Feeding Manila in
Peace and War, 1850–1945

FEEDING MANILA IN
PEACE AND WAR, 1850–1945

Daniel F. Doeppers

THE UNIVERSITY OF WISCONSIN PRESS

Publication of this volume has been made possible, in part, through support from the Center for Southeast Asian Studies and the Anonymous Fund of the College of Letters and Science, both at the University of Wisconsin–Madison.

The University of Wisconsin Press
1930 Monroe Street, 3rd Floor
Madison, Wisconsin 53711-2059
uwpress.wisc.edu

3 Henrietta Street, Covent Garden
London WC2E 8LU, United Kingdom
eurospanbookstore.com

Printed in the United States of America

Library of Congress Cataloging-in-Publication Data
Doeppers, Daniel F., 1938– author.
Feeding Manila in peace and war, 1850–1945 / Daniel F. Doeppers.
pages cm. — (New perspectives in Southeast Asian studies)
Includes bibliographical references and index.
ISBN 978-0-299-30510-9 (cloth : alk. paper) — ISBN 978-0-299-30513-0 (e-book)
1. Food supply—Philippines—Manila—History. 2. Manila (Philippines)—History—19th century. 3. Manila (Philippines)—History—20th century.
4. Manila (Philippines)—Social conditions—19th century.
5. Manila (Philippines)—Social conditions—20th century.
I. Title. II. Series: New perspectives in Southeast Asian studies.
DS689.M2D64 2016
338.1′95991609034—dc23
2015009224

For
CAROLE,
who made everything possible

and for
MATTHEW and TRACY, AARON and HALI
ANNIKA, CONNOR, GRADY, JORDAN, and BOWEN

and the rich rewards of family life

Morning is heralded in Manila by little busy puffing trains roaring . . . into the provinces as if they were really big trains and had some real purpose in being thus early on the road. . . . Their receding eloquence blends into the rumble of other wheels, converging into the city and really having something to do: yellow market carts with red-striped bodies so crowded with baskets that some of these are made fast to the uprights with tough rattan thongs. . . . And vehicles faster than the carts drawn by bays and pintos and sorrels take the road—market lorries loaded to the gunwales with double rows of passengers, bales, bundles and baskets, all lunging along in their mechanical-porter fashion and claiming, at this hour, the handsome midway of main thoroughfares.

They are . . . like burdened porters . . . their gaze upon the ground. . . . They roll their careless journey on . . . down the merry vales where lingering mists conceal . . . and at last along the flat valley and over the placid streams with lotuses and hyacinths nestled on their purple-black surfaces. The sun, in its good time, will . . . touch them into effulgent blossom. But now they sleep, yielding listlessly to the small current, wholly imperceptible, provoked by occasional dugouts paddling by. Like the carts, these boats are market bound, and like the lorries too. Manila must be fed, must have its breakfast, and will pay for the feeding, even of its animals.

—WALTER ROBB, "The Sunrise in Manila,"
ACCJ 8.5 (May 1928): 3

CONTENTS

ILLUSTRATIONS

TABLES

GRAPHS

PREFACE

The Hoosier grandparent I never met was Freeman Wood. A photograph taken around the end of the nineteenth century showing the interior of his parents' grocery in the county seat, complete with sacks of flour and kibitzers, was an object of wonder in the family album. Freeman left the grocery business to become a stockman, a dealer in workhorses, milk cattle, and hogs. His ability to judge the health, capabilities, and value of livestock became legendary. A sort of *viajero* in Philippine terms, he traveled to auctions in nearby states during the Prohibition era to bid on animals with borrowed money. And he was robbed at gunpoint by gangsters who threatened his family—one of the hazards of that line of work. His purchases were shipped back to Lafayette by train and auctioned on the premises of his farm on the banks of the Wabash. The farm was sold on his death. Other than a few photographs and my parents' stories, I had little connection to his world until I began this project.

~

Major strands of inquiry in historical geography and *Annales* history have evolved from roots in the work of Paul Vidal de la Blanche and his group in France a century ago, and historical geography and environmental history continue to overlap as the "same basic field of intellectual inquiry," in the words of William Cronon. Within both there is an interest in important everyday human practices, especially those concerned with food production and ecological management, diet, and health, in particular those practices and preferences that are of long duration in regional cultural communities. These were vivid occasional themes in the seminars of David Sopher and Donald Meinig at Syracuse University and of Andrew Clark in Madison. Questions involved with these themes have enhanced my awareness of the rich diversity of life and of the "changing human place in nature," to quote Cronon again,

but this is the first project in which they have emerged on my research agenda.[1]
A word about that is in order.

Thanks to an early experience in Calcutta, I entered graduate school with
the idea of focusing on cities and urban life in some part of "Asia." Once
launched at Wisconsin and finding a meager literature, I made a professional
choice to explore the social and economic processes that have shaped the
evolving society and geography of metropolitan Manila in the Philippines, in
short, to go deep. One of the megacities of Southeast Asia, Manila has been a
rewarding choice. I intended to focus on the whole of the twentieth century
but found that the records of ordinary individuals were much better for the
second half of the nineteenth century. I mined some of these in order to get
at migration behavior, the changing migration hinterland of Manila, and the
question of provincial and linguistic background in one's "assignment" within
the labor markets of the city. I was stunned by how restricted was the zone
from which the city drew most of its male and female migrants.[2] At the same
time I began to acquire a sense of the density of human and material flows
to and from the metropolis—with a population of only a quarter million in
the 1890s but already a relative powerhouse in extensive portions of Luzon.

This project began to take shape with an intense tracking of cargoes delivered to the city by coastal shipping starting in the 1860s. The records for rice
and other food commodities proved abundant. I became more aware that the
movements of vessels and transactions in foodstuffs were important everyday
connectors between provincial places and the growing metropolis. With the
partial exception of rice, historians and geographers had for the most part not
gone beyond the commerce in export commodities to look at that important
component of the domestic economy concerned with urban provisionment.[3]

I realized that a project on provisionment could make contributions to
several important lines of research linking urbanization to rural, provincial,
and environmental as well as economic change. Finally, Martin Lewis and I
offered a joint research seminar on a similar topic. The present project began
to crystallize out of our discussions with a very stimulating group of graduate
students. It became apparent that there was a substantial contribution to be
made with the richly textured study this was becoming. So the study grew.
While further research and writing were taking place, our family came to be
immeasurably enriched by the daughters-in-law and grandchildren named in
the dedication. In all, an incredible and wonderful phase of life.

Having started, a number of things fell into place. Some of these were personal and fortuitous, but knowing them may help the reader to appreciate the
place from which I am coming. During dissertation fieldwork in Dagupan,
my wife and I rented quarters in a house that was bordered by fishponds. Late
in the afternoon, one of us would go to the street in front of the public market

where women would be arriving with bamboo poles strung with pond-raised *bangus* (milkfish). They were fresh and cheap, and we enjoyed them frequently. Also in May there were local people in the *bangus* fry business. I watched men with fine nets capture the fry and also buyers and sellers counting and averaging the number of fry per container. Our *panganay* (eldest child) was born in Davao City that year in the wee hours. When the sun came up, we heard the muffled explosions coming from people "fishing" with dynamite. In the hospital we encountered one of these fishermen—missing a hand. And from my very first visit to the Philippines I can picture small boys leading giant, deliberate carabao. On a later stay in the city our sons came to look forward to hot *pan de sal* (light buns) eaten fresh and warm in the late afternoon or dining out on the excellent thin noodle and seafood dish known as *pansit Malabon*. Such images and experiences came to mind as I dug into the material for this volume.

In 1985, on a project concerning the Great Depression, I spent six months in Manila interviewing senior citizens about their work and family lives during the 1930s and 1940s. Some of the interviews took place in neighborhoods where people were involved in the fish or hog businesses, while others had worked in the public markets or as ambulatory vendors. Some of their stories were vivid. And during that period and later, I frequently ate lunch with Lito or Rose or Pepot—all those lunches and shared insights into Manila foodways and life. They taught me more than they know. Thus the analytical and general findings of this project are leavened with the lived experiences of a wide range of ordinary Filipinos.

After four decades of research on the Philippines and Southeast Asia, I feel fortunate to have developed a network of professional colleagues in many countries. This work would be much the poorer without their interaction and critical assistance. The list is long, and I fear that I will leave someone out, but the reader will find testimony and insight here from former University of the Philippines vice-chancellor Oscar Evangelista, who facilitated my work in so many ways, as well as Rico Jose, Benito Legarda Jr., Helen Mendoza, Resil Mojares, Yoshiko Nagano, Ruby Paredes, the late John Schumacher, SJ, and others. Mike Cullinane has been an inexhaustible source on semantics and usage and the contrasts and similarities with Cebu. Al McCoy consulted on the metastructure of this work and urged me on. Matthew Turner and William G. Clarence-Smith offered helpful critiques of papers that became portions of the manuscript. My research assistants at various times over the years in archives, parish records, and neighborhoods included Rose Marie Mendoza, Loreto Seguido, Marian Manalang (now attorney Marian Manalang Labog), Cleofe Marpa, Cristina Bernabe, and Dennis Santiago. The names of these and others are strewn abundantly in the notes. Early on Norman Owen opened several personal research files to me and later generously provided an invaluable structural assessment of each chapter. His persistent urging perhaps kept this from becoming a posthumous work. Owen and Jan Opdyke were the perspicacious consulting editors who helped greatly to consolidate the manuscript. Two anonymous referees made valuable suggestions. My gratitude to each and all is great. Shortcomings of interpretation and fact are mine alone.

The greatest debts are to Oscar and Susan Evangelista and their children Sara, Alex, and Ami, who welcomed me into their home and lives in Quezon City. The youngest would hold my hand while I walked her into kindergarten while missing my own boys deeply. The Evangelistas' *lola* (grandmother), Araceli, knew more about life than most of us and quietly held the entire family together every Sunday in a common bond of sharing food. Then there were all those earlier visits to the archives in Washington and College Park and the unfailing hospitality of Sandra and Jim Fitzpatrick and their boys, Michael, David, and Ben. Few have more gracious and welcoming in-laws. As Manila's traffic and air pollution grew worse, I became one of a whole cohort of researchers who came to stay with Helen Mendoza. Her family compound and "ashram" near the Welcome Rotunda provided an oasis of hospitality and good sense. An opportunity to spend a semester of research and writing under the auspices of the Amsterdam branch of the International Institute of Asian Studies not only opened up the marvelous world of a stimulating group of Dutch scholars, including Peter Boomgaard, John Kleinen, Otto van den Muijzenberg, Rosanne Rutten, and Willem Wolters, but also provided rich comparative insights from the provisioning of London and Amsterdam by the sail vessels and fishing communities of the former South Sea, the Zuiderzee. Here the project mushroomed from an essay on the rice trade into the subject of metropolitan provisioning writ large. While most of this work was written in Amsterdam and Madison, small but significant portions were written while I was ensconced in an adobe cottage, a former schoolhouse, in the village of Portal, Arizona.

Other institutional debts are many. I thank Rosalina Concepcion and the Philippine National Archives for assistance during both sunny times and typhoons; the U.S. National Archives now in College Park, Maryland; the Ateneo de Manila University and University of the Philippines libraries and their dedicated staffs; the Archives of the University of Santo Tomas; the Rural History Centre at the University of Reading, the Royal Geographical Society, and the Public Records Office at Kew in the United Kingdom; the Koninklijk Institut voor Taal-, Land- en Volkenkunde (KITLV) in Leiden; the Netherlands Maritime (Scheepvaart) and Tropen museums in Amsterdam; and the Memorial and Steenbock libraries of the University of Wisconsin, which together have quietly become one of the great places to conduct research on the Philippines. All the maps were created at the Cartography Laboratory of the University of Wisconsin under the able direction of Onno Brouwer and Tanya Buckingham. Rich Worthington, Qingling Wang, Caitlin Dorn, and Kristin Gunther each made signal contributions to shaping the maps. In addition to the support cited above, I am thankful for the opportunity to spend several months as a guest of the Research School of Pacific Studies at the Australian National University reflecting on the interviews used in this study and also for the critical financial support received from the Social Science Research Council and Vilas Fellowship program of the Graduate School of the University of Wisconsin. And I am happy to acknowledge the sustaining intellectual and personal communities of Madison and the University of Wisconsin, including the Center for Southeast Asian Studies and Geography Department.

Finally, sincere thanks go to the following authors and publishers for permission to reproduce or quote from certain copyrighted materials.

- KITLV Press for my essay "Beef Consumption and Regional Cattle Husbandry Systems in the Philippines, 1850–1940," in Peter Boomgaard and David Henley, eds., *Smallholders and Stockbreeders: Histories of Foodcrop and Livestock Farming in Southeast Asia*, 307–24 (Leiden: 2004).
- Ohio University Press for my essay "Fighting Rinderpest in the Philippines, 1886–1941," in Karen Brown and Daniel Gilfoyle, eds., *Healing the Herds: Disease, Livestock Economies, and the Globalization of Veterinary Medicine*, 108–28 (Athens: 2010).
- Ruby Paredes, Madison, for permission to reproduce a memorial on her mother's use of flour-based products in family cuisine.
- The family of the late Ambassador Marcial P. Lichauco and his wife Jessie Coe Lichauco (Sta Ana, Manila) and daughter Cornelia Lichauco Fung (Hong Kong), for permission to reproduce the painting of Cornelia Lao Chang Co Lichauco and a family photograph of Faustino Lichauco. Both pictures were previously printed in Cornelia Lichauco Fung, *Beneath the Banyan Tree: My Family Chronicles* (Hong Kong: 2009).
- Fernando J. Mañalac, MD, Steubenville, Ohio, author of *Manila: Memories of World War II* (Quezon City: 1995) for material on the death of Salvador.
- Robert M. Sears, executive vice president of the American Chamber of Commerce in the Philippines for a selection from the art piece by Walter Robb, "The Sunrise in Manila," ACCJ 8.5 (May 1928): 3.
- Edwin Green, group archivist at HSBC Holdings plc, London, for permission to refer to "LOH II, 124, folder Ig2" from 1916, and to Wigan Salazar who unearthed it.
- Washington Sea Grant Program, University of Washington, for permission to reproduce the *bangus* (*Chanos chanos*) drawing from Ling Shao-Wen, *Aquaculture in Southeast Asia: A Historical Overview* (Seattle: 1977).

MANILA

Y

SUS CONTORNOS

$$\frac{1}{20,000}$$

**Referencias de la orilla
derecha del Rio Pasig**

1. Visita de Santa Rita
2. Id. de Candelaria
3. Mesoneria de San Fernando
4. Cuartel de Infantería
5. Antigua casa de Recureilla
6. Dirección general
7. Quemadero
8. Fabrica de Tabaco
9. Administración de Vino
10. Plaza presentada en el lugar incendiado en Nov.º de 1836
11. Corregimiento de Tondo
12. Lugar incendiado el 16 de Abril de 1840

Calles

a. Calle Cerrada
b. Id. de S. Nicolás
c. Id. de S. Sebastian
d. Id. de Candelaria
e. Id. de Longos
f. Id. de Bancusay
g. Id. del S.ᵗᵒ Cristo
h. Id. de Jaboneros
i. Id. de S. Fernando

j. Id. de Sáo
k. Id. de Antaloague
l. Id. del Rosario
ll. Id. nueva
m. Id. de S. José
n. Id. de la Escolta

Tondo

Nuevo Tondo

S. Anton

S. Sebastian

Quiapo

S. Miguel

Fabrica de Tabaco

B A H I A

de

M A N I L A

Campo de Bagumbayan

Cementerio de Manila

Casa del Capitan

Sepulcro de Pineda

Malate

S. Antonio Abad

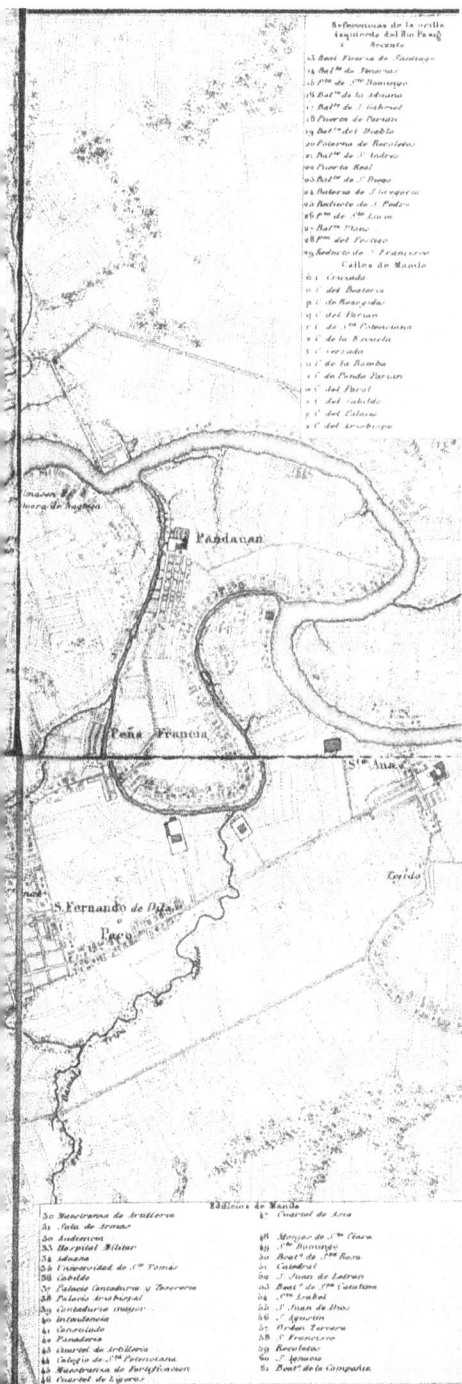

Manila in 1849. *Atlas de España y sus posesiones de Ultramar*, compiled and published by Francisco Coello, Madrid, 1848–68.

FEEDING MANILA IN
PEACE AND WAR, 1850–1945

INTRODUCTION

Why Provisionment?

URBAN PROVISIONMENT MATTERS like few other things. In 1944–45, the people of Manila experienced mass starvation. The wartime Japanese occupiers decided to defend their position in the Philippines from American and Allied reinvasion. They sent reinforcements. In line with the ongoing imperial war policy, they set about provisioning their forces in the field by taking foodstuffs from the local inhabitants. This included collecting rice before it could be distributed in the city. Japanese sentries were shooting farmers in dugout canoes transporting a few bags of rice into the city at night. Other "smugglers" at great risk brought in "contraband" domestic rice hidden under other cargo on river barges. In addition, many urban consumers traveled to nearby provinces to trade possessions for rice. But all this together was grossly insufficient to provision the city.

By August 1944 affluent families found starving beggars at their doors. By mid-October it was common to encounter adults desperately begging for a little rice or soup. At the same time squads of Japanese soldiers began to go door to door in residential neighborhoods. Ostensibly searching for hoarded rice, they took all sorts of foodstuffs, including household chickens and livestock. Now there was real hardship and starvation. People, not nameless refugees but known local residents, simply died on the streets. In November the mayor organized pushcart details to remove the dead. Soon every day saw trucks carting away the emaciated corpses. By December ragged bands of starving people were looting warehouses in full daylight.

Only the rich and well connected could afford rice near the end. Many others subsisted on homegrown sweet potatoes or swamp greens. Soon they had only squares of cassava bread or broiled coconut pulp. Others ate fried rice bran. In the great neighborhoods near the sea, thousands resorted to eating partially rotted rice recovered from ships sunk in the harbor. Ragged cadaverous people could be seen on the streets picking up a few bare crumbs—some were the final surviving members of poor working families. Tens of thousands starved to death or fled the city, evacuating to places where they hoped to find a survival ration. This starvation resulted from a catastrophic dismantling and breakdown of the metropolitan provisioning system.

These topics—collectively the operation of the city's provisioning system, developed over many generations, and its major perturbations—form the subject of this work. Provisioning systems are critical to understanding how the great cities of the region have been able to grow so large, and they form a theme that has been sorely neglected in Southeast Asian studies. What were the products ordinarily consumed by the people of the metropolis and how did the mix change over time? Where and how were these foodstuffs produced or manufactured and through what social intermediary and transport systems were they brought to the consuming population in the city? What exactly changed as a result of "global" commercial interactions? Interlaced with these questions is a concern for metropolitan-hinterland relationships, as well as the transforming effect of these relationships on both city and countryside.

~

This work constitutes a first scholarly exploration of the critical analytical problem of provisioning the "megacity" in Southeast Asia and, to a degree, a serious social history of one of the world's dozen or so largest cities, with an urban population of over 12 million within the designated "Metro Manila" territory but approaching 21 million when the rapidly growing outer suburbs are included.[1] Third World megacities present humankind with enormous challenges. Over the past decade scholars, strategists, and economists have, for diverse reasons, come to the realization that provisioning these vast conurbations represents one of the most formidable problems of the twenty-first century. In *Planet of Slums* (2006) popular urbanist Mike Davis offers scenarios of millions of urban poor crowded into sprawling favelas without water or services, resorting to terror, and restrained by massive acts of repression that make their urban territories into contemporary war zones. Provisioning megamillions in a few dozen megacities is one of this century's main social challenges.[2]

Megacities? In 1939 geographer Mark Jefferson observed the widespread existence of "primate" cities, pinnacle cities in national systems of urban places that are at least several times the population size of the second- and third-largest metropolitan areas combined. Many of these primate cities are port-capitals located in the former colonial world. Today, grown to great size, the largest of these Third World megacities are both celebrated and vilified for their many contemporary roles.[3] What isn't widely understood is how such conurbations actually grow and are sustained.

Provisionment was the essential precondition for the dramatic growth of historically modest colonial port cities into modern megacities of 10 to 20 million people. Without food security, great masses of humanity could not and would not have alienated themselves from the land and crowded into potentially lethal urban concentrations, cut off from most of the essentials for human survival. Together the seemingly small, ephemeral matters of grains, vegetables, proteins, and water are an essential foundation of the modern metropolis.

There were significant economic implications to these developments. It has generally been held that the chief impetus for economic change in former colonies was their participation in a growing system of world trade centered on one or more of the industrial economies of the day. This was especially evident during the period of "high colonialism" (1870–1930) when specialized "commodity export economies" were created in the subordinate territories, each exchanging raw materials for manufactured goods. This perspective suggests that the production, handling, and shipping of commodity exports formed the primary engine in the transformation of the colony's economy and society. Such a view has a certain analytical power.[4] But much colonial era change began, in fact, with the growth of a large city or cities with ramifications that were primarily internal rather than external. A systematic focus on the evolution of the food supply of metropolitan Manila provides a platform from which to observe these internal relationships in action and thus to view economic transformation from a new perspective.

Feeding Manila is not just about the city. In a sense, the countryside made the city by feeding it. At the same time, the city transformed the countryside, stimulating an interacting hinterland with various specialized forms of food production and processing. While a substantial part of export commodity product left the archipelago unceremoniously by the nearest significant port, the products of provisionment were more typically sent to the metropolis. Feeding the city involved changing forms of commercial organization and transport technologies and massive environmental transformations. The personal and commercial interactions coincident with provisioning Manila helped create the nation by placing more and more of it in frequent direct interaction.

Dramatically interrupting the flow of process and event is the fact of mass starvation in the city in 1944–45, the subject of the final chapter. This resulted from a catastrophic dismantling and breakdown of the provisioning system in wartime. Deurbanization by death and flight was the necessary outcome. It reminds us that, although food may be retailed through public markets, grocery stores, shops, and street vendors or distributed via some rationing system, it comes from soil and sea, from the efforts of farmers and fishermen, production organizations, and places large and small with various environmental characteristics. It tends to arrive via intermediaries, processors, and various systems

of transport and distribution. These subjects—the operation and major perturbations of the city's provisioning system—are the focus of this work.

The challenge is to bring coherence to these themes and approaches, to
construct a view of change across time and geographical/environmental context, and to bring to bear significant comparisons drawn from the larger
Southeast Asian realm. The result may be thought of as a work at the intersection of historical geography and economic history with attention to gender,
ethnic, and other social roles, as well as environmental questions. It provides a
sustained examination of provisioning an important Third World megacity
and of the evolving relations between Manila and the expanding array of localities and provinces on which it depended.

The structure of the text in four parts speaks for itself.

- The rice trade and its fluctuations due to natural and human events over
  nearly a century;
- The evolving supply and preference structures for "things eaten with rice":
  vegetables and fruit, fish, and meat (including the great epizootic waves that
  twice brought large animal destruction on a "biblical" scale);
- The provisioning of the city population with potable water and milk (with all
  their public health and public access implications), plus the foreign-induced
  rise of wheat flour products and coffee (versus cocoa) in the urban dietary;
- The provisioning catastrophe of the Japanese occupation.[5]

What may be less familiar is the degree to which such a work requires the
evidence and communication of historical photographs and themed graphic
advertisements, as well as maps and graphs. These are first used as discovery
and analytical devices and then presented as aids to reader comprehension.

To be clear, the "Manila" at issue here refers to the entire functional metropolis as it was at any particular time rather than the formally designated city
or the administrative structures therein.[6] As the country's capital, the early
concept of Manila was legally restricted to the zone demarcated by the great
defensive wall built around it in the late sixteenth and seventeenth centuries.
Locally this district is known as Intramuros (inside the walls).[7] In the nineteenth
century the multiple Filipino and Chinese cosmopolitan parts of the city,
including the most important commercial and production spaces, lay across
(north of) the Pasig River outside the wall. These were divided into a series
of municipal-parish territories known in Spanish as *arrabales* (from Arabic),
towns lying near but outside the walled city, on the outskirts.[8] Over time a
number of *arrabales*, both north and south of the river, came to be included in
the formal city. Numerous other areas and nearby town-parish centers became
integrated in the informal metropolitan expanse.[9] The formal administrative

city ran the public markets and slaughterhouse—major sources of revenue—and was important in public sanitation and disease control innovations. But the critical unit of analysis adopted here is the entire urbanized metropolitan area. Ignoring fluctuations, the population of the metropolis increased rapidly from fewer than 75,000 at the start of the nineteenth century to approximately 900,000 by late 1941.[10]

Provisioning Manila both affected and was effected by ecological change. In the past century human action has transformed the Philippine archipelago from a still heavily forested land to one in which substantial parts of the country have been stripped and have developed a flash-flood ecology where heavily concentrated rainfall from a tropical storm is no longer retarded by the canopy, roots, and buttresses of forest cover. Rather, the water gathers speed and mass as it moves downslope, setting off more and greater landslides, which cover fields, settlements, and offshore coral reefs with muddy sediment and debris, sweeping away transport and housing infrastructure and all too often burying animals and people. Due to the increasingly swift downslope run, groundwater is not recharged at former rates. The impact of the dry season is magnified. What were once consistent springs and creeks increasingly flow only seasonally or dry up altogether, leading to desiccation in areas with otherwise sufficient rainfall. This outcome has resulted from massive human population growth, "mining" nature's forest capital, and farming the slopes. The seriousness and insidiousness of this process at the societal level is considerable and by no means limited to the Philippines.[11]

Provisioning Manila played an important role in these ecological changes. A number of areas came to specialize for a time in supplying the growing city with firewood, charcoal, and timber.[12] Moreover, forest regeneration was often hindered by grazing cattle for the beef supply of the city, by the frequent use of fire to improve browse, and by the spread of monocrop farming. The demand for firewood also resulted in the removal of coastal mangroves, destabilizing coastlines. In another ecological change of massive scale, the great brackish nipa palm swamps, once exploited for the production of alcoholic beverages, vinegar, and roofing materials, were cleared and transformed into commercial fishponds to feed the metropolis.

Finally, food matters in its own right. Although it may reach one's table through a public market, grocery, or street vendor, it does not come from there. The human society of metropolitan Manila, or indeed any city, cannot exist for long without some fundamental and regularized division of labor and exchange. Tragically, when this axiomatic proposition was put to the test in 1944 and early 1945, a significant portion of the urban population vanished. Yet, as noted, the subject of provisioning the metropolis has been almost wholly neglected.

Only rarely are city and countryside and their mutual interactions investigated together, as they are in *Nature's Metropolis*, William Cronon's exceptional work on Chicago and its hinterland. In his phrasing, "urban and rural . . . [have] created each other, . . . transformed each other's environments and economies, and . . . now depend on each other for survival."[13] In the historical and geographical literature on Southeast Asia, food crops (with the exception of rice) and stock raising for domestic trade have been generally ignored.[14] This lacuna is puzzling because the commerce in foodstuffs for the great urban markets is one of the more substantial and enduring dimensions of modern economies.

At the same time provisionment is a persistent theme in geography, one involving ecological management, sociopolitical competition for essential resources, the spatial ordering of comparative locational costs, organizational change, and other topics.[15] Among historians and historically minded economists, the feeding of great capital cities has emerged as a major set of issues in important works on Beijing, Tokyo, and Paris, to name a few. In these studies, the emphasis has been on assessing the development and efficacy of policies and institutions designed to assure an adequate stream of basic grains to the population of the capital. Charles Tilly used these concerns to "observe the basic processes of state-making as they touch the everyday lives of ordinary people." Most recently this interest has been extended to the organization of meat supply and to analyzing an "oscillation between state intervention and market liberalization."[16]

Another prominent theme concerns the interaction between "global" and local phenomena. From the early agricultural and food imports of the "Columbian Exchange" to the essential and ongoing international trade in rice, the rise and social proliferation of wheat flour coming from California and Australia, the egg supply coming from the Canton River delta, and the decades of live beef imports from Cambodia and Australia, the Manila dietary in our period was deeply involved in both the global and the local.

In this work the record of imports gives us feedback on how some global food elements were judged and appropriated by families. For example, the rapid rise of "Germanic" beer consumption during the 1880s led to the astounding success of "local" San Miguel beer in the 1890s and beyond. Also a nutritional survey revealed the growing popularity of bread, a foreign item competing with rice for breakfast among the families of "workers" in the 1930s. And a memorial by the historian Ruby Paredes provides documentation on the way some "global" foods were categorized and intimately judged.

∽

This study covers the period from the 1850s through the catastrophe of 1944–45. The starting point was chosen to coincide with or closely follow several important institutional changes in the Spanish colonial system. In Manila in the 1810s, creole power holders were engaged in a fierce defense of their right to broadly control "the leases and taxes for supplies and meat (*abasto y carnes*) within the larger urban area." In peninsular Spain, the abolition of administratively fixed prices and the emergence of freedom of trade in all articles of food, drink, and fuel—a free market system for agriculture—occurred in the 1830s. The economic historian Jaime Vicens Vives writes of a resulting "agrarian revolution."[17] It took some time for free market ideas to be implemented in the archipelago, but by the 1850s centralized attempts to control urban provisionment had largely been replaced. The former special trade privileges and provisioning requirements of the Spanish provincial governors—"de facto trading monopolies" Owen calls them—had been curtailed in law, if not always in practice.[18] And the continuing attempt to enforce a government monopoly on alcohol was about to end. The choice of a starting point was also dictated in part by the diminishing frequency of attacks by maritime raider-slavers, which opened numerous coastal areas to more secure settlement and commercial participation. The two major public markets in Manila were built in 1851, and the single legal city abattoir had already been in service for some time. In regard to food, a commercial system with a number of "modern" characteristics was at hand.

At the same time the availability of historical sources improves significantly as we move into the second half of the nineteenth century. This is particularly true of the shipping cargo records, which allow us to quantify what had until now been impressionistic. Products brought to Manila by coastal shipping were elaborately recorded in the official press from 1861 to 1881 and in less detail for another decade. The data appear to be internally consistent and reasonably comprehensive for vessels arriving in the official Port of Manila. I selected four "sample" years at roughly decadal intervals and examined the records for food commodities in each. The results far exceed any previous evidence deployed on provisioning Manila. There is nothing of comparable detail in the twentieth century until some statistical reports of the 1920s allow us to see trends over time. These records have also been mined to get at the changing sociology of commerce. Since only a few of the merchants are well known, a sample of personal careers is briefly offered where the data allow.

Newspapers and magazines became prominent vehicles for the dissemination of commercial advertising during our period. A wide reading of the print media provides an abundant record of attempts to establish product and "brand" recognition for imported industrial products such as flour, beer, canned milk, and coffee, as well as plows, steam engines, and refrigerators. Such advertising carries messages about targeted consumers, as well as provisioners. Many commercial messages were clearly playing off family social aspirations, as well as group identity, especially among mestizos and expatriates. Later various foods and processing practices were baldly labeled "old-fashioned" or "modern," thereby further promoting social and material change.

In some cases items of fundamental importance—such as fish, the central flesh protein source in ordinary diets—escaped nearly all attempts to count and record them. But here we can delve into a rich literature on changing fishing methods and rely on fish corral registration records. For food, health, and numerous other topics the *Philippine Journal of Science* proved to be of great value. Meanwhile, the tax records of the late Spanish administration provide a wealth of information about mills, vessels, livestock buyers, and the like. Foreign consular officials were centrally concerned with commercial conditions, and the reports of British, Dutch, and American consuls form a useful critical record deeply exploited here.[19]

In studying provisionment one deals with many of the infrastructural requirements for urban growth: public health initiatives that attempt to guarantee sanitary meat, vegetable, water, and milk supplies; the development of transport, including rail and all-weather roads; and local delivery networks.[20] In numerous cases contending interests were involved in the implementation of these, for example, in the long bioscientific and policy struggle to end lethal outbreaks of rinderpest stemming from the import of diseased water buffalo and cattle. And there was conflict between the practices of mestizos and other Filipinos and the commercial efficiencies of some Chinese middlemen. Other policies were also important: Spanish neomercantilism, which made imported rice more expensive in times of shortage; American rules against the continued immigration of Chinese as laborers; and foot dragging on agricultural irrigation schemes. Such questions are pursued as they arise, but it is impossible to explore all of them, and this work is not intended as an institutional history. In the end a study like this will raise far more questions than it can reasonably answer. Hopefully others will pick up some of these threads and go further.

The strategy is to follow the commodity while at the same time recognizing that all food is interconnected. The study does not assume that there is but one story of change; rather it disaggregates the topic in order to uncover important differences. Since the surviving record is highly varied, the approach must be eclectic, involving different sources and methodologies. Few sources are available for the whole period. What I have done is to discover and exploit each of these new sources in order to go deeper than before, even if that deeper view is sometimes discontinuous. The strengths and shortcomings of the sources are discussed in the relevant chapters.

The analytical methods employed in this project are as much geographical and cartographical as historical. Graphs are used to calibrate rates of change over time and to spot covariation. Maps are employed as devices of discovery as well as communication.[21]

This research also opens windows into the lives of ordinary Manilans. Provisionment involved legions of producers and urban distributors. *Aguadores* and *lecheras* carried drinking water and milk into the homes of a string of regular clients. Urban fishermen caught and fish dealers and door-to-door venders distributed aquatic products. *Viajeros* journeyed to the provinces to purchase animals, fruit, and vegetables. *Cocheros* drove and cared for the horses that delivered provisions and carried consumers. Storekeepers and merchants transformed the formerly episodic commerce in various food commodities into everyday specializations. Oral accounts help capture the operations of these otherwise voiceless provisioners.[22]

The subject of provisionment penetrates many arenas. In Manila it sheds some light on a major issue in Philippine historiography: the background of the revolution of 1896 in Manila's hinterland. By the late 1890s Philippine colonial society had slipped its moorings. Indigenous forces carried out an early and resilient revolution aimed at evicting Spanish authority from the archipelago. The causes were complex. Historical demographers have now identified this period—roughly 1875 through 1905—as a time of repeated crises in human mortality, which brought the national rate of population growth down from perhaps 1.7 percent per year to less than 1.0 percent. Ken De Bevoise rightly settles on multiple cholera epidemics as a principal cause—"by far the most terrifying of the diseases"—in play. He points to "the obvious helplessness of healers, priests, and physicians" in the face of these "agents of the apocalypse." These waves of human mortality were exceeded by great waves of water buffalo mortality due to bovine rinderpest disease, choking rivers with rotting carcasses and bringing wet rice production to a standstill. These experiences could not but intensify the sense of popular disquiet and expose the impotence of the institutions of the late Spanish colonial state. Epizootic rinderpest, a rising incidence of El Niño–related droughts, doubts about drinking water quality, and attempts to keep cholera at bay all contributed to a growing sense of personal insecurity that was part of the revolutionary climate of the times.[23] Even worse conditions obtained in occupied Manila during the latter phases of World War II and are recalled in the final chapter.

This volume offers a view of change across time and geographical context and brings to bear comparisons from the larger Southeast Asian realm. It may be thought of as a work at the intersection of historical geography, economic history, and gender, ethnic, and other social roles, as well as several environmental questions.[24] It provides a sustained examination of provisioning an important Third World metropolis and of the evolving relations between metropolitan Manila and the expanding array of localities and provinces on which it came to depend. It begins to meet the call for studies in which "the nature, importance, fluctuations, and implications of provisioning are viewed in depth."[25]

PART I

The Rice Trade

I

The Manila Rice Trade
in the Age of Sail

RICE HAS BEEN AT THE CENTER of the everyday diet of Manilans ever since the city was instituted in the sixteenth century, through all the centuries that it served as the capital and chief trading center of the Spanish colony and after. No sizable portion of the Filipino and Chinese population of Manila preferred millet, wheat, or maize to rice.[1] The growing metropolis of the last century and a half is unimaginable without the rice trade. There have been accounts of the expansion of rice-producing zones and also of Philippine participation in the international trade in rice, but despite its importance the historical geography of the domestic rice supply of the city is a subject still in its adolescence.[2] Questions concerning the domestic shipping and milling of rice and the changing pattern of seasonal cycles in urban supply have been given still less attention. These questions in the age of sail form the core of the present chapter.

Simply stated, rice made Manila possible. This is not too much to say for Manila was chosen as the site of the imposed Spanish capital in the sixteenth century because it was one of the few good port sites in the archipelago that could be supplied with an ongoing stream of rice from nearby production areas. Rice has remained the primary staple food of Manila's population for almost all of the last 440 years. Further, growing demand from the city resulted in the mass conversion of land from grassland and frontier grazing or forest to rice production, especially in nearby Central Luzon. In this sense, rice made Manila, and Manila in turn made rice the predominant crop in its more immediate hinterland.

Filipinos employ no common word as encompassing (or vague) as the English word *rice*. Indigenous terminology refers instead to *palay*, *pinawa*, and *bigas*, that is, to the state of the grain. *Palay* is unhusked rice—harvested or

growing in the field. Writers of English nationality often used the Malay/
Indonesian cognate *padi* or *paddy,* meaning the same thing. Once the grain
had been husked, Tagalog speakers referred to it as *pinawa* or possibly *habhab,*
rice that is husked but unpolished. This was sometimes known as "brown rice"
in English, a reference to the color of the bran still adhering to the grain. At this
point, the grain is in a form that, with cooking, makes human consumption
possible though not especially palatable. International traders usually meant
something similar when they spoke of "cargo rice," rice that would withstand
the dampness and potential injury of water transport. However, Philippine
Spanish-language accounts now called it *arroz,* in some cases adding *corriente,*
or "plain," or *limpio,* meaning that it had been "cleaned" by having the husks
knocked off. In the shorthand of advertising, the Spanish-language Manila
press often referred to *pinawa* as simply *corriente.* The more the bran has been
removed from the grain, the whiter the product. At the point that nearly all
the bran has been removed—following a second pounding when using mortar
and pestle—Tagalog speakers call the grain *bigas* meaning "hulled white rice"
(*beras* is the Indonesian/Malay cognate). *Bigas* is the generic term in use in
the markets and rice shops of Manila today. In colonial Spanish, however, it
was still *arroz,* although qualifying terms might be added, for example, *arroz
blanco.*[3]

A perennial grass grown as an annual, rice is one of the handful of plants
that feed the great bulk of the world's human population. Without rice that
population would be much smaller. Domesticated lowland rice as a species
(primarily *Oryza sativa indica*) and the lowland Southeast Asian population
are mutually dependent. The unusual physiology of wet rice allows it to survive and flourish in water, producing more calories per acre than other grains
and more protein per acre than roots and tubers. Grown in ponded fields,
such rice solves one of the main problems of tropical agriculture—too much
rainfall, resulting in rapid leaching of soil nutrients. Wet rice relies less on soil
fertility than on nutrients washed into the field by sheetwash or irrigation
water and especially on nitrogen-fixing algae scum in the pond water. Fertilization, such as it was, relied then on the defecation of a few bovines browsing
on standing rice stalks after the harvest. The field bunds and level surfaces
minimize soil erosion, and the pond water hinders many of the weeds that
complicate dry land agriculture.[4] As a result, wet rice cultivation is sustainable
over very long periods. For at least five centuries scattered nuclei of wet rice
fields arrayed in wetlands around Manila Bay and the nearby lake Laguna de
Bay, as well as favored places in Ilocos, Bikol, and Panay, have been in continuous production. In that sense they are very old cultural-historical areas.

The indigenous inhabitants of the Philippines were not used to producing or
collecting surpluses large enough to provision the new Spanish regime. Getting

them to do so was one reason for the conflict and bloodshed that attended the initial imposition of Spanish rule.[5] In the first two or three centuries the Spaniards provisioned their capital settlement in several ways: by imposing a tribute-taxation system and demanding tribute in kind, as well as other forms of extraction; by encouraging the establishment of large landed estates; by encouraging commerce; and by dealing with Chinese and Japanese traders who brought foodstuffs from abroad. The Spaniards attempted to sustain their imposed capital by demanding tribute in kind to be paid in rotation among several nearby clusters of villages. For more than two centuries, part of the population routinely paid an annual tribute in kind. Such payments in *palay* or *pinawa* were still coming to warehouses in Manila in the 1830s.[6] But near the city, in the words of Linda Newson, "[T]he local population became rapidly integrated into the cash economy and from an early date paid tribute in cash rather than commodities."[7] In the 1590s and 1600s the requirements were for cash plus one hen. A system of tributes and head taxes paid in cash fit the reality of urban dwelling and also forced farm villagers to sell some of their product. A bit farther away, at roughly 10 to 25 kilometers, villagers were required to pay in rice, chickens, and silver. Luis Alonso Alvarez argues that sales from royal warehouses in the city were an important form of provisionment.[8]

Many of the landed estates that were created nearby were held initially by individual Spaniards in their own right or as front men for the several religious orders. Most of these became friar estates and in several cases were enlarged over time at the expense of the nearby inhabitants.[9] Given the early shortage of labor, many of these haciendas were involved in grazing, later transitioning to farming. The friars' ability to shelter their indigenous tenants from the corvée labor demands of the state became an important recruitment tool. According to Dennis Morrow Roth, the estates' religious owners became the "primary suppliers of agricultural commodities to Manila." From the extracted rents, the religious supported their missionary and educational training activities and contributed materially to provisioning the city.[10] "As in colonial Mexico," he writes, "the estate owners possessed considerable market power and occasionally were able to control supplies and prices to their advantage. Subsequently, the religious lost much of this dominance as a result of the commercial revolution of the nineteenth century."[11]

Rather quickly there also developed a commercial system of supply. This involved production in the vicinity of the city and also in China and even Japan for a time. An increasing number of people came to make a living growing and/or bringing foodstuffs to the city and providing credit and commercial organization for these activities. Feeding Manila "created chains of relationships that affected the economic life of producing communities. . . . Subsistence was replaced by production for exchange."[12] But at the end of the eighteenth century, the council of Manila still retained the right to set the price of food-stuffs brought into the city.[13]

~

Food supplies have to come from somewhere, and the where and how make a difference. In the middle of the nineteenth century, rice was delivered to the city by small watercraft on a daily basis. Some of these vessels came from Bulacan Province, just northwest of the city, arriving via Manila Bay. More came via a network of natural intracoastal waterways, which reached their confluence at Malabon, an important rice supply center just northwest of the city. Another supply route came from the east from Laguna Province and the Marikina Valley via the town of Pasig and other nearby places along the Pasig River. Both limbs of this system together constituted an "inner zone" of rice supply. By the second half of the nineteenth century even larger flows came from an "outer zone" where rice for Manila was gathered and transported by sea from more than 40 small ports, each drawing on its own microhinterland.

Dagupan, a small city in Pangasinan, eventually emerged as the single most important shipping point of Manila's domestic rice supply. Dagupan's fortune was to be located at a point where vessels of modest draft could enter from the sea and access the great quantities of rice that could be assembled there via the inland waterways of the northern Central Plain.[14] Dagupan was a preeminent beneficiary during both the sail and initial steam vessel eras, as the growing flow of rice from Pangasinan and northern Tarlac coursed through its riverside wharf area—the *pantalan* (figure 1.1).

A dynamic look at the rice provisionment system during the age of sail is made possible by an analysis of the detailed reports of individual maritime cargoes arriving in Manila. The use of such data is laborious, but sample years at decadal intervals, virtual trade cross sections, give us critical vantage points from which to comprehend the changing system of supply after the Philippines had begun to be transformed by participation in the growing world markets for export commodities.[15]

Our first vantage points, in 1862 and 1872, come during the age of sail navigation and hand-milling technology. This period also saw the waning of the so-called mestizo era in Philippine commerce. By the 1870s, the Philippines was sliding into a transition from a frequent net exporter of rice to coastal China to a regular net importer of rice, mainly supplied from the Mekong Delta. According to Benito Legarda Jr., one product of the processes of economic change in the Philippines was the reallocation of some labor and entrepreneurial activity away from rice production.[16] At the same time, resident Hokkien Chinese were becoming strong participants in the commercial supply of rice to the city.

FIGURE 1.1. Coastal sail craft and great mounds of *palay* in sheaves at the Dagupan *pantalan* (wharf). (Bureau of Agriculture, USNA II, RG151-FC-85A-4, box 85)

By the time of our third and fourth benchmarks, 1881 and 1891, the rice trade had entered a new era of steam navigation and was about to encompass railroads as well, with profound effects on transportation, milling, and the seasonal timing of deliveries to the city. Control of the rice trade was passing to Hokkien Chinese and European merchant houses, though not completely. The archipelago was now a massive net importer of rice.

What follows is an examination of the evolution of the supply and transport of rice to the city—by far the most important dimension of urban provisionment.

THE MANILA RICE TRADE AT MID-CENTURY

During the third quarter of the nineteenth century, Manila was operating largely within an "eotechnic" complex—a technological assemblage utilizing primarily wood, wind, and water variously as material, motive force, and transport medium.[17] This applied as much to the rice trade as to other sectors of the economy. Nearly everything moved by water, and a production location very far from a navigable waterway was a substantial commercial handicap. Even so, much of the work of agriculture, milling, and local transportation was done by hand or animal power. There was a steam-powered rice mill operating in Manila during the early 1850s, but it did not prosper. Until nearly the end of the nineteenth century most rice destined for human consumption entered the city as *pinawa*, clean (as opposed to white polished) rice, having been stripped of its outer husks by hand pounding or, in some cases, by means of a *gilingan* rotary hand mill. There were two geographical systems of domestic rice supply to the city, an inner one from the watersheds centered on Manila Bay and Laguna de Bay and an outer one from a dispersed group of production zones in Pangasinan, Zambales, the Ilocos coast, Panay, and Camarines Sur, the last two being strictly opportunistic suppliers.

Cascos and the Inner Zone

Defined as the area integrated by waterborne transport in the drainage basin around Manila Bay and the large shallow lake known as Laguna de Bay, the inner zone included much of the Tagalog and Kapampangan language areas. Movement by small boats on creeks, rivers, estuaries, bay, and lake had been important for a long time. Farmers in this zone largely practiced long-season, single-crop rice culture and shared a diverse family of seed stock, as well as important harvesting and crop-handling techniques. The majority of rice entering commerce here in the second half of the nineteenth century was initially germinated in seedbeds and transplanted 30 or 40 days later into well-prepared pond fields. Transplanting significantly raises yields per hectare over direct seeding in part because it facilitates better control of weeds and makes for a more uniform maturation, but it is highly labor intensive during certain periods of the crop cycle.[18] In Central Thailand, a switch from broadcasting to transplanting has been related by one author to expanding commercial opportunities and rising population densities. Certainly the rising demand for Thai rice in Singapore and among Chinese workers in Malaya was for the transplanted varieties.[19] These causes might hold in a general way for Central Luzon as well, although the farmers in at least some localities in the Manila area were already transplanting when the Spaniards arrived in the sixteenth century. Even in 1900, cultivation practices were not uniform in the inner zone. Although transplanting was common, planting by broadcasting seed directly remained in use in western Bulacan throughout our period, as well as in the Cavite uplands and parts of Laguna. More commonly the broadcast (*sabog*) method was used for a supplemental, second rice crop grown without a secure water supply. If the second "crop" was a ratoon crop allowed to grow from the stubble and rootstock of the first, this went without mention.[20]

Most Philippine lowland wet rice varieties are basically swamp plants. To grow them, farmers need a means of soil preparation that results in a deep slurry—even when they are starting out with a hard surface at the end of the dry season. The plowing and repeated harrowing of muddy fields was largely accomplished by the labor of the water buffalo, a very well adapted draft animal. The water buffalo, or carabao (*Bubalus bubalis*), is the critical plow animal where agricultural soils are deep mud. They are stronger than cattle and work well in mud—unlike cattle, horses, or mules—even mud so deep that they virtually slide on their bellies. They thrive on poor village browse and are easier and cheaper to maintain. They are also slower than cattle and must be given frequent opportunities to cool themselves by immersion in mudholes or water. According to one account, a properly managed carabao was started out on light work in its third year and in a few cases was still working at 30, although an average of 18 was more common.[21]

For a long time lowland rice farmers of the Central Plain raised a single annual crop timed to take advantage of the rains. Only rarely did these farmers use early-ripening (short-season) varieties; rather, in normal times they took advantage of the full rainy season (*tagulan*) so as to obtain a greater yield. If one includes the east coasts and far south in the consideration, then the Philippines has quite a variety of rainfall regimes, but in Central Luzon both the rainy season and the dry season are normally substantial and well defined. In terms of cropping security, it is the sufficiency of rainfall during late May through October or early November that counts most.[22] In a normal year this was the rice season in much of Central and northern Luzon. But many years were not "normal." There might be too much rain, not enough, rain at the wrong time, death of draft animals, human cholera, or war. In any of these cases the harvest would have been short, and so would the amounts of domestic rice forwarded to Manila.

In the mid-nineteenth century *palay* grown in the inner zone was harvested with a sickle, dried in the open, and threshed by foot, carabao trampling, or by being beaten on a stone or rack and winnowed. Rice that was not immediately needed or threshed was often piled in large cylindrical stacks for further drying (figures 1.2, 1.3). Later it was threshed, winnowed, and placed in the family granary or sent into the commercial system unhusked.[23]

Palay bound for the Manila market from the inner zone was often collected in one or another of several commercial towns in the vicinity of the city. There it was stored, usually partially milled, and sent into the city by small watercraft on a daily basis. These watercraft did not pass through the statistical apparatus of the Port of Manila at the mouth of the Pasig River because they proceeded directly up the tidal creeks and canals—*las vias fluviales*—to the commercial small-craft landings and public markets in and around the Binondo commercial district (map 1.1).

Malabon was the most notable of the towns collecting and supplying this rice. In the nineteenth century, Malabon and its parish were commonly known as Tambobong, a Tagalog word meaning "granary" or "storage structure for rice." In their gazetteer of 1850, Augustinian friars Manuel Buzeta and Felipe Bravo note that the merchants of Tambobong carried on an extensive commerce in

FIGURE 1.2. Threshing by means of carabao trampling, early twentieth century. Threshing was also done by beating the stalks on something or trampling by foot. (USNA II, RG350-P-Ac-4-4)

FIGURE 1.3. Hand threshing by beating the stalks on a stone and winnowing with an open-work tray (*bilao*) or by means of a hand-cranked winnowing machine (*background*), Bulacan, early twentieth century ethnographic photo. Note the high stacks (*mandala*). (USNA II, RG350-P-Ac-4-14)

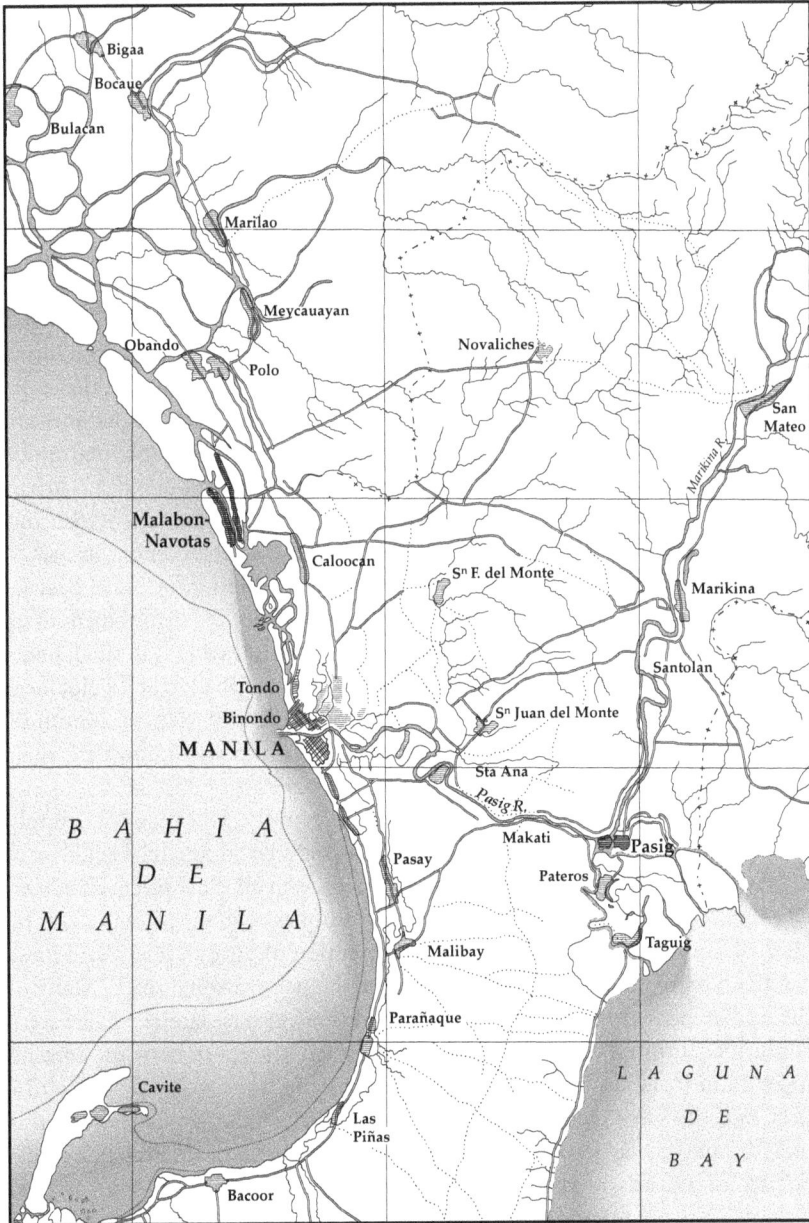

MAP 1.1. Water Routes and Nodes in the Inner Zone Rice Supply System, 1885.
Redrawn by Rich Worthington from "Provincia de Manila, 1885," 1:100,000
blueprint map, John E. T. Milsaps Collection, Houston Public Library, Special
Collections, series D, box 1, folder 5.

unhusked rice with nearby provinces, such as Bulacan and modern day Rizal. In the 1880s, Bulacan and Nueva Ecija and also Pampanga and Bataan were supplying Malabon with *palay* for subsequent processing in the 65 or 70 *fábricas de arroz* located there—presumably hand-operated rice mills—which employed 600 or 700 workers on a daily basis. More than 60 storage structures lined the waterway in the 1890s (map 1.2). British Consul Pauli called it "the depot for the productions of Bulacan and Pampanga, from whence they are forwarded to Manila or direct to shipping in the bay."[24] Following considerable population growth, its municipal territory was divided in 1859 to create the administratively separate municipality and parish of Navotas. Arranged along parallel beach ridges, Malabon and Navotas lay astride the natural intracoastal waterway that connected the towns of southern Bulacan with the commercial district of Manila. For functional purposes the two jurisdictions are better thought of as one. Their efforts to supply and support the metropolis were sufficient to create in Malabon-Navotas one of the more significant urban places in the archipelago. Taken together, they were fourth, just after Cebu and ahead of Legaspi and Cavite, in a ranking of Philippine urban centers in 1903.[25]

Beyond Malabon, near the outer end of the estuary and intracoastal passage in Bulacan, the small town of Bocaue was also notable for the involvement of its inhabitants in buying and husking rice and then transporting it in small boats to Manila for sale in the public market. In between, at Obando and Polo, local women trafficked in rice and carried it to the city.[26] Several of these communities, notably Malabon-Navotas and Obando, were also well known for their roles in supplying the city with fish. But their critical function in the rice supply system was equally vital in the mid-nineteenth century. This inner rice supply zone was an arena of considerable indigenous-mestizo commercial activity.

Nature's endowment of inland waterways allowed all these places to be easily connected to the city. Maps of the period emphasize this fact (map 1.1). The maze of minor deltaic channels drew the attention of cartographers, and these are much more prominently shown than are the poor seasonal roads. Many of these channels lead to Manila Bay itself, but others traverse laterally in a watery landscape, forming a convenient and protected transport network coming together in the southeast at Malabon-Navotas and ultimately via Tondo district into the city's commercial heart. A problem in the last portion of this inland waterway lay in how to get through the swamps. In 1823 the Economic Society of Friends of the Country awarded a gold medal to one Doroteo Punzalan Estrella for opening a channel that gave "a new and more convenient direction to the river of Tondo." Likely this refers to the Canal de Maypajo, which connected Malabon and the adjacent Dagatdagatan (lagoon) to the Vitas-Pritil area of Tondo with landings in the Bankusay neighborhood. Coello's map depicts this route as it was circa 1850 (map 1.2).[27]

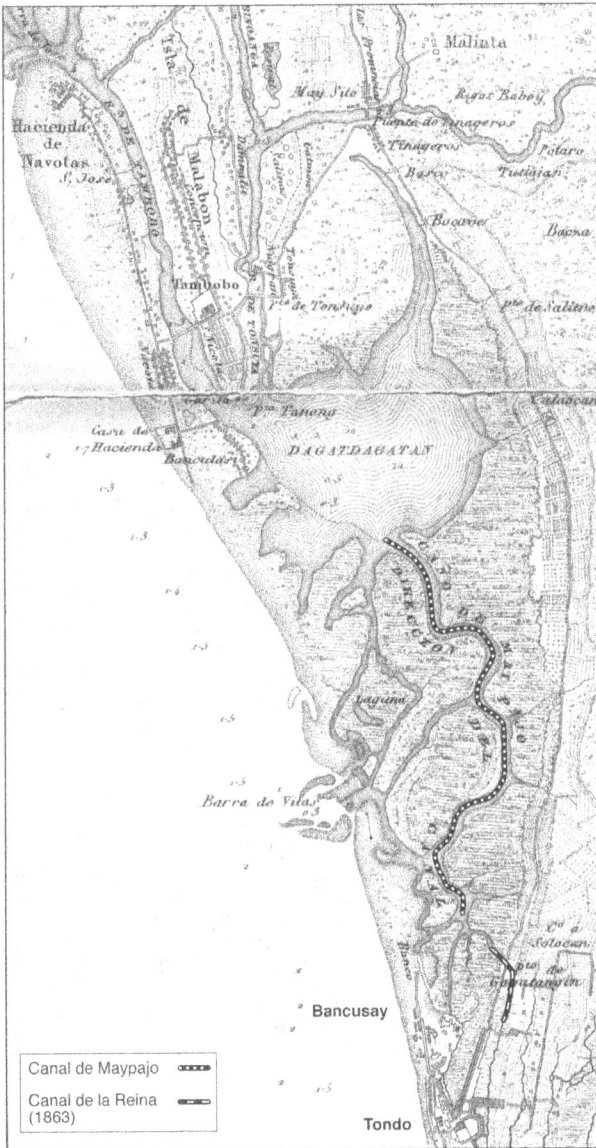

MAP 1.2. Malabon-Navotas and the Intracoastal Waterway into the City, 1849. This remarkable map depicts the course of the Rio de Tambobo (now the Navotas River), Dagatdagatan, and the Canal de Maypajo into the city. Substantial quantities of rice were delivered via this route. Rice storage structures and the homes of *casco* owners lined both sides of the river. "Parte de Costa de la Bahia de Manila," inset on the "Islas Filipinas, Primera Hoja Central," 1849, sheet 25, *Atlas de España y sus posesiones de Ultramar*, compiled and published by Francisco Coello, Madrid, 1848–68.

Then, in 1864, the Queen's Canal, the Canal de la Reina, was cut through the commercial fishponds and swamps of Tondo proper to provide a secure and direct connection to the commercial district. Passage was free. Its primary purpose was to bring rice and other food products into the city. It was connected to the Estero de Binondo, and together these gave ready access to Divisoria, the principal public wholesale and retail market of the city.[28] By contrast, a private toll was charged for use of the Canal de Maypajo. As a result, vessels tended to follow the coastline instead. But in even moderate seas the canal was the safer route.

Cascos were the workhorse river vessels that delivered most of the freight from inner zone production areas. These were long, flat-bottomed watercraft propelled by poling and sometimes by sail (figure 1.4). Cargoes included rice, firewood, palm vinegar, freshly cut sugarcane for immediate consumption, and semiprocessed *pilón* sugar, even watermelons—almost anything of value and bulk that was not immediately perishable. The efforts of two or three polers moving along the side-mounted running boards of these vessels became a notable subject of artists and photographers. Poling against the current could be hard work, but the major loads bound for the city generally moved downstream with the languid current. The shallow-draft *cascos* worked well on sheltered

FIGURE 1.4. *Cascos* were the primary bulk transport vessels on the rivers and inshore waters of the inner zone. Here crewmen move firewood into the central city by poling. Rice also arrived this way. (John Bancroft Devins, *An Observer in the Philippines* [Boston, 1905], facing 56)

intracoastal waterways or Laguna de Bay but were not suitable for open water in less than ideal weather. Although they were often equipped with movable arched cargo covers, loaded *cascos* rode low in the water and the cargo could easily become soaked in rough conditions. *Cascos* were built in various sizes. Many had a substantial capacity, though probably somewhat less than the 800 to 1,500 *cavans* routinely carried by *pancos* and *pontines* (smacks and yawls)—the standard coastal sail craft of the day. Agustin de la Cavada reports 440 *casco* and other boat departures from the interior province of Nueva Ecija during 1870—all bound for Manila and carrying 75,000 *cavans* of *palay* and rice, as well as other cargo.[29] There were also smaller river vessels. But most shipments to Manila from the inner zone remain a mystery because few reports of *casco* arrivals and departures have been found.

Casco ownership was concentrated in the hands of the same local people who owned the storage structures in the rice supply system. A plurality of the vessels used to connect Nueva Ecija and Bulacan with the city was based in Malabon-Navotas. Over 300 *cascos* have been located in the tax-registration records of the 1890s. More than half of these were based in Malabon-Navotas and Manila itself—100 and 69, respectively. Other *cascos* were based in landings distributed about the bay.[30] Tax licenses for *cascos* based in Cavite routinely mention that the vessels are for use around Manila Bay and on the rivers of Pampanga. Most of the *cascos* of Malabon are described as being for rent.[31] Clearly Malabon-Navotas combined a large number of available *cascos* and many rice storage and hand-processing facilities. Business arrangements involving *casco* rentals remain obscure, but we can see the eventual introduction of formal insurance and the operation of a judicial system that could see to the compensation of merchants when their food cargoes were damaged by the negligence of the operators of leaky vessels.

Ownership of the 300 *cascos* was moderately distributed, with almost 90 persons owning 1 to 3, 18 with 4 to 10, 2 with 13 to 15, and 2 with more than 20 (table 1.1). The last two were based in Manila where many of their *cascos* were used as lighters, ferrying cargoes and ships' provisions to and from ocean-going ships. *Casco* ownership was most concentrated in the city. Luis Rafael Yangco was the largest and best-known owner, an entrepreneur in shipping people and cargoes to and from Manila over relatively short distances from the shores of Laguna de Bay to Subic Bay in southern Zambales. In transport terms, he became a major figure in feeding the city. Yangco's career began with distributing drinking water by *banca* (native boat). In the 1870s he arranged to construct a number of lighters to service domestic and international vessels. By the 1890s he was operating 23 *cascos* and several small steamers. On his death in 1907, he owned a fleet of more than 25 small coastal steamers, as well as machine shops and repair slips in Malabon and Manila.[32] His 10 main

TABLE 1.1. Major Casco Owners in the 1890s

Owner	Base locale	Number owned
Luis Rafael Yangco	Manila, San Nicolas	23
Enrique Rodriguez	Manila, Binondo	22
Roman Sta. Maria	Malabon, Sta. Rita	15
Pedro Naval	Navotas, Calle Real	13
Doña Maria Santos vda. de Macario Lichauco	Manila, Quiapo	10
Romulo Mercado	Pampanga, Sesmoan, San Nicolas	10
Mamerto Rivera	Malabon, Sta. Rita	9
Julian Andres	Navotas, Almacen no. 60	9
Florencio Andres	Navotas, Balite Centro	9
Francisco de los Santos	Navotas, San Roque	7
Segundo Mercado	Pampanga, Sesmoan, San Nicolas	6
Doña Luisa Naval	Navotas	6

SOURCES: PNA, Contribución Industrial: Manila, 1892, 1896 (1), 1896 (3), 1896–97; Bataan, 1892–97, 1893–96; Bulacan, 1893 (2), 1893 (4), 1894, 1894 (5), 1895–96, 1895–98 (1), 1897; Cavite, 1892–94, 1894–96, 1895–97, 1896 B; Nueva Ecija, 1893–97 (R2); Pampanga, 1892, 1891–97, 1893–98, 1893–97 (1), 1893–97 (2), 1894–97, 1880–98 (altas y bajas).

warehouses in the 1890s were located along or very near the Pasig River *muelle* (wharf) in San Nicolas district. Yangco also served as *capitan municipal* (municipal executive) of Binondo and in 1894 was appointed a *regidor* (councilman) of the city, a position earlier reserved for Spaniards and creoles (conventionally Spaniards born in the Philippines). Several other owners can be identified. Enrique Rodriguez was a general landing and transfer agent based in San Nicolas. In Navotas, Pedro Naval and Julian Andres were important *casco* owners. Andres also maintained a very large rice storage facility, and both men were important fishing entrepreneurs, as we will see.

Another Manila *casco* owner in the 1890s was Maria Santos, widow of Macario Lichauco, carrying on his water transport business. This enterprise was operated by her sons, Crisanto and Faustino Lichauco, the latter subsequently famous as a major supplier of beef to the city.[33] After the Philippine-American War, Faustino Lichauco began building *lorchas* (cutters) and using them to unload oceangoing vessels in the Manila roadstead. Luis Yangco and Faustino Lichauco appear repeatedly in the business of provisioning Manila.

The flat-bottomed *cascos* were well adapted to the commercial need for reliable and inexpensive bulk transport in a day of modest infrastructure, but their passage could be perilous in choppy waters. Also the main river channels of western Bulacan were brigand lairs in the days before the mass conversion of the landscape into fishponds.[34]

∿

East of the capital, another network of commercial centers was arrayed along the Pasig River, the artery that connected communities and production zones around the interior lake with the markets of the city. Again *cascos* and also narrow sail and outrigger vessels called *paraws* were critical in this commercial flow. For a long time the key commercial town was Pasig, strategically situated at the intersection of the south-flowing Marikina tributary and the main channel of the Pasig River (map 1.1). The deep alluvial Marikina Valley was celebrated for its lowland rice. Pasig town performed functions somewhat like those of Malabon. In normal times, the rice trade here could be lucrative. In 1850 the town's commercial intensity was marked by having two official periodic market days (*tianguis*) per week rather than one. Husking rice was one of the major tasks carried on here by local women, and the clean rice they produced was a major item in the active river commerce that Pasig folk carried on with Manila. In 1870 a visitor wrote, "Pasig is a big and handsome pueblo; the people are relatively wealthy since they handle the intermediate trade between Laguna and Manila." It was a place where river and lake vessels could readily be engaged. When the new province of Rizal was inaugurated in the early twentieth century, Pasig was its principal market center.[35]

Just east and south of Pasig town, several channels connect the lake to the Pasig River. Along one of these channels, at Pateros, Buzeta and Bravo note the operation of 20 rice-milling establishments that they call "work places for husking rice by machine." In 1850 Pateros people were carrying on a substantial commerce, partially milling and carrying rice to the metropolitan market every day by boat, a trip of several hours. There would be no adequate road connection until the twentieth century. Thirty years later officials commented on the storage and cleaning there of *palay* produced in other locales, especially in various parts of Laguna Province. Still later Montero y Vidal calls the rice-milling establishments of Pateros "*pilanderías*" (places where the pounding of rice takes place).[36] The Pasig River arterial also connected places very near at hand. From Pandacan, a settlement almost contiguous to the built-up city in the 1870s and 1880s, women bought rice in nearby municipalities for local milling and eventual sale in Manila. In the city, a steam-powered mill for cleaning rice (and sawing lumber) was in operation at Malacañang from the 1830s. But, say Buzeta and Bravo, while "some Europeans have lately introduced [milling] machines, [Filipinos] do not find that the grain becomes as polished in these as by the methods that they employ," namely, pounding.[37]

Beyond the Pasig River settlements was the littoral of Laguna de Bay and, to the south, Laguna Province. Western Laguna was well suited to rice cultivation.

This was the location of large friar estates. By the nineteenth century these were leased in parcels to persons known as *inquilinos*, who then sublet the land in small sections to Tagalog cultivator families.

The religious owners increasingly became rentiers rather than managers. In this zone were numerous water control devices used to enhance rice production. Many of these were already more than a century old. Among the Dominican haciendas, Biñan became a trading center from which rice was sent to Manila via lake and river.[38] Farther east Santa Cruz had emerged as the important provincial market center—with two market days—by the mid-nineteenth century. From this wet rice zone, much of the local grain surplus was traded to the rice-deficit interior uplands rather than mainly to Manila. Increasingly this trade allowed several interior municipalities to specialize in producing coconut oil and alcoholic beverages from local palm stands. These products were forwarded to the city along with some commodities brought overland by pack train from Tayabas.[39] Along the Laguna lakeshore was a string of large estates from San Pedro Tunasan and Biñan southward through Santa Rosa and Calamba, mostly under Dominican ownership. Directly west, in Cavite, were the huge estates of Imus and San Francisco de Malabon (now General Trias). Both had installed extensive irrigation works.[40]

Returning to Pasig and turning east, one entered the administrative and military district of Morong centered on the municipality of the same name. Here, just before 1850, an earthen dam was constructed on the Morong River together with a 500-meter irrigation ditch. Half a century later, a local official organized farmers to construct and operate a temporary dam to facilitate irrigation of a second crop of wet rice during the dry season. Until after World War II, "this dam was constructed every year during . . . February and demolished either by flood or by the farmers themselves in [late] May or early . . . June when *palay* was about to be harvested and water was no longer needed."[41]

So during the middle and late nineteenth century there was an active trade by a host of small cargo craft carrying *palay* from nearby growing areas such as Bulacan, the Marikina Valley, and western Laguna to storage and husking or milling sites located just outside the main urban area. These sites supplied clean and also white rice, *pinawa* and *bigas*, to the city on a daily basis. This "just-in-time" arrangement for milling and delivery had the great advantage of supplying a commodity that was fresh and minimizing spoilage in a day before mechanical dryers.

Rice also occasionally arrived in the city by boat from lower Pampanga and Cavite. The available statistics are but tantalizing fragments counterbalanced by other reports of rice shipments *to* Pampanga as the production of sugar expanded there. Cavite was known for supplying Manila with small quantities of high-quality rice, but so far there is little record of Cavite rice arriving in bulk.[42] In the agricultural survey of 1886–87, both Pampanga and Cavite reported sending some rice to Manila, but the amount now sent from Bulacan was said to be "insignificant."[43] In the 1860s and 1870s, the relative share of Manila's requirements supplied by the provinces of the inner zone was diminishing.

The pattern of specialized hand-milling centers arrayed about the city was certainly not unique. There was something similar in the penumbra of Hanoi in northern Vietnam and no doubt other places, but, unlike the *hang sao* of Tonkin, the artisanal rice-milling laborers of Luzon attracted little notice.[44] In northern Vietnam, the *hang sao* in such tiny, specialized milling centers often raised pigs as a way to capture some economic value from the stocks of bran and bits of broken grain that their labor created, and manure from the pigs was welcomed by local farmers. In the Tagalog area, many households raised a pig, and these formed part of the meat supply of the city, but no special connection between hand-milling centers, swine raising, and attendant manure use has yet emerged from nineteenth-century data. The Tagalogs were living at a lower density on the land than the Vietnamese of the Red River delta and were also somewhat less devoted to swine raising.

By contrast, the rice bran generated by milling was an important component in horse feed, and in the 1880s and 1890s the nearest hand-milling center at Pandacan became the stabling area for the 134 horses of Manila's horse-drawn streetcar system, the *tranvía*.[45] Likewise, *darak* (rice bran), *zacate* (forage grass), and raw *palay* itself were carried into the city by small boats for use as horse feed. So there was a relationship between the by-products of hand milling and the maintenance of the considerable horse population of the city. *Darak* was also a good feed for chicks and, in mixture, for hogs as well. Another specialized economic use of these by-products was seen at Pateros, where thousands of ducks were bedded down each night in special sheds on a thick layer of rice husks from which fresh eggs were collected each morning. Raw *palay* itself formed part of the diet of mature ducks. Rice husks were also used at Pateros as packing and insulation in the process of partially incubating duck eggs into *balut*, an important snack, as we shall see.[46]

This was the inner zone rice supply and *casco* navigation system as it operated in the mid- to late nineteenth century. The same provisioning system also brought domestic fruit, vegetables, palm vinegar, chickens, and eggs to the metropolis.

Coastal Shipping and the Outer Supply Zone

Rice also came to the city in large quantities from the outer zone. Linked to Manila by coastal sail shipping was a series of production zones more distant than those around the bay and lake. By the mid-nineteenth century central

Pangasinan, together with what would later become northern Tarlac, came to form the main concentration of outer zone rice production for the city. In this broad alluvial plain a considerable area of soil is clay loam, very well suited to wet rice cultivation because of its ability to absorb and retain moisture. Here fine silt is frequently deposited by the overflow of the Agno River, improving soil texture and providing nutrients.[47] As in the inner zone, the rice farmers here largely practiced single-long season cropping. June rainfall was more secure in central Pangasinan than in Manila or Nueva Ecija. At the same time, October rainfall was marginally chancier.

The rice system in central Pangasinan differed from that of the Tagalog zone in some important ways. Here and in the adjoining Ilocano folk region, many farmers planted awned, or "bearded," varieties of rice that were little used in the inner zone. Awns are slender bristles growing out of the end of the husks. In central Pangasinan these varieties were primarily grown without active irrigation simply by trapping rainfall and sheetwash in the field, a system of rice culture known in Tagalog as *sahod* or *sahod-ulan*. Harvesting practices provide another point of difference. In Pangasinan, "[T]he spikes are cut off one by one with a hand instrument consisting of a handle and blade." This instrument, called a *rakem* in Ilocano, has a blade set sideways on a small handle. Its use as a harvest instrument in Luzon was specific to people and places raising the awned varieties. In Ilocos Sur the small initial bundles of awned rice were tied together in groups of six to make a *manojo*—a bundle graspable by hand. In Pangasinan the system was similar. As historian Rose Cortes remembers it from the 1930s, all the harvested *palay* was made into small bundles and placed unthreshed in the family storehouse. As the need arose these small sheaves were brought out, further dried, threshed, and pounded. This routine was little changed from that reported in the 1830s.[48]

The gathering and storage or trading in sheaves was practical because the awned grains tended to remain firmly attached to the stalk. In Pangasinan in the 1850s and long after, rice entered commerce in *manojos* unthreshed. Henry T. Lewis reports that awned varieties simply store better unthreshed and with less spoilage than the beardless varieties but that the bundles first need to be set out "to mature in the sun for a few days" (figure 1.5).[49]

As commercialization proceeded there were efforts to intensify production by manipulating water to extend the growing season, to make it more reliable, and even to facilitate limited second cropping. Rosario Cortes places the construction of "dams, irrigation canals, and ditches" in the 1830s in the eastern region of Pangasinan on the Angalakan and Toboy-Tolong rivers. These efforts finally resulted in irrigation dams in ten municipalities in the eastern portion of the province—versus only two in the west. Predictably, the diversion of

FIGURE 1.5. The northern rice system: sheaves of freshly harvested *palay* are set out to dry prior to storage in the family granary, Dipilat, Vintar, Ilocos Norte, circa 1901. (USNA II, RG350-P-Ac-3-3)

upstream water for seedling plots sometimes led to controversies over water rights and protests by downstream farmers.[50]

East-central Pangasinan is also where the noria waterwheel irrigation device was introduced some decades later. This device is particularly useful where the river is well incised—precisely the case with much of the Angalakan. Norias were designed to harness shallow stream flow to power large vertical wheels to lift bamboo tubes filled with water. As each tube passed the zenith of the wheel's rotation it poured water into a high trough or bamboo pipe leading to the field (figure 1.6). Usually, such a noria was capable of irrigating two or three acres. The norias were an appropriate technology—lightly constructed from readily available materials, then dismantled and taken out of harm's way during the typhoon season. Pangasinan was the main place where they were used. Scores were eventually employed along the Angalakan from Pozorrubio and Manaoag to Mapandan (map 1.3). They provided water to get seedbeds started and deal with the effects of drought at the start of the main rice season.[51]

These mechanical water lifts were an innovation of the 1860s in Pangasinan, but they were also widely used for irrigation in other parts of Southeast Asia. In central Vietnam their design was a high art. They were common among the Minangkabau of Sumatra and were reported in Malaya (Negri Sembilan),

FIGURE 1.6. A noria lifts irrigation water near Pozorrubio, Pangasinan, early twentieth century. A temporary barrage deepens the river flow to power the wheel. (Bureau of Science photo, USNA II, RG151-FC-84D, box 84)

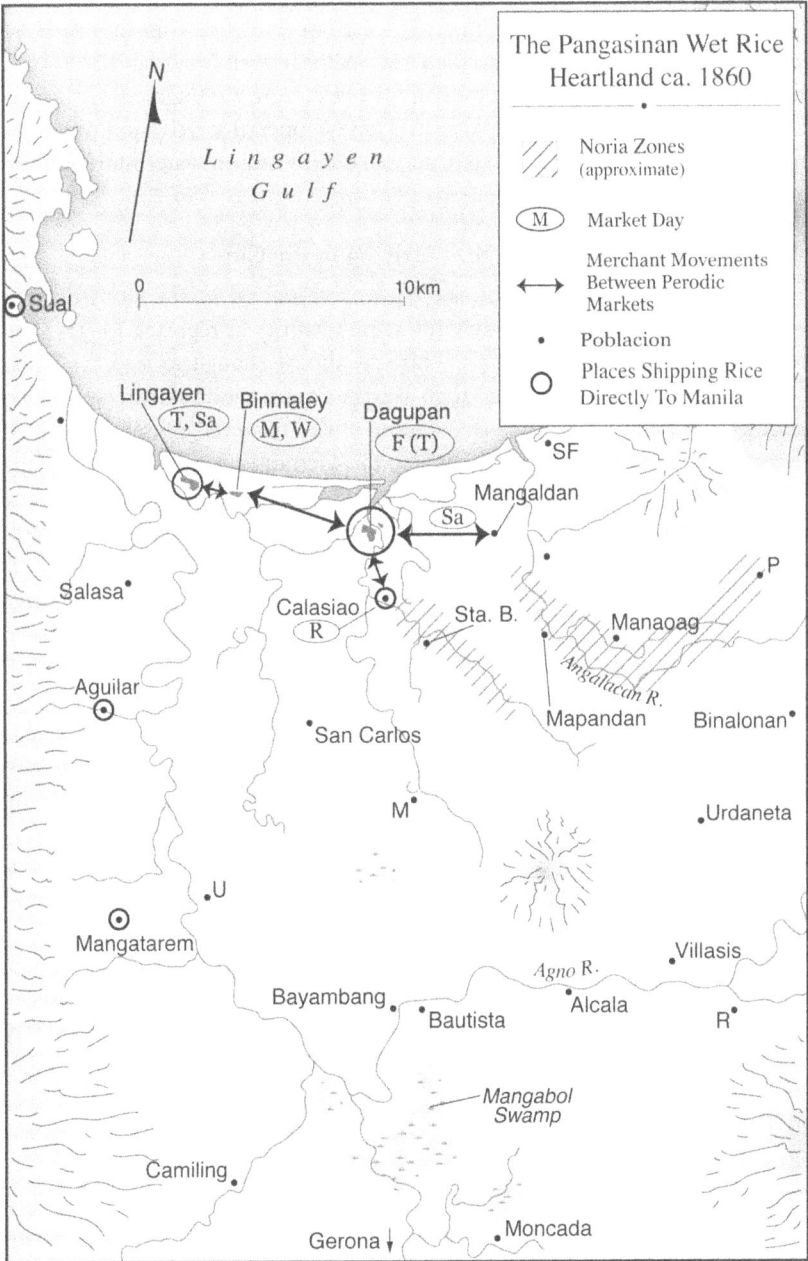

MAP 1.3. The Pangasinan-Tarlac Surplus Rice Zone, including Major Periodic Markets and Noria Irrigation Areas

Cambodia, and northern Thailand. In fact, norias had been employed in India, the Middle East, and China for some two millennia. Yet they seem to have been introduced in Pangasinan via Spain or even Spain and Mexico.[52] Farther north, in Ilocos Norte, rainfall for rice culture was augmented by local irrigation systems known as *zangjeras*. These cooperative irrigation organizations and infrastructures were used to make rainy season water more reliably available.[53]

Rice Cargoes Arriving in the City

Some 400,000 *cavans* of clean rice equivalent arrived in the city from the outer zone in 1862, approximately the same quantity as reported for 1853 but considerably more than in 1854.[54] This was almost enough to feed the entire population of the city for a year at an average consumption rate of two *cavans* per capita.[55] It easily exceeded the volume of rice from the densely populated inner supply zone surrounding the capital. Certainly it did so prior to the development of Nueva Ecija in the later nineteenth century. Primarily the cargoes consisted of ordinary clean rice, *pinawa*. The general Spanish term *arroz* was used to record 87 percent of the total volume, but sometimes *arroz corriente* (rarely *arroz ordinario*) was employed (table 1.2). At this time *arroz corriente*, meaning the standard form and variety, likely referred to *pinawa*. Only twelve shipments were recorded as including *arroz blanco*, and reports of commerce in *arroz blanco* are also rare.[56]

Unhusked *palay* comprised more than 10 percent of the total when calculated on a clean rice equivalent basis. Since husking reduced the volume and cost of shipment by approximately half, one wonders if the absence of husking in some supply localities reflects a fear of spoilage in transit during the rainy season or, in the case of central Zambales, some combination of proximity reducing the cost of transport and a local labor shortage. Whatever might have been the case, the decision on form was often related to shipping cost.

Just two special rice types are singled out in the record, both in minuscule quantities: *malagkit* from Zambales and Pangasinan and *mimis* from Ilocos Norte and Pangasinan. *Malagkit* refers to a diverse group of glutinous sticky rices used in sweet confections, while *mimis* denotes an especially white, nice-smelling, and expensive variety. Del Pan describes it as a variety of "exquisite flavor"—"sweet, clear, and white"—and Carro calls it the best in Ilocos. In 1917 Hill wrote that neither type yielded enough to attract commercial growers in Central Luzon. *Mimis* began to pass out of use in the twentieth century because its grains broke too easily under mechanical milling. Today *milagrosa* has largely replaced it.[57]

Widespread use of standard units of measure constitutes a sign of a penetrating commercial system and the power of the state to regulate it. This can

TABLE 1.2. Form and Quantity of Rice Arrivals from the Outer Zone, 1862, 1872, and 1881

Form and type	1862 Cavans	1872 Cavans	1881 Estimated cavans
arroz	342,275	260,062	581,819
arroz corriente	8,621	49,898	13,310
arroz blanco y corriente	—	49,303	2,897
arroz blanco y segundo blanco	1,872	21,621	3,800
palay	41,273	24,735	19,173
malagkit	478	190	552
Total	394,703	405,972	621,551

SOURCE: Calculated from the daily record of arrivals in the *Gaceta de Manila*.

NOTE: *Palay* is stated in *arroz* equivalent terms. For 1881 ten cargoes were reported only as *con arroz* or *con palay*, and two serious misprints in the *Gaceta* have been corrected using values reported in *El Comercio*. Twenty other cargoes could not be located. These have been estimated in reference to the record of shipments for the particular vessel or vessel type consigned to the particular merchant: *con arroz* 22,750 *cavans* and *con palay* 2,050.

be seen in the case of rice entering the Port of Manila, since almost all cargoes were expressed in terms of the 75-liter *cavan/kabán*. Our sample year 1862 was the first in which the size of the *cavan* was legally standardized in all provinces, although in fact there were several in the Visayas and Mindanao where this was not immediately followed.[58] This relatively high level of standardization was decidedly not the case with a number of minor food commodities entering the port in one of several alternate systems of volume description. The exact conversion of these is problematical, but they accounted for no more than 1 percent of total arrivals of rice.[59]

～

Ten years later, in 1872, the internal rice trade centered on Manila appears to have changed little. Again about 400,000 *cavans* of clean rice equivalent were recorded as having been received from the outer zone (table 1.2). This lack of expansion in the flow has several possible explanations. The previous year witnessed a moderate El Niño in the eastern Pacific, and, although it failed to produce the usual drought in East Java, Manila and Central Luzon were nonetheless extremely short of soil moisture at the start of the 1871 agricultural cycle. This may have delayed and compressed the growing season and thus affected the harvest at the end of 1871.[60] This harvest, in turn, provided the domestic stocks available for shipment to the city during most of 1872. Further, the first quarter of 1872 saw a brief revolt in the Cavite garrison near the city followed by secret trials, public executions, and exilings. Although we are without specific evidence, the resulting fear may have had a dampening affect on the commercial activities of mestizo merchants.

One change during the decade was the growing quantity of clean rice described more specifically than *arroz*. Extrapolating from incomplete data, we might estimate that *arroz blanco* now made up about one-third of clean rice deliveries and *arroz corriente*, or *pinawa*, about two-thirds, showing a trend toward more complete milling. At the same time, unhusked *palay* declined from more than 10 to only 6 percent of the whole. Little can be inferred concerning the methods of milling, but British consul William Gifford Palgrave reported in 1878, "The work of cleaning and husking is entirely done by the cultivators and their families, chiefly the women." Only hand labor was employed according to him, since local people "object" to machine milling, preferring instead the mortar and pestle.[61]

Annual Cycles

The rice system that fed Manila operated on a seasonal calendar, and the schedule of grain arrivals had implications for the annual cycle of grain availability, price, and seasonal urban employment in milling and handling. The southwest monsoon and its associated rainy season drove the timing of the annual crop cycle, the flow of interisland sail navigation, and, ultimately, the march of prices and much else besides, including the relative seasonality of human births and mortality.[62] In the 1850s, January–May was a well-developed commercial season in Pangasinan. This overlapped with and followed the harvest of the previous year's rice crop. It was also the dry season and northeast monsoon, during which coastal vessels operating from Dagupan and other ports could safely navigate to Manila or Ilocos. Rice trader and British vice consul Jose de Bosch describes an intense rhythm: light bridges replaced, roads repaired by corvée workers, and commerce humming with shipments to Dagupan and other downriver ports. Much of this ceased when prevailing winds reversed and the rains set in. Only the protected harbor at Sual remained in ordinary, if desultory, service (map 1.3).

The general rice harvest in Pangasinan and the Ilocos coast began at the end of the rainy season in November, peaked in December, and in some locales and years lasted well into January. In Nueva Ecija and Bulacan in the inner zone, the harvest centered on December and January and continued into February.[63] In 1861 significant rice shipments from Zambales and Pangasinan began to arrive in the city in mid-December. Arrivals were flat during January 1862, then built to notable peaks in March and May. More than 54 percent of the total outer zone supply arrived during the March–May quarter that year. Arrivals fell off sharply in June as southwesterlies, the gathering rainy season, and the passage of typhoons made navigation problematic. Shipments picked up again in the fourth week of November.[64]

Provincial reserves of rice left over from the previous harvest fell inexorably, approaching the vanishing point as the new crop season wore on if the price offered was attractive or the previous crop was short, or both. The flow of rice arriving in Manila from the outer zone during the December to May dry season averaged more than 50,000 *cavans* per month, while during the June to November *tagulan* the monthly average was 14,000 *cavans*. September–November was the lowest period. It is a fair hypothesis that some of the slack in the outer zone supply during this season was made up by deliveries from the inner supply points, including Malabon and the Pasig River towns.[65]

A decade later the volume of shipments followed a similar but smoother cycle. Rice was already flowing into the city at a moderate rate during December 1871 and continued to rise to a high, flat plateau during February, March, and April 1872. There followed a regular decline, reaching a low during August through November. Overall, 86 percent of the calendar year total had been landed in Manila by June 30 in 1872 versus 74 percent by the same date in 1862. Again, these year-to-year peculiarities may be due to weather events affecting the maturation and harvest of the crop, on the one hand, and the navigation of sail craft from provincial ports on the other.

Transport by sail craft took place on a schedule largely determined by nature. It is not hard to understand why a week or even two might pass during the rainy season without a single significant outer zone rice shipment arriving in the city. Vessels venturing into Lingayen Gulf and down the Zambales coast during the southwest monsoon often faced difficult seas and dangerous winds. In May it took 3 or 4 days for a cargo of Pangasinan rice to arrive in the city. In June and July 1872 nine vessels from Dagupan and nearby ports took from 14 to 23 days to complete the same voyage. Some of these sought shelter in Bolinao and others stopped farther along the Zambales coast or at Mariveles at the entrance to Manila Bay. Several needed to repair or replace damaged sails. Shipments from the north dwindled to nothing.

This phenomenon had an impact on the annual cycle of prices. In ordinary years the price of rice moved inversely to the level of supply on hand and the flow of arrivals—reaching its highest point as domestic shipments dried up and stocks dwindled in the rainy season months prior to delivery of the new harvest. One critical question has to do with the store of rice on hand at the start of the rainy season. Relying, as it did, on a "just-in-time" system of maritime delivery in an age that lacked mechanical dryers, Manila probably did not have a great cushion of commercial reserves.[66] Still, it has proved difficult to gauge the capacity of the storage structures for urban supply, whether the *tambobong* of Malabon or bodegas in Manila. A century earlier religious estates often kept a significant supply of rice in storage on their properties. Nicolas P. Cushner writes that in times of need this would have given them "considerable influence on local and city markets."[67] Whether that was still true in the much more commercially active mid-nineteenth century is a question waiting to be answered.

Some evidence on the state of rice reserves in Manila may be gathered from structures such as "the cavernous rice granary" on the ground floor of a house built by a rich *inquilino* in Marikina in the 1840s. This was one of several kinds of *bangán* (grain storage structures) built of stone and lime.[68] Some of these were inside the great house, as above, some leaned against the outside of the house with their own roofs, and some were freestanding. In the 1870s through the mid-1890s, before the railroad was constructed, urban commercial warehouses were concentrated "in the center of commercial movement" next to the riverside wharves and Estero de Binondo in San Nicolas and Binondo districts.[69] By the 1890s shipping entrepreneur Luis R. Yangco alone owned several blocks of warehouses near the river. During 1891 his *almacenes de deposito* took in almost 39,000 *cavans* and 67,000 *sacos* of *arroz*.[70] Still, timely deliveries were critical.

In Manila during 1861–62, the cycle of prices for Pangasinan rice started at 1.50 pesos (1 peso, 4 reales) in late October 1861 and remained flat through early June 1862. The price then rose to a high plateau of 2 to 2.125 pesos from July 4 through October 8, declined to 1.875 pesos in late October and continued downward, momentarily reaching the very low price of 1 peso on January 7, 1863. Urban households that could afford it were well advised to make a bulk purchase of rice before June.[71] Most working families could not afford a bulk purchase.

In all, 1862 was a nearly normal year for the seasonal fluctuation in rice supply and price. In half the years during 1850–72, the annual rainy season price rise in Manila was under way in June or July and the higher prices lasted through September, October, or November, when abundant deliveries from the outer zone resumed. The most frequent gross deviation from this pattern came when the new harvest was poor, meaning that the rainy season price plateau failed to come to a timely end and the resulting shortage kept the price in the city high during the first half of the following calendar year.[72]

It is important to keep in mind that from at least the 1830s Manila was connected to good commercial intelligence on the accessible coastal and riverine areas of China and Southeast Asia. A change in the somewhat unpredictable market for rice on the China coast could dramatically affect the quantity of Philippine export shipments, the local price of rice, and the annual price cycle. It was surely no accident that two of the three major arrivals of foreign rice in 1862 entered Manila in early November, though normally a September arrival would have been optimal in order to catch the highest price of the year.[73]

Spanish policy, like the policies of the indigenous mainland kingdoms, was to keep rice available and affordable. The chief instrument was the potential to ban exports when demand and supply pushed prices over some threshold. In 1855 that threshold was set at 2.25 pesos per *cavan* for Manila and 1.75 pesos for the provinces. The authorities in the Philippines imposed such export bans on a number of occasions, including nearly all of 1851 and June–August 1855. In extremis, duties on imported rice could be waived as well, as during the rainy season of 1857 when supplies were imported from Singapore and Batavia. By contrast, Dutch policy in Indonesia at this time was to avoid such measures even in the face of famine.[74] In the Philippines during the 1850s and 1860s, imports of rice were modest. Only during 1858 and 1859 did they exceed exports. Rather, the Philippines remained an episodic net exporter of rice to the China market, sending 10 million kilos or more in 1853, 1855–56, 1860, and 1865, an amount approximately equal to 40 to 45 percent of the outer zone supply to Manila calculated for 1862 and 1872. By the end of the mid-1870s, however, the situation had reversed and net imports became routine.

The Evolving Geography of Outer Zone Supply

In 1862 rice arrived in Manila from 45 ports in the outer zone (map 1.4). Since little of the crop was shipped far overland, each port represents a microregion of production, with its diminutive hinterland integrated by river and creek or coastal short haul.[75] A few provincial ports such as Dagupan, Lingayen, San Narciso-Alusis, and Vigan captured the trade of somewhat larger zones due to their strategic location at the mouths of productive river basins and small deltas. Pasacao was exceptional in receiving its supply overland—from the rice bowl of Camarines Sur. In Owen's view the cost of this overland shipment was no greater than the cost of reaching the nearest alternative markets overland in Albay. Of the many points of shipment in 1862, Dagupan was by far the most significant. Together the top ten ports were responsible for at least 55 percent of the total volume—and probably much more if cargoes reported only by province could be allocated more specifically.[76] In sum this was a highly dispersed system of supply nodes, characteristic of an absence of efficient overland transport, as well as of the flexibility offered by shallow-draft sail craft. Such vessels sometimes brought rice directly to Manila from interior loading points in Pangasinan, bypassing the downriver ports (map 1.3).

Ten years later one sees only incremental change in the pattern of outer zone supply. In each year during this era, more than half the total flow from the outer zone came from the Pangasinan ports.[77] One important trend was some narrowing of the outer zone. Pasacao, a minor supplier in 1862, sent no recorded shipments in 1872. In the expansion of export crop production, the Bikol region lost its self-sufficiency in rice and the direction of net flow reversed. During the same period, production of rice in this region may have

Manila's Rice Supply, 1862 Volume by Port of Origin*

▲		71,400
▲ ●		22,000–26,000
▲ ●		9,500–14,500
▲ ●		2,600–6,000
▲ ●		100–1,550

Port Totals in *Cavans*

△	80,000
△	13,000–15,000
△	6,200–7,600
○	1,500

Province/ Island Total Unaccountable by Port

Shipments primarily: ▲ clean rice/*arroz* ● *palay*/unhusked

Shipment Data Not Available

*clean rice/*arroz* and equivalent

0   50   100 mi
0   50   100   150 km

Batanes Islands

Batan

Lacag
Paoay
Currimao
Vigan/ Pongol
Candon
Bangar

Luzon

Sarapsap
Anda
Bacnotan
Bolinao
Bani
Agno
Sual
Caba
S. Tomas
S. Fernando
Dasol
Aguilar
Calasiao
Dagupan
S. Cruz
Manga-tarem
Lingayen
Masinloc
Palauig
Iba
Botolan
S. Felipe
S. Narciso (Alusis)
S. Marcelino
S. Antonio

★ Manila

N

Taal

Marinduque

Mindoro

Pasacao

Sibuyan Sea

Burias

Tablas
Romblon
Romblon
Sibuyan

Ticao

Sorsogon

Calamianes

Masbate

Ibajay
Kalibo
Capiz
Batan

Catbayog

Samar

Panay

S. Jose de Buenavista
Iloilo
Bacolod

Leyte

Negros

Cebu

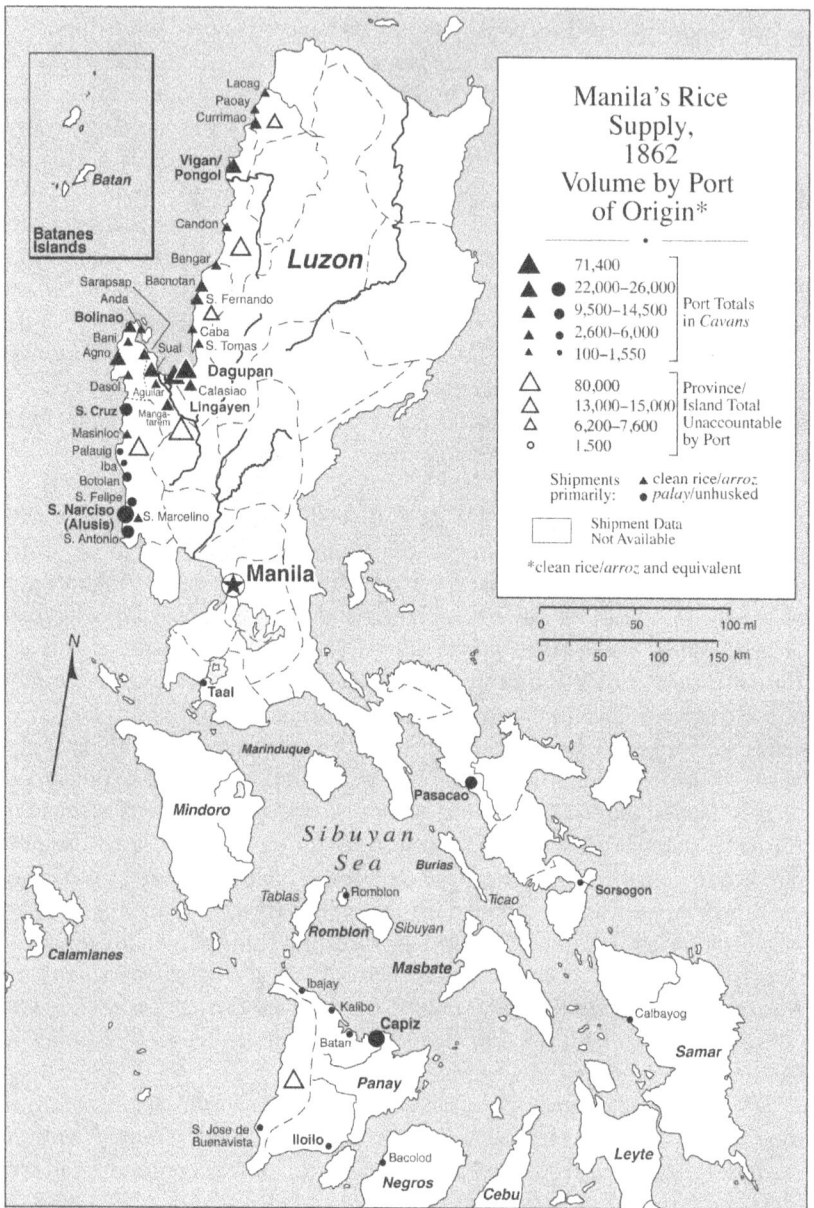

MAP 1.4. Outer Zone Ports in Manila's Rice Supply, 1862 (volume by port of origin)

TABLE 1.3. Rice Shipments to Manila from the Ten Leading Ports of the Outer Zone, 1862, 1872, and 1881

1862		1872			1881		
Port	*Cavans*	*Port*	*Cavans*	*Percent*	*Port*	*Cavans*	*Percent*
Dagupan	71,417	Dagupan	201,972	49.8	Dagupan	490,340	78.9
S Narciso Z	26,131	Lingayen	34,305	8.4	Lingayen	26,793	4.3
Lingayen	24,493	S Narciso	26,053	6.4	S Tomas LU	14,050	2.3
Capiz	21,843	Vigan	20,584	5.1	S Narciso	13,820	2.2
Pasacao	14,578	S Antonio	20,582	5.1	S Antonio	13,782	2.2
Vigan	13,711	Sual	14,453	3.6	Sual	10,324	1.7
S Antonio Z	12,580	Agno	10,208	2.5	Capiz	6,975	1.1
Sual	12,280	Sta. Cruz	7,466	1.8	Sta. Cruz	6,767	1.1
Sta. Cruz Z	10,918	Capiz	7,210	1.8	Subic Z	6,044	1.0
Agno	9,567	Alaminos	6,561	1.6	S Felipe, Z	5,662	0.9
Total	217,518		349,394	86.2		594,557	95.7

SOURCE: Calculated from the daily record of arrivals in the *Gaceta de Manila*.
NOTE: Percentage is not calculated for 1862 because numerous shipments were reported only by province, thus understating the volume by port. For 1881 cargo size is estimated where the arrival report was *con arroz* or *con palay*. Shipments of *palay* are included in *arroz*-equivalent terms. The few cargoes not stated in *cavans* are excluded. Percent refers to total outer zone volume. Z = Zambales, LU = La Union.

increased by less than 10 percent. Shipments to Manila from Panay declined by half during the 1860s, and those from the small ports on the Ilocos coast by one-quarter. Manila export price quotations on Ilocos rice ended in 1865, but, despite a vanished regional self-sufficiency marked by growing shipments to the region from Dagupan, the quantity actually shipped from Ilocos to Manila declined only gradually and still accounted for nearly 11 percent of Manila's outer zone supply in 1872.[78]

This set of regional trends is mirrored in the changing composition of the list of ten leading provincial ports engaged in shipping rice from the outer zone to Manila (table 1.3). The loss of supply to the city from some regions was more than replaced by the increased shipments through Dagupan. The growth in tonnage received from Dagupan was phenomenal—even allowing for the understatement of true totals in the earlier year. By 1872 the Dagupan *pantalan* had become the point for collecting and forwarding half the entire outer zone trade (figure 1.1). Together the top ten ports now accounted for 86 percent of the flow. The origins of the domestic supply of rice to the city narrowed during the decade bracketed by our sample years. It is possible that this pattern was partly shaped by the specific geography of the 1871 drought, but it is a change that fits ongoing trends.

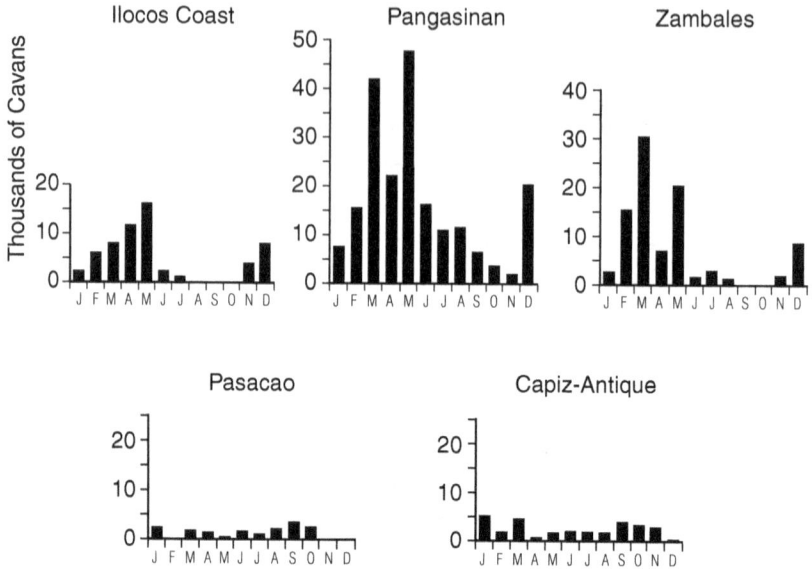

GRAPH 1.1. The Seasonality of Rice Shipments to Manila by Production Zone, 1862

Each collection zone had its own characteristics in the supply system. Panga-sinan was the single major source of domestic supply to the city (graph 1.1, map 1.5). Its production zone included both the Pangasinan cultural-linguistic core in the delta and the new settlements of Ilocano migrants on the plains to the southeast and southwest. With the looping Agno River as the highway, Dagupan became the way station for rice transported from all over central Pangasinan, as well as what became northern Tarlac and northwestern Nueva Ecija.[79]

Outside Pangasinan most of the surplus grain generated in Zambales came from the pioneer Ilocano settlements in the south along the Santo Tomas River valley with its main port at San Narciso-Alusis.[80] Along with those from Ilocos, shipments from Zambales completely ceased during the late rainy sea-son. The minor supply zones of Capiz-Antique in Panay and Pasacao in Camarines Sur exhibit much flatter seasonal patterns of shipments to Manila but with a notable rise during September and October as a few merchants took advantage of their ability to reach Manila by sail during the upswing in the annual price cycle. They were also taking some advantage of an early harvest cycle. Still, Capiz and Camarines Sur were important rice production zones increasingly committed to supplying more immediate regional and interregional markets, markets that were growing rapidly with increased spe-cialization in export commodities.

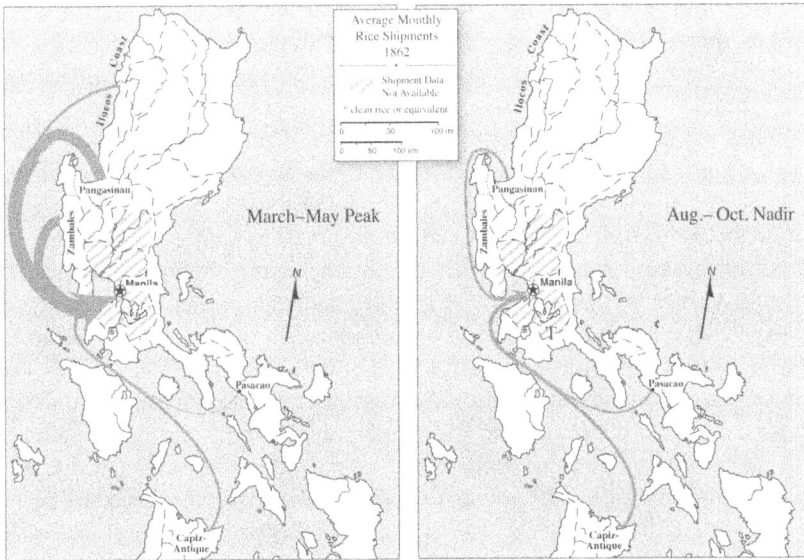

MAP 1.5. Average Monthly Rice Shipments to Manila during the March–May Peak Flow and the August–October Nadir by Production Region, 1862

A notable feature of the Manila rice trade was that the rice arrived in the city unmixed from numerous microareas of production. It was relatively simple to maintain the commodity's regional identity and then use that as a commercial descriptor implying a certain range of qualities. In the Manila marketplace, rice from the major regional source areas was differently priced. In a broader context, Legarda cites commercial circulars from Hong Kong in November 1860 placing the highest premium on Ilocos rice, followed by rice from Pangasinan and Saigon, then Java, and finally Siam and Arakan.[81]

International arrivals of rice formed a small part of the Manila supply system in our base year, 1862. The aggregate amount reported in the official summary foreign trade record, 27,516 kilos, or less than 500 *cavans* of clean rice, appears to be a considerable understatement. Three arrivals are noted in the *Gaceta*. The first, 15,640 *sacos* arriving in February from Siam, would by itself have exceeded the official import total for the year. The other two arrived from Saigon and Bangkok, respectively, at the moment of maximum depletion of existing stocks in early November, but no specific quantities are given. In gross national terms, however, these imports were outweighed by the direct export of rice from Sual to Hong Kong and Macao. There were also small exports of rice direct from Manila to coastal Fujian Province. The total exports in 1862 are listed at about 31,000 *cavans* of clean rice (almost 1.8 million kilos), depending

on the conversion factor used. When viewed in the context of internal shipments, 1862 was a year of significant but not unusually high net exports—a
response to the extraordinarily high prices bid for rice along the China coast
that year.[82]

The third quarter of the century ended with a historic transition—the Philippine role as a supplier in the international rice trade was drawing to a close.
Rapid population growth plus a lack of emphasis on rice production, compared
to expansion in the production of major export commodities, was leading to
a chronic net national caloric deficit. As a result, eight foreign vessels delivered
rice to Manila in 1872, compared to three a decade earlier. The timing of
foreign shipments is now familiar—concentrated in the rainy season or at its
end—three in July, four in August, and another in November. All of this was
well timed to coincide with the annual season of dearth in internal shipments.
It was taking on the routine of a well-established import trade. The point is
that as the Philippine commercial economy expanded around the production
of commodities for export and as the price of domestic rice failed to rise in
proportion to the general level of export commodity prices, it is not surprising
that rice production failed to keep up. This is a central finding in the work of
both Legarda and Owen.

At the end of the third quarter of the nineteenth century, the Manila rice
trade was moving through an epochal transition toward net imports as a regular part of the provisioning system and a narrowing of the principal sources of
domestic supply. Increasingly, the city and the broader rice trade based there
were relying on Pangasinan and Central Luzon as the chief domestic production region. As supplies in other outer regions were outstripped by population
growth, the growing caloric deficit was made up by local maize production,
imports of rice, or both.

～

Within this era of coastal sail navigation, Dagupan, a small city in Pangasinan,
emerged as the single most important center of supply to the capital. Dagupan's
good fortune was to be located at a point where vessels of modest draft could
enter from the sea and access the great quantities of rice assembled via nature's
network of shallow inland waterways. Conversely, the flows from Ilocos, Bikol,
and Capiz/Panay became increasingly reoriented to their immediate regional
markets.

Finally, typhoons aside, this was a moderately fortunate era for agricultural
weather. For the entire period from 1850 through 1863, there were no intense
El Niños to harm rice production by causing severe droughts. Likewise, through
the mid-1870s there were no wars or great epizootics to destroy standing crops
or great numbers of plow animals. With a few exceptions, these were reasonable times from the point of view of Manila's rice supply, but a new era was
about to begin.

2

Paleotechnic Marvels and Rice Production Disasters, 1876–1905

THE NEWLY PRACTICAL TECHNOLOGY of steam power affected the expanding Manila rice trade in two stages: in changing the characteristics of maritime shipping and in the nearly simultaneous introduction of the revolutionary forms of railroad transport and power rice milling. As elsewhere, the railroad with feeder short lines tended to open up the broad interior to integration in a more far reaching commerce—the inner Central Plain of Luzon just as had occurred in the Great Plains of North America, the Ukraine, Manchuria, and Argentina. It brought great economies of scale and efficiency in overland movement. Everywhere this "paleotechnic" complex of smooth rails and reliable steam engines led to great geographical and commercial concentration in the principal foci of the new system of movement.[1] Manila was the principal place whose reach was being extended, the headquarters of increasingly well organized commercial networks, the prime beneficiary. The changes in transport, production location, and milling were paralleled by important transitions in the commerce in rice (as we will see in chapter 4). These were important innovations.

At the same time this was a disastrous era, not least for rice production. The formerly rapid expansion of the human population of the archipelago was greatly slowed during 1875–1905—the result of the diffusion of human disease, problems in the food system, and warfare, all of which left conditions disrupted and people vulnerable to a degree that was not the case in the third quarter of the nineteenth century. One result of being increasingly integrated in international maritime trade was the arrival of the bovine disease rinderpest, also known as cattle plague, which proceeded to kill perhaps 85 percent of the carabao and cattle, twice. Each of these epizootics severely affected rice production by removing the work animals. On top of this, El Niño–driven droughts became frequent after two generations of very few. Two decades into the multiple difficulties of this era Spanish colonial society came off its moorings, resulting in two rounds of revolutionary fighting, the siege and near famine of Manila in 1898, and the subsequent full-blown Philippine-American War. All this affected rice production and the rice trade, domestic and international. This chapter opens with the last net export year for Philippine rice, addresses the new transport and milling technologies, and proceeds to examine the causes of the string of unprecedented domestic rice deficits.

TECHNOLOGY

It took more than two generations for steam applications to come into general use. A power dredge was employed to keep the lower Pasig River wharf area open in the 1830s. Steam-powered gunboats played a significant role in turning the tide against coastal slave raids and extending imperial control over the maritime Muslim Sulu world.[2] And steam-powered vessels increasingly connected Manila not only to the China coast, Singapore, and beyond but also to Iloilo and Cebu and a few small provincial ports such as Batangas and Guagua (Pampanga). Nevertheless, as late as the early 1870s, the rice supply system of the city remained almost entirely a sail- and muscle-power affair. Bulk purchase orders for rice still went to Dagupan in longhand form delivered by sail until 1873. Most of the rice milling was done by hand or was assisted by animal and water-powered devices. Clean but generally not polished rice was delivered to the city by coastal sail craft and riverine *cascos*. The domestic system of delivery from the outer zone was almost wholly driven by nature's calendar with arrivals dropping to very low levels from June or July through November as prevailing wind direction and storm tracks turned unfavorable and dangerous.

By the early 1880s an internal and international telegraph system delivered market information and commercial orders almost instantaneously and steam-powered vessels breached the long-standing pattern of a deep rainy season nadir in shipments. In the following decade the country's first railroad line opened through the heart of the principal zone of surplus rice production, greatly lowering the cost of bulk shipment to the city.[3] At the same time an important new area of domestic supply was being established in and around Nueva Ecija Province and northern Bulacan. While the former riverine access was constricted due to siltation set off by land clearance erosion, new production localities were increasingly made economical by roads and railways in the early twentieth century (figure 2.1). As all this happened, the relative contributions of the inner and outer zones of urban supply reversed. During the same era, after 1876 the demand for rice in the archipelago routinely exceeded the domestic supply. In the later 1880s a tide of much needed rice was imported from Saigon, some in polished form. Starting in 1891 several large steam-driven

FIGURE 2.1. Threshed *palay* in sacks transported by cart in Bulacan around 1910. In the northern system, rice was often moved in sheaves unthreshed. (John and Kate Evans Collection, USNA II, RG200-S-PE-album 3-74)

rice mills opened along the new railroad in Central Luzon and began sending significant quantities of polished domestic rice to the city. The international complex of iron, coal, and steam power had arrived. A British engineer wrote, "[The new] railway gave a great stimulus to the husking and pearling industry, which was taken up by foreigners. There are now important rice steam power mills established . . . along the line from Calumpit towards Dagupan which supply large quantities of cleaned rice to Manila and other provinces, where it is invariably more highly appreciated than the imported article."[4]

The last quarter of the nineteenth century opened with an important change in the annual cycle of domestic rice supply to the city. Suddenly in 1881 the exaggerated seasonal low point of arrivals from the outer zone was moderated (graph 2.1). Gone was the predictable deep monsoon dip lasting four to six months. Gone, too, were most of the desultory two-week gaps in rainy season arrivals. In that year 45 large cargoes of domestic rice arrived aboard nine different *vapores*, as steam vessels were called. From July through September the little steamer *Camiguin* delivered rice to the city every six days. Operated by the British firm Smith Bell, it made the round trip to Dagupan against the monsoon winds in less than half the time it took coastal sail craft to make the trip in favorable weather.[5]

Now when abundant supplies of Pangasinan rice were available, steamers could deliver them in all seasons. But 1880–82 were unusual years when relatively little imported rice was needed to fill out domestic supplies. After 1876 most years required comparatively massive amounts (graph 2.2). Steamers made

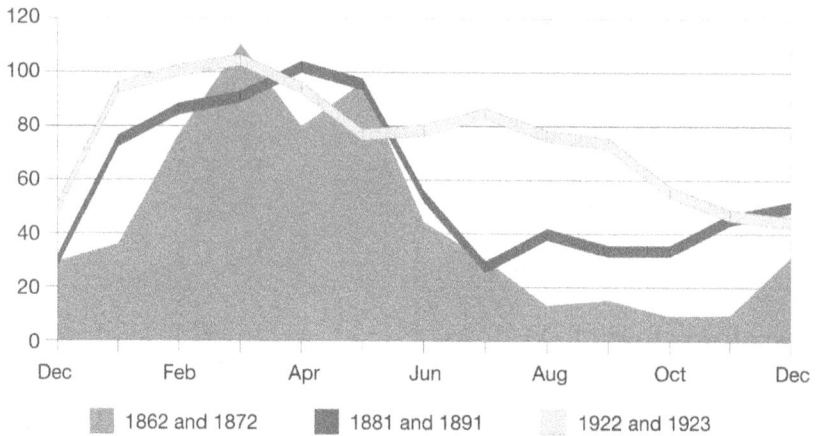

GRAPH 2.1. The Changing Annual Cycle of Domestic Rice Arrivals in Manila during Three Eras: 1862 and 1872, 1881 and 1891, 1922 and 1923. In each case two years have been averaged and the mean of the three top months made equal to an index of 100. (Compiled from daily reports, *Gaceta de Manila*, 1862, 1872, and 1881; *El Comercio*, 1881 and 1891; SBPI, 1922, no. 5, 87; SBPI 1923, 97; and SBPI 1924, no. 7, 99.)

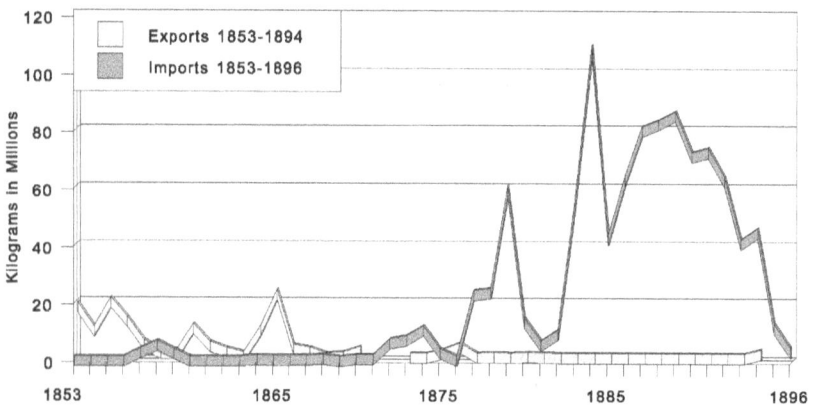

GRAPH 2.2. Philippine Rice Exports and Imports, 1853–1896, total annual shipments in millions of kilograms. Import data for 1868–72 are approximated by export data from Saigon-Cholon. (Compiled from Benito Legarda Jr., *After the Galleons: Foreign Trade, Economic Change, and Entrepreneurship in the Nineteenth-Century Philippines* [Madison: 1999], table 12; *Balanza*, 1877–94; *Census 1903*, 4:87; Manuel Azcarraga y Palmero, *La Libertad de Comercio en las Islas Filipinas* [Madrid: 1871], 244–45 [for 1868–69]; and Frank H. Hitchcock, *Trade of the Philippine Islands* [Washington, D.C.: 1898], 20.)

possible a substantial modification of the traditional calendar of domestic rice supply, but the exact shape of that calendar depended on agricultural weather conditions and commercial market forces. Growing demand from Manila was calling forth an enlarged domestic flow, but this was not entirely related to the population dynamics of the city. It was especially brought on by Manila's expanding role in the supply of both domestic and imported rice to regional populations engaged in export agriculture.

The geography of provisionment continued to change as well. According to the commercial press, Pangasinan ordinary (*corriente*), Pangasinan second-class white, and *palay de Factoría* from Nueva Ecija and northern Bulacan constituted the everyday varieties in the Manila marketplace in 1881.[6] These three, together with the more expensive *arroz blanco para mesa* (polished form ready to be cooked), were the only varieties that were always available whenever summary prices at wharfside were reported. What the newspapers missed was *pinawa* and *palay* supplied through Malabon. Still, Pangasinan rice was ascendant within the domestic supply system (table 2.1).[7] Partly this was due to the new production generated by the streams of Ilocano pioneers settling in areas of the inner Central Plain and connected to Dagupan by river.[8]

Not all harvests in Pangasinan were bountiful. The crop brought in at the end of 1882 for use in 1883 was short. Farmers had left much unharvested because they were afraid to venture out of their homes due to a lethal cholera epidemic. In this case, the cholera began to cause significant mortality in Manila during late August. The process of spatially contagious diffusion brought the

TABLE 2.1. Rice Arrivals from the Outer Zone by Region of Origin, 1862, 1872, and 1881

	Percentage of rice arrivals in Manila by volume		
Production area	*1862*	*1872*	*1881*
Pangasinan	51.3	64.0	85.7
Zambales[a]	22.7	21.2	9.3
Ilocos	14.6	10.8	3.0
Capiz-Panay	7.2	3.6	1.5
Pasacao-Bikol	3.8	—	0.2
Other	0.4	0.5	0.2
Total percent	100.0	100.1	99.9
Total *cavans*	392,287	405,569	621,551

SOURCE: Calculated from the daily record of arrivals in the *Gaceta de Manila*.
NOTE: All figures include *palay* in *arroz*-equivalent terms.
[a] Includes Bolinao and nearby municipalities later transferred to Pangasinan.

epidemic to the rice areas of Tarlac at the end of September and then to the
heart of Pangasinan in early November, just at harvest time. In Dagupan and
neighboring Binmaley the mortality was unprecedented—over 1,000 burials
each that month compared to the usual average of fewer than 100. Central
Luzon again experienced a very poor harvest in 1884, and this was followed by
the crushing drought of September–October 1885. Both agricultural disasters
were likely caused by the spectacular explosion of the volcanic island of
Krakatau off the west coast of Java in 1883—an event that left so much volcanic dust in the atmosphere that midlatitude sunsets were brilliant red for several years after.[9]

In this era the emergence of an important stream of rice from Nueva Ecija
in the inner Central Plain was a major development. New as an abundant item
of commerce, *palay de Factoría* was also more highly valued than *palay* from
other sources. By 1881 *palay de Factoría* was more consistently available than
rice in *palay* form from elsewhere. Production of rice for the Manila market
was expanding rapidly in Nueva Ecija.[10]

In the evolving geography of supply, Zambales continued to be a major
provider of clean rice in the 1880s, especially from the Ilocano settler enclave
in the southern part of the province. Still, in 20 years the Zambales share of
outer zone deliveries fell from more than 20 percent to less than 10 percent.
From Ilocos, La Union still supplied about 3 percent of the outer zone total.
But northern Ilocos was now sending rice to the new settler communities in
Cagayan, and Ilocos Sur had become a net importer—a symptom of its economic decline. Its former flow to Manila had evaporated.[11]

The changing geography of supply is mirrored in the increasing concentration of flow through Dagupan (table 1.2) and in the reduction of shipments
from other places. Only 27 ports sent rice to Manila from the outer zone in
1881, as opposed to about 45 in previous decades. Likewise the number of ports
shipping at least 10,000 *cavans* to the capital also fell. The urban rice supply
from the outer zone was becoming more concentrated in the rapidly expanding
production and trade of the Pangasinan-centered watershed of the northern
Central Plain. Furthermore, this concentration was happening during a year
of extraordinary harvests when one might have expected the opposite.

In 1891 both Ilocos and Zambales appear more prominently in the supply
of rice to Manila, which was locked now in a chronic national shortage.[12]
Ilocos as a whole was no more self-sufficient than in the 1880s, but the regular
arrivals there of ordinary rice from Pangasinan and even Zambales allowed
some of the local higher quality grades to be sent to the Manila market. Some
of this was shipped from Currimao in Ilocos Norte and from La Union—this
almost 30 years after an end to Ilocano rice surpluses had been noted.[13] At the
same time, Zambales was at last connected to Manila by steam-powered craft—

reducing the cost of shipment. The commercial emergence of the great natural harbor at Subic is seen in the record of 1891. Located at the southern end of Zambales, it became an important regional center for gathering rice and other commodities for shipment to Manila. Suddenly Subic ranked second in terms of the number of rice cargoes sent from the outer zone. A decade later it would be described as the most commercially vibrant of the southern Zambales towns, with large warehouses for rice and a regular connection to Manila by means of the Yangco line of steamers. Like Taal in coastal Batangas, Subic became the focus of an indigenous commerce that largely excluded Chinese participation.[14] At the same time shipments from Capiz to Manila declined to a trickle as more rice from there was absorbed in Bikol, the Visayas, and even Batangas.

On the international front, the level of rice imports was miniscule during bountiful 1881 and 1882. Only one international vessel was reported to have landed rice in Manila in 1881, although three others came from Saigon with unspecified cargoes. The pattern for 1891 forms a total contrast as it came during the great Philippine rice deficit of 1884–92. From Saigon, at least 53 steamship arrivals and one sail vessel brought rice to Manila. Further, a vessel from Hong Kong and two from Kobe, Japan, also brought rice. So dense was the flow from Saigon that some European tramp steamers became dedicated to the run.[15] The high volume of imports reached a hiatus during 1895–98.

Rail Transport and the Rise of Power Rice Milling in the 1890s

The opening of the Manila-Dagupan Railroad permanently ended the monopoly of water transport in the movement of domestic rice supplies to the city. Built largely with British capital and opening in segments from early 1891 to November 1892, the new railroad began regular operation through central Pangasinan–northern Tarlac and lower Pampanga-Bulacan, two of the major production zones of the Central Plain.[16] For more than a year, however, there was a break at the Pampanga River crossing where passengers and cargo were floated across. Finally, the last major bridge was completed in 1894, facilitating through shipments along the entire line. A spur at the northern end of the line linked the railway to the Dagupan wharf. With this the railroad took over a good portion of the transport of rice to Manila from the small armada of privately owned coastal vessels (graph 2.3).

Unlike sail vessels, the railway could normally continue operations year-round, occasional typhoon damage excepted. More important, rice from nearby interior areas along the railway line could now be brought to Manila at about "a quarter of its former cost." In response, maritime shipping charges were reduced by a reported 40 percent during 1891–93 even before the railroad was fully in service.[17] This was a transport revolution with important implications for the geography of employment in rice processing and shipping.[18]

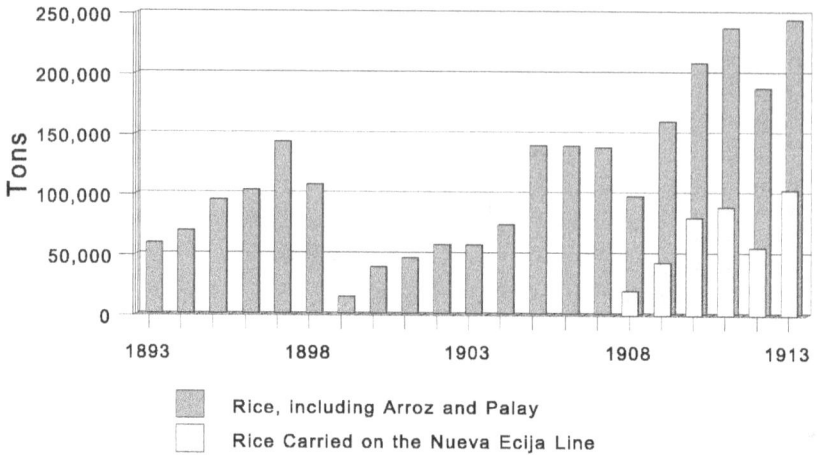

GRAPH 2.3. Rice Tonnage Hauled on the Manila-Dagupan Railroad, 1893–1913. "Rice" here includes both milled rice and *palay*. The portion carried on the Nueva Ecija branch is also shown separately for 1908–13. (Compiled from RPC 1904, pt. 3, following 228; and Manila Railroad Co., *Report of the General Manager*, 1912, 1913, 1914, and 1918.)

The railroad privileged Manila. Whereas the construction of most sailing vessels took place in dispersed provincial locations, railway equipment was largely imported from Britain through Manila. While sailing vessels were often based in provincial ports and drew locally for crew and captain, railway operations were concentrated in the Tondo and Caloocan yards in the metropolitan area. The proliferation of steam-powered vessels had already begun the process of change, and now centralization in Manila and concentration along the railway lines emerged as major themes.

The 1890s also marked a major transition in domestic milling. Actually, a variety of powered rice-milling arrangements had been developing for some time. These included animal-drawn mills, waterwheel mills, and a growing number of small steam-powered mills—later replaced with small gasoline-engine mills. Each of these forms had its own geography. What was truly new was the rise of large-scale power mills producing for the city and the broader commerce conducted from there. Foreman thought "the Manila-Dagupan Railway gave a great stimulus to the rice husking and pearling industry." He mentions important new "steam power mills established at Calumpit" along the river at the Bulacan-Pampanga border and at "other places along the line from Calumpit towards Dagupan" (map 6.2). Even before the outbreak of the Philippine Revolution, these were supplying "large quantities of cleaned rice to Manila and other provinces."[19]

The first of the big steam-powered mills was opened at Calumpit by Warner Blodgett and Company, and its product was advertised for sale in Manila in late 1891 as "*arroz de la máquina*."[20] Calumpit was an excellent site because it effectively linked river and railroad modes of transport and was on the Manila side of the railway bridge that took the longest to complete. A new mill opened at Bautista in southern Pangasinan the following year. These were soon followed by others at Gerona (1894) and Moncada in northern Tarlac and at Dagupan (map 1.3). Initially each had the capacity to process 400 to 500 *cavans* per day.[21] Run by British managers, these mills became centers of indigenous employment. All the big early steam-driven mills in Central Luzon were located along the railway line.

These mills were freestanding capital investments not integrated with investor attempts to grow rice on a large scale. Most were owned by Luzon Rice Mills, Ltd., a business entity set up and staffed by Smith Bell & Company with British and, reportedly, some Philippine capital participation. Smith Bell was a leading commercial house, number one in the export of Philippine sugar in the 1880s and number one in abaca in 1886. It would not be surprising to discover that the HSBC and other British banking institutions played a role in financing the construction of these mills. Certainly these banks long sought to promote the export of British capital equipment, and in the early twentieth century they did grant loans on these mills. A major block of shares in Luzon Rice Mills was owned by Smith Bell's erstwhile competitor, Warner Blodgett, later known as Warner Barnes. Otto van den Muijzenberg calls this practice "overlapping investorship" and notes that it was common among Europeans and other investors in Manila "to spread risks as well as opportunities for profit" and was often used in mercantile partnerships in the days before bank loans were common.[22]

The mill at Bautista was apparently the first of the new *fábricas* erected by Luzon Rice Mills. It was described as a "huge, corrugated-iron mill with [an] ugly chimney and clustering of godowns [bodegas] where the paddy is stored." With the opening of the railway and this mill, Bautista emerged from obscurity as a place where *palay* was taken to be sold and processed. It grew rapidly for a time as a locus of various Chinese businesses and also of Filipino alcoholic beverage distribution.[23]

During 1895–96 *arroz* from the mills at Bautista and Gerona was merged into a single category in the Manila price listings of *El Comercio* and in advertisements placed by Smith Bell. Both mills were managed by a succession of Smith Bell career men. Their product was the most expensive regularly quoted Philippine polished rice. While the first phase of the Philippine Revolution apparently had no major effect on the facilities supplying domestic rice to the city, the Philippine-American War that followed destroyed the Warner Blodgett mill at Calumpit. The others were not destroyed, and indeed the British man-agers of rice mills and the railroad cooperated with the forces of the nascent Philippine Republic. By contrast, the railway was mostly out of service during June and July 1898 and from February 1899 through May 1900.[24]

~

The opening of these larger power mills was presaged by more than two decades of innovation with small mills. There were micro-steam-powered rice mills operating before the Revolution, possibly more than 60. Research has so far revealed 2 medium-sized mills and 21 smaller units in Manila's provisionment hinterland between 1893 and 1898. Most of the small steam mills were found in Bulacan province (13) and Manila itself (3). A majority were polishing mills. Others are more vaguely described as *pilanderías* or *fábricas de arroz*.[25]

The two medium-sized mills were found in Manaoag (Pangasinan) and Balayan (Batangas), neither connected to railway lines. These were projects of affluent entrepreneurial landowners. In Manaoag, Tomas Rous was the owner of both a medium-scale rice mill and one of the larger alcohol-distilling oper-ations in the province. Rous got his start in these industries from his father, a French immigrant named David Rous. Over three decades the family lived in various places in eastern Pangasinan. Investing in a more advanced mill in the 1890s was part of a family pattern. The other medium-scale mill was owned by the Martinezes of Balayan—leaders of one of the major political factions in that sugar town and also an affluent family with multiple commer-cial interests.[26]

In the 1890s most of the small power mills appear to have been owned by indigenous Filipinos and mestizos, including the Cojuangco family in Paniqui, Tarlac. Chinese entrepreneurs owned 5 of the 21, including 3 located in Malo-los (Bulacan). The small-capacity steam mills tended to remain in local hands and were not adversely affected by the rise of big mills; rather, their numbers continued to grow. There were 250 by 1912, cleaning rice mainly for local use.[27]

In the transition period of the 1890s, there were also small mills powered by water flow or animals. Surprisingly, rice mills driven by waterwheel appear to be a late-nineteenth-century development, coincident with the advent of steam milling, perhaps even in part a response to the growing commerce in *bigas* as opposed to *pinawa*. This commerce was largely created by the big steam-powered mills in Saigon-Cholon and now along the Manila-Dagupan Rail-road. Twenty-five water-driven mills in 13 provinces turned up in the 1890s records. Most were in the hilly southern Tagalog zone—Tayabas (13), Cavite, and Laguna—and also northern Pampanga. Apparent Filipinos paid the tax on all the waterwheel rice mills in Tayabas and most of the others as well.[28]

There were also a few husking mills powered by water buffaloes in the 1890s—a mode of power more frequently employed in artisanal sugar milling.

One of these was established by David Rous in Binalonan, Pangasinan—a *molina de sangre* powered by carabao and operated by Chinese workers. Foreman describes a mechanical arrangement in which a carabao plodded around in a circle pulling a sweep that rotated a vertical central shaft. On this shaft were "pins which at each revolution caught the corresponding pins in vertical sliding columns." These columns, or pestles, were thus raised one after the other and fell of their own weight on the rice set on mortars below. Various reports locate such mills in use in Candaba (3), at Pagsanjan and Calamba in Laguna, and in nearby Naic, Cavite, as well as Bulacan and Tarlac, in the 1890s. Chinese owned several of these. The governor of Pangasinan reported 17 similar mills in his province in 1908, but one hears little of them thereafter.[29]

Imported Agricultural Machinery and Advertising

The rise of steam power in rice milling should be seen in the context of the technological and marketing efforts of the manufacturers. Among the elements of a gathering visual cacophony in late-nineteenth-century Manila newspapers was a steady stream of ads for iron-milling machinery, steam engines, pumps, and plows. Most of the advertised manufacturers were British—the initial world leader in this technology. Rice-milling equipment was usually ancillary to their core business due to the rapidly expanding sugar industry, the tonnage of Philippine sugar exports having doubled between the mid-1860s and 1874 and doubled again by 1881.

The most graphic advertisements (at least before the mass marketing of beer got under way) were placed by the British manufacturing firm of Ransomes, Sims, and Head of Ipswich. This company offered grain mills powered by an innovative mobile steam engine capable of running on chaff as well as fuelwood, charcoal, and even cogon grass—a major economic advance over engines requiring expensive imported coal.[30] The graphic shows a workman pouring a generic sack of grain, possibly meant to simulate *palay* in this case (figure 2.2). The grain was poured into one of two hoppers that fed into horizontally rotating millstones geared in turn into a vertical flywheel powered by a belt from the mobile engine. This was state of the art machinery of the early industrial era, and Ransomes, Sims, and Head was a premier manufacturer.

Ransomes made its way against the persistent British agricultural depression in the late nineteenth century by aggressively developing export markets in agricultural-commodity-producing economies worldwide. An integral part of its business plan was to create high-quality graphics in its catalogs and then use them in advertising in its far-flung markets. The catalogs were printed in several languages, including Spanish, English, and German. Ransomes' precision eye-catching ads appeared dozens of times in Manila's *El Comercio* during 1879–82. Its local agent was nonetheless based close to the sugar industry in

MAQUINARIA AGRÍCOLA.

Gran fábrica de los Sres. Ramsomes Sims y Head.

Ipswich y Lóndres.

Esta gran fábrica, establecida en 1786, ha obtenido de diversas Sociedades de Agricultura y de las prin-
cipales Exposiciones internacionales del mundo 40 medallas de oro, 108 de plata, 157 primeros premios 40 se-
gundo 30 medallas de bronce y 94 certificaciones de mérito.
ARADOS de surcos rectangular, trapezoidal, volteado, pulverizado, de hierro forjado, con cuchilla circu-
lar, anglo americanos, Aguila, de vuelta, dobles y triples ó sea de dos y tres puntas, y de otras muchas clases
GRADAS ó RASTRILLOS, de hierro forjado, costiculadas y de cadena.
SEGADORAS, para yerba, henos, trigo etc.
MAQUINAS DE VAPOR LOCOMOVILES, alimentables con carbon, leña, estiércol, paja, cogon,
tigba y cualquier otro combustible, de 8 á 20 caballos de fuerza de uno y dos cilindros.
MAQUINAS de vapor horizontales con calderas tubulares.
MAQUINAS de alta presion, fijas.
MAQUINAS elevadoras para minas, túneles etc.
TRILLADORAS movidas por fuerza animal y de vapor.
MOLINOS perfeccionados para moler trigo, centeno, cebada, arróz, etc.
BOMBAS centrífugas, fijas y portátiles, movidos al vapor.
MAQUINARIA para aserrar.
QUEBRANTADORES de piedras ó minerales.
Autorizados por los fabricantes, recibirán pedidos, y darán cuantas esplicaciones se soliciten, en Iloilo.
Innes y Keyser.

FIGURE 2.2. A steam-powered flour and feed mill advertised in Manila by the British manufacturer Ransomes, Sims, and Head, 1881–82. In Ransomes' catalog this is identified as a mill for wheat, rye, barley, and rice. The catalog drawing has been substituted here for the identical but smudged newspaper image. (*El Comercio*, January 24, 1882, 4; *Descriptive Catalogue of Agricultural Machinery Manufactured and Sold by Ransomes, Sims, and Head, the Orwell Works, Ipswitch, July 1879*, folio 1390H, Rural History Centre, University of Reading)

Iloilo. Ransomes enjoyed its chief Philippine sales in the in the early 1880s but ceased sales operations there during the sugar crisis, following the liquidation of its agent firm, Innes and Keyser, in 1887. The immediate cause of this termination was due to the death of Keyser and the poor health of Innes.[31] Not all machinery agents were foreigners. In the 1890s, Jose Leoncio de Leon of Bacalor (Pampanga) and Manila represented Clayton and Shuttleworth and advertised machine threshers as well as steam- and water-powered mills.[32]

The advertisements find resonance in the annual tally of machinery imports. Little was reported in the 1860s, but finally in the late 1870s a few threshing machines and one for polishing rice appear, followed by four machines for cleaning *palay* and six "rice mills" in 1880. Most of these came from England.[33] These numbers pale in the face of the flow of machinery for sugar. By the early 1880s, iron-milling equipment powered by steam was increasingly available, and one company was offering iron gearing for waterwheel mills as well.

The most advanced sectors of the commodity export economy were now well into the machine age, and the import of coal and delivery of firewood to power some of these engines were large businesses. Little wonder that this spilled over into the processing of rice destined for the city. The overall machinery trade suffered "quite a collapse" in the sugar trade depression of the later 1880s, leaving "large stocks" of unsold imported machinery on hand. In the early 1890s, the machinery advertisements in *El Comercio* were less exuberant. Ransomes had withdrawn, but the products of G. Buchanan and Company of London were abundantly advertised by its agent firm, Smith Bell. Products featured in the Buchanan ads during the 1870s included rice mills and copper stills. By the 1890s, they focused exclusively on the sugar industry with steam-, animal-, and water-driven mills. Still, the company's long association with Smith Bell makes it a possible choice to have supplied the equipment for the *fábricas* of the new Luzon Rice Mills. At the same time, Frederick Sawyer was advertising an "economical mill for rice" capable of hulling and cleaning a ton of *palay* per hour. At the end of the 1890s, Englishman Fred Wilson and his partners were selling *molinos de sangre* and servicing all kinds of steam-operated mills and other machinery.[34]

Large or small, the new steam rice mills were not heavily concentrated in Manila. As in the 1850s, there was a cost advantage to milling rice near the site of production and then paying to ship a lighter, more concentrated commodity. The establishment of a decentralized geography of steam-powered rice milling in the 1880s and 1890s was an innovation of broad significance that antedated the emergence of similar patterns in the Mekong Delta and central Burma by 15 or 20 years.[35] Still, the Philippines lagged behind the three massive rice-exporting river deltas on the mainland in investment in large power mills producing polished white rice rather than simply "clean" rice. When a chronic rice deficit emerged in the Philippines in the later 1870s, the mills of Cholon, the ethnic Chinese rice supply section of Saigon, began to find a regular market there.[36]

Disasters in the Health and Rice Production Systems

Simultaneous with the advances in grain transport and milling, serial disasters struck. These included epidemics and epizootics for man and beast and the

onset of a period of more frequent El Niño atmospheric events, which brought
an increased incidence of drought to Philippine rice production regions. We
start with human disease, an important cause of the notably slowed population growth alluded to at the outset of this chapter.

Beriberi Comes to Manila

Despite substantial savings in female household labor, the rise of machine
polished rice had a downside. In some cases the switch from consumption of
twice-pounded *bigas* to machine milled and polished rice had important
health consequences. Ken De Bevoise has linked the wide use of polished
Saigon rice to the appearance of beriberi in Manila. On the heels of a cholera
epidemic and panic, the first widespread outbreak of beriberi settled on the city
in late 1882 and early 1883. It could be deadly. Having progressively lost neuro-
logical control of their extremities, victims could eventually die in excruciat-
ing pain. In the 30 years before the mystery of this condition was fully solved,
the problem became increasingly acute in many Asian places and situations.
Following the discoveries of Pasteur and Koch in Europe, a contaminant was
suspected. Much scientific effort was expended exploring this "poison hypothesis." In the urgent search for cause and cure, researchers in a number of Asian
colonies and countries came increasingly to share their reports and discoveries
through the effective media of scientific journals, flyers, and conferences.

Eventually beriberi was understood to be due to a dietary deficiency of
thiamin, or vitamin B$_1$, normally found in rice bran—not to a toxin.[37] Thorough power milling was capable of removing all the bran. The lack of bran in
household stores of polished rice became critical when the consumption of
fresh green vegetables, *kamote* greens, legumes, many fruits, pork, and other
sources of thiamin was interrupted for an extended time or when people were
too poor or restricted to consume them. It is precisely such "protective" foods
that the poorest households found difficult to fit into their dietary budget.
This could be deadly when infants were fed only a gruel of polished rice or by
nursing mothers already suffering from beriberi. Further, in the Manila beriberi outbreak of 1882–83, fright kept some affluent Filipinos in their homes.
Patients confessed that they had lived "for months . . . solely on rice out of fear
of contracting cholera." Ken De Bevoise's well-sleuthed reconstruction of this
event ties it to machine-polished white rice.[38]

Polished rice thus became a serious problem in the health system. Philippine consumer preference provided the demand, and relatively fast and well-
ventilated ships now minimized the risk to polished rice of spoilage in transit—
a risk that earlier had been managed by leaving the bran in place.[39] Finally, a rise
in value of the milling by-products—bran and broken "brewers" rice—made
highly milled white rice increasingly feasible economically. From De Bevoise's

perspective, the onset of beriberi was causally linked to the downside of expanding Philippine participation in commerce and foreign trade, since Saigon was the source of considerable polished rice in Philippine commerce during the 1880s and early 1890s.[40] But this was not a simple matter. Rice imports were minimal during 1882. Further, as late as 1890 only about 19 percent of Philippine imports of Saigon rice was polished *bigas*. De Bevoise may be right about the cause of the 1882–83 beriberi epidemic, but if so, it began in a year in which imported rice made a quite restricted contribution to feeding the city and in a decade when only a modest percentage of the imported rice arrived in polished form. Nevertheless, beriberi became a significant cause of human suffering and mortality. We will return to this subject in chapter 3.

Disasters for Rice Production

Aside from human disease and warfare, two sorts of bio-environmental crises were behind the increasing and frequent shortfalls in domestic rice production—drought and work animal mortality. The droughts were primarily caused by atmospheric pressure anomalies, now known in general as El Niños and involving the Pacific Ocean tropics from Peru to Indonesia, the Philippines, and beyond. A second critical factor was the arrival and diffusion of rinderpest. In the Philippines this epizootic disease began in 1886 and twice in fifteen years killed the great majority of the carabao and cattle in Manila's rice supply areas. In each case it took years for their numbers to recover. The multiyear loss of work animals brought rice production to a near standstill in affected areas. As a result of these causes, the high rate of rice imports in most of the 1880s and early 1890s became even greater during 1899–1905, 1908–12, and 1915–18. Increasingly, the country produced less rice than its usual consumers required. Imports replaced episodic exports as the predominant feature of Philippine participation in the international commerce in rice.

The Legarda Thesis on the Rice Deficit

Both the Philippines and Indonesia (Dutch East Indies) became chronic deficit producers of their major food staple in the 1870s. Both remained dependent on imported rice to augment domestic supplies for a very long time.[41] For the Philippines, economist Benito Legarda Jr. explains this switch from frequent surplus to deficit as the result of general population growth in concert with the higher returns often available from the production of other crops, especially export commodities such as abaca and sugar. Likewise, he points out that the export market for rice in China, formerly serviced in part by the Philippines, was increasingly met by cheaper rice from the Mekong Delta.[42] In ordinary times the Philippine rice deficit may simply have been a sign of increasing commercial specialization in crops with a higher rate of return in international markets—an economically rational response in the context of declining shipping costs and widening market integration.

As a result and starting in the fourth quarter of the nineteenth century, Philippine rice monocrop regions were increasingly seen as among the most impoverished of commercially active agrarian production areas in the archipelago. Frederick H. Sawyer, writing in 1900, said he believed that "the cultivation of rice is the lowest use that the land and the husbandmen can be put to" and that it was being given up in favor of raising more remunerative export crops.[43] A decade later, a colonial official reported that "the average production of rice per acre . . . is below the amount required to cover the average cost of production. . . . This . . . is made possible by the utilization . . . of the labor of many women and children for which there is no competitive market."[44] Legarda argues in the case of the rice-producing Ilocos coast that the onset of deep poverty came with the loss of cotton as a complementary dry season crop—the result of imports of factory-made cotton textiles and yarns. In any case, there was some relative reallocation of labor and entrepreneurial activity away from rice production as well as gender-selective (male) outmigration.[45]

Legarda's views on the rice deficit are surely correct in general and have been widely accepted. However, the growing deficit in rice production did not closely parallel the rate of population growth. Legarda gives considerable attention to the expansion of the Philippine population from the late eighteenth century, when it was growing about 0.4 percent per year accelerating to a then extraordinary 1.6 to 1.7 percent rate by the 1860s, and he acknowledges that it fell to an average of only about 1.0 percent or less during 1875–1905.[46] With the rate of population growth reduced, this was a time when Philippine rice production might have caught up. But far from a gradual rise in line with the rate of population growth, net rice imports shot up dramatically in the late 1870s, practically vanished during 1880–82, established a variably high level during the rest of the 1880s and early 1890s, receded, and then took off to completely unprecedented heights during the first decade of the twentieth century. Population growth does not explain the stunning speed and erratic magnitude of this phenomenon—even allowing for some statistical smoothing. A fuller explanation must take account the catastrophic loss of work animals during the two great epizootics caused by rinderpest and of the much higher frequency of El Niño droughts after 1876 as opposed to the several decades immediately before.

A critical factor in the advent of the prolonged rice deficits was the arrival and diffusion of rinderpest beginning in 1886 with a few animals imported from the mainland. By the end of the second year it had affected most of the areas supplying rice to the city, including Central Luzon, the Ilocos coast, and Capiz. Some 80 percent of the major work animals perished during the first two years. A European businessman serving as the Dutch consul reported, "Bands . . . of starving and miserable people are found in several provinces pillaging and murdering on occasion. . . . At the highest point of these problems . . . an epizootic broke out for six months that killed two-thirds of the farmers' beasts . . . and the government couldn't do anything to stop it. The cadavers infested the air and the rivers; we bless providence that the epidemic didn't attack our species."[47] This outbreak is depicted symbolically on graph 2.4. The second major outbreak centered on 1900–1902 and was even more widespread than the first because it was embedded in the disruptions caused by the United States' imperial conquest. While the effect on agricultural production may have been less critical in coconut or abaca areas, these epizootic waves were disasters for rice production because the wet paddy soil must be intensively plowed and harrowed or trampled into fine slurry. The carabao population eventually recovered in each case due in part to further imports

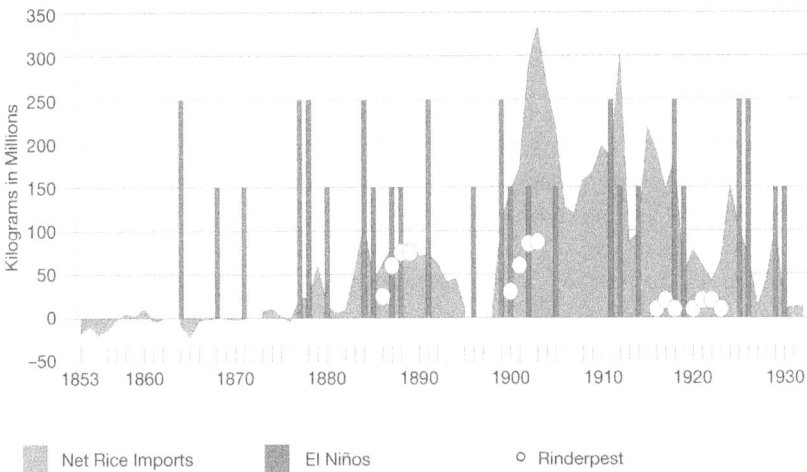

GRAPH 2.4. Philippine Net Rice Imports, ENSO/El Niños, Rinderpest Epizootics, and Severe Droughts, 1874–1932. Net imports are shown in millions of kilograms. El Niño southern oscillations (ENSOs), representing a variably high risk of regional drought, are shown by vertical bars representing the "very severe," "severe," and "moderate+" categories. Severe droughts in Manila are defined as years with rainfall at least 25 percent (550 mm) below the 100-year average (2,057 mm). The degree of bovine mortality from rinderpest is depicted impressionistically. ENSO designations are from W. H. Quinn and V. T. Neal, "The Historical Record of El Niño Events," in Raymond S. Bradley and Philip D. Jones, eds., *Climate Since A.D. 1500*, rev. ed. (London: 1995), 623–48. Their work was later extended. See Mike Davis, *Late Victorian Holocausts: El Niño Famines and the Making of the Third World* (New York: 2002), 271–72.

of animals from the mainland, but for several years domestic rice production was devastated. During these crises, many farm families switched to growing roots and tubers on a subsistence basis and cultivating with a hoe. The result of this animal disease was an immediate human food crisis and, in some cases, a land tenure crisis as well. It is not an independent event that the greatest three-year period of rice imports was 1902–4 when the mass loss of work animals and a significant drought episode coincided. In Central Luzon considerable rice land was still out of production for want of carabao five years after the second rinderpest wave.[48] The negative impact on rice growing contributed to the often desperate nature of the decades leading up to the Philippine Revolution. (Rinderpest is taken up in greater detail in chapter 8.)

Lacking an adequate time series of Philippine *palay* production in the nineteenth century, we turn first to an examination of the fluctuating level of imports as a rough reciprocal guide to production fluctuations and of their relation to the experience of El Niño–driven drought. Then the tradeoffs between importing more rice at such times and switching to domestic maize are taken up.

The sudden expansion of rice import tonnages in 1877–79 and again in 1883–94 went well beyond all precedent. They exploded again in the early twentieth century and, at one level or another, remained a standard feature of the Philippine economy for many years (graph 2.5). Inexpensive rice from the great river deltas of the Southeast Asian mainland took over supply of the Philippines' former markets in China. Increasingly, this rice also became available to make up for any local deficit in the archipelago. (The story of how Vietnamese tenant farm labor was organized and exploited in the Mekong Delta to produce very cheap rice is beyond our scope.) Philippine rice producers had thus lost their occasional but important export market. At the same time, higher returns were often available from export crops other than rice.[49]

At the moment of the shift from net exports to imports during 1872–76, the price of rice in the Yangtze Delta was particularly low and, by extension, on the South China coast as well—too low to draw out any Luzon supplies. At the regional level in 1876, rice imports at Canton were up by 50 percent over the previous year, and the Philippines was again, briefly, a marginal net exporter. When the price was high in the Yangtze Delta in 1878, supplies were low in both places due to the same cause: the agricultural effects of the El Niño southern oscillation (ENSO) drought of 1877–78. And when there was a probable surplus in Luzon under the excellent crop conditions of 1881, the price in China was again quite low. In this decade, Philippine and South China coastal supplies were often closely in synch—both affected by the same weather phenomena.[50] The big profits in the external rice trade in the late 1870s came from importing. This was as much because of rapidly swelling export production in

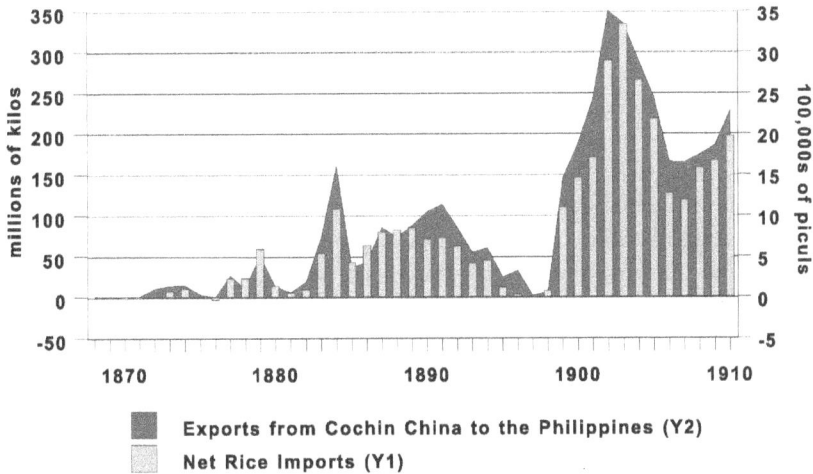

GRAPH 2.5. Net Rice Imports (*left scale*) versus Rice Exports to the Philippines from Cochin China (*right scale*), 1868–1910. The two scales are not perfectly equivalent, but a rough comparison of the profiles reveals few anomalies. The form of rice is not distinguished, but during 1890 it was 1.5 million *picos* of white rice, 5.3 million *picos* of *arroz corriente*, and 2.1 million *picos* of *palay*. During January–November 1902, it was 604,000 tons of white rice versus 35,000 tons of *palay*. In 1910 it was all white rice. (Compiled from graph 2.2 in this volume; Albert Coquerel, *Paddys et Riz de Cochinchine* [Lyon: 1911], two tables at the end; and "Arroz," *El Comercio*, 19Jan1891 and 29Nov02.)

the Mekong, Chao Phraya, and Irrawady deltas as from dynamics internal to the Philippines.

After generations of thinking dominated by a consciousness of the sudden shattering effects of typhoon winds and floods, Filipinos awoke in the 1990s to a perception of drought. Strong Pacific ENSOs frequently produce serious drought in Philippine food production regions.[51] A well-developed El Niño emerges off the coasts of Peru and Ecuador when the sea temperature rises near the usually cold surface of the Humboldt Current and stays abnormally high. This brings on an extreme set of interlocked events in the oceans of the world. It happens when atmospheric pressure differences fail to drive the trade winds from east to west with sufficient force to displace the heated water in the upper layer of the current.[52]

The causes of this complex phenomenon remain elusive. One result is high convection rainfall over the warmed mid- to eastern Pacific Ocean surface, often with some pattern of drought farther west in Indonesia and/or the Philippines and beyond. Since important droughts in the Philippine rice-growing regions are typically caused by this trans-Pacific, even global, atmospheric phenomenon, I take the incidence of the more powerful ENSO categories as potentially special years of low rainfall in Luzon and/or the Visayas.[53]

One approach is to look at the record of annual rainfall totals.[54] In the 75 years from 1865 to 1898 and 1900 to 1940, there were eight years with total rainfall of less than 1,500 mm at Manila, more than 25 percent (at least 550 mm) below the long-term mean. Of the eight, three were associated with the most intense El Niños, and another was a second-year continuation of such an event. Two more of the eight drought years at Manila were linked to the moderately strong ENSO category.[55] In addition, the 1911–13 moderate+ El Niño produced severe droughts in Cebu and Central Luzon especially during June and October 1911 (addressed in chapter 3). Because of storms in other months, however, Manila's total for that year was above 1,500 mm. Thus, there were four intense El Niños in this 75-year period, three of which produced four years of extreme drought at Manila. There were 13 moderately strong El Niños in the same period, which produced two drought years at Manila—three if the 1911 drought in Central Luzon and Cebu is included. In 57 other years (some with weaker El Niños) there was only one annual drought at Manila. The probability of each of these sorts of years producing a precipitation shortfall of at least 550 mm in one year all the way across the Pacific Ocean was then .80, .15, and .02—a very strong relationship.[56] In addition, there was the catastrophic explosion of the Krakatua volcano, which perhaps accounted for an eighth great drought in this era.

With regard to Central Luzon's single-crop rice culture, a rainfall "problem" was caused by a late start to the rainy season, producing negligible rainfall in May and a poor total in June. In the absence of active irrigation, this typically brought on a delay in field preparation and transplanting, thus shortening the growing season or forcing farmers to depend on still chancier rainfall in early November. A delay at the start of the season could also lead to the use of rice strains with shorter growing periods and lower yields. The end of the season also presented a common problem—as when the rains tapered off early in October or, worse yet, in September and October, leaving the plants short of moisture just as the rice grains were filling out near the end of the crop cycle. In these regards a year that approximated the mean Manila–Central Luzon monthly rainfall pattern was probably fine. The problem of drought for grain maturation and human economy was much more likely to come at one or both ends of the rice-growing season rather than in the middle.

How much rainfall is enough? This depends on when it comes. Climatologists have answered the question of rainfall requirements by taking into account both direct evaporation and evapotranspiration directly from the growing plants and devising a formula for approximating potential evapotranspiration. The amount of direct evaporation declines steeply as the rice plants leaf out and shade the surface water, but most of this gain is offset by increased transpiration from the plants themselves. In addition, a certain amount of water that is surplus to this loss (up to perhaps 100 mm) may be stored in the paddy soil and thus carry over. The amount of moisture needed for a month of normal rainy season rice growth, taking into account potential evapotranspiration and soil storage, varies from 190 to 200 mm per month for the latitude and temperatures of Central Luzon and Ilocos. Anything less quickly becomes grossly insufficient to replace the moisture lost to evapotranspiration from the plant and through evaporation. Without active irrigation, a month of rainfall measuring less than 100 mm, or even two weeks, can be disastrous.[57] The record for Manila is replete with at least 27 individual months below 100 mm during the main June–October growing season. Eighteen of these severely dry months were Octobers.

It is in the nature of monsoonal circulation systems to be highly variable, so while 1903 was extremely poor in the Manila area, with low rainfall in June, August, September, and October, there were no months below the standard index at Dagupan in the same year. While there could be considerable variation within Central Luzon, there were also many relatively dry Junes and Octobers when the entire plain was affected. Such a year was 1911—resulting in near record rice imports the following year. Even then the harvest in some peripheral areas was up. Again, the vagaries of tropical storm tracks could quickly alter conditions, and in some years, such as 1907, it was a combination of too much rainfall in July and August plus too little in October and November that halved the crop in Nueva Ecija and parts of Tarlac.[58]

There were also a small number of annual cropping cycles in which problems arose when the rains failed to *stop* in a timely way and continued through the harvest months of November and December. Such a situation made it difficult to successfully harvest and dry the crop. This could lead to plant rust, rot, and loss. In the 75 years from 1865 to 1940 (less 1899) there were six Novembers at Manila that recorded 300 mm of rainfall or more. This hazard could be general, but it was more likely to occur across the southern portion of the Central Plain.[59]

The frequency of El Niño conditions varies both over time and in the geographical pattern of drought and harm to human welfare. The record is very strong that the El Niño/ENSO that produced the record of much-studied droughts on Madura Island off the coast of Java also produced a closely related record in the Philippines taken as a whole, though not always in the Central Luzon rice basket in particular.[60] The historical record of rice imports, then, is a fair guide to the record of food production shortfalls in some parts of the then commercially integrated sections of the Philippine archipelago.

The incidence of Pacific El Niños was variable. The stronger events are symbolized on graph 2.4 with tall to medium columns. During 1847–63, there were no El Niños classified as moderately strong or more (o for 17 years). By contrast, the period of rapidly rising rice imports, 1871–91, witnessed a concentration of intense El Niños that remains unusual in the entire four-century record (6/21). Even worse, this was followed by ten more during 1897–1920 (10/24), or one every 2.4 years. The intersection of short and long intervals in recurrent ENSO phenomena produces just this sort of complex pattern. Again, the relationship on the graph is not always straightforward because not all El Niños produce significant droughts in the Central Luzon rice-growing areas or cause them to occur in the months when they would do the most damage to production. Further, a drought-causing El Niño that began in the November–December season in one calendar year could cause a poor harvest only near the end of the following calendar year and create a market for increased imports primarily in the year after that. One may conclude that the historical suddenness of the onset of very high imports and the ongoing magnitude of these shipments are correlated in more than a general way with the onset of a multiyear period of frequently problematic agricultural weather.[61]

In a season of serious shortfalls in domestic rice production the deficit was often made up in some part with rice brought from the Southeast Asian mainland. This left urban consumers and others with a choice between the domestic and imported staple. City folk were advantaged because most of the imported rice was landed and handled in Manila. But this juxtaposition does not exhaust the options. Sweet potatoes, yams, and taro root were also carried in the public markets at one level or another. But the main alternative to domestic or foreign rice was maize. Before recounting various spectacular droughts and hard times, this is the place to consider the urban maize option.

Coping with Shortfall: The Maize Option

Maize, or corn, became an increasingly important part of the national diet as the later nineteenth and early twentieth centuries wore on and the rice deficit continued to grow.[62] But in the Philippines its production and mass human consumption have usually remained highly regionalized, and Manila was not located within one of the emerging maize concentration areas. In Cebu maize was grown along with millet in the early nineteenth century. Under local conditions corn outproduced and over time replaced millet as the principal staple. But in nineteenth-century Manila, as far as maritime arrivals from the outer zone are concerned, maize remained almost a null category. In late 1861, a shipment of 8,000 *mazorcas* (ears) was received from Vigan. Five cargoes from Cebu, Vigan, and La Union brought altogether less than 500 *cavans* of maize to the city in 1862. This was "whole" maize, not processed into grits or flour, which would have been more immediately subject to weevil infestation. Ten years later only a single small cargo was logged. This constitutes almost the entire record of coastal maize shipments to the city. In our 1881 sample year, a great year for rice production, maize is nearly invisible in the commercial record, never quoted when commodity prices in Manila were summarized. From diverse occasional sources and an almost nonexistent maritime flow, we may conclude that maize was not critical to feeding the city.

But it was not absent. The American consul Alexander R. Webb wrote in 1888, "I have never seen . . . it shelled for sale on the market, nor is corn meal made, as far as I know. It is apparently raised . . . for roasting, and when green it is peddled about the streets," fresh or boiled or roasted and ready to eat (figure 2.3). Evidently, the consul did not encounter the street vendor delicacies of leaf-wrapped *suman de maiz* or *ginatan de maiz*, corn kernels cooked with coconut milk.[63]

It is worth noting that as few Mexican women ever came to the Philippines almost none of the indigenous Mexican maize-based cuisine was introduced to the archipelago or through it to Southeast Asia and China more broadly

FIGURE 2.3.
A vendor roasting fresh maize over charcoal, a typical way corn entered the diet of rice consumers except in hard times, weekly market, Malasiqui, Pangasinan. (Photo by the author, 1969)

despite the fact that the plant was broadly transferred. As the ethnobotanist Robert M. Zingg put it, "[C]orn was taken completely out of its characteristic Mexican complex and fitted into the prevailing rice patterns."[64] Since dried maize kernels are hard, fitting them into the rice patterns meant fracturing them so they could be more readily boiled. This product is sometimes called rice-milled maize.

As Consul Webb inferred, maize was widely, if sparsely, grown on scattered small plots on hilly or other well-drained land in the provinces surrounding the city. In this zone, it was sometimes planted as a dry season crop after rice was harvested and sometimes was produced in two or three crops per year on its own plot. By itself it could provide a "more balanced diet than many root crops." If it was sowed intermixed with rice, as in the *tenggala padi* system employed in parts of Malaya, no reports have come to my attention.[65] Maize was nearly always quoted in the weekly market price reports for Malolos (Bulacan) and Batangas in 1861–62, and it was sometimes reported to be in transit to the city from the inner zone—as when four *bancas* loaded with maize were recorded in a sample week of observation on the Pasig River in 1853. An important agricultural survey conducted in 1886–87 estimated maize production of about 600,000 *cavans* in all of Central and northern Luzon compared to 2.2 million *cavans* in Cebu alone.[66]

Maize became more significant whenever the rice harvest was short, and this crisis substitution had been going on for generations. So it is not surprising that as a result of the drought of 1911–12 the colonial executive recommended a policy of encouraging "the diversification of crops," and corn emerged as a critical crop in Central Luzon and Laguna.[67] Certainly the acreage planted to maize nationally underwent a major expansion during 1911–14. Experts said that the productive "Moro white" variety, "a stable hybrid between Mexican June . . . and a native white variety," enjoyed the greatest use as a human food.[68] Despite all this, milled dried maize and cornmeal did not become major staples in the Manila dietary. Hominy, to say nothing of several varieties of cornbread from the American South remained novelties despite demonstration events and the wide distribution of recipes during the early period of American control.[69]

Nevertheless, starting in the 1880s and during the prolonged period of rice shortfall, a sprinkling of advertisements in *El Comercio* announced maize for sale from the latest crop. The grain is variously described as shelled, or shelled and machine ground, and as superior or cheap. In a number of cases maize is advertised as a feed for horses and poultry. At the same time, small hand-cranked cornmills were offered for sale by several Manila stores.[70]

Finally, a new source (the "Arribos" column in *El Comercio*) briefly allows us to track domestic commodities entering the city by "water, land, and rail."

At last modest arrivals of maize come into view: about 700 to 800 *cavans* a month plus hundreds of thousands of ears in July and September–November 1894 and March and April of the following year. At the same time, La Castellana, a major store on the Escolta, was advertising new "maize for fattening horses, chickens, pigs, etc., in whole grain and milled . . . fine or coarse."[71]

A similar shortfall in domestic rice production relative to population growth in Java during the first two decades of the twentieth century was accompanied by a dramatic increase in the production and consumption of cassava—like maize, an American plant of the Columbian Exchange. But, whereas pioneer farming in Java now led to the uplands where cassava was the most calorie-productive crop on fields without benefit of irrigation, the same conditions did not apply on Luzon. Dried cassava chips (*gaplek* in Indonesian) were known, but they did not become a major foodstuff of the Luzon lowlands or metropolitan Manila.[72]

For human consumption in Manila short of a crisis, the dietary choice remained overwhelmingly between various grades of domestic and imported rice rather than between rice and maize. In the view of Tagalog Manilans above the line of destitution, rice-milled maize, long a staple among ordinary folks in Cebu City, and even maize-rice blends were for poor Visayans. Of course for desperate people there were options other than switching to a fall-back food. There was the sale or mortgaging of assets, if one possessed any—land, house, carabao—and in the city there were social support networks, pawnshops, and moneylenders. There was also begging, thievery, banditry, pillaging, and flight—things that tore at the social order, as we saw in the 1887 report of Consul Hens. They would occur again under starvation conditions in the city in 1944–45.[73]

In twentieth-century Manila the everyday human consumption of maize in grain form remained minimal except in years of marked drought. It continued in frequent use as a fresh vegetable and became more widely consumed in special products such as the iced confection known as *halo-halo* or popcorn. But, despite its growing use, all this hardly softened the general perception that corn was something better fed to pigs, an undesirable substitute for rice.

Disaster

The several disasters in rice production during our period were seared into the memories of the vulnerable. We pursue them in part to get a sense of the vital commercial role Manila came to play in supplying rice to other parts of the country. And we pursue them to get some sense of coping behavior at the individual and community level. Because Manila was relatively well provisioned, the center of the largest food system, the worst effects of the great droughts often played out in the provinces. The severe droughts of the 1870s and 1880s give a graphic sense of the relationships among droughts and yields, as well as of imported rice and maize as alternative staples.

The powerful El Niño of 1877–78 brought a drought that ruined crops in many parts of the archipelago. This ENSO event was so strong that it caused droughts not only in Southeast Asia and China but also in northern India, eastern Australia, the watershed of the Nile, and Brazil. The Dutch consul's report confirms that 1877 and 1878 were "years of drought, insufficient harvests and consequently misery and mortality."[74] In the Manila area, the month of October 1878 saw only half the rainfall normally required to finish the annual rice-growing cycle. Less than ten kilometers south of the urban area, the parish books of Las Piñas report "drought" and a "very poor harvest," followed by "famine" and what Peter Xenos calculates to have been a clear episode of "crisis mortality."[75] Proximity to the capital did not save the starving. As far as the records allow us to see, Manila's parishes did not suffer this fate to the same degree, but the median wholesale price of clean rice in the city that year reached the extreme level of three pesos as against a more usual median of around two. Farther away famine also struck at the northern end of the Central Plain in Pangasinan, Nueva Vizcaya, and no doubt elsewhere.[76] Besides bringing immediate human misery, the production shortfall that year wrote a definitive end to the long-standing pattern of intermittent surplus rice production and export.

Outside the Manila area, Iloilo and Cebu were important ports that were still small cities in the late nineteenth century but second and third in size in the national urban hierarchy. Each was the center of its own regional food supply and distribution system. In Cebu (map 1.4), the British vice consul described the effects of the 1877 drought: "The natives continue extremely poor. . . . The bulk of the population finds difficulty in paying the poll-tax. . . . A partial failure of food crops occurred in the past year, and maize has been sold as high as . . . three times its usual price, and this has aggravated the distress of the [people]. Unusually hard measures have been used by the Government during the year to collect arrears of tribute." When the rice crops "failed" during the droughts that accompanied these "very strong" El Niños, "it was feared that a famine would break out. . . . Timely arrival of the grain from China, Japan, Saigon, and Siam . . . [helped to] alleviate the situation."[77]

But rice was of less importance in Cebu than in many other places. For 25 years British businessmen there had been observing that "maize [has] largely displaced rice as the ordinary food of the people." Steam-powered mills for maize were starting to appear, but it was usually milled by hand in a circular stone hand mill (*gilingan* in Tagalog).[78] Cebu's maize supply came from the immediate area and also in shipments from across northern Mindanao. Beyond this system of maize production and regional commercial supply, one gets the sense of both chronic and periodic regional and urban caloric deficits met with imported rice from the Mekong Delta.[79]

Unlike Cebu, Iloilo was set within a rice-rich ecumene. Considerable rice was produced in both Iloilo and Capiz provinces on Panay (map 1.4). Further, in the 1860s at least, the port of Iloilo received many cargoes of *palay* from various landings in nearby Negros, and it sent out about half as many to Bikol, Leyte, Cebu, and southeastern Negros during the same period. It was the center of its own regional food supply and distribution system.

The El Niño of 1877–78 hit very hard in Iloilo. Large quantities of rice were imported directly from abroad, and "extremely large" amounts were brought from Manila, having originated in Pangasinan and some other provinces. Still, there was "great distress among the poorer classes, and many deaths . . . occurred from starvation," while others died of starvation-related diseases. Tragically, poor families had "already in many cases sold their little all to save themselves from starvation" the year before. A vice consul observed, "One man has sold his house, another his buffalo, and so on, but this year [1878] there will be nothing left for them to sell."[80] Drought was the triggering condition.[81]

Maize yields were not a major factor in feeding Manila, but a decline in the harvest of rice or maize outside the areas of domestic supply to the city could readily affect the tonnage of rice imported through the city.

Another extreme event began in 1884, this one linked to the explosion of Krakatua, a volcanic island off western Java, which the year before had spewed great amounts of volcanic dust into the atmosphere and strongly affected rainfall in Indonesia, Thailand, and the Philippines. Rainfall in Manila during October 1884 was disastrously low. This was followed by further significant drought from August through November 1885 in a year that set the record for low rainfall during our period at less than half of the 100-year mean.[82] Philippine rice imports during 1884 exceeded 100 million kilos—setting the nineteenth-century record. It is remarkable that the price response was not more extreme, but by this time an international commercial system was fully in place with experienced professional operators buying clean and milled rice and transporting it in rapid steam-powered vessels over an increasingly well-traveled sea-lane between Saigon and Manila.

Nevertheless, a catastrophe of biblical proportions ensued. It was by total chance that this severe natural perturbation coincided with the onset of a foreign trade depression caused by the first international sugar glut. These shocks were quickly followed by the death of most work animals in the main Philippine rice-growing areas and by cholera in the human population. As the price of rice rose, it was followed by an immediate cut in ordinary urban family budgets for fruits and vegetables, as well as meat and fish. There were dietary and health consequences even when people were not technically starving. Further, we may expect that this economic depression and drought produced a substantial loss of ownership of agricultural property by landholding farmers. Mestizos have been accused of acquiring an outsized share of good agricultural land.[83] But such acquisition has not been studied in relation to drought cycles, and so far this severe depression episode has not been thoroughly studied either.

In the 1890s, a frequently difficult rice situation was made worse by colonial protectionist policy set from afar. In Iloilo the level of rice imports was down by half in 1891 "owing to a fair crop of paddy in this district, and the consumption of maize caused by the enhanced price of foreign grain." The price of imported rice was "enhanced," in part, because of an import duty applied under Spanish policy starting in early 1891. From the perspective of a foreign merchant in Iloilo, this duty "of about 4 dollars [Mexican] per ton on all rice imported . . . falls very heavy on the poorer classes." Again, in 1896, the crop on Panay was so poor "that in some of the inland villages the priests had to ask help from the public charity to supply funds to feed the starving [inhabitants]." But, judging from the declining level of direct imports, destitution and the import tax were dampening effective consumer demand for rice.[84] It appears that the same was true in Manila. The period of high imports reached a hiatus during 1895–98. In order to make certain that the situation was not exacerbated by the opportunistic export of domestic rice, the Spanish authorities boosted the standing export tariff from 25 centavos to 2 dollars (Mexican) per 100 kilograms.[85]

Manila Food Crisis during the Siege of 1898 and Its Aftermath

Not all the serious urban food supply problems in this era were the result of disease and weather. The Philippine Revolution against Spain broke out in 1896 and reached a hiatus the following year. In 1898 American president William McKinley sent a naval squadron followed by expeditionary troops to establish a new regime of imperial control. Knowledge of the provisioning-related events of the Revolution and the subsequent Philippine-American War is fragmentary but suggestive. The distinguished author Nick Joaquin colorfully insists that Spaniards caught in Manila during the Filipino and American siege of the city in 1898 were reduced to consuming dogs, cats, and rats.[86] The siege lasted roughly from the destruction of the Spanish fleet on May 1 until the capitulation on August 13. During this time the city was progressively cut off from its provisionment hinterland. Victor Buencamino later recalled that the food supply was so interrupted that demand for meat in some quite affluent households, including his own, was met by eating their carriage horses one by one. Some city residents fled—those who could afford it—to Hong Kong, Pasig, or rural estates. Eventually, many European foreigners and Manila Chinese were evacuated by neutral vessels. So the civilian population of the city declined. Earlier it had expanded with the arrival of Tagalog refugees fleeing Spanish attacks at Lemery, Nasugbu, and elsewhere in western Batangas in 1896.[87]

A month after the siege began a French officer visited the city's major public markets. He reported that "baskets full of fish and delicious fruits" were available to the crowds of Tagalogs and Chinese. But as the Filipino insurgents drew closer (June 9), he reported, "The noose is tightening." The precautionary evacuation of foreign civilians was under way. "The last of the Chinese are ready to leave," he wrote, and would be transported to Hong Kong. By June 20, market prices had doubled, and fresh food was becoming rare. Toward the end of the siege (August 7–11), the troops were "living on biscuits."[88] A Spanish resident reported, "[W]ant had become acute because the bakers and other traders of prime commodities had not foreseen the situation. . . . Food became scarce, then ran out. Abuses arose." By late July, "Bread was made of wheat flour mixed with rye and very soon it would be substituted with rice. Beef had long ago disappeared, and carabao meat was given as substitute, [and] when even this is exhausted it would be substituted by horse meat. The situation [was] afflicting."[89] Another observer noted, "Sickness and disease prevailed among the Spanish soldiers to an alarming extent."[90]

The effect of severe food shortages on the less affluent majority of the city population during the siege of 1898 is largely undocumented. Indeed, many of the cited examples of privation among Spaniards—a shortage of bread and beef or having to eat rice—were hardly relevant to Filipino families. Still, a cessation of maritime trade due to the blockade would have caused great unemployment, and a shortage of fresh food would have affected everyone. In the second quarter of 1899, well after the capitulation, a flood of imported rice began, but serious disruption of the ordinary patterns of provisionment continued for several years.[91]

Residents on the urban fringe resumed their provisionment roles within a week, using the surviving carabao to till the fields of Mandaluyong, San Juan del Monte, Caloocan, and Malabon. Then, in February 1899, American troops attacked the surrounding Filipino forces, striking outward from the city and launching the Philippine-American War.[92] Several important agricultural supply areas became venues of conflict. Rice milling and forwarding by rail remained interrupted. The disruptions caused by warfare and the effects of the second wave of rinderpest can hardly be separated here. In a short time the population of work animals was destroyed, and in the midst of the very strong El Niño of 1899–1900, rice production collapsed.[93]

In several areas a desperate guerrilla war ensued. The invaders' response to the struggle in Batangas and Laguna provinces included concentrating the population in towns, allowing people out to harvest rice only under military guard, and burning thousands of rural homes and hamlets outside the garrisoned town centers.[94] Further, American forces embargoed the Batangas ports from December 1901 until May 1902. All this resulted in major interruptions in domestic production, as well as the flow of ordinary food supplies to the city: hogs, onions, garlic, lentils, and citrus fruit. And, as imports from China, including food commodities, had already declined by two-thirds in 1899, in 1901 there was only a trickle.[95]

In the immediately following years, rinderpest, drought, and the effects of conquest stalked the land. Many fields were left unplanted. But very large quantities of rice were now entering the port of Manila from Saigon. In early 1902, normally the season of high domestic arrivals, an armada of small steamers was kept busy shuttling back and forth between Manila and Saigon (graph 2.5). Considerable rice was also purchased in Hong Kong, Siam, and even Calcutta.[96]

Following a moderately strong El Niño, which had begun late the previous year, the 1903 rice season in the Manila area was almost as dry as in the disaster year of 1885. The price of rice was unprecedented. Not only did drought affect rice production in 1903, but the numbers of carabao and other bovines fit for work in Central Luzon had just recovered from the first rinderpest epizootic when they became sufficiently numerous to sustain another. They were then completely devastated by a second epizootic, which developed around the turn of the century, an event made more widespread than the first by military movements from region to region and island to island. This time it took more than a decade for carabao numbers to recover. The symbols on graph 2.4 are meant to be indicative in only a general way. What one can see is that drought and the death of work animals in the second epizootic and military and economic disruption combined to produce very large caloric shortfalls. Unprecedented imports of mainland rice followed during 1902–5. The three principal ports, but especially Manila, were centers for the reception and distribution of imported rice during each of these events.

As if all this were not enough, the droughts also sometimes triggered infestations of locusts, which caused further local and regional crop damage. Typhoons hitting the principal surplus rice zones in Central Luzon could also cause catastrophic damage.[97] More rarely, volcanic eruptions buried everything in a certain area, including crops, fruit trees, and villages. Massive eruptions, some distant and some nearby, ejected sufficient material into the atmosphere to cause global cooling and shorter growing seasons for a time.[98] The combination of ENSO droughts and rinderpest epizootics goes far toward informing our understanding of the vicissitudes of domestic rice production and the specific timing and magnitude of rice imports.

Manila's Domestic Rice Supply at the Start of the Twentieth Century

With the construction of the Manila-Dagupan Railway and steam-powered hulling and polishing mills along its right of way, more and more rice was diverted away from coastal and internal waterway vessels. Both Dagupan and Malabon-Navotas were affected. This trend began just before the Revolution. It was made more drastic by various wartime restrictions placed on coastal shipping by the American military (starting in July 1899). In Dagupan the change was not instantaneous, and it remained a notable, if slowly growing, commercial and transport hub. Since the railroad did not immediately go far up the Ilocos coast, Dagupan retained its role in servicing places like Vigan and connecting them by sea to the Manila Railroad. But Dagupan and a major portion of its riverine hinterland were on this railway line, and coastal shipping of rice from Pangasinan to Manila declined. By 1908 there were big new rice mills at Dagupan and Mangatarem in addition to the one at Bautista and a growing number in Tarlac.[99] All of them moved their product southward by rail, diverting rice that might once have coursed through Dagupan's *pantalan*. Near Manila, swampy Malabon-Navotas was bypassed by the railway—the same railway that now serviced a considerable part of Malabon's former hinterland.

The rice trade system of the 1880s through 1905 and beyond was a triumph of the paleotechnic means of production involving power mills, steamships, and railroads. But, although the crop could be milled and brought to Manila much more efficiently than before, this was an intermittently disastrous era for production. It was a prolonged period during which rice production and society were famously beset. As Benito Legarda Jr. has observed, "We economists try to attribute as much as possible to endogenous factors interacting within the economic system, but in this case the exogenous factor of El Niño cannot be ignored. How puny economic forces look beside the forces of nature."[100]

3

The Manila Rice Trade
to 1941

THE TONNAGES OF RICE received in Manila eventually grew to something far beyond the requirements of the urban and nearby populations for the city was integrating an expanding hinterland. It had become the principal center of the large import trade in rice and also for organizing the shipment of both domestic and imported rice. On balance this represented a reversal of commercial flows. In the worst of the drought and rinderpest years, the reversal was nearly complete. It changed the economic role of the city and the nature of its integration with a growing number of provinces.

At the same time an important new area of domestic supply was emerging in Nueva Ecija and northern Bulacan in the south-flowing watershed portion of the Central Plain. From a benchmark of circa 40,000 *cavans* in 1870, Nueva Ecija was providing Manila with more than 800,000 *cavans* of rice annually by the early 1920s. Increasingly railroads and roads facilitated the opening of these promising rice production localities in the eastern Central Plain. As this happened, the relative contributions of the former inner and outer zones of supply were reversed. Less and less of the domestic rice flowing into Manila now moved by water. The new patterns of the rice trade were professionally managed, smoothed out, and increasingly routine. Although Manila and the Philippines remained reliant on foreign sources to make up for the ongoing rice deficit, the magnitude of the shortfall declined over several decades to a much lower level—with a few jarring exceptions.

THE CENTER OF GRAVITY IN
RICE PRODUCTION SHIFTS EAST

In the first decade of the twentieth century, a branch of the main trunk railway line was extended northward through the eastern portion of the Central Plain into northern Bulacan and the heart of Nueva Ecija. Here commercial rice agriculture was rapidly expanding and displacing former frontier stock raising and lowland forest.[1] Rice here was initially more productive per hectare on virgin land than in some longer-established areas, but it was not without problems. As it happens, central Nueva Ecija averaged about 77 percent of the typical rainfall of central Pangasinan and with that came increased risk of shortfall in June and October. It became more common here than elsewhere in the region to use some form of active irrigation on each end of the rainy season since the yield of many rice varieties responded directly to the opportunity for a longer period of growth.[2] As this province was opened, settlers often built local irrigation systems using an earlier technology. The construction of larger, modern water management systems got off to a slow start. A result of this is that the agricultural and economic experience of the Philippines was quite different from that of Thailand or Java where state-led irrigation projects were central. After a first larger project was constructed in Tarlac in 1913 (watering 6,400 hectares), it was another decade before others of similar scale were attempted. Another feature of this new area was that large tenanted rice estates became much more common than had been the case in central Pangasinan.[3]

Still, considerable land was available, and small farmers streamed into northern and central Nueva Ecija from Ilocos and into the southern municipalities from the Tagalog area. In wooded areas, it took some time to clear and level the land during which families needed some support. In the longer run, it proved difficult for them to deal with the aggressive expansion of the ill-defined borders of various estate entities (as it had also with the more rapacious of the friar estates in the eighteenth century), to master the complex requirements to acquire the land and water rights to which settlers were entitled (in the early twentieth century), or to deal with the thugs of land-grabbers (in the 1920s).[4] One of the estates in Nueva Ecija was formed in 1857 with an elephantine grant of 16,740 hectares to the brother of the Spanish provincial governor. This was Hacienda Esperanza centered on Cuyapo. Not precisely defined, it sprawled over what eventually became three municipalities in Nueva Ecija, as well as parts of northeastern Tarlac and southeastern Pangasinan, and sparked numerous complaints.[5] The larger parts of the original grant, by 1930 under several owners, brought the total to 17,910 hectares. At least 16 other estates exceeded 500 hectares in 1930. In many cases, these parcels were parts of still larger aggregations under the control of one or another extended family.[6] All these paled before the 27,080-hectare Buenavista Estate in northern Bulacan owned by an entity of the Catholic Church.

Another distinction of this new area was the increased use of mechanical rice threshers known as *trilladoras*. These became much more common in this eastern zone than elsewhere in the country as modern American threshing machinery became popular with estate owners (figure 3.1).[7] During the great rice shortage of 1919, the director of the Bureau of Agriculture, Adriano Hernandez, saw a solution in the wider adoption of "modern machinery" with the bureau taking the lead in demonstrating "such machinery and modern methods."[8] At the same time, the threshers were hated by tenant farmers because their use meant a loss of family income from hand or foot threshing. Because the mechanical threshers did not perform well using the awned *palay Iloko* varieties, these were gradually given up in Nueva Ecija. The Bureau of Agriculture also worked on the introduction of improved seed and, for better or worse, on "reducing the number of varieties planted." Although this resulted in more efficient milling, a diversity of varieties would have provided better protection against a widespread plant disease.[9]

The new railway line into Nueva Ecija had a concentration effect identical to that along the Manila-Dagupan main line a decade earlier. Entrepreneurs were quick to spot the opportunity to erect sizable steam-powered rice mills. Two principal collection and milling centers emerged. The smaller of these was Gapan, the traditional source of *palay de Factoría* in the nineteenth century and still the general market center for the province in the early twentieth. By 1895 the upriver locale of Cabanatuan was crystallizing into a major center in the rice supply system. Early investors there erected a *tranvía*, or local tramway, more than a kilometer long for "carrying *palay* and some passengers." Reached by the steam railroad a decade later, bulk transport charges to Manila dropped sharply. By 1913 there were six power mills in the province with the three largest in Cabanatuan and nearby Sta. Rosa.[10] These were mills on the same scale as those already operating at Bautista and Gerona. A branch railway line was opened into Cuyapo in the northwest corner of Nueva Ecija (1908) and was subsequently extended to Rosales and San Quintin in eastern Pangasinan (1912–18), both of which then emerged as milling centers. In 1911 Pangasinan was still producing 35 percent more rice than Nueva Ecija.[11] But by 1918 there were 18 modern rice mills in Nueva Ecija and more planned. Rice agriculture in the province was expanding rapidly.

As it happened, the railroad line to Cabanatuan opened just in time. Land clearance nearly always results in increased rainwater runoff. No longer impeded by the preexisting vegetation, the already substantial rainy season volume and velocity of surface water increased the transport of sediment. Much of this larger sediment load then settled in the existing watercourses as the stream velocity slackened at the end of each rainy season. Whereas a dense traffic of *cascos* and smaller vessels had carried rice from Gapan and Cabanatuan to Malabon and Manila in the late nineteenth century, by the early twentieth higher rates of siltation made river passage problematic.[12]

FIGURE 3.1. A steam-powered *trilladora* threshing rice on a Central Luzon estate (Murcia, Tarlac) about 1914. The thresher was popular with large estate owners but not with share tenants, who lost income because of it. (USNA II, RG350-P-Ac-4-6)

The magnitude of rice shipments from Nueva Ecija to Manila finally approached those from Pangasinan in the second decade of the twentieth century (graph 2.3). When it did, this was in part because of the increasing diversion of Pangasinan rice to the expanding provincial market in Ilocos, especially Ilocos Sur. With the advent of rail transport to Manila, rice grown in northern Tarlac or northwestern Nueva Ecija was no longer shipped north by river and onward via Dagupan. As Nueva Ecija surged ahead in the 1920s and became clearly number one in supplying domestic rice to the city, some Pangasinan rice fields were converted to sugarcane, coconuts, or mangos.[13] Averaging the available data for 1922 and 1923 reveals the structure of supply (table 3.1). By this time Nueva Ecija accounted for almost 42 percent of total domestic arrivals in Manila and was supplying by itself more than the entire former outer zone.

For the first time, a quantitative picture emerges of the flow from Bulacan, especially the northern portion of that province. Also in the inner zone and after years of running a rice deficit, Pampanga began to record a surplus again starting about 1910. In the early 1920s, it was providing 6 percent of the supply forwarded to the capital. During the poor market for sugar in the 1890s and early 1900s, some land in Pampanga was converted from sugar to rice production. By the early 1920s, the acreage dedicated to sugar had recovered, but land planted to rice had expanded even more.[14] At the same time arrivals in the city from the outer zone as a whole had advanced barely 23 percent over 1881. As before, Pangasinan or the combination of Pangasinan and Tarlac was predominant within the context of outer zone supply. The emergence of Cagayan as a surplus producer was a new development. With these changes brought by the opening of new lands and by new transport technologies and infrastructure, the concept of "inner" and "outer" zones loses its analytical usefulness.

Equally revolutionary from the vantage point of forty years earlier was the small role now played by coastal and interisland shipping in the recorded supply of Manila: about 4 percent of total domestic receipts in 1922–23. The balance arrived in Manila by rail. This would change during the later 1920s and 1930s with the increased use of trucks in rice hauling. The total tonnage of milled and unmilled rice carried by the railroad peaked at the end of World War I, but it remained moderately strong through 1922–23, recovered in 1927, and then tailed off as trucks took over more of the trade from the provinces nearest the city. As power mills proliferated, more and more rice arrived in the city as *bigas*—milled, polished, and ready for consumption. *Palay* as a percentage of all rice carried continued its long decline.[15]

The use of motorized trucks in the rice transport system was made possible by an aggressive and persistent program of road building and the establishment of a network of local road maintenance workers, *camineros*, who labored to keep them in good repair. The Spanish administration was later credited

TABLE 3.1. Rice Arrivals in Manila by Province and Zone of Origin, 1922–1923

Province	All		Former outer zone		Former inner zone	
	Cavans	Percent	Cavans	Percent	Cavans	Percent
Nueva Ecija	802,600	41.5	—	—	802,600	68.6
Pangasinan and Tarlac	438,900 / 250,900	22.7 / 13.0	689,800	90.1	—	—
Bulacan	248,700	12.9	—	—	248,700	21.3
Pampanga	118,300	6.1	—	—	118,300	10.1
Cagayan	40,900	2.1	40,900	5.3	—	—
Zambales	27,200	1.4	27,200	3.5	—	—
Capiz	2,900	0.1	2,900	0.4	—	—
Other	5,100	0.3	5,100	0.7	—	—
Total	1,935,500	100.1	765,900	100.0	1,169,600	100.0

SOURCES: SBPI, 1922, no. 5, 87; 1923, no. 6, 97; 1924, no. 7, 99.

NOTE: The data are an average of 1922 and 1923 in *cavans* of 57.5 kilos. The table includes amounts arriving by railroad and coastal vessel. It does not include the unreported amounts arriving by truck or rivercraft.

with having laid out routes for public roads, thereby clearing the way for the subsequent regimes to use them as public domain rights of way. But on Luzon there was only a very limited endowment of more or less all weather roads from the late nineteenth century, and these mostly connected Manila to the nearest parts of Cavite and Bulacan. The reason, according to a British engineer and longtime resident was that under some governors-general "scarcely a cent was expended on roads or bridges . . . [for] the provincial governors simply pocketed every dollar." Owen simply says, "Those seeking an explanation for the rise of the Philippine export economy [in the nineteenth century] must look elsewhere" to factors other than the expansion of a well-built infrastructure of roads and bridges.[16] By contrast, state investment in road building received a high priority in the early twentieth century. This program of infrastructure development was the particular obsession of W. Cameron Forbes, who became secretary of commerce and police with responsibility for public works in 1904 and then governor-general in 1909. Quick action followed. From 1908 to 1910, the total of first-class roads was increased from 394 to 915 kilometers. Road construction became popular with elected Filipino politicians and their constituents. Initially these roads facilitated the delivery of rice to nearby ports and railway loading docks and mills. In the longer run, with the introduction of trucks and buses, roads came to link millers and wholesalers in the nearer provinces directly to the city. From the late 1930s through 1941, trucks delivered approximately 65 percent of the rice brought to Manila from Bulacan and Pampanga provinces but only 10 percent from Nueva Ecija, Tarlac, and Pangasinan—28 percent overall. As Dick and Rimmer point out, the railroad still had the competitive cost advantage in long haul, full carload units and express service.[17]

From the perspective of the consumer, the high price of milled rice in Manila during the 1920s was not a triumph. Given the shortage of merchant shipping during and just after World War I, as well as the local effects of the world flu pandemic during the harvest season of 1918 and the disastrous regional drought of 1918–19, it is no surprise that rice prices matched and exceeded the general inflation of World War I. But, while the prices of imported manufactured goods deflated rapidly after 1920, the price of rice stayed relatively high through 1929. Something similar happened in Java during the same years.[18] Political forces in the Philippine legislature evidently placed a higher value on protecting large-scale producers from the competition of rice imports than on assuring affordability for consumers. In the Philippines, higher domestic prices may have stimulated production and thus had something to do with the declining trend of rice imports. But this era saw important productivity advances in rice production in Japanese-dominated Taiwan, while in the Philippines fertilizers and fertilizer-responsive rice varieties continued to play only a minor role.[19]

The annual cycle of deliveries was further smoothed during the 1920s. With domestic deliveries coming primarily by railroad, arrivals shot up at the start of the year and stayed high through April. Thereafter there was a predictable dip, but the volume of shipments then remained more or less level from May through August or September before declining to a predictable nadir in November–December. The low point was now 40 percent or more of the average for the three top months (graph 2.1). This represents a more even cycle of deliveries than before, something that could be expected with the transition to large provincial mills, substantial warehouses, and rail transport. Of course there were still year-to-year perturbations. Only a small part of this smoothing was due to new irrigation systems and a change from one to two rice crops per year. In Nueva Ecija and Bulacan, irrigation before the 1960s led mainly to greater security against drought during the rainy season and to growing varieties that took longer to mature rather than to double cropping as one might have expected if the systemic response had been more involutionary.[20] Timing of the annual price cycle remained as before with the peak wholesale price almost always found within the period July to November.

As noted, the construction of large irrigation schemes was slow to emerge. Despite a sizable project in 1913, it was a long decade before others of similar scale were completed in Luzon. Finally, the Talavera River scheme in Nueva Ecija (9,500 ha) and the Angat River project in central Bulacan and Pampanga (23,100 ha) were completed in 1923 and 1926, respectively. The Angat River scheme was extensive, delivering water (north side) to farms along the river from San Rafael to Baliuag, downstream to Calumpit, and farther south (south side) to Quingua (Plaridel) and a long string of other municipalities in Bulacan Province all the way to Hagonoy.[21] Again, such irrigation works helped make rice growing more secure for farms near the river during the rainy season. However, they did little to make irrigation water available for a second crop during the dry season. That would require a substantial nearby water storage capacity—nearby because of the gross loss of water to evaporation in transit. A leading newsmagazine pronounced this "A Record to Be Ashamed Of," compared to over a million hectares of rice fields serviced by modern irrigation systems in the rather different physical geography of late colonial Java.[22]

Marshall McLennan points to a form of "rent capitalism" that emerged in large parts of the Central Plain—an economic form within which landowners have less interest in advancing productivity than in converting hacienda arrangements to formal contract sharecropping and perfecting mechanisms for extraction. Benedict J. Kerkvliet calls this extending capitalism and curtailing formerly prevailing practices such as gleaning, food loans, and rights to fish and cut firewood or raise vegetables in the dry season, as well as the hacendero's sponsorship of tenant weddings and baptisms. In Nueva Ecija and other nearby Central Luzon provinces this amounted to terminating many reciprocal obligations of the landlord toward tenant families.[23] Rice production during the last decades before World War II was marked by the end of the frontier in the Central Luzon Plain and by an increasing number of estate owners taking up permanent residence in metropolitan Manila, leaving the management of their rice lands in the hands of overseers. Accompanied by rapid population increase, the result was an infamous squeeze on tenant farmers. These were real processes of change in the lives of the producers of the domestic rice supply.

In the settler areas of eastern Pangasinan, Nueva Ecija, and other nearby provinces there emerged a profound sense of injustice vis-à-vis estate owners, the legal apparatus as it applied to land rights, and the Philippines Constabulary (PC).[24] Continuing a much older tradition of peasant millenarian revolts, in 1931 this disquiet led to a midnight attack by 70 local Ilocano farmers, both men and women, on the PC barracks, municipal offices, and land records archive in Tayug municipality in eastern Pangasinan.[25] Soon it gave rise to more radical protests and then revolutionary organizations among Tagalog and Kapampangan farmers in Nueva Ecija and neighboring provinces. These changes and the revolutionary movement that grew out of them in Central Luzon constitute a critical subject in its own right, one that lies beyond our focus.[26]

~

As elsewhere, implementation of the new transport modes opened up deep interior spaces and brought concentration to the big mills and railway loading stations in some localities. It brought concentration to Manila as well, since the city was the major center of distribution to the rest of the country of both imported and Luzon-grown domestic rice. Regional growth in rice-consuming populations, inexpensive transport charges along the main railway lines leading to the city, and concentration meant that many of the small ports of supply of the 1860s and 1870s were now gone from the metropolitan rice network or had been rendered insignificant. Interisland steamships now routinely carried rice *from* Manila to regional and provincial distribution points in the Visayas and Bikol.

Power rice milling, however, did not become as geographically concentrated as one might have suspected, for new *kiskisan* mills (a rice mill operated by rubbing/friction) continued to operate on a small scale and in dispersed locations—well suited to the needs of the producers. Over time, steam power was abandoned in favor of the small internal combustion engine.[27] As to timing, in early 1922 the Pacific Commercial Company sponsored a series of expensive graphic advertisements for Bandera brand mills powered by small engines. Nearly two decades later *kiskisan* mills in Pangasinan and Nueva Ecija had an average capacity of 4 to 6 *cavans* per hour. Ads suggest that this rate was achieved with internal combustion engines of ten horsepower or less. Larger mills had capacities of 10 to 50 *cavans*. By 1929 Macleod and Company was advertising that over 1,000 small rice-milling outfits were in operation (figure 3.2).[28] In the early postwar era, small-capacity local mills continued to process up to 30 percent of Philippine *palay*.[29]

The small-capacity power mills could have certain advantages vis-à-vis their larger competitors. Siok-hwa Cheng chronicles the proliferation of small mills in Burma during the first decades of the twentieth century. Because their typical milling lot was small, the operators could avoid mixing grain varieties of different shapes and hardness, resulting in a more efficient physical milling process. This produced a lower rate of broken grains and thus higher-value output. Because of this the small millers "could often afford to pay higher prices than the big mills for consignments of particularly good and uniform grain." Further, as they were embedded in local production areas, they had little need for a network of buyers. They also had a readily accessible pool of labor as rice farmers looked for other work during the dry season, which was also the milling season.[30]

FIGURE 3.2. "Mill Your Palay by the Macleod Method." Milling and polishing rice with mortar and pestle is picturesque but old-fashioned and inefficient. Using a small mill powered by an internal combustion engine is the modern way according to this artistic ad. (Macleod Machinery, International Harvester, PFP, February 22, 1930, 9)

By 1936 there were 35 large mills in operation in Nueva Ecija alone—double the number in 1918. These were especially concentrated in Cabanatuan and Gapan, each with more than a quarter of total provincial milling capacity. The balance was scattered among at least ten other communities.[31] Where such mills were established, they became the leading purchasers of locally grown *palay*. Some of the polished rice they produced was sold in local markets, but more was forwarded to Manila, the central market for rice.

Chinese ownership of the major provincial rice mills and their growing control of interregional rice marketing were changes that accompanied the massive development of rice production in the eastern Central Plain. The vertical integration of a number of Chinese commodities and rice dealers to include ownership of large provincial rice mills and the construction of substantial on-site storage facilities was largely in place by the early 1920s. By the 1930s it was said that rice storage facilities in Manila itself were ample. Smith Bell and Warner Barnes had largely left the business. By contrast, the Filipino Chinese distillery magnate Carlos Palanca Tan Quin Lay now entered the rice-milling, warehousing, and trading business by creating the Manila Rice Mills Corporation and buying out the former assets of Mariano Velasco, another Filipino Chinese.[32]

REDUCING BERIBERI MORTALITY IN MANILA AND BANGKOK

Diseases connected to the food supply did not disappear in the early years of the twentieth century. As it happens, the experience with beriberi in Bangkok became important in unraveling the mystery of this disease in Manila. The common thread turned out to be the steadily broadening consumption of rice thoroughly polished by machine. In Bangkok beriberi first became common in 1900–1901 and increased alarmingly through 1910. Admissions for beriberi at the Bangkok police hospital quadrupled to over 4,500 per year during this period, and the number of deaths rose from 14 to 282. But an odd pattern emerged from scientific scrutiny of the Bangkok data. It seems that a number of beriberi sufferers were sent home to their families in the provinces. Miraculously, many of them returned after a few months apparently cured. As a result, medical scientists became convinced that beriberi had something to do with eating a diet primarily composed of rice polished in the new steam-driven mills. Somehow the act of uniform high polishing was removing some substance "essential for the maintenance of the normal metabolism of nerve tissues."[33] The important new *Philippine Journal of Science* became the vehicle for communicating the inquiry: "Beriberi was, so far as we can find out, unknown in Bangkok until white, steam-milled rice began to be retailed locally." The first outbreak in the Bangkok jail was in 1890, approximately ten years after two local steam mills began producing white rice. The delay in onset, apparently, was because the product of these early mills was shipped to Europe. "During the nineties [however], attracted by the immense profits which were being made by owners of steam rice mills, many [new] mills were erected, and soon large quantities of white rice were thrown upon the local market." By 1900, with many or most of the hand mills cast aside, people began to suffer from beriberi.[34]

There also developed a distinct geography of this disease. It tended to be concentrated along the Menam River artery and along the "banks of the large, navigable canals which join this river with the adjacent streams throughout the flat, alluvial plains in the neighborhood of the capital." The great river and intersecting canal system constituted the principal means of commercial transport, and along these waterways "steam-milled rice from Bangkok is freely hawked." The population living away from these arteries was observed to be much less likely to be affected.[35]

It was 1911 before the cause of beriberi was reasonably well understood by public health officials in Manila. When it was, experiments began on the use of rice bran, or *tikitiki,* as a medicine useful in preventing and curing the disease. Although beriberi could be ended in institutions where the diet was controlled, it would have been a very hard sell to convince urban Filipino families to go back to consuming home-pounded *bigas—pagputi ng uwak* (when the crow turns white) is the local saying. Still, as a matter of public policy, the American director of health, Dr. Victor Heiser, and the colonial secretary of interior, Dean C. Worcester, sought to discourage the mass consumption of machine-polished rice by recommending a tax. As Heiser saw it, such a tax would not affect the poor, since they would purchase unpolished rice (*pinawa*) and pound it themselves, leaving some of the bran intact, and it would not materially injure the better off since they would typically consume a more varied diet. The secretary rather distantly remarked that such a tax would "impose no hardship worth mentioning." The tax was not approved by the elected Philippine Legislature.[36] Beriberi was not ended in Manila, but even before World War I some were being cured of it by a local formulation made from rice bran.

Although affluent urbanites had sometimes died of beriberi during cholera scares, it came to be primarily a medical condition of very poor mothers and infants. In a dietary study of more than 100 Manila working-class families in the 1930s, a rough calculation was made of the ratio of vitamin B_1 intake to something called "daily total energy exchange." The result depicted most of the families in the study as being on the borderline of risk except for those with the highest incomes. These spent the most on food per person and consumed more fats, as well as fruits and vegetables, than the others—a common finding of such studies in many cultures. Conversely, it was precisely the lowest income at-risk families, those with the least to spend, that had to maintain their caloric intake in order to work and so spent the lowest percentage of their dietary peso on meat, fruits, and vegetables. Six cases of active adult beriberi were manifest among the sample population (666)—all in mothers and all in families with dangerously low ratios of estimated B_1 intake. The study pointed to elevated levels of metabolism during pregnancy and lactation, which increased the personal requirement for B_1. When the intake did not increase to compensate, the woman tended to develop beriberi, starting with tingling and numbness in fingers and legs.[37]

During 1909–12, a time that included major rice shortages, there was a multiyear surge in the Manila death rate from beriberi. Still, the rate fell steadily. A major reason for this decline was the emphasis given this condition in the district mother and infant public health facilities known as "puericulture centers." These were spreading the message about symptoms and cause and directly dispensing large quantities of a concentrated extract of *tikitiki* as a prophylaxis.[38] Studies of rural family diets in the Philippines in the prewar period indicate that worker families in Manila were marginally better fed.[39]

Rinderpest

Between epizootic peaks rinderpest continued to kill Philippine bovines in local outbreaks. For a time, however, the restricted numbers of animals imported helped to keep the disease under some control. But beef prices doubled in the last few months of 1913 due to the shortage of domestic cattle. Agitation by cattle traders and elite consumers proved effective, and bovine imports from Indochina and China were resumed, amounting in 1915 to perhaps 16,000 head. Predictably, the third rinderpest epizootic wave was set off early in 1916 and reached almost a score of provinces by the end of that year. Because of more effective intervention and perhaps more regular disease exposure, this third wave was more prolonged but not as intense as the first two. Annual national bovine mortality peaked at 35,000 in 1922. The intervention of government veterinary scientists finally stemmed this iteration and broke the cycle (an analysis of this event is included in chapter 8). The third wave of bovine mortality did not have a major effect on rice production and thus did not seriously interrupt the declining trend of rice imports.

Drought and Rice Production Shortfalls

The coincidence of droughts, carabao loss to rinderpest, and the Philippine-American War produced the most profound Philippine deficit rice production cycle during 1902–5, including four of the five largest years for rice imports in the era preceding independence. In an expected parallel, the same ENSO event likewise produced a significant rice shortfall in Java (1901–2). As we have seen, this irregular drought pattern continued. In fact the period 1897 through

1920 included at least ten El Niños classified as moderately strong or more and occurring on average once every two and a half years. The moderately strong 1911–12 El Niño produced severe droughts in Central Luzon and Cebu. In Manila and Nueva Ecija (San Isidro), the rainfall totals for October 1911 were grossly insufficient—10 and 58 mm, respectively—and still less in November. Rice yields plummeted in Pangasinan (down by one-third) and Nueva Ecija (down by two-thirds), severely affecting the domestic flow to Manila.[40] So Philippine rice imports were again at fantastic levels during 1910–12, especially 1912, when they reached 300 million kilos for the second time.[41] Imports were above 200 million kilos again in 1915, part of the response to the generally poor agricultural years of 1915 and subsequently 1916 and 1918. Thinking in comparative terms, these were also very poor years in Java, with net rice imports for 1917 and 1918 both among the five highest of the prewar era. Thereafter, the Luzon and Java patterns diverge.[42]

The Philippine rice deficit eased during the 1920s, although there were further import spikes in 1924 and 1929.[43] Again these were caused by drought, but they were not as severe as in preceding decades. In the economic depression of the early 1930s, imports of rice to Manila declined to levels not seen since the early 1880s, but this was a sign of declining incomes rather than a close approach to self-sufficiency.[44]

Any year of much elevated imports reveals not only an important Philippine shortage but also a tradable surplus within the international rice system. In July 1919 a bountiful domestic *palay* crop seemed to be in prospect later in the year. This was the season when residual stores of rice from the previous crop year were increasingly depleted and arrivals of imports typically picked up. But the 1918 crop had been short due to a particularly strong and extensive ENSO event that affected not only the Philippines but also mainland Southeast Asia and eastern India. Such rice stores as were available from the deltas of the mainland were high priced. At the same time the French colonial government banned the export of rice from southern Vietnam for fear of serious domestic shortages. Abundant imports for the Philippines could not be found, certainly not at a "reasonable" price. So the ENSO of 1918–19 was not matched by a corresponding surge of imports (graph 2.4).

By the end of July 1919 the rice situation in the Philippines was serious. High retail prices pushed it out of the diet of a great many Tagalog and other families. As one provincial report has it, "Batangas people are using corn as a substitute for rice during this crisis. While many only mix corn with rice, the majority of the laboring classes [are using] corn as their staple food." Fortunately a plentiful maize crop was harvested, and the price promptly declined. Similarly, maize was now critical in the upper Marikina Valley near the city.[45] The early efforts of the Department of Supply to distribute rice to be sold at a fixed lower price had not reached ordinary people. The situation was not a famine, but in Manila and the provinces a great many people were being forced to rely on maize, a food for which they had not developed a taste. Despite the fact that many Filipinos in the Visayas and Cagayan Valley ate maize every day, Tagalogs and others usually felt poorly served when forced to eat it. In the midst of this situation some professional rice dealers in Manila sought to sell some of their stores in the even higher priced Hong Kong market. A mass circulation English-language newspaper called these dealers "enemies." And since most were Chinese, they were blamed for holding back supplies in order to drive local prices higher and were also accused of giving customers short measure. They were threatened with deportation. The government moved quickly to make both "hoarding" and exporting rice illegal. Still, the situation continued to worsen before improving.[46] Growing out of this experience was a policy climate favoring everyday government intrusion in the rice market.

MANUEL QUEZON AND THE
END OF AN OPEN MARKET IN RICE

The great success of some Chinese in the rice business brought a reaction. When rice shortages occurred, as in 1919, this concentration of non-Filipino market power inflamed nationalist passions—even as the situation was made worse by legislation favoring large rice producers and dealers. After the Philippines became a commonwealth, with a much greater degree of self-government, the National Rice and Corn Corporation was established—an attempt by politicians and the state bureaucracy to intervene in rice trading, not just in emergency situations but every day. The NARIC, as it was called, was almost explicitly intended to challenge Chinese control of this aspect of the national economy. Its public and perhaps contradictory objectives were to keep rice affordable and to assure a reasonable return to domestic rice farmers.

Where rice forms a central item of everyday food consumption, few things undermine the credibility and moral position of administrative authorities faster than a situation in which it is unobtainable or exorbitantly priced. The same was said of bread wheat flour in European societies. To one degree or another, such authorities keep an eye on the rice situation in the capital city and elsewhere. They may be especially concerned with "food security" in the capital because their grip on power would surely be tested if things were allowed to get out of hand. Food or bread riots, became a major form of popular political expression in eighteenth- and early-nineteenth-century Europe.[47] Governments in Southeast Asia, whether run by foreign colonialists or indigenous authorities, were (and are) sensitive to this issue. In the Philippines, we have seen how the Spanish authorities imposed a ban on rice exports during 1851 and part of 1855 and went even further to waive import duties on rice during a seasonal supply shortage in 1857. In 1895 the Spanish authorities suddenly boosted the standing export tariff on rice from 25 centavos to two dollars (Mexican) per 100 kilograms in order to make certain that none left the archipelago as imports fell off.[48] Intervention in the commerce in rice was hardly a new idea.

The American authorities were forced to confront similar issues. They started by affirming a ban on rice exports. In the Philippines rice exceeded 20 percent of all imports by value from 1902 through 1905 and again in 1912. In the same period, the total was routinely above 10 percent. In 1903 the Philippine Commission moved to defray the costs of importing and distributing more than 20 million pounds of rice to be sold "to the inhabitants of those provinces in which the rate was excessive." Along the way, the government found itself frozen out of the rice market in Saigon by the same "syndicate" that was allegedly responsible for the general price escalation and so ended up securing its requirements from other venues, even as far away as Calcutta. Very little of the imported rice was given away. Rather, paid work for some of the needy was created using funds made available for road construction—a method of alleviation ultimately used throughout the twentieth century.[49] In 1911 an El Niño drought reduced rainfall to zero during the crucial final month of the growing season, and together with earlier typhoon damage this resulted in a disastrous crop (graph 2.2). On the advice of an ad hoc committee, the governor-general directed the Bureau of Supply to purchase large quantities of the grain in Rangoon and Saigon. "Thereafter," he reported, "wherever it appeared necessary, the government controlled the price . . . by placing on sale in the public markets government rice at a reasonable price." The Philippine Legislature subsequently approved this action. Another export embargo was enforced from 1919 through 1927.[50] So, although the rice market was generally allowed to proceed under the direction of private traders, there was a history of exceptions adopted in an attempt to avoid public suffering and the political cost of shortages.

Enter Manuel Quezon and the new Philippine Commonwealth government. No sooner had they taken power than they were faced with consecutive years of poor rice harvests (1935–36). Hoarding was said to be rampant on speculation of rising prices, and there were popular protests and demands for rice in some provinces, including Pangasinan. Within a month of the idea, legislation authorizing a National Rice Commission, and through it the NARIC, had won the approval of the legislature. Victor Buencamino, a prominent veterinary, former importer of beef cattle, and sometime Quezonian troubleshooter, was appointed vice president and manager of NARIC when business began in 1936. Buencamino later recalled that President Quezon came to the "realization" that the rice dealers, mostly Chinese, "were in a position to starve the population" if it suited their business purposes. True or not, Quezon

decided "to set up a government machinery that would effectively wrest control of rice distribution from alien hands."[51] In the near absence of an analytical literature on Philippine rice prices, these ideas had been brewing within
the Bureau of Commerce bureaucracy and among some members of the public for at least a decade.[52]

The stated objectives of the NARIC were both to assure consumers that
rice would remain affordable and to guarantee a reasonable return to domestic
rice farmers—a difficult policy objective. The chairman of the NARIC board
was Vicente Singson Encarnacion, a onetime member of the elite Philippine
Commission (1914) and secretary of agriculture and almost certainly a close
relative of one of nineteenth-century Manila's leading rice traders, Petronila
Encarnacion, as we will see.[53] Buencamino had considerable ties to the owners
of big rice estates in Nueva Ecija. At President Quezon's insistence, the NARIC
was given a significant advantage vis-à-vis the private traders, as the rice it
imported was exempted from both import duties and sales taxes.[54]

According to Buencamino, the NARIC's first action was to import rice on
its own account. It also purchased some domestic rice in order to get started.
It then attempted to establish a "reasonable" retail price, starting with Manila.
Approximately 2,000 retailers in the city were required to post bond guaran-
teeing that they would adhere to the now official price, while the NARIC
undertook to ship rice stocks to locales where the official price was not being
respected. Buencamino asserts that at first "certain retailers sold their stock to
hoarders, mostly aliens, at above the official price. These speculators in turn
shipped their rice to provinces still in short supply and thereby made a killing." Over the course of nine months, more than 150 dealers were "suspended,"
whatever that meant in practical terms. In October 1936, Philippine troops
were sent to several locations in Pangasinan "to prevent the shipment of rice
to other provinces as a result of the order fixing the price of rice at ₱0.28
per ganta."[55]

By 1937, the NARIC was routinely involved in domestic rice purchasing.
The intention was to flatten the annual price cycle for the consumer and also
for the farmer, who was often forced to sell at harvest time at the bottom of
the market cycle. The plan evolved to use bonded provincial warehouses where
farmers could deposit *palay* and receive a loan of 70 percent of the market
value while awaiting a seasonal price increase. Alternatively, the farmer could
sell to the NARIC for the stated floor price. Without standing facilities of its
own, the NARIC increasingly bought or leased warehouses and mills, especially in Cabanatuan, Nueva Ecija.

This period, April 1936 through December 1941, was the start of "guided
marketing" in the view of rice economist Leon Mears. The impact of this set
of programs was real but limited. With limited facilities and funding, the

NARIC's purchases of domestic *palay* are estimated to have been less than half of 1 percent of the Philippine harvest. Apparently the agency's activities were regionally limited as well, for in late 1937–38 the NARIC announced that it was shelving plans to construct a distributed set of rice warehouses and instead was building both warehouses and a modern rice mill in Cabanatuan. Mears infers that the program of *palay* loans and purchases required greater sophistication than was possessed by most ordinary farmers and therefore "maximum benefit at the farm level undoubtedly went to large landowners." On the other side, the financing of rice purchases abroad was adequate; the annual retail rice price cycle in Manila was held to 20 percent between highs and lows until 1941, and corruption was apparently minimal, at least before the war.[56] The fact that things ran even fairly well during this period shows Quezon's attention to close oversight. Still, it is not surprising that such controls led to black-marketeering.

One would like to know more. What were the negative, as well as the positive, outcomes of this policy? At what price beyond the later recovered value of sales were rice purchases, storage facilities, and price policing delivered? Economist Ian Coxhead comments that the NARIC's "duty-free imports, subsidized storage, and price ceilings" may actually have driven a number of private traders "out of the rice business, thus creating or exacerbating a situation in which a few large traders could actually exercise market power during supply crises" as happened later during the Marcos era. Depending on the amount of duty actually paid on rice imported by private traders, the fact that NARIC rice entered duty free would have constituted an economic if not a formal "import restriction" on ordinary traders.[57] Still, the bulk of the rice trade centered on Manila remained in private, especially Chinese hands, and the mounting problems of everyday tenant rice farmers in Central Luzon remained unsolved only to burst forth in revolutionary rebellion. The Chinese were removed from legal participation in the rice trade during the Japanese occupation, but they reestablished their predominant role as they helped to revive the trade in rice after the war. The changing social profile of those active in the Manila rice trade forms the subject of chapter 4.

4

Changing Commercial Networks
in the Rice Trade

YEARS AGO, in one of the core texts in Southeast Asian social and economic history, Edgar Wickberg proposed a thesis about commodity marketing in the Philippines, including rice marketing, that has remained the standard ever since. Simply stated, Hispanized mestizos, of mostly Filipino-Chinese extraction, dominated for a century, roughly 1750–1850, but then were increasingly pushed out of the trade by a substantial influx of immigrants and sojourners from South China whose methods of acquiring *palay* from the growers proved more competitive. "In the new [post-1850] Chinese economy," Wickberg wrote, "the most important new activity was that of commercial agent, or middleman. . . . In addition, the Chinese partly reclaimed their position as provisioners of urban areas from the mestizos and *indios*." In this transition the mestizos ended up producing export commodities through the control of land and labor. This thesis has lasted because it clearly identifies the major shift in trade and its elegant simplicity.[1]

Wickberg was primarily interested in delimiting Chinese and Chinese-mestizo roles over time rather than in the rice trade per se, so he omits details that a more extensive and quantitative historical analysis reveals. First, he tends to overlook (since it is not his focus) the critical participation of other groups in the trade, particularly Spaniards (especially Basques) in the early stages and Britons later on. Second, the record shows that Manila-based mestizos active in the rice trade tended to shift to other commercial enterprises, not just export agriculture. Third, Wickberg does not attempt to analyze a larger trend in rice marketing, namely, that it became a more professional career, increasingly concentrated in fewer hands. Finally, his subtle discussion of Chinese mestizos has led others to mistakenly reify this category and depict it as immutable.[2]

In the 1870s both the Philippines and Indonesia (Dutch East Indies) moved from being frequent exporters to regular net importers of rice. In the Philippines the opportunity for simple profit on whole shiploads of imported grain attracted the foreign export houses to the domestic rice trade as well. In the 1890s British interests pioneered the establishment of large steam-driven rice mills, physically linked to the new railroad running through Central Luzon, itself the creation of mostly British capital. By the second decade of the twentieth century, however, the same British-managed mills had largely ceased to be profitable, and by the 1920s Hokkien Chinese were operating most of the larger mills and becoming predominant in the wholesale supply of rice to the city.

Our benchmark samples of cargoes entering the Port of Manila from the outer zone provide a basis for statistical analysis. Each entry gives the name of a consignee (*consignado*), which allows us to assess the participation of individuals, families, and ethnic groups and to calculate the degree of concentration in the commercial flow. Our focus here is on the commercial dimension of provisioning the city, the structures within which networks of commercial agents operated.

Gathering Rice for Manila in the 1850s and 1860s

In the mid-nineteenth century the province of Pangasinan alone accounted for half of the entire documented flow of rice to the city from outer zone. During 1862 some 237 shipments of rice (excluding a few partial cargoes of less than 100 *cavans*) arrived in Manila from ports in Pangasinan. These arrivals were consigned to 97 individuals and firms, an average of just 2.4 shipments each. Of the consignees, some 86 operated on a small scale, handling 1, 2, or 3 cargoes each. Six received from 4 to 7 cargoes, and 5 handled more than 10. Arriving cargoes varied considerably in size. Most were less than 1,000 *cavans*, 72 were in the 1,000 to 1,499 range, and 13 fell between 1,500 and 2,400. It is tempting to surmise that small operators received small cargoes, but this was not necessarily the case. Many cargoes (33) received by single-shipment consignees equaled or exceeded 1,000 *cavans*—a full load for many coastal sail vessels. This fragmented pattern was not unique to the Pangasinan trade; for example, in the commerce in rice between Vigan and Manila in the same year, 3 consignees handled 2 cargoes each and 20 received 1. Consignment to the master of the vessel, or *arraez*, indicates that a particular cargo was not organized and financed by a Manila-based merchant but was transported to the city on speculation and sold off the deck; a number of cargoes were so listed. Clearly the overall pattern is one of remarkable commercial dispersion, with many limited operators and a few medium- to medium-large-scale merchants.

The Pangasinan-Manila rice trade was not dominated by the foreign export houses, such as the American commercial houses Russell & Sturgis and Peele

Hubbell or the British firm Smith Bell. These were among the leading entrepreneurs in the accelerating commercial development of the Philippine economy from the 1820s to the 1870s, but they lacked any major comparative
advantage in the domestic rice trade. It was not that they shunned the rice
trade, but beyond occasional speculations it was never a primary business
focus for them.[3] As Norman Owen reminds us, although there were many
commercial opportunities in the Philippines, the American firms lacked the
personnel and capital resources to tackle them all, remaining concentrated on
abaca (Manila hemp) and sugar exports.[4] Some individual members of these
firms were more active. George Peirce, an agent of Peele Hubbell based in
Legaspi during 1863–67, made ten "adventures" on his personal account, buying 1,000 *cavans* of rice at a time in Manila, arranging transport, and selling it
in various port towns in Albay. Most of his speculations sold out within four
to six months—confirmation of Bikol's growing rice deficit—and he estimated
a return of 23 percent on this activity.[5] Thus the export houses had some effect
on the supply of rice to the provinces where they were active, but they were
never central to bringing the crop to Manila.

Spanish and what would in the twentieth century be called "Filipino" businessmen and -women were the major participants in the Pangasinan-Manila
rice trade. The scale of their operations is set out in table 4.1. Five consignees
received more than 10,000 *cavans* in 1862, and together conducted 32 percent of the trade from that province to the city. Five medium-scale operators

TABLE 4.1. Major Participants in the Pangasinan-Manila Rice Trade, 1862

Merchant / consignee	Cargoes from Pangasinan	Cavans from Pangasinan	Average size of shipment	Cavans from other provinces	Total consignment cavans
Don Francisco Mortera	13	14,656	1,127	5,882 LU	20,538
Don Narciso Padilla	14	13,963	997	2,526 P	16,489
J. M. Tuason y Cia.[a]	24	13,690	570	—	13,690
Señores Aguirre y Cia.	13	12,085	930	1,400 LU	13,485
Da. Cornelia Laochanco	13	10,610	816	—	10,610

SOURCE: Calculated from the daily record of arrivals in the *Gaceta de Manila*.
NOTE: The next five ranking consignees were Don Jorge W. Petel / Señores Petel y Cia. (8,469);
Fabian Vinluan, *arraez* (5,623); Doña Tomasa Laochanco (5,186); Don Juan Reyes (5,173)[b]; and
Antonio Ferrer, *arraez* (4,230). None received rice from other provinces. Vinluan's average shipment size was 1,874 *cavans*; the others' ranged from 860 to 1,210. Names and titles are as listed in
the *Gaceta*. LU = La Union, P = Pasacao.
[a] Variously attributed to José Maria Tuason (father, d. 1856) and Severo Tuason (eldest son), a
partner in the firm.
[b] There were also shipments to a Juan de los Reyes (2/2,400).

handled a further 14 percent of the flow. Only 3 of the 10 largest-scale merchants were also active in a production region outside Pangasinan. The remaining half of the trade was dispersed among 87 others.

Who were these merchant traders? The list in table 4.1 reflects the multiethnic nature of Philippine commerce and urban society of the day. The Tuasons and Laochancos (also spelled Lauchangco) were Chinese mestizos. José Maria Tuason was lord of the huge "*mayorazgo*," an estate founded in 1794 that entailed the Santa Mesa–Diliman and Marikina haciendas on the edge of the city, and his company owned much of the Hacienda de Maysilo in Malabon and Caloocan as well.[6] The estate founder, Chinese mestizo Antonio Tuason, had made a fortune in the galleon trade, and other members of the family had been prominent traders. In the 1850s, they constituted an import-export firm dealing in imports of cloth, Spanish wines, and even Javanese rice in years of shortage. Along with Fernando Aguirre, J. M. Tuason was an investor in and comanager of the Banco Español Filipino, founded in 1851.[7]

Francisco Mortera and the Aguirres—a Basque name—were Spaniards or creoles.[8] In the mid-nineteenth century the Aguirres were engaged in transporting bulk leaf tobacco from Ilocos to Manila for the government monopoly and also in maritime commerce with places like Aparri, Taal, Negros, Misamis, and even Palau. They operated a steam-driven sugar refinery in the city, exported semirefined sugar, and imported coal. The rice trade was just one of their interests.[9] Narciso Padilla was also sometimes listed as a "Spanish merchant," but Legarda clearly treats him as a "Filipino" entrepreneur. He had been involved in the international rice trade with China in the 1830s–50s. The Padillas were substantial landowners in Lingayen and Binmaley, Pangasinan, but Narciso was based in the business district and legal community of Manila. By 1860 he was serving as a *regidor* (city councilman).[10] The Aguirres, Tuasons, and Padilla all operated moderately large scale, Manila-based, export-import businesses. They were affluent and well connected, part of a growing Philippines-based entrepreneurial class. British Consul Farren observed about this time that the enterprise of Spaniards in the Philippines was "chiefly . . . directed . . . to monopolies, government contracts, and the carrying and coasting trade."[11]

Among medium-scale rice traders in 1862 the G. van Polanen Petel company was the only non-Spanish European firm active in the Manila trade. Sometimes regarded as French, according to Otto van den Muijzenberg the Petels were Protestants of Dutch nationality, with roots in Philippine commerce since the 1840s.[12] A host of other operators, including Fabian Vinluan and Antonio Ferrer, were indigenous Filipinos or mestizos. Chinese accounted for 10 of the 97 consignees of Pangasinan rice arriving in Manila in 1862, but they handled less than 6 percent of the flow (11,500 *cavans*). None handled more than 2,000 *cavans* or operated in other provinces—understandable given the restrictions then on movement beyond their province of registration. Most of their shipments were from Dagupan, which was emerging as the major locus of Chinese commercial activity in the province. By contrast, mestizos dominated the trade from the provincial capital of Lingayen. There is little here to suggest that Chinese were about to take over the rice trade.[13]

~

An on-the-scene report allows us to see how rice was moved commercially. Export rice merchant and British vice consul Jose de Bosch left a vivid account from Pangasinan at the start of the rainy season in 1856. Wet rice was grown in substantial quantities in all jurisdictions of the province, he said. Following the harvest, some *palay* was kept aside for household consumption, "and the rest is either sold to brokers, or taken little by little to the *tiangues*, or market days, in the towns or to Dagupan and Lingayen in canoes [*bancas*], for trading purposes."[14]

At the time the principal markets in Pangasinan were still periodic rather than everyday; all were located in the town-parish centers of communities on the lowest margin of the river delta close to the gulf, the zone of highest population density. These periodic markets formed an integrated system, with synchronized schedules allowing mobile merchants to attend most of them on a regular basis (map 1.3). The Thursday market in Calasiao was said to draw the largest number. Dagupan was not initially one of the most populous nuclear centers, but as it became more important in the rice trade a second official market day was added.[15] Other periodic markets, not so well attended, served people living farther inland. Historian Rosario Mendoza Cortes emphasizes the complementarity between the coastal zone, where the most important early markets were located, and rice production locales in the interior. In exchange for rice, coastal settlements produced salt, nipa "wine" (*tuba*), and nipa roof shingles. De Bosch singles out Camiling, now included in northern Tarlac, for special mention as one of the great rice-producing municipalities of the Pangasinan river transport zone (map 1.3).[16]

At first most of these merchants were mestizos, but small numbers of Chinese began to arrive. Within two years the newcomers were ensconced in stone shop houses in Binmaley and Dagupan and on the ground floors of the more substantial homes of Lingayen. The Chinese initially encountered hostility, but by 1856 they were "joining the mestizos in all kinds of commercial businesses," said de Bosch.[17] By the later nineteenth century everyday shops and a permanent commercial district had crystallized in Dagupan, and some merchants based there were sending out strings of carts to service the interior markets.

De Bosch reports that a large part of the rice that entered commerce did so through the activities of broker-agents known as *personeros*. Usually mestizos, plus a few Spaniards, *personeros* worked on behalf of a principal wholesale rice merchant or speculator in a complex system involving cash advances. A *personero* might receive 1,000 silver dollars to advance in small amounts, usually indirectly through one or more layers of intermediary brokers operating on a smaller scale. Although brokers and merchants would gladly purchase rice directly when the opportunity was presented, this trade was then primarily organized on the general lines of the Asiatic advance system, which established patron-client relations and deflected upward some of the risk of production. In various forms, it was widely used in the smallholder export rice industries of mainland Southeast Asia, as well as in abaca (and some sugar) purchasing elsewhere in the Philippines.[18] Depending on the season, advances could also provide agricultural credit. De Bosch continues, "With the money [now] advanced to the brokers, [the *personero*] receives orders to purchase grain at a certain fixed rate—and as he bargains and collects it, it is his duty to report operations to the merchant who employs him. If the latter hears of a rise or fall in the price of the article, at Manila, or in China, he immediately writes to his brokers, giving them a new collection price. . . . As the acquisitions of rice are affected, the brokers forward it to Dagupan, or Sual, or ship it direct for Manila from the towns where the purchases are made, if accessible to coasters."[19]

A later letter allows us to see the initiation of a transaction between a Manila speculator and a provincial *personero*.

Don Juan Bta. de Arrechea   Manila, 13 March 1871
Lingayen, Pangasinan
. . . I have received your pleasing letter of the 9th brought by Elizalde, which responds to my letter of the 4th that included $4,000. As before, you will receive $3,000 more. Starting tomorrow I will [get] another $3,500, and you can set the wheels in motion, prepare your good buyers. In these circumstances, he who falls asleep misses the good business. . . . With an eye to Christ, let's get to work and see if God will help us. Your letter was very timely. . . . By the way! it really costs to send letters there. . . .

You can purchase a lot of rice by paying 2 cuartos more than the other buyers. There is no hurry to buy. But if I understand [correctly] in paying 2 cuartos more than the others, you will have plenty and within 12 days you will be pleased to have been a little more audacious than the rest, because although 2 cuartos doesn't seem like much, he who gives shows himself to be confident, saying he is generous and a friend of the poor. . . . You buy cheap rice, and we will figure out how to get it here soon enough.

Antonio de Ayala[20]

Arrechea bought rice as directed, and over the next six weeks Ayala received in Manila at least 2,000 *cavans* carried on vessels departing from Lingayen and

Dagupan. Two years later Pangasinan was connected to Manila by telegraph. In theory this might have removed some of whatever information advantage was held by larger traders and lowered the risk of smaller ones.

The principals to this particular transaction were both Spaniards of Basque background operating in a network of trust. Both had lived in the Philippines for decades. Arrechea—originally from Navarre—was engaged in the construction of sailing vessels and in the trade linking Pangasinan with Ilocos and Zambales, putting down substantial business roots in the region.[21] Antonio de Ayala (1804–76), born in the Basque province of Alava, came to the Philippines while his uncle was archbishop of Manila. In the city, he worked for Domingo Roxas, a famous creole entrepreneur, and then joined him in 1834 as a partner in forming Roxas y Cia (Casa Roxas). Subsequently, Ayala married the daughter of Roxas, Margarita Roxas de Ayala, apparently his first cousin.[22] Casa Roxas and Antonio de Ayala went on to found the famous Ayala distillery, and they owned large haciendas at Calatagan and Nasugbu in Batangas. Ayala also invested in agricultural lands near the urban area in San Pedro de Makati and in Panay and became a major stockholder and counselor of the Banco Español Filipino (1851–60, 1863–69), later known as the Bank of the Philippine Islands.[23] This Pangasinan rice venture would have been just a short-term commodity speculation for him.

According to de Bosch, in addition to a layered network of agents and brokers, the wholesale merchant also needed "some means of conveyance of his own such as lighters, large boats, carts, sheds for storing merchandise at Dagupan or Sual, and sometimes also at the growing localities where grain is purchased, as for instance, when he buys paddy, and has to get it cleaned and husked, in order to diminish the cost of transit. Husking [by pounding] a *cavan* of white rice [*bigas*] costs from 15 to 16 cents, and a *cavan* of brown rice [*pinawa*] from 12½ cents. There are pounders who readily engage to husk rice for the above prices, at all the towns of the province, and who will even go from one place to another for that purpose, if necessary."[24]

In this way rice was drawn downstream from interior farming communities to Dagupan, Lingayen, and other centers and sent on to the metropolis. Increasingly Dagupan became the way station for rice arriving from all over central Pangasinan and present-day northern Tarlac along nature's watercourses.

A different linkage effect was occasionally seen when the price offered by the agents of city merchants was so attractive that rural producers sold too much of the harvest, leaving their families vulnerable. El Niños of various strengths were recorded in 1864 and 1866, and one or both may have produced a drought, so urban rice dealers offered "very good prices" in the first half of 1866, realizing that there would be a "scarcity of production in the provinces." Attracted by the high returns, Pangasinan farmers "imprudently sold even what was necessary to their own subsistence, the result of which was a scarcity during the months of August, September and October so great that in some villages where resources had been exhausted, the inhabitants nourished themselves on herbs and roots."[25]

A Decade of Change

The growing geographical concentration in the flow of rice to the city was matched by a trend toward greater concentration among commercial agents. In 1872 the total number of consignees handling rice cargoes from Pangasinan had shrunk by one-third, with the contraction coming wholly among small-scale operators. The share of the top five had doubled to 39 percent of the entire bulk of outer zone rice arrivals in Manila (tables 4.1, 4.2), and the membership had changed completely. The next five likewise experienced a turnover of four of its members, although there was little change in the quantities handled. Fabian Vinluan, the Pangasinan native and boat captain, was the exception—the only consignee from 1862 still commercially centered in Pangasinan and one of two still active in the Manila rice trade.

❧

TABLE 4.2. Major Participants in the Pangasinan-Manila Rice Trade, 1872

Merchant / consignee	Cargoes from Pangasinan (Dagupan)	Cavans from Pangasinan	Average size of shipment	Cavans from other provinces	Total consignment cavans
Agapito Siap	74 (48)	58,950	797	Capiz	59,531
Don Isidoro Lopez Cordero	47 (42)	35,844	763	—	35,907
Tomas Puson	35 (18)	23,525	672	—	23,525
Francisco Sy-Quiaco	19 (18)	20,986	1,105	—	20,986
Doña Petronila Encarnacion and Vicente R. Sy Quia	18 (18)	15,208	845	1,833 Ilocos 575	17,616

SOURCE: Calculated from the daily record of arrivals in the *Gaceta de Manila*.

NOTE: The next five consignees were Don Alfredo Camps (8,379); Señores Inchausti / Don José Joaquin de Inchausti (6,371); Don Vicente Genato (5,515); Fabian Vinluan, *arraez* (5,110); and Juan Carballo / Caraballo (3,940). None received 1,000 *cavans* from outside Pangasinan. Cargoes containing fewer than 100 *cavans* of *arroz* are excluded except for total consignment *cavans*. "Don" or "Doña" is included when the term is used more than once in the printed record.

What became of the top rice traders of 1862? In the evolving commercial environment they evidently withdrew from the rice trade because they found other lines of investment more attractive, not because they lacked resources. Doña Cornelia Laochanco (1819–1900) was one of the top Manila consignees in 1862.[26] Born to Chinese mestizos in Binondo, Manila, in 1836 she was married to Tomas Ly-Chauco, an immigrant Hokkien from Tongan in the Xiamen (Amoy) marketing and dialect zone. Widowed in 1857, while still in her thirties—and still residing in Binondo, though now with five children—she continued in business. Ultimately she married twice more, in each case to an immigrant businessman from the Xiamen area, and outlived both husbands.[27] Cornelia is known to posterity as the progenitor of the prominent extended family known as Lichauco, whose entrepreneurship in the food supply of Manila over multiple generations is remarkable.

Remarried and evidently not short of resources, Cornelia Laochanco nevertheless received only one cargo of rice from the outer zone in 1872, having reoriented her interests. She invested in urban properties developed as *accesoria* (simple apartment building) rentals, especially in Binondo. In the 1870s and 1880s, she operated a *fardería* and warehouse complex in the Tanduay section of Quiapo, Manila, where partially refined brown (*pilón*) sugar was spread out on mats to dry in the sun and then sacked by degree of purity (color) for sale and export, resulting in a product that was more valuable per pound but still not fully refined and thus dutiable abroad at a lower rate. She sold this sugar to local British merchant houses rather than trying to export it directly. Such *farderías* typically employed scores of laborers, often Chinese; her family records that "at one time she employed 500 persons."[28] During the cholera epidemic of 1882, her warehouses (*camarines*) were temporarily converted into a 150-bed emergency hospital. Doña Cornelia also bought and sold gold from Paracale in Camarines Norte, loaned money on agricultural lands, and acquired rice and sugar land in Arayat, Pampanga. Later in life, she maintained a pleasant seasonal home along the Pasig River in Sta. Ana, on Manila's outer fringe. She is remembered in the family as a formidable businessperson and matriarch—"Lola Grande"—but she did not reenter the rice trade (figures 4.1, 4.2).[29]

In 1862 Francisco Mortera (d. 1884) had topped the list of urban consignees of Pangasinan rice and was also carrying on a considerable trade with neighboring La Union. Ten years later he had been joined by a relative, Ramon Mortera. Each handled only one rice cargo from Dagupan; both were now primarily buying in La Union (2,200 and 7,700 *cavans*, respectively), receiving shipments from seven different La Union ports between them—still important players, but no longer in Pangasinan. After the government tobacco monopoly was ended, Ramon dealt in bulk leaf. In the 1890s, he owned six substantial buildings in Binondo and Tondo, as well as two *accesorias* in Sta. Cruz

FIGURE 4.1. Cornelia Laochanco (1819–1900; a.k.a. Cornelia Lao Chang Co Lichauco), a leading Manila rice dealer in the 1860s. The portrait is a copy by Fernando Amorsolo of a damaged nineteenth-century original. (Reprinted by permission of the family of the late Ambassador Marcial P. Lichauco and his wife Jessie Coe Lichauco and daughter Cornelia Lichauco Fung)

FIGURE 4.2. A *fardería* in San Miguel–Quiapo in the 1890s.

district, and was doing well enough to serve as a counselor of the Banco Español Filipino.[30]

By 1872 the Tuasons were concentrating on banking and their extensive
urban properties and suburban estates. They were the biggest creditors of the
commodity firm Russell & Sturgis when it failed in 1875. The Aguirres continued to import coal and general merchandise, but their company was in liquidation in 1875. Possibly business reverses had something to do with Aguirre's
withdrawal from the commerce in rice, but the move fits a more general pattern of commercial reorientation. Padilla's active business career ended in 1865.
In local lore his daughter Barbara is said to have made a fortune transporting
rice, but her name does not appear in our sample years.[31] Jorge Van Polanen
Petel was managing a tea estate in West Java by 1869. The firm bearing his
name was still commercially active in the city in 1872 but not in the rice trade;
it was now operated by Jean Philippe Hens, a Belgian, and in the 1880s was
importing salt as well as manufacturing and exporting tobacco products.[32]

∽

Who were the new rice traders? By 1872 Agapito Siap (or Sy-Siap) had managed to establish a level of commercial dominance undreamed of a decade
earlier, handling alone almost 15 percent of the total outer zone flow of rice
to Manila. Although there were still scores of persons acting as *consignados,*
including persons of diverse backgrounds, it was lost on no one that Chinese
now handled 27 percent of the entire outer zone rice trade with the city. Less
noticed was the fact that Agapito Sy-Siap and Francisco Sy-Quiaco, together
with six or seven others also surnamed Sy, accounted for almost 23 percent of
the outer zone trade with the capital on the 1872 list. The commercial cargoes
handled by the Sys came primarily from Dagupan, the most rapidly expanding shipping point in Manila's domestic rice supply system.

Clearly Hokkien Chinese in general and the Sys in particular were doing
something effective. Chinese merchants were culturally adept at establishing
and maintaining commercial and credit trust among themselves, even over substantial distances. In the Philippines they established simple town and village
stores, often linked, and regularly attended the increasingly dense schedule of
periodic provincial markets. Through these venues they offered less perishable
food supplies (dried fish, noodles, onions, garlic, and rice), imported factory-
made textiles, and hardware items as material advances against a later exchange
for newly harvested rice or other cash crops.[33]

These exchanges were mostly not true barter but rather what Willem Wolters calls "cashless transactions" on which accounts were kept. W. G. Huff calls
them "bookkeeping barter."[34] In essence the merchant advanced goods and
kept a record of the value owed. Later the producer presented *cavans* of rice
(or other specified commodities) and was credited with their value, thereby

offsetting some or all of the value advanced earlier. On both sides of the ex-change, the merchant held the edge in assigning a price. The system was valu-able to the producers in giving them a convenient way to obtain valued basic goods for their efforts. It also worked well in a provincial economy that was chronically short of coins. By contrast, Spanish merchants in Pangasinan largely eschewed the import and retail aspects of commerce. Mestizos were active at the periodic markets, selling native textiles and imported manufactures, but it was becoming more challenging for a mestizo or Spanish *personero*, liv-ing on a materially better scale, to assemble competitively priced cargoes of rice for the city. In 1872 a major change in commercial networks was in view, less than 20 years after the initial arrival of Chinese—"mere agents, depending on other [Chinese] of greater consideration established at Manila," according to de Bosch—in the province.[35]

Despite growing competition from the Chinese, a new group of Spaniards and Filipinos had also acquired major positions in the Pangasinan-Manila rice trade by 1872 (table 4.2). The Spanish and creole traders included Isidoro Lopez Cordero, the Señores Inchausti (Ynchausti), Vicente Genato, and Alfredo Camps.[36] The Inchaustis were best known for their role in commodity export commerce. Spanish Basque in background—originally from Guipuzcoa Province—José Joaquin de Inchausti (1815–89) arrived in Manila in the second quarter of the nineteenth century and founded the Inchausti Company in 1854.[37] By the standards of the day this company became a big business, espe-cially in trade, shipping, and alcohol production. Inchausti interests eventu-ally included abaca, sugar estates, and mills in the provinces and in Manila the important Tanduay distillery, rope manufacturing, and substantial urban prop-erty. The Pangasinan-Manila rice trade was never a central preoccupation.[38]

Vicente Genato was the son of Manuel A. Genato, a Spanish creole and active importer and commodity trader, who had taken over a prominent pro-visioning firm begun in the 1840s. It was still in business—and still a Spanish company—a century later. By the 1890s, when Vicente took over its manage-ment, Genato interests in the commodities trade centered on coffee planting and export (just as blight devastated this crop), as well as sugar and hemp. The subsequent Genato Commercial Corporation maintained a long-term focus on provisioning by importing food products.[39] Doña Petronila Encarnacion, as the *Gaceta* listed her, was a Filipina mestiza from Vigan, unusual among those active in the Pangasinan rice trade in her ongoing connections in Ilocos Sur. Together with her immigrant Chinese husband, she would remain active in the rice trade through the 1890s. Tomas Puson may also have been a mestizo.

In 1872 the northern European and American commercial houses were still minor players in provisioning Manila from domestic sources, but they were prominent in the growing rice import trade, as regular arrivals of whole cargoes of foreign rice fit the way they liked to operate. They could purchase bulk rice in Saigon without the need to maintain a network of *personeros* or make advances in coin or material goods and could obtain advantageous access to information on supply conditions and price.

At the end of the third quarter of the nineteenth century, the Manila rice supply was in an epochal transition: almost simultaneously Hokkien participation in domestic commerce rapidly expanded and European and American commercial houses built up the import trade. Meanwhile, Cornelia Laochanco had moved into urban sugar processing, while other former Manila rice traders similarly moved into banking, cigar manufacture (after the end of the monopoly in 1880), property development, and foreign trading. Some developed rural production estates, but once established in the city few larger families left, although a branch might become established in the province to manage rural holdings. In the Manila rice trade mestizo buyers were increasingly cut out—in Iloilo and Cebu as in Manila—although a few prominent mestizo traders, such as the Nable Joses and Petronila Encarnacion, continued to flourish for years.[40]

The duality of urban business and rural commodity growing is seen through the next two generations of Lichaucos. Following the premature death in 1889 of Cornelia's son Macario, two of his sons, Crisanto and Faustino, took turns managing the family rice lands in Arayat. Both were involved in shipping in the 1890s and in managing their mother's fleet of *casco* river vessels. Fleeing the Spanish regime in 1896, the brothers relocated to Hong Kong and were active in the foreign affairs of the nascent Philippine government during the Revolution. Returning to Manila, Crisanto eventually took up management of his late father's estate, the Hacienda "El Porvenir" in Tayug, eastern Pangasinan.[41] At the same time, his brother Faustino made a fortune supplying Manila with beef imported from Cambodia and Australia and later invested in fishing trawlers. The larger family genealogy is replete with such pairings; for every hacendero, one finds another relative with urban professional and business interests. Following Cornelia, several family members inherited or initiated investments in urban properties, including rental *accesorias*; others were publishers or businesspeople. The commercial competition from Chinese immigrants was real, and eventually there was some movement of mestizos away from commercial pursuits in the provisioning system, but the transition was far from simple or complete, although from a provincial perspective it might look more one-dimensional.[42]

IMPORTS AND INNOVATIONS IN THE AGE OF STEAM

Concentration in the commerce in rice advanced rapidly.[43] One major change in the domestic rice trade after the 1860s was the rising scale of operations by the major wholesale dealers. Another was the rising consistency of their dedi-cation to the rice trade as a career enterprise. Where nine of the top ten dealers of 1872 were apparently new to the business, this pattern of opportunistic involvement and sudden turnover diminished thereafter, and substantial carry-over became the rule. In all, six of the top eleven *consignados* in the Pangasinan to Manila rice trade in 1881, or their business successors, were on the same list in 1891. Despite impressive continuity, this was not an ossified oligarchy of trade. Federico Cosequin disappeared from his commanding position follow-ing 1881. The American firm Peele Hubbell was forced into liquidation. The wealthy creole Roxases appear for the first time on the lists of large-scale par-ticipants in the rice business, but Pedro P. Roxas was soon to become absorbed in developing the new San Miguel Brewery.[44] Just as some small ports and microhinterlands fell out of the system of metropolitan rice supply, so, too, did many small-volume consignees vanish from the business (table 4.3). The volume handled by the three largest consignees in 1881 was 256,000 *cavans*, up from only 42,000 in 1862 and now accounting for 41 percent of total arrivals from the outer zone.

Networks of Hokkien businessmen continued to expand their position; by 1881 Chinese consignees were handling 52 percent of the volume of rice reach-ing Manila from the outer zone. Seven of the top ten *consignados* were Chi-nese, and four were named Sy (table 4.4).[45] In a little more than 25 years, Hokkien rice merchants had achieved a position of considerable commercial prominence based on domestic purchasing operations through networks of provincial stores using a system of carefully recorded cashless transactions, as well as relations with strings of urban and provincial shops whose operations

TABLE 4.3. Increasing Commercial Concentration in the Flow of Pangasinan Rice to Manila, 1862–1891

Rice cargoes per consignee	1862	1872	1881	1891
1–3	86	50	26	16
4–9	6	8	6	7
10–19	4	2	8	7
20–49	1	2	3	3
50–74	—	1	2	—
Total Consignees	97	63	45	33

SOURCE: Calculated from the daily record of maritime arrivals in the *Gaceta de Manila* (1862–81) and *El Comercio* (1891).

NOTE: Totals are not adjusted for increasing size of cargo. The record for 1891 omits rice carried as secondary cargo.

included retailing milled rice. As Isabelo de los Reyes tells us of Malabon in the late 1880s, "The Chinese bring in rice from Saigon which they give, on credit, to the tiendas of the mestizas."[46] The Hokkien would consolidate their position in the domestic rice supply system in the twentieth century, but in the 1880s and 1890s ongoing Spanish and mestizo competition and the deployment of emerging steam technology by British firms still delayed this development.

Among the largest rice merchants in the 1880s and 1890s were Alejandro Nable Jose and Smith Bell and Co., the British export and import firm. Nable Jose, from a well-known Dagupan-Manila shipping and rice-trading family, was based from the mid-1870s through the 1880s in Binondo in the Manila central business district, where he held various official positions, including *gobernador-cillo* (mayor) of the Gremio de Mestizos (the official organization of mestizo residents). At the same time in Dagupan and Calasiao his brother Mariano Nable Jose was licensed to trade in "products of the country," including rice.

TABLE 4.4. Major Participants in the Pangasinan-Manila Rice Trade, 1881

Merchant / consignee	Cargoes from Pangasinan (Dagupan) [by steam]	Cavans from Pangasinan	Average size of shipment	Cavans from other provinces	Total consignment cavans
Federico Cosequin	59 (59) [9]	93,759[a]	1,600	8,210 LU	101,969[a]
Alejandro Nable Jose	65 (61)	85,117[a]	1,310	5,450 LU, Z	90,567[a]
Smith Bell & Co.	29 (29) [27]	63,450[a]	2,187	—	63,450[a]
Encarnacion–Sy Quia:	29 (28)	33,300	1,148	—	33,300
Vicente R. Sy Quia	16 (15)	18,130	—	—	—
Petronila Encarnacion	13 (13)	15,169	—	—	—
Sy De	23 (22)	29,764[a]	1,285	—	29,764[a]
Domingo Uy Quince	20 (11)	25,010	1,251	6,135 LU, Z	31,145
Peele Hubbell & Co.	19 (19)	24,009[a]	1,273	1,300 I	25,309[a]
Joaquin Sy Tay	17 (17)	19,659	1,156	5,000 Z	24,659
Agapito Siap	18 (18)	16,575	921	—	16,575
Vicente Yu Chauqui	14 (14)	14,980	1,070	—	14,980

SOURCE: Daily arrival listings in the *Gaceta de Manila* supplemented with information from *El Comercio*.

NOTE: Frank (Francisco) Heald was the eleventh-ranking *consignado* (13,922). I = Ilocos, LU = La Union, Z = Zambales.

[a] One or more shipments was recorded simply as *con arroz*. In each case, an estimate has been made in reference to the record of shipments for that vessel or vessel type consigned to that merchant.

Fuller records would likely show that Alejandro went to Manila while his father Donato and brother Mariano handled the rice-buying and shipping activities based in Dagupan.[47] When Donato died in 1885, Mariano inherited the family business venue and pier (*pantalan*) on the Dagupan riverfront and became the leading Pangasinan owner and builder of ships used in the coastal trade. In 1890 he registered seven sail vessels; in 1896 his vessels included three highly maneuverable *pailebots* (cutters), two workhorse *pontines* (yawls)—all in the 50- to 60-ton class—and a single, much larger *bergantin-goleta* (brig-schooner) of 214 tons. Mariano also operated a small steamer in various years, as only a shallow-draft boat could pass over the sandbars and enter the river port of Dagupan. Throughout the 1880s, Alejandro continued to function as a preeminent *consignado* of Pangasinan rice cargoes arriving in Manila.[48]

For Smith Bell, on the other hand, large-scale involvement in the domestic rice supply of Manila was a novelty. Along with its fading American competitor Peele Hubbell, Smith Bell had been centrally involved in exporting abaca and sugar for decades.[49] No doubt they were drawn into the rice trade by opportunities for simple speculative profit on the importation of whole cargoes of Mekong Delta rice. For some time, they had been supplying rice purchased in Canton and Hong Kong to deficit areas in the Visayas. In years of bumper domestic harvests, Smith Bell also looked to Dagupan for supplies; in 1881 Smith Bell and Peele Hubbell were the leading advertisers of Pangasinan rice for sale in the city. Another ranking consignee in 1881 was Frank Heald, a British citizen and founding member of the Manila Jockey Club. He began as a clerk with Peele Hubbell in the late 1860s and by 1874 was buying rice in Pangasinan for the company and on his own account. He remained active in the Dagupan rice trade, by 1886 was said to be doing well, and would stay on after Peele Hubbell folded, trading in Pangasinan and alternating with his brother Ernest as the British vice consul for Sual—suggesting that one brother always remained in Pangasinan while the other handled arriving cargoes in Manila.[50]

By 1891, after several decades of rapid turnover, the participation of major *consignados* in the Manila rice trade had settled down, with considerable carryover from a decade earlier (table 4.5). Smith Bell and the Nable Jose family were now by quite a margin the leading participants in the Pangasinan-Manila trade. Peele Hubbell had been forced to liquidate but was succeeded in some of its assets by the new British company known as Warner Blodgett—owners of a new steam-powered rice mill at Calumpit.[51] Alejandro Nable Jose continued to be extremely active in this trade until his death in Binondo in 1890.[52] His eminent commercial position was immediately assumed by his widow, Luisa Lichauco (d. 1909), (at least in name) a daughter of Cornelia Laochanco and sister of Macario Lichauco (d. 1889), a business colleague of Alejandro's father, Donato (d. 1885).[53] A year later Mariano Nable Jose brought the steamboat

Dagupan from Hong Kong, and Luisa handled its cargo and passenger arrangements in Manila; the extended family remained a substantial business team.[54] Mariano also acted as a transport contractor outside his own family business, with his vessels used on occasion for the wholesale rice dealers Julian Siap and Sy Tay.

Members of the Roxas family, famous creole entrepreneurs and landowners, were also among the leading rice merchants of 1891, as was Luis R. Yangco (1841–1907), a major mestizo entrepreneur in coastal shipping.[55] All these traders and transporters would subsequently be thought of as Filipinos, as would the mestiza Petronila Encarnacion, an important supplier of rice and other commodities to the city for at least 20 years. She began her commercial career in Vigan, married there in 1853 (she later claimed to have brought 5,000 pesos to the marriage), and baptized her children there through at least 1860. Her husband was an immigrant Chinese merchant named Vicente Romero Sy Quia, who was born near Xiamen, in coastal Fujian, in 1822 and arrived in Manila in his teens. By 1848 he was working in Vigan, where he was employed in a preexisting Chinese business. His Philippine marriage was unusual in that

TABLE 4.5. Major Participants in the Pangasinan-Manila Rice Trade, 1891

Merchant / consignee	Cargoes from Pangasinan [by steam]	Estimated cavans from Pangasinan	Cargoes from other provinces [by steam]	Total shipments	Total estimated cavans
Smith Bell & Co.	38 [24]	62,000	1 ST	39	63,000
Luisa Lichauco	45 [12]	57,000	3 Z, ST	48	60,000
Joaquin M. Sy Tay	23 [1]	24,000	7 [2] I, Z	30	33,000
Sy De	18	18,000	4 Z, ST	22	22,000
Pedro P. Roxas	13 [4]	17,000	1 N	14	18,000
Julian Sy Yap (Siap)	16	16,000	1 ST	17	17,000
Sy Socsoy	16	16,000	1 Z	17	17,000
Joaq'n Uy Diongco	15	15,000	1 Z	16	16,000
Francisco L. Roxas	7 [7]	14,000	—	7	14,000
Warner Blodgett	11	11,000	3 [1] B, IN	14	15,000

SOURCE: Daily arrival listings in *El Comercio*.
NOTE: Luis R. Yangco was the eleventh-ranking consignee with only 6 cargoes from Pangasinan but 13 from Subic and Mindoro for an estimated total of 29,000 *cavans*. Estimates are based on the assumption that rice cargoes in sail craft averaged 1,000 *cavans* each, while those in steam vessels averaged 2,000. The results are only suggestive. B = Bolinao, I = Ilocos, IN = Ilocos Norte, LU = La Union, N = Nasugbu, ST = Santo Tomas, Z = Zambales.

only about 3 percent of Chinese registered in the 1890s had formally converted to Catholicism and legally married in the country.

Sy Quia and Encarnacion extended their business operations to Manila. Both were *consignados* for rice cargoes there in 1872 and were extremely active in the 1880s, and Petronila is listed in the surviving registers of the Binondo Gremio de Mestizos from 1873 through at least 1884. Although the Ilocos region declined as a source of rice, Encarnacion remained involved there with other commodities—and probably as a supplier of rice from Dagupan—into the 1890s. At the time of Sy Quia's death in 1894, he was listed in the highest tax category for resident Chinese, 1 of only 34, out of more than 20,000 Chinese registered in the city. The couple became considerable Manila property owners and their estate was said to have been worth a million pesos.[56]

In Ilocos Sur, Petronila Encarnacion's registered commission agents during the 1890s purchased "the fruits and products of the country" on her behalf in Tagudin and Sta. Cruz. In Vigan, two women named Benita and Juliana Encarnacion had adjoining homes and were licensed to buy and sell local commodities.[57] It is quite likely that these women were of the same powerful political clan, siblings or close cousins and business associates of Petronila. Benita Encarnacion was the mother of the future elite politician Vicente Singson Encarnacion, born in Vigan in 1875. Thus the Sy Quia–Encarnacion couple nicely bridged cultures; through Petronila they were deeply rooted in the business landscape of the region, and through Sy Quia they linked to the rice-trading Hokkien Sys.

On the Chinese side, five merchants—four of them named Sy—continued to hold important positions in the rice trade of 1891.[58] Their work and commercial capital were critically important in feeding the city, but they could not be described as predominant. None appears to have been operating on the scale formerly achieved by Agapito Siap at his peak in 1872 or by Federico Cosequin in 1881, although Joaquin Martinez Sy Tay, a brother of V. R. Sy Quia, had increased his business by at least 50 percent. In none of the tax records for his several properties, however, is he listed as Chinese, which, along with the acculturated form of his name, suggests that he was moving into a more "mestizo" identity, casting his lot with the Philippines.[59]

Besides ongoing Spanish and mestizo competition, Chinese ascendancy in the rice trade was also delayed by the capital and organizational requirements of technology. In 1881 all but 2 of Smith Bell's 29 domestic rice cargoes landed at Manila were transported there by steamer, principally the vessel *Camiguin*. In this bumper year, the company also carried unstated cargoes—probably rice—from Dagupan directly to the Visayas by the same mode. One of Smith Bell's advantages was that the range of products and geographically widespread operations of the company allowed the vessel to be profitably redeployed on other routes when there was little need for rice transport from Dagupan to Manila. Larger merchant firms, presumably with good access to capital markets (possibly through one of the British overseas banks), could better afford to operate more expensive steam-powered vessels than their smaller rivals could, and they accrued the advantages of faster delivery, some economies of scale, and possibly enhanced market information.[60]

In 1891 Smith Bell, using the now venerable *Camiguin*, maintained its advanced position, but in August Mariano Nable Jose brought a steamer from Hong Kong. Named the *Dagupan*, it was swiftly employed by Nable Jose and Luisa Lichauco on the Manila-Dagupan run, matching the *Camiguin* trip for trip.[61] The Roxas interests also owned a number of steamers that became part of the nucleus of the new Compañia Maritima, formed in 1889 with Pedro Pablo Roxas as manager, and they made frequent use of these vessels in the domestic rice trade. Luis R. Yangco, with scores of *cascos* and 28 small steamers, by 1907 had become wealthy servicing the routes linking Manila to the ports of Laguna de Bay, Manila Bay, and the Zambales coast. Starting in late 1891, he had begun using the steamer *Vigia* on regular runs to the new port at Subic, as well as to San Antonio in Zambales.[62]

During the 1890s Spanish protectionists attempted to head off the British and others by banning foreign-owned (non-Spanish) shipping in Philippine domestic commerce, and Spanish firms did emerge as important operators of steamships in Philippine waters. Such an official climate would have discouraged attempts by Chinese to acquire steam vessels for the rice trade, a situation that changed little during the early years of the American administration. Steam technology in transport was increasingly important in the domestic rice supply of Manila, but Philippine Chinese rice merchants were followers rather than leaders in its employment.

◇

Meanwhile, with the opening of the Manila-Dagupan Railroad and the big steam hulling and polishing mills along its right of way, a great deal of rice was diverted from the downriver ports and away from coastal vessels. It is probably no accident that the Chinese community in Dagupan experienced the least demographic stability of any significant Philippine provincial city during the 1890s.[63] Mariano Nable Jose continued in the coastal shipping business for a time, registering nine sailing vessels and a steam launch with the new authorities in 1904–5. Two of his vessels were now based at San Fernando in nearby La Union, which was not reached by the railway until the age of highways and trucks. None of these vessels had been in his fleet in 1898, and each was half the tonnage of his earlier vessels. As the railroad took over much of the shipping of Pangasinan rice to Manila, Nable Jose switched to carrying rice to Ilocos, as well as passengers and cargo between San Fernando and Dagupan.[64]

Our sample year 1891 falls toward the end of the first long period of heavy rice imports. There were some 54 arrivals from Saigon that year. Spanish ship-owner Francisco Reyes received 14 cargoes, all on a Spanish-flag steamship. Yap Tico, a successful Hokkien importer-exporter, brought 12 cargoes on German steamers. Each carried an estimated 214,000 *picos* (approximately 13 million kilos). The combination of Augusto Saavedra and the Mensagerias Maritimas accounted for another 12 cargoes on the French ship *Volga*. These were followed by Smith Bell (4) and 3 further Chinese, as well as several single cargoes. In total Chinese organized 20 of the 54 known arrivals from Saigon.[65]

The Rice Supply in the Twentieth Century

The catastrophic deaths of the nation's plow animals, the El Niño–related drought cycles, and the disruption and damage of imperial conquest brought about the collapse of domestic rice production in the first years of the new century, creating an inflationary situation in which the Chinese rice dealers realized their greatest business success. With the profits of this era, Wickberg believes they "probably emerged more solidly placed in the rice industry than before."[66] But British business did not suddenly disappear. During the previous decades, Smith Bell had played a supporting role in importing Saigon rice, but it became much more aggressive in the surge at the start of the new century, continuing a long association with Denis Frères, an export firm in Saigon. During the short period for which we have shipping data (November 1901 to early March 1902) 25 small steamships and tramp steamers arrived from Saigon-Cholon with rice—2 a week, every week. Half were organized by Smith Bell (8) and Warner Barnes (4), British owners or managers of large domestic rice mills. A further 8 cargoes of Mekong Delta rice were brought by Spanish firms, especially Tabacalera (6), which had not been active in this commerce a decade earlier.[67] Spain had lost control of the country politically, but a few Spanish businesses were nevertheless doing well, even taking on new roles.

Some Philippine Chinese were also making fortunes in rice wholesaling and distribution in this inflationary environment, but in 1901–2 most of them were importing through British or Spanish commercial houses. Hokkien merchant Yap Tico, who in 1891 had imported twelve cargoes of Saigon rice (estimated at 214,000 *picos*), handled just two cargos in this period of even greater demand. Did he cut back because of the competition of the European trading houses? In Cholon it is probable that Yap dealt with one of the Chinese milling firms. We see elsewhere how an international trade in rice, initiated from abroad, might develop directly through ethnic business contacts without European or American trading companies as intermediaries. Go Bon Tiao, later known by his baptismal name, Pedro Singson Gotiaoco, was a late-nineteenth-century Hokkien businessman in Cebu who began by "peddling oil," presumably kerosene.

He moved on to "selling rice consigned to him by a Vietnamese merchant"—
surely a fellow Hokkien based in Cholon—became "a trusted vendor," and
"soon, with enough capital, began selling his own rice." In the 1930s Chinese
rice dealers from Saigon continued to come directly to Cebu City—and no
doubt Manila—bringing thousands of sacks of rice that were unloaded and
taken to the bodegas of the consignees, one or another of the local Chinese
wholesale rice dealers.[68]

Advertising notices placed in *El Comercio* by Manila wholesalers during
1901–2 add nuance to this picture. Smith Bell ran a standard message featuring
both streams of its supply: *segunda blanco de Saigon* and first- and second-class
white and *corriente* from its mill at Bautista-Bayambang. Smith Bell's main rival
in advertising frequency was M. Yap Siocco, who was offering Saigon rice at his
store on Rosario Street in the central business district, just as he had been
doing since the 1880s.[69] F. M. Yap Tico, an occasional advertiser, was from
Xiamen rather than the more common origin locale of Jinjiang and had begun
in Iloilo with a general merchandise importing and exporting business. Alfred
W. McCoy describes him as the only Chinese in Iloilo in the decade before the
Revolution who was "equal in stature to the major European or mestizo merchants." Expanding his operations to Manila and Cebu, he imported rice
directly from Saigon (documented in the records of 1891 and 1901–2), as well
as textiles, and exported Philippine commodities to China, Hong Kong, and
Japan; he operated a small steam vessel as well. A son, Yap Seng, became the
manager of the firm in the early twentieth century; by the 1930s it was generally
acknowledged to be the largest rice dealer in the Visayas, an effective business
organization that expanded both laterally and up the hierarchy of commercial
centers rather than down from Manila.[70] Other dealers, such as Yu Biao Sontua, and Lucio Lim Pangco, also appear occasionally in Manila advertisements.

In the change from pounding by hand to power milling, the trade in rice
became more capital intensive. By the 1920s the real economic power in the
domestic rice supply system lay with the owners of the bigger mills and their
attendant storage facilities, now predominantly Philippine Chinese, many of
whom were also involved in the rice or broader commodities trade in Manila.
They largely superseded the British firms, which had initiated the construction
of big domestic mills 25 years earlier. A clue to this transition was provided when
the Dutch consul reported as early as 1895 that the new European-managed
mills could not work continuously "due to a lack of sufficient grain." Luzon
Rice Mills, which built a number of the early mills, paid a dividend in 1907
but none during the nine years that followed; its general manager, Smith Bell,
explained this poor performance as resulting from "the difficulty of securing enough paddy [*palay*] to keep the mills constantly working." By 1909 and

probably earlier, the company had mortgaged its facilities at Bautista, Calumpit, and Dagupan to the Manila Branch of the HSBC.[71] The Dagupan mill had ceased operations altogether by 1916, although Luzon Rice Mills remained a major player in the Manila rice trade in 1919.[72]

A similar transition took place in Cabanatuan, Nueva Ecija. There were at least two sizable rice mills in that town by 1907, one owned by W. W. Weaver, an American. Weaver's mode of operation was "to buy *palay* for cash, mill it and ship it to Manila and have a straight miller's profit." But he could not generate a sufficient flow of *palay* to make the mill a success. Neither could Lizarraga Hermanos, the Spanish company that owned the second mill. Both were out of the business in a few years, replaced by Chinese entrepreneurs who were better connected to networks of local buyers and to the growers themselves.[73] In Dagupan another non-Chinese mill had failed even earlier. Faustino Lichauco organized a partnership in 1901 to carry on a "rice-cleaning business at Dagupan, and for the purchase and sale of *palay* and rice." Unfortunately, this was the period when many rice paddies were out of production due to the second rinderpest epizootic; the business proved unprofitable and was closed in May 1904.[74]

By 1935 there were 42 power rice mills in Nueva Ecija alone. Where such mills were established, they became the leading purchasers of locally grown *palay*. Within Nueva Ecija, the largest commercial flow of *palay*—some 43 percent of the total—was to the mills situated in Cabanatuan. Once milled, 89 percent of the rice found its way to wholesalers in Manila. Indeed, many mills were owned by the same Manila rice traders or large-scale general commodity traders who had provided most of the finance within the system. In 1930, 7 of the 8 largest mills in Cabanatuan were owned by Chinese.

Many tenant rice farmers found it necessary to obligate some portion of their share of the eventual harvest in return for loans from the estate owner. The loans were crucial to family budgets in a one-crop rice system because there was so little remunerative work during the long dry season. By contrast, most of the estate owners sold *palay* to the millers or their agents for cash, though some was sold on preharvest contracts. A sample survey in 1930 concluded that the Nueva Ecija *palay* sellers received an average price equal to 87 percent of the wholesale price in Manila that year.[75] A later study demonstrated that nominal monthly wholesale *palay* prices in that province closely paralleled the swings in wholesale polished rice prices in Manila. But in fact most of the *palay* passed into the ownership of the millers early in the calendar year, when prices were lowest following the harvest.

Following passage of a 1931 law, most of the significant provincial mills were "operated in conjunction with . . . bonded warehouses." These facilities provided the operator with a "steady and sufficient volume of *palay* to keep his

mill in operation." Most growers lacked such warehouses, and if they did not
sell the harvested and threshed crop outright early in the calendar year they
tended to place it in the bonded warehouse of the miller with whom they had
an ongoing relationship. To promote this relationship, the miller loaned empty
sacks and often helped finance farm operations. There were also storage charges,
which could be waived if the grower eventually sold to the miller operating the
bodega. Further, the miller often made credit or cash advances on the *palay*
thus stored under bond. It was usually in the growers' interest to delay the sale
of at least a portion of the crop until prices had commenced their usual advance in the late rainy season months before the new harvest, but at this point
the grower had little bargaining leverage, as with such arrangements there was
only a modicum of competition among the mills in a given locality.[76]

Some Filipinos still owned and operated small *kiskisan* mills serving their
own estates and local communities, just as they had owned small steam- and
water-powered mills in the 1890s. By the 1930s they also owned a number of
small- to medium-capacity bonded rice warehouses. There were a few other
exceptions to Hokkien dominance in the interprovincial rice trade and large-
scale milling. One was the sizable Cabanatuan and Manila milling and rice
business of Cantonese businessman Kwong Ah Phoy in the 1920s and 1930s.
Among Filipinos, Basilia Huerta viuda de Tinio invested in a substantial rice
mill in Cabanatuan and fitted it out with British steam-powered machinery,
although rather than trying to manage this mill, she chose to lease it to a Chinese businessman.[77] Other Filipinos entered the business more directly but for
the most part at a below-average scale of operation.[78]

~

By patient building over 60 or 70 years, by staying in the business in all seasons in good years and bad, by operating networks of local businesses employing carefully recorded but often cashless transactions, and by offering a limited
range of retail consumables and other goods, the Hokkien Chinese came to
dominate large-scale commercial rice milling and conduct the greatest share of
the trade, especially the interprovincial rice trade centered on Manila. Kwok-
Chu Wong emphasizes yet another aspect of their dominance: the marketing
power of the Chinese rice dealers in the city itself. The myriad small Chinese
stores acted as a rice distribution network and made it increasingly difficult for
competitors not attached to such networks to reach the main body of urban
consumers (figure 4.3). Despite the disruption of revolution, war, and blockade, several new power mills were built in Manila early in the twentieth century. By 1904 three large mills operating there were owned by Chinese. All
three were located near the Tutuban railway station in Tondo—the point of
mass entry of domestic rice arriving by rail—with its adjacent mills and storage bodegas extending onto Dagupan Street. The Tutuban Rice Exchange was

FIGURE 4.3. A basic element in retail provisioning: a Chinese *tienda* offers polished rice (by grade in pyramidal mounds), as well as garlic, onions, canned goods, and other items, 1920s. (H. V. Rohrer, U.S. Bureau of Foreign and Domestic Commerce, USNA II, RG151-FC-85B-21)

founded on this street in 1922 on the premises of the Philippine-Chinese Rice Merchants' Association. Three years later an estimated 80 percent of the rice arriving in the city by rail was traded there.[79]

Many sources on the Tutuban Rice Exchange emphasize its anticompetitive practices, which included employing the Hokkien language and honoring secret bids scribbled in such a way that open auctions failed to develop. No sales prices, volumes, or parties to the transactions were posted; only with difficulty could prospective buyers find out the bids of others or the price reached on the previous sale. This may well have disadvantaged Filipino or even Cantonese buyers, but it was not a system with which they were culturally unfamiliar. The *bulong* (whisper) system of bidding was widely used in Manila by Filipino wholesalers of both fish and hogs, including when they were selling to Chinese buyers. The Tutuban exchange was at least a substantial success in bringing buyers and sellers together on a regular basis in one place. Its emphasis was on trading rice just then arriving in boxcars, which allowed independent millers to avoid the costs of unloading and warehousing prior to sale in the city.[80]

In Wong's analysis, it made more sense for specialized Chinese rice traders to expand upstream into milling than it did for affluent Filipino rice land owners, lacking wholesale and retail networks, to move downstream into large-scale milling.[81] For example, Yu Biao Sontua, whose career began in Leyte, did very well trading in rice and other commodities at the end of the Spanish period, and during the disastrous times that followed he made a major killing on rice imports. Based in Manila by 1901, Yu built a business network with interests in abaca, stevedoring, and interisland shipping, with branches and agencies in Leyte, Samar, Masbate, Cebu, and Surigao. In Yu and in Yap Tico, Wong sees prime examples of companies "that had their origin as commissioned merchants of western business houses, or *cabecillas*, [and] gradually moved . . . into rice imports, abaca exports, sugar trading, and general merchandise trading on their own accounts, . . . running some interisland shipping services and remittance businesses at the same time."[82] One part of Yu's interests by 1915 was a rice mill at Bautista, Pangasinan. He was still a major player in the Manila rice trade in 1919, but the great wave of deflation that followed World War I caught him overextended.[83] He eventually became insolvent and was forced to sell the Bautista mill.

~

Importing rice in the 1920s was quieter than during the first dozen years of the century, but there were some changes. An important Philippine business reference for 1926, while failing to mention rice trade in its profiles of Smith Bell and Warner Barnes, notes that both Siy Cong Bieng and Company and Siu Liong and Company (G. A. Cu Unjieng) were now major importers of rice from Saigon and Rangoon. Both companies were established during the 1890s;

Siy Cong Bieng also owned rice mills in Cabanatuan and Rosales and still owned a mill and bonded warehouse in Cabanatuan in 1935.[84] Elsewhere in Southeast Asia there were similar transitions, mainly to Chinese ownership, within the rice trade.[85]

Complementarity and mutual dependence among European and Asian commercial networks, while it lasted, had turned heavily on differential access to market information.[86] For a long time, northern European networks had a strong advantage in information on the economies of industrializing countries, the ultimate consumers of many tropical exports. On the other side, Asian networks became adept at understanding and working in the context of local needs and conditions while relatively few northern Europeans were willing to pursue their careers in the "provinces." There were, of course, a few exceptions in the Philippine rice trade. In addition to a number of Spanish Basques, several British businessmen were willing to live and work if not in the villages and townships at least in some of the small urbanized ports, such as Dagupan and Lingayen, or were posted there as buyers and managers with the new inland rice mills.

Over time, however, there was declining complementarity as European and Asian networks became more directly competitive. The increasing quality and abundance of business information on international markets and growing contacts with foreign millers and manufacturers worked to the advantage of the Asian—in this case Chinese—networks. The advent of both international and internal Philippine telecommunications aided in this development.

British networks were still able to hold on and prosper for a time because of their superior access to steam technology and perhaps to their access to credit as well, as shown in the number of bank loans to Luzon Rice Mills managed by Smith Bell. British international banks, lacking reliable sources of information on who might be adept and trustworthy within Asian business communities, favored the enterprises of their compatriots, who were often precocious in the use of new technology: steam power in cargo vessels, milling, and the railway.

The Hokkien Chinese networks ultimately won this competition because of their strength at both ends of the trading system. Increasingly, the Hokkien advantage in forging *poblacion* (county seat) and even village-level buying and selling connections with the local rice producers put them in a position to control acquisition of the crop, which tended to starve the power mills owned by others of sufficient *palay* to maintain profitable operations. They succeeded in the rice trade because it played to their competitive strengths, and it allowed these businesses to save on the transaction costs that would have accrued had they remained specialized at one or the other end of this commerce. The breadth and strength of the wholesale and retail networks developed by Chinese businessmen over time made it difficult for others to find a market among metro area consumers. Arriving Hokkien migrants who went to work in the rice trade needed to learn the local language of commerce. At the same time they needed to emphasize and act on their narrower linguistic, place of origin, "clan," and even shipmate identities as crucial links to the sources of commercial training, market information, credit, and the like. In this sense, the operation of the rice trade tended to sharpen social and ethnic segmentation in the metropolis and provinces.

At roughly the same time, however, the Chinese were shunted aside in the Philippine sugar trade. Whereas *palay* keeps relatively well and does not require immediate milling, harvested sugarcane must be milled immediately or lose its value. Commercial power in the sugar trade passed to the owners of the modern centrifugal sugar-milling *centrals*. As a result, the Hokkien from Xiamen (Tongan) who had been important in the sugar trade in the nineteenth century were increasingly bypassed in the twentieth, when a successful operation was more affected by international power relations and import quotas.

PART II

Ulam:
What You Eat with Rice

5

Vegetables, Fruit, and Other Garden Produce

THE CONCEPTION OF A MEAL in Tagalog society starts with boiled rice (*kanin*), and the main element of what one eats with it is represented as the *ulam*.[1] Generally speaking, fish or fish products form the principal *ulam*, but vegetables (*gulay*) and meat (*karne*) are also included. In addition to vegetables and fruit, the animal protein available in Manila in the nineteenth and early twentieth centuries was composed of fish and shellfish, fowl and eggs, swine, and bovines, with occasional goats, sheep, wild deer, dogs, and other animal products. This and the following chapters take up the changing vitamin-carrying and protein diet of the people of Manila, including the production, capture, and commercial systems supporting that diet and the degree to which various localities and provinces became specialized in their supply.

DIET AND NUTRITION

Too little is known about who actually ate fruits and vegetables other than affluent Filipinos, resident Chinese, and those other foreigners who developed a taste for native foods. But fruits and vegetables were certainly coming to the markets of the city, many on a seasonal basis. Photographs capture the delivery of stems of bananas and other heavy produce by carabao cart and small boat and of vegetables in baskets carried by boys using balance poles (figure 5.1). Tomatoes, onions, and lemons were often delivered in a small basket called a *buslo* or *canastrillo*. The flow is hard to document because of the great variety of comestibles involved and the various modes of arrival, but traces can be found if we look for them.

We can surmise that the urban poor ate less of all these—fruit, vegetables, meat, and even fish—because expenditures for rice took more of their limited dietary budget, especially during hard times.[2] The consumption of fruits and

Figure 5.1. Bananas (*saging sabá*) arriving in Manila by carabao cart circa 1900. This cart, or *karreton*, was an essential form of year-round bulk transport during the era of limited infrastructure. (USNA II, RG151-FC-85D)

vegetables was and is one of the most income-sensitive aspects of family diets in the Philippines, particularly for those with incomes that barely cover the cost of daily carbohydrates and are insufficient for the vitamin-rich foods essential to more robust health and growth. Low-income urban families always look first to filling energy foods—to rice or rice substitutes. For poor workers in our era, the percentage of the food budget spent on vegetables and fruit could be modest in the extreme. Diets also varied with the seasons. For example, only the affluent and the securely employed ate mangos in February, early in the season, but as the flow became a veritable tide in late March and April prices plunged and many others joined in.

A dearth of vegetables and fruits apparently developed as Manila grew during the nineteenth century. During a January visit in 1842, Charles Wilkes had reported, "Vegetables are in great plenty, and consist of pumpkins, lettuce, onions, radishes, very long squashes, etc.; of fruits they have melons, chicos, durians, marbolas [*mabulo?*], and oranges." But thirty years later, a Belgian visitor wrote, "The nearly complete lack of vegetables is a major privation to Europeans" in the city. And in 1888 the Dutch consul welcomed plans to construct a railway into the hills east of the Marikina Valley because it would "make possible the transport of vegetables and healthy fruits which are in general lacking in Manila."[3] Evidently the supply had not kept up with population growth.

Ken De Bevoise points out that a diet of rice and fish alone is deficient in "vitamin A, vitamin B$_2$, vitamin C, folic acid, iron, calcium, and magnesium and [is] low in fat."[4] Most people in the lower reaches of Manila society ate bananas routinely and also nipa vinegar (*suka*), fish sauce (*patis*), and *bagoong*—tiny fish pickled, fermented, and aged in rich brine to make a loose paste. On occasion, at least, they may have made *tinola* soup, delicious tamarind-sour *sinigang*, or their equivalents with a few minnows caught in the ditches of the city (as one sees yet during hard times) or a bit of fish—Alegre even mentions *tinola* made with captured frogs—green papaya or sweet potato, and chile pepper, *kangkong*, or *malunggay* leaves. *Kangkong* is a delicious swamp green, an excellent source of vitamins A, B, and C, calcium, potassium, and iron. It is commercially available and also scavenged or grown in Manila as a subsistence food and is important in the swampy interstitial spaces and edges of a number of Southeast Asian cities. *Malunggay* is a common urban tree whose delicate compound leaves are notably high in vitamin A, niacin, and riboflavin when eaten fresh. It is particularly good when cooked in a little water with mongos.[5] De Bevoise remarks, "The addition . . . of regular but modest portions of mongo beans, bitter melon, eggplant, chili peppers, sweet potato [*kamote*] shoots and leaves, bananas, papayas, and pineapple would have provided a nutritional intake meeting all modern minimum daily requirements," though remaining low in fat.[6] But how regularly were these consumed?

A dietary survey of more than 100 families of urban "workingmen" was conducted in Manila's Paco district from October 1936 to March 1937. It reported that "bananas, tamarind, tomatoes, and onions" were the most commonly consumed fruits and vegetables. Predictably, it also found that families with the least disposable income to spend on food also spent the lowest proportion of their dietary budget on fruits and vegetables—about 8 percent—while the mean expenditure share for the total sample of worker families was 14 percent.[7] The poorest families bore the greatest nutritional deficit, which increased their vulnerability to cholera, diarrhea, intestinal parasites, and tuberculosis and other respiratory diseases. There were also notable socioeconomic differences in physical stature.[8]

The prosaic indigenous bananas were routinely underappreciated by observers, whether varieties of ordinary bananas or the short, thick, cooking bananas known as *saging sabá*. John Super tells us that once they were introduced in Mexico, bananas quickly became the "staple for the poorer classes," especially when mixed with tubers and grains. They required little labor to grow and produced in all seasons, yielding vitamins and minerals, as well as considerable calories. In their native Philippines bananas were the most common element in ordinary nineteenth-century diets after rice or maize.[9] Important questions include how these fruits and vegetables were raised, in which environments and locations, and by whom.

The Metropolitan Garden Ring

One can still see remnant orchards in the hills of Antipolo to the east of the Marikina Valley, but since most former sites of intensive gardening and fruit cultivation are today almost completely filled with humanity and concrete, it takes a special concentration to imagine the garden belt that nearly encircled the metropolis in the nineteenth and early twentieth centuries. The vegetable gardens essential to nineteenth-century Manila were eventually displaced outward; the specialized betel gardens of Pasay succumbed to the mansions and polo grounds of the well-to-do (as well as to the replacement of betel with cigarettes), and the irrigated fodder (*zacate*) fields in San Juan del Monte, Pandacan, and elsewhere were displaced by subdivisions or abandoned with the slow decline of horses in the city. Little is left to remind us.

Many ripe fruits and soft vegetables do not travel well. The fruits get bruised, and the sugars break down; when picked green for better transport, they often fail to develop their full complement of flavors and textures. This modern problem was more acute in the days of slower animal and water transportation, so locales with the right ecological characteristics that were highly accessible to the urban market and not in immediate demand for urban expansion often were selected for growing such produce. Unlike rice and hogs, which were often transported over substantial distances, many vegetables and fruits were grown in the immediate area, needing only a brief journey to market. (Shipping from much beyond Laguna and Bulacan chanced the perishable nature of the products and raised the duration and cost of transport.) Decisions about which crops to grow and what practices to use on a given plot might be those of a farmer whose effort and time were important considerations or they might be forced on a farmer by a landowner more concerned with return per unit of land than per unit of labor. It is at the intersection of these considerations that local producers, whether Filipinos, Chinese, or colonials, worked out the changing calculus of metropolitan land use that led to a "ring" pattern around Manila.

This is hardly a new discovery—in the 1820s agricultural location theorist Johann Heinrich von Thünen identified the rings of land specialization that tended to form around semiisolated commercial towns using ox and horse transport.[10] What is relatively new is scholarship demonstrating that nearby specialized market-gardening zones formed not just around the rapidly industrializing cities of the West in the nineteenth century but also around large cities in the tropical colonial world, as Kenneth Kelly demonstrates for alluvial land next to Calcutta.[11] Meanwhile, railroad building started in the Philippines in the 1890s, leading to a sharp decline in transport costs for areas with favored access; trucks followed. This innovation allowed field crops, such as rice, to be delivered to Manila at prices that undercut the local comparative advantage, which made general farming next to the city less attractive than before, another impetus to the emergence of specialty crops in areas immediately accessible to the city.

By "Manila" I mean the entire functional conurbation rather than simply the designated "city." The formal territory of the latter changed over time, and in our period seldom included the entire urbanized area. Multiple parts of Manila, including the most important commercial and production spaces, lay near but outside the ancient walls of Intramuros, which defined the *"leal ciudad."* These *arrabales*—outskirts or suburbs—were divided into a series of municipal-parish territories. Numerous other close-in areas and town centers were added to the metropolitan expanse over time.

Agricultural activities in the several *arrabales* and the immediate set of areas beyond them are described in nineteenth-century records. Just inland from the coast there emerged a mixed-use zone that included fruit and vegetable culture and continued in fragmented fashion around the landward side of the city. As the urban seaside was bracketed with fish traps and ponds, so the land side became the location of clusters of gardens on extensive areas of gently sloping or nearly flat terrain with alluvial and volcanic soils.

Buzeta and Bravo's gazetteer of 1850–51 provides considerable detail. Father Buzeta actually lived for a time in Malate, then on the southern margin of the urbanized area, and he wrote with feeling when he described adjacent Pasay as a place of "delightful *jardines* and fruitful kitchen-gardens filled with fruit trees of several species and vegetables that are carried to the market in Manila on a daily basis." Farther south was Parañaque where, just inland, there were beautiful gardens with fruit trees such as lemons, oranges, and bananas, as well as uncommon specialty products that also came from gardens here. The pre-Hispanic Filipinos enjoyed fruit trees, but a new element arrived with the Spanish conquest. Spaniards had once been leaders in western horticultural techniques, which they transmitted to the Philippines via Mexico; in the process, orchards became regular features of the landed estates that bracketed the city in the eighteenth and nineteenth centuries.[12]

On the eastern margins of the city, the cluster of Sta. Ana, Mandaluyong, and San Juan del Monte is depicted as producing considerable fruits and vegetables, and from Pandacan along the river came freshly cut sugarcane stalks—the ordinary sweet of the day. To the northeast, parts of Sampaloc district were still little developed, yet, it was said, "All the houses have their gardens with

fruit trees and different vegetables."[13] Coello's map of the city in 1849 (the fron-
tispiece to this volume) shows these gardens. Beyond Sampaloc the jurisdiction
of Caloocan then included much of the large territory that became Quezon
City; vegetables and legumes were part of the produce from *sitios* and estates
here.

Tondo, on the northern margin of the city, also played a part in the daily
supply of vegetables and fresh maize and produced good oranges as well. Farther out, Malabon is mentioned for its vegetables and fruits. The Franciscan-
administered parishes of Polo and Marilao in the same direction supplied
produce that was carried into the city by small boats on a regular basis. All these
gardens, fruit trees, and *zacate* paddies constituted a patchy mosaic—depending
in part on the variable quality of the soils—that by the mid-nineteenth century approximated a ring (though one much smaller than the "inner zone"
described for the rice trade).

Beyond these most immediately accessible jurisdictions lay some well-watered
alluvial areas with easy access to Manila by dugout *banca* or *casco*. Here, too,
commercial production of fruits and vegetables came to play an important part
in local economic specialization.[14] All sorts of essential perishables came from a
special cluster of settlements to the east, along the channels of the Pasig River.
From Pasig municipality came fine fruit and vegetables; from Pateros and Taguig
came watermelons in addition to an everyday flow of duck eggs and rice.
Through these places passed *cascos* loaded with fruits and vegetables and rafts
of coconuts on their way to the city from Laguna. Along the western shore
of the lake south of Taguig were friar haciendas leased to mestizo *inquilinos*
and sublet to Tagalog farm families. Here the gardens and tree crops of Biñan
elicited great respect from Buzeta and Bravo: mangos, *nangkas* (jackfruits),
papayas, *atis*, oranges, bananas, and many other fruits are named. Some of this
was true of the neighboring hacienda communities of San Pedro Tunasan,
Calamba, and Cabuyao as well.[15]

Stretching north of Pasig town the alluvial and rice-rich Marikina Valley
became a prime locale for supplying Manila with fruit. In the 1880s José
Montero y Vidal singled out this valley as an important source of Manila's
vegetables.[16] In Cavite the giant Imus Estate was famous for its outstanding
mangos—90,000 in 1884, as well as a thousand pineapples. This fruit was sent
to Manila by small boat via nearby Cavite el Viejo.[17] Young alluvial soils based
in part on volcanic ash, reasonably fertile and workable, and adequate rainfall
played a part in these developments, but so did indigenous practices of fertilization and soil manipulation. The further endowment of an elaborate network of natural waterways and canals, essential to the supply of rice by *cascos*
from the broad inner zone, was also critical to the delivery of fresh fruits and
vegetables from such places.

Suburban Estates

A lot of the vegetables and fruits that supplied the markets and households of Manila in the nineteenth and early twentieth centuries were grown on the immediate urban margins. Almost by definition this implies that many were grown on the holdings of the extended Tuason family, since these wrapped around the north and northeast sides of the city. Here, three enormous haciendas had been purchased in the 1790s and 1810s by the eldest son of Antonio Tuason, the Chinese mestizo who was possibly the richest person in the Philippines in the late eighteenth century. In 1900 these holdings still stretched from Tinajeros in Malabon through Balintawak in Caloocan, then eastward across to Diliman and Santa Mesa and on to the Marikina Valley beyond. The Hacienda de Marikina alone encompassed 30 square kilometers. Two of the Tuason haciendas remained intact through five generations because they were entailed in a *mayorazgo* and could not be divided or sold. The disentailment of such estates was a crucial step in the program of agricultural modernization undertaken in Spain in the 1830s, but in the colonial Philippines the Tuason *mayorazgo* lasted through the end of the century.[18] Within the Tuason estates there were sections with substantial vegetable gardens.[19] The Hacienda de Sta. Mesa included territory extending from Sta. Mesa in eastern Sampaloc northward to include the Diliman Estate, now in Quezon City. In the 1890s, there were widespread vegetable gardens on these lands, from which local youths carried vegetables into the city in baskets slung on balance poles.

As the city grew, did cultivators on the Tuason Estate decide to switch from rice to more labor-intensive gardening in response to the growing market opportunity or did they make such a transition because they could no longer pay the rent asked based on rice production alone? If Tagalog farming families were turning to vegetable gardening in order to cope with rising rents, this might reflect demands from the lessees of the Tuason Estate. Benito Legarda Jr. offers a caution about soil quality, however. Around 1910, shortly after his family secured the present Legarda Estate, several members rode across it on horseback to make a quick survey. They found that a lot of the estate was not in intensive production and where it was *zacate* was at least as notable as rice. The upland portions, away from alluvial soil concentrations, had only thin clay soils on rock and were not suitable for vegetables.[20]

~

Outside the Tuason lands legumes and vegetables were grown on other portions of the city margin, including the alluvial riverside in Pandacan and Sta. Ana. These often included eggplant and *ampalaya*, or bitter gourd, grown on trellises. Long, thin eggplants were easily grown and widely available in the markets. One or another variety of bananas and papayas, the fast-growing Latin American tropical fruit, were also common. *Sitaw*, the high-quality, very long green bean grown all over the archipelago, a good source of vitamins A and B, goes unmentioned in the 1896 Estadistica, however. Tomatoes—generally small and picked before fully ripe—were raised in eight localities of Navotas and also on land near the lake at Taguig.[21] Using small watercraft, people in both places could readily move such perishables to the great public markets of Manila. In 1905 a labor survey noted that the "high price [of agricultural wages] in suburban Rizal province is accounted for by the fact that the tilled land is occupied by vegetable gardens selling in the Manila market." Nearly every nearby Rizal municipality produced bananas, and it was common to see them being delivered by carabao cart. The encircling band of vegetable and fruit production areas was patchy, but it was expanding.[22]

The Chinese Connection

There is little in the available sources identifying vegetable growers as anything other than (Tagalog) Filipinos, but this silence conceals the fact that in developing a regular food supply for the imposed urban center in the early colonial period, the Spaniards had attracted many market gardeners from South China. By the mid-seventeenth century there were perhaps 2,000 Chinese market gardeners and numerous orchards along the Pasig River stretching from the city to Laguna de Bay.[23] Even after the expulsion of many Chinese in the mid-eighteenth century this tradition remained, to be renewed by the new wave of nineteenth-century immigrants.

Chinese methods of leafy vegetable gardening generally worked well in the Philippines. In the 1920s Hokkien horticulturists north of Tondo, between Maypajo and Caloocan, and also in Paco were known for intensive soil preparation by means of hoeing and the use of raised beds. Both techniques assist in soil aeration and drainage. For fertilizer they made intensive use of *lumbang* cake, a residue left over from the extraction of *lumbang* lubricating and illuminating oil.[24] Such vegetable gardening involves incessant nutrient recycling by laboriously working biomaterials into the soil. It is likely that they made use of lime produced locally by burning seashells or limestone.[25] A domestic guano fertilizer industry had begun as well, and its product was used by suburban *zacate* growers, although it was mainly used on sugar plantations. Manila's Chinese gardeners developed a reputation for hand watering three to five times on sunny days to avoid wilting, using pairs of buckets slung on a balance pole. Tagalog gardeners at this time apparently watered less frequently.

With crops grown in rapid succession, the local gardens of Chinese farmers produced cabbages, onions, Chinese cabbage (*petchay*), lettuce, Chinese celery (*kinchay*), mustard greens, tango herb, spinach, and peppers, with one crop seeded in rows and another broadcast in between on the raised beds. Some of the leafy crops matured in a little over 40 days; others could take four months.[26]

Horticultural production by local Chinese market gardeners was certainly important in provisioning the city, although we lack a basis for estimating how much of Manila's supply they accounted for at any particular time. We know that an appreciable, if indeterminate, amount of Manila's bananas and vegetables still came from Tagalog dooryard gardens rather than dedicated commercial growers.[27]

At the same time, however, the gardens of Manila were not totally comparable to those elsewhere in urban East Asia. For a start, there was no integration with fish farming, as there was in many gardens of the Canton River delta. Leafy garden wastes were used in Cantonese polyculture ponds as a food source for grass carp, and the rich pond bottom detritus was brought up periodically and worked into garden soils. In the Philippines, by contrast, fishponds mainly produced *bangus* (milkfish) in this era, a species that lives in brackish water and does not eat greens. The ecology of *bangus* ponds did not lead to integration with gardening.

Perhaps more significantly, whereas Chinese and Japanese cities were immediately surrounded by an intensive gardening zone, the excrement of the urban population was routinely captured and transferred to these gardens. In the case of large cities with a dense network of waterways, such as Shanghai, Suzhou, and Canton, this zone of intense fertilization could be extensive. Around prewar Shanghai innumerable small plots of vegetables were mixed with rice fields to a distance of 30 miles—a day's travel by small boat. This was exactly the same radius "where the city's night soil is cheaper and more readily available than in outlying districts" and where the overall vegetable yields were among the highest in the country.[28] A similar system was employed around Hanoi in northern Vietnam, where residents of several villages specialized in removing urban night soil and selling it as fertilizer. According to Pierre Gourou, women carried this matter up to 15 kilometers using a balance pole and buckets.[29]

There was nothing like this among insular Southeast Asian farmers or in Filipino market gardens. When the new American authorities set up a "pail conservancy system" for removing excrement from Manila, it was gathered in half barrels in midden sheds, carried through the city on mule-drawn freight wagons, loaded on barges, and summarily dumped into Manila Bay (figure 5.2). A few years later a new sewer did the same—delivering the effluent a mile out in the bay.[30] In neither case were these wastes processed or distributed for their fertilizer value. Some use of dove and chicken droppings is reported, but the mainland Chinese method of recycling human wastes was noxious to Filipino gardeners. Horse manure was used in various Spanish crop yield experiments in the 1890s, but if it was used for fertilizing suburban vegetable beds this has gone without comment.[31] The vegetable gardening ring encircling Manila, then, was the result of a market proximity advantage and the existence

FIGURE 5.2. Workers in the "pail conservancy" sanitary organization remove barrels of night soil for dumping in the bay. (*Lipang Kalabaw*, October 26, 1907)

of pockets of alluvial and volcanic soils; human night soil as fertilizer apparently played little role.

Betel

By the 1880s, suburban Pasay had developed into a quintessential microspecialty production area, with gardens known as *ikmuhan* or *buyales* producing the fresh leaves of the climbing pepper plant (*Piper betel* L.) included in the popular mouth stimulant known as betel. Pasig was another nearby place where several Chinese mestizos grew rich producing betel leaves for the city.[32] The chew (*ngangà* or *buyò*) also incorporated slaked lime and a wedge of areca palm nut. Wrapped in the leaf, it was chewed like tobacco, often with a bit of tobacco leaf added. Chewing betel is a very old practice throughout Southeast Asia, including Vietnam, and subsequently in South India, Taiwan, and most recently highland New Guinea. For a long time, "the amenities of friendship and hospitality [were] expressed in a ceremonial chewing of the betel nut."[33] Areca nuts were also widely grown in and around Manila, where "One frequently sees the native cottages enclosed on two or three sides by rows of areca palm" each grown for its hundreds of nuts.[34]

By the tens of thousands, the heart-shaped betel leaves were carefully arranged in layers on winnowing trays (*bilao*) and delivered fresh to the public markets of the metro area. From the point of view of freshness, it was a prime advantage to grow the leaves close to the major body of consumers. As early as the 1860s, Manila was not just consuming but also supplying betel leaves to provincial towns located within easy reach by boat around Manila Bay. In the 1880s, leaves were sent to Bulacan, Pampanga, and Nueva Ecija. In the late nineteenth century, the horticulturists of Pasay were still raising enormous quantities of betel leaves. Stretching southward, Parañaque and Las Piñas were also heavily committed to betel growing, and in 1896 on the rural Makati estate owned by Pedro P. Roxas betel was intensively grown on parts of 39 out of 40 tenanted parcels.[35]

At the beginning of the twentieth century Frederick Sawyer was particularly taken with the cultivation techniques of the Tagalog betel gardeners of Pasay: "[Betel] is grown in small fields enclosed by hedges or by rows of trees to keep off the wind. The soil is carefully prepared, and all weeds removed. As the tendrils grow up, [tall] sticks are placed for them. The plants are watered by hand, and leaf by leaf carefully examined each morning to remove all caterpillars or other insects. The plants are protected from the glare of the sun by mat-shades. . . . . The ripe leaves are gathered fresh every morning, and taken to market, where they find a ready sale at remunerative prices."[36] Increasingly the leaves were grown on raised beds to promote drainage and give a superior matrix for root development. In Laguna, soil for betel vines was sometimes fertilized with tiny shrimp.

In the city betel selling was a common part of the street scene. In the 1850s more than 240 women sold *buyò* in the commercial section north of the river. Some were vending from the doorways of houses or shops, others were ambulant. Nearly every street and plaza had half a dozen such vendors, known as *buyeras*. Compared to textile vendors, the *buyeras* were far more likely to come from poor circumstances and be operating with a microloan. The work was highly social, since male customers often lingered, making good looks and a pleasing personality a definite plus for sales (figure 5.3).[37]

Although cigars had been around for generations, in the twentieth century the smoking of well-advertised brands of cigarettes gradually replaced the general custom of chewing betel. Such cigarette smoking likely resulted in a net loss to health, although betel chewing itself could sometimes lead to cancer of the mouth. According to Francis A. Geologo, a significant moment in the transition away from betel came with the catastrophic mortality brought by the world influenza pandemic of 1918–19, during which annual Manila mortality rose by 69 percent in 1918.[38] The use of betel was attacked as backward, and the practice of chewing and spitting became one of the targets of hygiene campaigns. Even so, Filipino dealers in Manila's Divisoria Market remained at the center of the commerce in betel.[39]

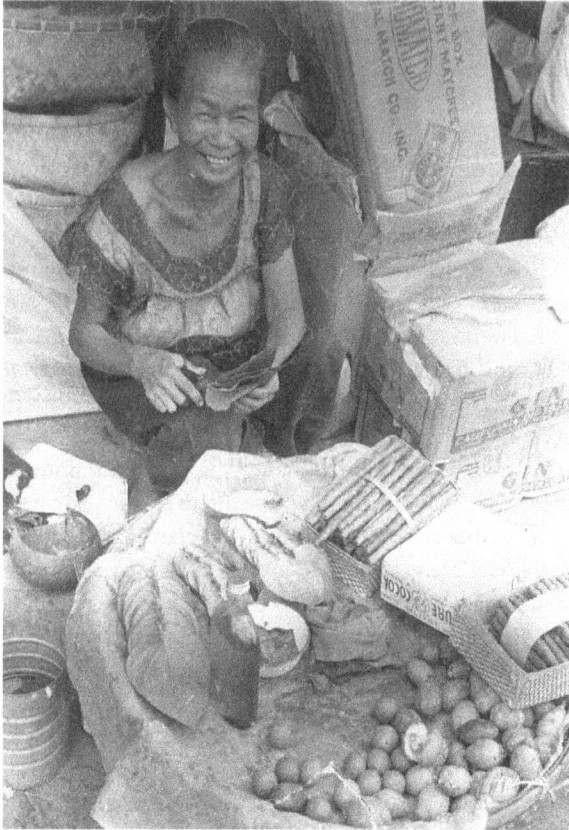

FIGURE 5.3.
La buyera—a betel
vendor—in the
Dagupan public
market displays the
makings of her
trade, as well as the
winning personality
that was a hallmark
of it. (Photo by the
author, 1969)

By the end of our period the geography of production had changed. As the gardens of Pasay were converted into urban infrastructure, Pangasinan, Laguna, and Ilocos Sur became the main leaf-growing areas. For a time, however, war reversed the decline of betel. During the Japanese occupation many cigarette smokers turned to betel chewing when tobacco cigarettes became unobtainable. Leaf growing expanded in central Pangasinan along the Angalakan River with the use of *noria* waterwheels for irrigation.[40] Likewise, as familiar medicines ran out, many turned to betel for its formerly well-known medicinal effects: healing, soothing, and calming, including calming the pangs of hunger. Betel was also administered to animal wounds.[41]

Zacate Fodder

The local access advantage that produced a ringlike pattern of horticultural production also extended to fodder for the many native horses—and a few larger imported horses—widely used for local transportation in nineteenth- and twentieth-century Manila, especially for pulling carriages and light taxi vehicles. The local breed of horse evolved in the archipelago in the sixteenth and seventeenth centuries from larger imported stock and, though small, was well adapted to the available forage supply. Since they lacked the physique needed to pull or carry heavy loads, such horses complemented rather than competed with carabao as plow and draft animals. Hundreds of native horses in pack trains carried sacks of copra or coconut oil in kegs from the hills of Laguna-Tayabas to Santa Cruz, Laguna. Citing a report from 1823, Bruce Cruikshank calls attention to the overland trade from several Tayabas locales, especially Mauban, and writes, "Much of their income comes from portage from the Camarines [through Tayabas] . . . to Santa Cruz de la Laguna and on to Manila." The coconut oil was produced by a human-powered press made of timber, one of many processing activities undertaken by provincial workers for the urban market.[42] Another was the production of firewood, long Manila's leading cooking fuel.[43] Pack trains were also used to transport oranges from eastern Batangas and coffee beans from Lipa (figures 5.4 and 5.5).

Although most city streets were poorly surfaced, they were hard enough to make horses practical in nearly all weathers. In Manila horse-cart taxis were used rather than the human-powered rickshaws that were common in Chinese

FIGURE 5.4. A packhorse train moves sacks of copra overland to Laguna de Bay for shipment to Manila, early twentieth century. Oranges and coffee were transported in the same manner. (Bureau of Science photo, USNA II, RG151-FC-85A)

Figure 5.5. Provincial processing for the city: Extruding coconut oil with a side-screw press, Laguna, 1891. (PNA, Contribución Industrial, Laguna, 1863–95, doc. Pagsanjan, S9)

and Japanese cities, as well as Singapore and other colonial cities of the day. The Japanese consul in Hong Kong commented in astonishment that there were more horse carriages in Manila than in any other city in Asia (figure 5.6).[44] Greg Bankoff reports that there were more than a thousand carriages and gigs (*kalesas*) in the city and nearly as many *karromatas* and carabao carts in the 1880s. In the 1930s, there were still several thousand horse-drawn vehicles in the city, nearly all pulled by male horses, many of which had quite long working lives.[45]

Outside the city, the typical roads were essentially seasonal tracks, quagmires in the rainy season, during which the carabao cart was the most reliable land vehicle. But by the end of the Spanish era public works, such as metaled roads and graceful arched bridges, made it possible to venture out of the city in a horse-drawn vehicle to the town centers of nearby communities in Cavite and Bulacan. Spaniards and northern Europeans and American men sometimes rode (and raced) horses as well.

FIGURE 5.6. A great density and variety of horse- and mule-drawn vehicles on the Escolta in the Manila central business district, 1898. All these animals had to be fed. (Colquhoun, photo 091055, Photo Archive, Royal Geographical Society, London)

As the city expanded there was growing spatial separation between home and workplace, market, or school, which often made some sort of urban transportation system essential. During the last two decades of the nineteenth century, Manila was served by a system of horse-drawn streetcars known as the *tranvía*—slow but useful and well patronized. The *tranvía* company alone had 134 horses. In 1905, the new city administration was feeding 600 horses in two corrals, half of them kept for the use of the personnel of various insular bureaus. And many carriage horses were kept by affluent households. Almost 9,000 were enumerated in the city in the census of 1903. Although the number declined during the 1910s, it picked up again with the massive urbanization and suburbanization of the 1920s and 1930s, standing at an estimated 25,000 shortly before World War II.[46]

Walter Robb wrote, "Manila must have its breakfast, and will pay for the feeding, even of its animals."[47] As in cities everywhere, Manila's horses depended on food brought to them—in particular *zacate,* one or another of several long, freshly cut native grasses, rather than cured hay. Von Thünen located the production of hay for stall-fed animals in the inner ring, along with vegetables and fruit.[48] So it was with *zacate* for Manila's horses. Planted in irrigated paddies, it would not have been grown on soils of a quality to support horticulture.[49] It was bulky, worth little per unit of volume, and needed to be delivered fresh. Transport very far would have made it prohibitively expensive.[50]

A week-long survey of Pasig River traffic in 1853 recorded 108 small *bancas* carrying *zacate* into the city; extrapolating from this number one could project 5,600 *banca* loads for the year.[51] In the 1880s and 1890s, there was scattered but extensive *zacate* production in a broad inland band around the city from the northeast to the southeast. In various locales along this arc, men made a living growing *zacate,* which they then carried into the urban area for sale. Pandacan was first among the outer city districts in the number of *zacateros* in 1880s with 156; Paco/Dilao and Sampaloc were next. In the 1920s, as in the 1850s, a good deal of the daily *zacate* ration was delivered to dealers in the city by dugout *bancas* moving along the *esteros.* Smaller quantities were carried in loose sheaves slung on balance poles and sold by ambulant vendors. In the 1930s bales of *zacate* were found in sheds located in the parts of Tondo where *kalesas* were based, notably Gagalangin. The urbanized area was expanding rapidly, and by the 1930s suburbanization had displaced *zacate* production in almost all of its late-nineteenth-century locales; *zacate* and forage lands were now found farther out, in Rizal, Laguna, and Cavite provinces.[52]

Zacate, the dietary mainstay of urban horses, was usually supplemented with vitamin-rich rice bran (*darak*), a local milling by-product. *Darak, palay,* and sometimes maize and rice straw were used as horse feed in combination or rotation, depending on the price. On rare occasions, copra meal was advertised as a horse feed meant to replace grain, but it does not appear to have seen wide use.[53] Other by-product livestock feeds, such as used brewer's grain, were produced in the city but not used for horses.

Zacate and *darak* were not an adequate diet for the larger horses imported from Europe and later from the United States and Australia. Bankoff notes the interest of Spanish officials in developing better quality horse feeds, including the introduction of some from Cuba. One of these was the Mexican plant *teosinte,* promoted by the Spanish agriculture and botany school in Manila. Grown in the metropolitan area in the 1890s in irrigated, manured, and sometimes limed plots, *teosinte* could be cut once a month during the dry season. It was found to be a better dry season fodder than most.[54] *Palay* was minimally adequate as a substitute for oats but was often relatively expensive. So for large horses various fodders were imported from the American Northwest, Australia, or India. Little progress was made during the first decade of American occupation in developing a local fodder source. William Clarence-Smith describes several regimes in Southeast Asia engaged in cross-breeding programs intended to "improve" the quality of native horses, which usually resulted in "equids that were vulnerable to prevailing diseases and that rejected local fodder."[55]

A 1930s example gives practical insight into the use and feeding of horses. San Juan district native Liberate Tuaño bought a *karretela* with his older brother's assistance and became a rig driver. His day typically began at 5:00 a.m.: cleaning the stall, hitching, picking up passengers in suburban San Juan, and driving them downtown to Quiapo. He then spent the day making produce deliveries for one of the Chinese stores in the Quinta Market. Returning to San Juan in the evening, Tuaño still had to feed the horse. At intervals he bought *zacate* from a local dealer. By 1938 the same elder brother (his *kuya*) had taught him how to drive a motor vehicle, and he was glad to make a change. Now working out of neighboring Mandaluyong, he made deliveries for a Filipino store that sold molasses to horse feed dealers. Part of his work involved mixing water with the molasses. Later *darak* was added. His deliveries went to *zacate* dealers in outer surrounding places such as Binalonan, Taytay, Marikina, and Cavite.[56]

PROVINCIAL GARDENS

Some vegetables and fruits that travel relatively well had ecological requirements that were not well met near the metropolis. In the nineteenth century these products tended to be supplied from Batangas, the Ilocos coast, and Laguna; in the twentieth century they also came from the eastern plains in Nueva Ecija and parts of Pangasinan, while midlatitude vegetables and white potatoes, not Manila's customary fare, were grown in the Benguet highlands. Batangas farmers originally took advantage of the fertile loamy soils that developed from the disintegration and breakdown of alluvially deposited volcanic materials to develop several horticultural and arboreal microspecialties: onions, garlic, mongo beans, and even peanuts and soybeans.[57] Northeastern Batangas became home to a famous concentration of citrus groves, while the Vigan area of Ilocos Sur sent important quantities of onions and garlic.

Such crops were often collected and brought into the urban marketing system by traveling buyers known as *viajeros*. Only oral "records" document their work. One *viajera* was Natividad Samio de Gamboa (b. 1918), a native of San Nicolas district in the city. A high school graduate, she married a sweepstakes agent in 1936. By then she was selling vegetables in a market stall. Her (Tagalog) parents had some land in Pangasinan, and she began going there to buy vegetables. Before long she began traveling throughout an extensive territory. For more than 20 years, she went to different places buying vegetables: Pangasinan, Tarlac, and Bataan in the hot season, Cavite and Laguna in the rainy season. Using the railway lines where possible, her sacks of vegetables were loaded into special produce cars at night. The sacks were delivered at Manila's

Tutuban Station the next day. "From Tutuban in peacetime, we used our own
carabao *karreton* [freight cart] and delivered to customers in different markets
of the city," she recalled. After the war, "if you had 100 sacks, then just hire a
truck from Manila." Increasingly she specialized in the *kamote*, even leasing
an "hacienda" in Balanga, Bataan, where *kamotes* were grown for her under a
local manager, or *katiwala*. In those days, "the Chinese were our good customers. . . . They were waiting for *kamotes*. Later, after Liberation, they started
going to the province to get them directly. They gave money [advances] so that
during harvest, they would get it."[58]

The parents of Filoteo "Lolo Feling," Tuason (b. 1906) also made their living
in the commercial supply of foodstuffs. His mother was a vendor of chicken
eggs in the Quinta Market. His father was a *viajero* dealing in eggs, fruits, and
vegetables. According to Lolo Feling, his father's commercial travels in the
1920s took him around a territory that included Laguna, Bulacan, Tayabas,
and even Mindoro.[59] When his trip was to Laguna and he was able to purchase
quite a bit, he would join other *viajeros* and hire a *casco* for the return trip. The
varied produce of five *viajeros* was enough to fill the vessel. If the available
produce were less, he would return on his own aboard the *Napindan*, a steam
"paddleboat" that landed behind the Quinta Market (figure 5.7).

FIGURE 5.7. Vendors looking over produce arriving on the Pasig River quay of the
Quinta Market circa 1904. (USNA II, RG350-P-E-19-7, folder 8)

Many *viajeros* worked through and were dependent on particular wholesale distributors. During the 1910s and 1920s, the father of carpenter Arturo Bautista, was a *negociante* in Tondo, a businessman dealing in fruits brought by *viajeros* from the provinces. The produce included watermelons and mangos among others. He recalled, "Retail vendors came to our house to get goods, and on occasion we sold in the market. *Bateles* [sailing vessels] brought the goods."[60] Others used cascos to bring produce to the Binondo Estero landing at the Divisoria Market. Later the trade in some products was taken over by Chinese operators who came to concentrate on a particular product group, assuring themselves of favorable buying opportunities by extending credit, hybrid seeds, and other inputs required for specialized production.

Usury and Market Vendor Indebtedness

Small-scale credit arrangements and usury were (and are) an integral aspect of retail provisioning and public food markets. Always some vegetable and fish vendors lacked the revolving commercial capital needed to pay their wholesale suppliers each morning. Unable to purchase fresh stock, such retail vendors would have had little to sell. Certain individuals known as *kapitalistas* or *usureros* made a handsome return providing short-term loans to needy vendors. In my experience such individuals make loans to regular clients with very little ceremony; often a nod is enough.

Like her father before her, Saturnina Salazar de Abreu made such loans to vendors at the Divisoria Market in the 1890s and early 1900s, reportedly charging 10 percent per day. She also ran a gambling and numbers operation. As a result, she was widely known as Lola Supot, roughly "grandmother bag lady." She also came to own considerable city real estate and rice lands. Raised in her big house in Tondo was her grandson, Victor Buencamino. From there he was launched on a career as a veterinary, as a businessman importing live cattle and selling sides of beef, and eventually as the administrator of the NARIC. He knew well how market vending worked.[61] Some other market *kapitalistas* were Chinese, and even a few resident Indians took advantage of the opportunity.

In the 1930s, market vendor loans were often for 100 days at a nominal rate of 20 percent total interest with daily payments. But the interest was the same every day. No account was taken of the declining principal of the loan, so the effective rate was far higher. Like the *buyeras* of the nineteenth century, vegetable venders were likely to be undercapitalized. In response to exposés, the legislature eventually set up an Anti-usury Board with powers of investigation. Further, the Philippine National Bank formed a small loan department and made at least 2,000 loans at 8 percent interest to small-scale retailers such as market vendors. These actions helped, but what was needed were ethical small-loan institutions operating on site in a rapid nonbureaucratic manner within the context of market, factory, stevedoring organization, and office. Properly licensed moneylenders in the 1930s were limited to interest charges of 14 percent on an annual basis—too low to for them to stay in business with unsecured small loans—leaving the field open to loan sharks.[62]

Onions and Garlic

Onions are an Old World domesticate, probably grown in the Philippines for many centuries. Garlic is a related plant (both are alliums) with a long history in Southeast Asia as a comestible and folk medicine.[63] Both remain in everyday household use. The present Tagalog term for onions is *sibuyas*, derived from the Spanish *cebollas*. This borrowing might tend to put the matter of origins in doubt, but in the nineteenth century *sibuyas* was still used concurrently with the older Tagalog term *lasuna* (*lasona* in Ilocano).[64]

The use of onions and garlic—and also tomatoes—in household cuisine was widespread in Luzon in the 1890s, and no doubt well before. Felice Prudente Sta. Maria points out that in the Philippine context the Spanish *guisa*, or "sauté," came to mean "frying garlic, onion and tomato in pork fat." By the early twentieth century, this procedure had become popular "as the first step in cooking many dishes."[65] Garlic and onions kept fairly well and could be sold in stores rather than amid the rapid turnover of perishables in the public markets (figure 4.3). In numerous locales in the 1890s small, generic, Chinese provincial stores were licensed to sell rice, garlic, onions, candles, and other items. Slightly larger stores sold the same things plus cotton thread, coconut oil, and kerosene. Both were called *tiendas de sarisari*. In some localities the list could also include flour, betel, Chinese noodles, vinegar, salt, and firewood.[66] Most of the large-scale onion wholesalers in the early twentieth century were also Chinese, based near the Divisoria Market.

Although some onions and garlic were imported, principally from China, they remained domestic products as well. Onions are grown in the dry season because they rot after exposure to tropical downpours. No place in the immediate Manila area offered protection from occasional unseasonable rains, so farmers of Taal-Lemery on the Batangas coast and the Vigan area in Ilocos Sur, which have protective mountains or hills just to the east, seized the opportunity. Much of the traceable domestic flow of onions in the nineteenth century came from these two places.[67]

The geography of the city's onion supply in 1862 was sharply differentiated by season. Most of the arrivals from December through February came from Ilocos, while from April through November nearly all came from Batangas, which had the advantage of quicker and cheaper access. Batangas was close enough to be able to continue a reduced level of shipments during the rainy season. (We may infer that Vigan found its opportunity in shipping onions that had been dried and stored during the rains, followed by the new onion harvest in February–March until Batangas took over.) Onion and garlic shipments from Taal-Lemery were small scale. They were never the primary cargo but were always an extra, filling out a bulk cargo of something like coffee or sugar. Of 66 arrivals of onions from Batangas ports in 1862, most were of 100 *picos* (6,200 kg.) or less. Smaller coastal vessels from there could sail safely to and from Manila, making many trips throughout the year, and smaller vessels did not require great capital resources to enter the business.

Thirty years later the pattern of arrivals was little changed. Onions continued to be grown on a commercial basis around Taal-Lemery and were often shipped in the same cargo as garlic and handled by the same *consignado*. Evidently there was no principal buyer in Taal who could concentrate the flow of either. The high level and broadly distributed nature of indigenous commercial activity made the town of Taal one of the more generally prosperous in the province—and unwelcoming to Chinese businessmen. In fact, in 1895 no Chinese were registered in Taal and only 16 in its twin, Lemery, versus 200 or more in Batangas town and Lipa.

In the twentieth century, onions became an object of formal experimentation by the newly arrived American authorities, initially at the Singalong horticultural station on the edge of the city. Finally, in the 1930s, Bermuda onions were successfully grown, using seeds brought from the Canary Islands. This capitalized on the rising "global tropical exchange of seeds and plants" and the work of local agricultural experiment stations.[68] In this effort the Philippine Bureau of Plant Industry carried out dry season experiments using transplanted seedlings and thick rice straw mulch in freshly harvested and weeded paddy fields. Where it rained notably, the seedlings rotted. In other places good medium-sized bulbs were obtained 85 to 120 days after transplanting.

Domestic onion production in the 1920s and 1930s was still concentrated in Batangas and Ilocos, but some varieties were being taken up in the market gardens on the metropolitan fringe. By 1939 the Bongabon and Muñoz locales in eastern Nueva Ecija were becoming established as the premier dry season onion production areas, and the province now accounted for half the national production.[69] Eastern Nueva Ecija has been the chief locus of Philippine onion production ever since, but it remains a risky crop. As production became more capital intensive, utilizing fertilizer and insecticides as well as commercial seed selection, credit relationships—often personalized—doubtless became more critical, but we know little about this in the prewar era.[70]

Garlic was reported as being grown in Taal and Bauan in 1850. The town name Bauan, in fact, appears to be a version of *bawang*, the Tagalog term for this plant. Recorded shipments to Manila in 1862 came mostly from Taal-Lemery, although two came from Batangas municipality and one from Guimbal, Iloilo.

Some garlic arrived in Manila almost every month of the year. As with onions, this commerce was highly distributed. In the 1890s, Calaca was singled out as a major Batangas source, and a decade later a significant flow of garlic was coming from Tanauan, the center of domestic citrus production.[71]

By the 1930s, there were two areas of specialty production using contrasting methods. In Batangas, garlic "seeds" or cloves from China were planted directly following considerable plowing and harrowing. Taking advantage of proximity and thus freshness, much of the Batangas crop was picked green and marketed for its leaves with the bulb only half developed. Meanwhile, faced with the competition of Nueva Ecija's surging onion production, a number of farmers in Ilocos Sur had given up that crop to concentrate on garlic. In Ilocos, the crop was grown to maturity using local stock planted in late October through December in moist, but not irrigated, paddies following the rice harvest. Fresh rice straw was used to slow evaporation and retard weed growth. The crop reached maturity in a little more than three months and was harvested from February through April.

Other Provincial Vegetables

Mongo beans, called *balatong, munggo*, or mung beans, were a small-volume specialty lentil crop, another of the commercial specialties of lowland farmers in Batangas.[72] About 80 percent of the minuscule recorded shipments to Manila in 1862 came from there. In 1870 coastal shipping delivered more than 170,000 kilos of mongos to the city. A decade later all the shipments were from Batangas Province, especially Balayan. Dry season mongos were relatively affordable in the city, costing approximately the same per liter as second-class white rice in the 1880s.[73] Like garlic and onions, mongos were often grown in fields following the main rice harvest. Along with peanuts they were planted so as to ripen during the dry season, since they tended to rot if harvested during the rains.[74]

In the twentieth century, mongos became a common dry season crop in Asingan and elsewhere in Pangasinan; like other legumes they were an excellent soil restorer and forage crop. In the decades before World War II, they became a common ingredient in the iced soda shop confection known as *halo-halo*, or *mongo con hielo*, increasingly sold by resident Japanese. Mongos were also sometimes boiled; mixed with chopped vegetables, garlic, and onion; and fried in small patties. By 1939 they were among the top 20 Philippine crops by value and were traded on the Tutuban Rice Exchange. Pangasinan and Iloilo became the primary producers, followed by Batangas.[75] Soybeans were also grown in the dry season in Batangas, upland Cavite, and elsewhere, though even more were imported via Amoy and Hong Kong.[76]

Sweet potato tubers (*Ipomoea batatas*) were of considerable importance in the national diet during this era. Originally introduced from Central and South

America, these are known in the Philippines as *kamote* or *kamote-bagin*, literally "vine potatoes" (as opposed to cassava, *kamoteng-kahoy*, "tree tuber").[77] The green-purple vine shoots are a nutritious vegetable and the tubers a filling starch food. They are mentioned as an important component of the diet of poorly paid office workers in the late nineteenth century and became a critical home-grown survival food for a broad spectrum of Manilans during World War II. By the nineteenth century the *kamote* had become the predominant staple crop in drier parts of the mountains of northern Luzon and, along with native Asian yams, in the typhoon-ravaged Batanes Islands to the far north. In the vegetable and fruit zone surrounding Manila, the *kamote* was a common house garden product. In many lowland places, they were planted after the principal rice season.

Some of Manila's supply may have come from nearby provinces, but few specifics have emerged. Twenty-six shipments were reported in Manila in 1862, almost all from Vigan or Ilocos Sur more generally. Transported in barrel-sized produce baskets (*cestos*), shipments ranged from a few hundred to a few thousand baskets.[78] Sweet potatoes were usually landed in mixed cargoes, and the commerce in them was highly dispersed. Despite potential difficulties with spoilage, these shipments were all composed of whole sweet potatoes, not dried sections. Some were likely fed to hogs, although at that time there was not much of a concentrated swine industry. This was a cheap foodstuff for the poor of the city, a bit of dietary variety for others. In the 1890s a visitor mentioned yams and cabbages as cheap vegetables that rounded out the diet of clerks and their families.[79]

Vigan was still a supplier in 1872, but the largest volume now was from the tiny nearby port of Sulvec. In the agricultural survey of 1886–87, at least 16 provinces reported *kamote* production.[80] With increasing commercial integration in the twentieth century, the domestic *kamote* in Manila often came from the southern uplands of the Mountain Province, where they were a local staple. Immediately following the very poor rice crop of late 1920, with a concurrent nose dive in tobacco demand, so many Pangasinan farmers planted *kamotes* that by March and April 1921 the province had a marked oversupply, and one could buy a whole cartload for only three or four pesos.[81] In the later 1920s 250,000 pounds a year came from China.

More difficult to track are other roots and tubers such as taro (*gabi*) and *ubi* (*Dioscorea alata*), the latter a violet-colored Asian yam widely grown for subsistence but also routinely sold through the public markets in small quantities.[82] In 1862 and 1872, baskets of *ubi* were recorded as arriving in Manila from Romblon, Batanes, Cebu, and Bohol. In April 1881 the rice dealer Petronila Encarnacion received a shipment from Caoayan in Ilocos. These and other roots and tubers can be preserved by drying thinly sliced cross sections, but there is almost no record of commerce in such products. Rather than being sent in bulk by growers, much of the prewar domestic trade was organized by *viajeros*.

Citrus Fruits

In 1850 Buzeta and Bravo singled out Biñan, Meycauayan, and a few other places as producing citrus, but this was still clearly uncommon. The real rise of commercial citrus growing in the Philippines resulted from a project in northeastern Batangas begun in the 1870s. Oranges had been part of an important transformation in the agricultural economy of Spain in the later decades of the nineteenth century. Along the Mediterranean coast a skillful farming population turned oranges into a major export crop.[83] Following the collapse of wheat cultivation in Batangas—where citrus had scarcely been mentioned by Buzeta and Bravo—the Spanish authorities required landowners there to plant a number of mandarin orange trees annually.[84] The result was that thousands of trees were set out in small orchards.

In a survey taken in 1886–87, Batangas was already producing 5.5 million *naranjas* (sweet oranges) plus 100,000 *kahels*, another citrus variety (often called Seville oranges).[85] By the 1890s, Tanauan municipality was the source of oranges sent to Manila in large quantities. Soon this innovation spilled over into neighboring Santo Tomas. The principal variety was the mandarin orange, known specifically as *sintunis* or generically as *dalandán* in Tagalog. To reach the urban market, pack trains of native horses carried the harvested fruit overland to Calamba for shipment onward by lake and river (figure 5.4). Despite the expense and slow pace of packhorse transport, the mandarins sold well in the Manila market.

When the protracted guerrilla war of resistance against the Americans finally ended, citrus production in Batangas increased significantly. Former revolutionary general Miguel Malvar was an enthusiastic orange grower in Santo Tomas.[86] Increased interest in citrus growing among landowners arose in part because in the wake of war and rinderpest they lacked the work animals to continue sugar production—and there was a worldwide sugar price decline. By 1909 individual orange groves were generally one to two acres in size and growing. Malvar and the governor were convinced that citrus cultivation could be further expanded. The governor, in particular, wanted rail transport extended to inland municipalities in order to facilitate this.

Batangas shipped 10,000 tons of oranges, 8,000 by rail, in the first full year of rail service in 1910. The northeastern part of the province now enjoyed great economies of scale in overland movement. Virtually none of these oranges came from foreign-owned plantations. Modest production expanded in San Jose and Lipa in addition to the core locales. The railroad and subsequent development of roads and truck transport helped get the fruit to Manila in good condition. Yet the requirements for fully successful citrus production are exacting; the mandarin that obtained such "superlative excellence" in Tanauan and Santo Tomas was judged to produce an "utterly inferior fruit" at nearby Calamba to the east, and "still worse on seemingly like soil at Lipa," only 15 kilometers to the southwest.[87]

The burgeoning Batangas citrus industry was devastated by the violent eruption of the Taal volcano in January 1911. The ashes fell mostly eastward onto the trees of Tanauan, causing immediate defoliation. The 1911 crop was down by 90 percent. A year later 50 percent of the citrus trees of Tanauan and Santo Tomas were dead, or nearly so. A report on the subsequent conditions by the government horticulturist P. J. Wester was pessimistic. Former head of a horticultural experiment station in Florida, Wester was horrified to discover that orchardists in Tanauan tended to leave the trees largely to their own devices, writing, "Little or no cultivation appears to be in vogue in a grove of full-grown trees. . . . No pruning worthy of the name is ever practiced and the dead wood in the trees is swarming with borers." Wester predicted that citrus would follow Batangas wheat and coffee into the dustbin of history unless a more active arboriculture, involving weeding, pruning, removal and burning of dead trees, and active fertilization, came into general use.[88]

More gently, other arborists predicted that income per tree could be increased if the trees were pruned to yield higher quality fruit. Instead of pruning, the local practice was to spread the crown of the tree by hanging "stones from the branches" or by placing them "in the crotch between main branches."[89] Citrus production at Tanauan partially recovered and then declined again. For a long time, the orchardists did not develop the quantity or uniform quality of citrus fruit wanted in the urban marketplace, leaving the higher quality segment to producers abroad.

Mandarins, oranges (sweet oranges), and pomelos were the principal citrus fruits reported in city markets and the arboriculture literature.[90] Over time the location of their production became more diverse. In the early 1920s, more than 22 million "mandarin fruits" were reported as having been harvested annually, 12 million in Batangas and 1 to 2 million each in adjacent Tayabas, as well as in Ilocos Norte, Cebu, and Pangasinan. At the same time, about 12 million "oranges" were harvested, with approximately 1 million reported from most of the same provinces plus Iloilo and Albay. The thickly pulped, grapefruitlike pomelo, known locally as the *suha*, was grown on about the same scale as the orange, with an estimated 11 to 14 million fruits harvested, though more came from farther south; the Bikol provinces of Camarines Sur and Albay accounted for half of the *suha* total and Negros a fifth.[91] Several other citruses, some of which apparently had evolved in the Philippines over many years, were also occasionally part of Manila's cuisine, but little is known of these in commerce.

For generations it has been the tiny domestic citrus *kalamansi*—which resembles a green ping pong ball—that is routinely available and affordable. *Kalamansi*, a.k.a. *kalamondin*, a hybrid (Citrus x microcarpa), was more widely grown and less seasonal than mandarins and oranges, available in local markets throughout most of the year. It was widely used to flavor *pansit*, fish, and other dishes and to make a lemonade-type drink that was enjoyed in its own right and sometimes prescribed for the sick—a good source of vitamin C. In 1903 it was said to come from "a small tree common in all gardens." Despite its popularity, it rarely appeared in official reports, but in the research orchard of the Lamao Experiment Station in Bataan, *kalamansi* did better and exhibited more disease resistance than many other citrus varieties on somewhat swampy land.[92]

Vegetables and Tubers from the Uplands

White potatoes are native to the equatorial Andean highlands and have also done famously well in the cool upper midlatitudes around the world, an important instance of the intercontinental movement of biologic materials.[93] Nutritionally, white potatoes compete with rice as a chief carbohydrate and calorie source, but until the recent fast food phenomenon, with its french fried potatoes, urban Filipinos were not much interested in this dietary alternative. Even today, white potatoes are far less widely grown than sweet potatoes.

Fewer than ten significant domestic shipments of white potatoes—called *patatas* or *papas*—arrived in Manila from the outer zone in 1862, mostly from Cebu. A few years later *patatas* were imported from Hong Kong and Amoy (1873), "*patatas de Benguet*" were advertised by a Manila store (1884), and small quantities of potatoes continued to arrive from Malaga, Spain (1889).[94] In the 1890s, Foreman reported walnut-sized potatoes grown in Cebu but also that a potato of excellent flavor and pinkish color was being cultivated on the Cordillera in Benguet: "In Manila, there is a certain demand for this last kind." More emphatically, Sawyer claims that the small potatoes grown on the Cordillera "are much prized in Manila."[95]

According to Martin W. Lewis, white potatoes (and cabbage) had long since been added to dooryard gardens in the uplands.[96] Various locales in the Philippine highlands proved suitable for their production, but these places were fairly isolated. The pioneer German Filipino family of Otto Scheerer grew midlatitude garden plants in what became the Baguio area during the late 1890s. To westerners in Manila hungry for familiar fresh vegetables, the uplands promised a veritable pasture of plenty. American officials, once U.S. rule was established, were keen to encourage upland vegetable culture. The Bureau of Agriculture began setting up a formal seed and plant introduction center in the fertile Trinidad Valley, with plans to grow a seed supply for temperate zone vegetables. In 1904 officials reported that "native planters were growing magnificent cabbages, turnips, tomatoes, potatoes, beans, etc." The bureau chief had "never seen finer English peas and cabbages in any country. . . . [The] Vermont [white] potato yielded at the rate of 100 bushels to the acre without fertilizer or irrigation, and native potatoes, which were planted earlier and suffered less from drought, did better. Pumpkins, carrots, squashes, beets, spinach, parsley, kale, eggplant, beans, radishes, lettuce, cauliflower, and nearly all ordinary vegetables grow to perfection in the dry season with some irrigation."[97]

The greatest boost to the distant vegetable supply of prewar Manila, however, arose from American insistence on building a very expensive mountain road into the uplands of Benguet, thus opening them to commercial garden development. Robert Reed has written tellingly of the fixation of some American members of the Philippine Commission on building a proper "hill station" in the manner of European colonial hill resorts from India and Indonesia to China, while Warwick Anderson has looked deeply at the intellectual relationship between presumed "neurasthenia" in "white" colonialists in the tropics and the value placed on a health respite in the more familiarly cool uplands.[98] In the days before air-conditioning, the main alternatives to the sweltering heat and humidity of the hot season were to seek sea breezes or go up in elevation.

Convinced that such facilities were critical to maintaining the physical vigor and mental health of Euro-Americans in the tropics, early U.S. administrators lavished great sums of public money on building the Kennon Road into the highlands and developing the town of Baguio. The road began not far from the northern terminus of the Manila-Dagupan Railroad. It was built along a poorly chosen route that frequently washed out. Nevertheless, its construction increasingly opened up the southern Cordillera to integration with the Manila-centered commercial system.[99]

Both Japanese and Cantonese laborers and entrepreneurs became specialized upland market gardeners growing numerous midlatitude vegetables. The more accessible and entrepreneurial upland people also came to grow a great abundance of cabbages. Government research stations played a role in fostering such diversification—introducing and testing new vegetables, fruits, and fodders. Some of these actions met with notable success, particularly the early attempts to introduce midlatitude vegetables.[100] Clearly much of the innovation was due to the efforts of individual farmers and to those commercial and transport agents who communicated market information and organized shipping.

Over time an important commercial flow of midlatitude vegetables began, and the number of market gardening communities expanded. Part of this commerce was within Benguet, provisioning workers in mines and lumber mills. Increasingly, however, it responded to the commercial demand from Manila.

Commerce proceeded in both directions. One early Baguio-based dealer was a Hungarian American army veteran, Joe Rice, who sent huge baskets of cabbages down to Manila and brought up rice from the lowlands; he also encouraged a modest commerce in highland strawberries. As the mountain road network was improved in the 1930s, upland native Bado Dangwa began operating a system of buses and trucks. He distributed seeds and was soon hauling great baskets of cabbages. Although truck transport from Baguio direct to Manila was now reasonably priced, at least some of the upland produce continued to move by train during the dry season.

Commercial vegetable culture spread among the indigenous upland Ibaloi, but the major market gardeners and dealers in midlatitude vegetables in the Trinidad Valley were Japanese and Cantonese—many first attracted to the uplands by road construction work. Because of their prominence in this business and the grocery trade in Manila, Baguio became the only city in the country where Cantonese Chinese outnumbered Hokkien.[101] Each group contributed important innovations. The Japanese began using small glass greenhouses to raise seedlings and introduced a variety of white potato developed in Japan. The Chinese introduced Shanghai cabbage, which was subsequently more widely adopted as ideal for rainy season planting, along with several beans and a delicious form of *petchay*.

Both groups quickly adopted the use of raised beds and greatly modified the local red clay soil by hoeing in humus from composted plant residues, ashes from pine needles and dried grass, manures, and commercial fertilizers. One observer reported gardeners using a mix of "six parts of stable manure or compost, three parts of soybean cake and fish meal, and one part of potash and phosphorus fertilizers." The result of this considerable labor was the development of an anthropogenic garden soil that was black and friable.[102]

By the early 1930s cabbages occupied approximately three-fourths of the upland vegetable area. *Petchay*, beans, strawberries, and green onions were next, followed by many others in small quantities. In 1935 Ah Gong's Sons upscale grocery in Quiapo carried fresh carrots and both red and white cabbage in addition to "Baguio potatoes." Apparently all this came from the gardens of Benguet (figure 5.8).[103] In the last phase of World War II, however, exhaustive scrounging by starving Japanese soldiers caused the loss of several well-adapted cultivars, a catastrophe for mountain people.[104] The war also the removed Japanese vegetable growers and dealers, leaving Cantonese as the primary traders organizing the commerce in upland produce.

Gardens and Orchards Overseas

For centuries some of Manila's foodstuffs have come in from overseas, especially from South China. Father Alcina even reports "little oranges" from China

FIGURE 5.8. Ah Gong Sons and Co., a longtime Cantonese grocery, makes a home delivery, 1929. The company's slogan was "Good Food Makes a Happy Family." (H. V. Rohrer, U.S. Bureau of Foreign and Domestic Commerce, USNA II, RG151-FC-84D, box 84)

arriving in Manila around Christmas time in the seventeenth century. But in the nineteenth and twentieth centuries, as domestic production failed to keep pace with both the growing urban population and expanding tastes, a large maritime trade from highly productive areas of coastal South China arose.

Imported Vegetables

Onions were a significant item of import throughout the nineteenth century. Small quantities of both green and dried onions arrived from China and from British "possessions" (presumably Hong Kong) in the 1850s and 1860s. Thereafter, they were folded into the general vegetable category, but it appears imports were still increasing. In the 1880s, upmarket stores in the city advertised Spanish onions "just received and very fresh" and also Bombay onions, known elsewhere as Bermuda onions. The advent of American control brought shipments from North America, and the quantity of onions imported continued to grow. By the 1920s and 1930s, it was on the order of 7 to 14 million kilos per year—several times the level of Philippine production.[105]

Most garlic was also imported. There are very few nineteenth-century records of imports—just a small quantity in 1855. Thereafter, garlic is not reported separately. But Xiamen, the main port in the region of origin of most

Chinese in the Philippines, exported 1.5 million pounds (3.3 million kilos) of garlic to unstated destinations in 1898.[106] In the late 1920s and early 1930s, Philippine garlic imports amounted to 2 to 3 million kilos per year, virtually all from China. Then, part of a general pattern of shrinking imports, only 1.5 million kilos were landed in 1934.

In Manila the wholesale dealers in garlic were now almost entirely Chinese, mostly Cantonese, the same individuals who also handled the wholesale trade in onions.[107] This trade fit well the growing position of Cantonese merchants in the grocery shops of early-twentieth-century and peacetime Manila. Robb reports that one reason the Cantonese grocers did relatively well in Manila was that while they competed with each other in retail trade they cooperated closely in buying and importing these fresh perishables.[108]

In the twentieth century, millions of pounds of white potatoes were also imported annually—circa 22 million pounds in 1927 and 1928—95 percent from Japan. American high-grade potatoes were "sold to the European trade," that is, to foreigners living mostly in the city, and large shipments went to U.S. Navy and Army forces stationed in the country. Potatoes were among the major casualties of the Philippine Chinese boycott of Japanese goods in late 1931—an expression of outrage over the Japanese seizure of Manchuria. In the mid-1930s recorded imports were 12 million pounds—again mainly from Japan.[109]

Perhaps the most significant agricultural import (after rice) in the late nineteenth century, however, was fresh green vegetables. In fact overall vegetable imports were many times larger than imports of fruit. The great majority of this produce came from South China, although some vegetables were also imported from other sources. Under the high tariff protection of the 1890s, there was a brief spate of fruit and vegetable imports from Spain. During the early decades of the twentieth century, vegetables accounted for 1 to 2 percent of all imports by value. Between 1917 and 1929, these were split in origin among Japan, China, and the United States. An American commercial agent maintained that "Filipinos relish cabbage, and when the price is sufficiently low they will purchase it readily." The United States supplied nearly all the cabbages imported in the late 1920s: 1.5 million pounds a year.[110]

The exact origin of the vegetables arriving via Hong Kong remains to be worked out, but likely they were from the Canton Delta rather than what became the Hong Kong "New Territories." Such produce may have been grown in the rich, symbiotic, fishpond-garden production system of Shunde and Nanhai in the northern part of the delta, but on balance this seems unlikely because this system was famously given over to the production of mulberry leaves and exportation of silk fiber through the 1920s.[111] More likely the export vegetables and fruit came from other counties in the delta. The Cantonese grocers of

Manila came from an area that included Xinhui and Kaiping, both among the possible vegetable sources suggested by historian Alfred Lin.[112]

∽

Thus there developed over time a maritime trade in green vegetables, oranges, onions, garlic, and eggs from the alluvial areas of South China, made even more productive by active human soil manipulation. In return substantial tonnages of Philippine sugar were shipped to China after the mid-1880s; in the 1930s, large quantities of Philippine mangos were added to this stream. To make the economics of the trade work, vessels needed paying cargo in both directions. As W. G. Huff said of Singapore's analogous trade, since these ships were already carrying Chinese migrants in both directions, vegetables and other foodstuffs "could be carried in relatively small quantities . . . at little additional cost."[113]

Sanitation and the Vegetable Supply

Whatever the public health concerns of the late Spanish regime—and it made aggressive use of ship quarantines and fumigation—the Americans happened to arrive in the Philippines during the great urban public health debates generated by scientific discoveries of the biological causes of food spoilage and contamination. These concerns were brought to a head by the "muckrakers" and reformers in American literary and political life and were in the minds of colonial bureaucrats, informing the policies pursued in Manila. As with health and sanitation in general, Dr. Victor Heiser was the point person, first as quarantine officer and then as the director of the Bureau of Health from 1905 through 1914. Barely half finished with a medical internship, he had been appointed to the U.S. Marine Hospital Service in 1898; his first duty was to treat soldiers returning from service in tropical Cuba and Puerto Rico. He was soon reassigned to deal with the medical screening of immigrants and was then posted to Naples, Italy, to advise governments and steamship lines on screening emigrants; in 1902 he was sent to the international tropical medicine meetings in Cairo. Heiser was one of the few Americans well prepared to serve the imperial enterprise as chief quarantine officer for the Philippines. Highly qualified and hard-driving but culturally tone-deaf and autocratic, he quickly gained a reputation as a skilled intriguer with a militaristic style. Nevertheless, he had a vigorous professional commitment to ambitious sanitation and inclusive health goals for the entire city and society, not just the resident foreigners, as was the case in some other colonial cities.[114]

The new public health authorities in Manila quickly zeroed in on a contaminated food supply as a prominent cause of diarrhea and cholera and took action.

March 3, 1902, the attention of the Chief Quarantine Officer at Manila [Heiser] was called to the existence of Asiatic cholera in Canton. Five days later came news of it at Hong Kong, from which large quantities of fresh vegetables were constantly being shipped to Manila. In the effort to ward off infection, the port authorities at Manila immediately placed an embargo on low-growing vegetables. This step was necessary because the Chinese were accustomed to sprinkle human excreta in liquid form on growing cabbages, not only for fertilizer, but also for protection against insect pests. If cholera were present in the vicinity, it was always possible that each fresh, crisp, tender leaf would enfold a myriad cholera germs.[115]

The master of the first ship to be turned back unceremoniously dumped his cargo of vegetables into Manila Bay, from whence they washed ashore in the poorest seaside neighborhoods. Despite the embargo, cabbage and lettuce were apparently not in short supply in Binondo, the most Chinese part of the city. This was taken as evidence of smuggling.

The new authorities moved quickly to repair or replace most of the former public markets with airy concrete and steel buildings that could be hosed down and where vegetables and other perishable foods could be inspected, disinfected, and controlled. These were utilitarian structures—nothing like the architecturally distinguished public markets erected by the French in Cambodia.[116] The rebuilt Divisoria Market, as well as the Quinta and Arranque markets, which were spaced in and around the city's central area north of the river, were reopened in late 1901. Each became the center of an everyday consumer activity space critically important to thousands of households. Together they accounted for 87 percent of total market stall collections in the city—more than half from Divisoria alone, since it also had a central wholesale function.[117] A number of small markets served the expanding outer neighborhoods.

A lethal outbreak of cholera in January 1908 led Heiser to temporarily ban the sale of a long list of vegetables, some fruits, and numerous locally prepared foods, as well as all street peddling of food and drink. Individual street fruit vendors had been a common sight; as late as 1906 the Philippine Commission had exempted them from the municipal license tax. But increasingly Heiser saw to it that perishables could be legally retailed only through public markets or groceries.[118]

The early gains in market sanitation were heavily concentrated in Manila, but by the 1930s the major public markets there were said to be unsanitary and mobbed by flies.[119] *The Critic* pointed out to Mayor Juan Posadas that the public markets needed a "thorough overhauling and cleansing" and that petty graft there was well known. The next mayor, Eulogio Rodriguez, announced that cleaning up the public markets was one of his top priorities. The new city health officer, Dr. Mariano C. Icasiano, said the same in 1940.[120] But along with the renewed problems of sanitation inside the public markets, there was now a similar problem outside. Unlicensed but convenient street markets known as *talipapas* had emerged (or reemerged?). There were at least three of these in Tondo and two in Sampaloc in 1939 and some in other districts as well. They were illegal but protected by city authorities and bitterly resented by fee-paying vendors in the regular markets. One reason that some Filipino vendors favored *talipapas* was that many of the public market stalls were already occupied by Chinese.[121]

Still, the great majority of perishable vegetables continued to be sold through the public markets. Heiser liked to believe that this public health initiative pleased nearly everyone. "The city liked it because of the income," he wrote, "the dealers because of the cheap rents, the housewives because of the wide choice . . . and the convenience of being able to buy all their supplies in one place." And the public health authorities liked it because of the enhanced sanitary control. During the worst of the cholera epidemics, the authorities required everyone entering market buildings to disinfect their hands.[122]

Imported Fruit

Most citrus fruits travel well, so when growing urban demand exceeded provincial production it enticed imports from China and elsewhere. Oranges and mandarins were imported from China in the 1850s and 1860s. From 1876 through the end of Spanish reports in 1894, fruits in general were being imported in significant quantities: 200,000 to 500,000 kilos per year. Primarily these were coming from China and, in some years, British "possessions." The American consul reported in 1884 that "we depend here on China for our supply" of quality oranges.[123] In the 1890s, when oranges emerged as an important Spanish export, some of them also found their way to the Philippines.

Oranges continued to be the main imported citrus—from 330,000 fruits in 1913 to 840,000 in 1920—and they still came at first mainly from China. In 1921 the total shot up to 2 million. In the prosperous late 1920s, however, imports of oranges from the United States became substantial—over 7 million individual fruits in more than 40,000 boxes arrived by ship each year. By the mid-1930s the United States was supplying 80 percent of imported oranges by value and was also doing most of the business in apples, lemons, and grapes.[124]

American exporters were advised to send only good quality, reasonably sized oranges that were likely to keep well. Low-grade oranges faced competition from domestic, Chinese, and Japanese oranges. Each September mass arrivals of the new American crop of Valencia oranges led to a temporary glut in the city. Navel oranges arrived on a different schedule.[125] By the 1930s, retail groceries catering to the affluent also tended to stock fresh grapefruits and canned grapefruit juice from Florida.[126]

Why did imported fruit sell so well and at higher prices than local fruit? In 1930 the official *Commerce and Industry Journal* reported, "Our own oranges from Batangas are sold in the local market as fast as they are gathered. . . . After the orange season, one cannot usually find locally produced oranges." At the same time the California oranges were cured and prepared for the market, passing through "an elaborate process of cleaning, washing, sweating, disinfecting, curing, and packing." As a result, the imported California oranges "sell from three to four times as much as the local product," and were available most of the year.[127]

Even before the turn of the twentieth century the import trade in fine fruit came to include midlatitude apples and grapes. Facilitated by the Suez Canal, which materially shortened the trip, rapid steamships, and special tariff protection, fresh fruit from Europe began to appear in a few Manila stores catering to the affluent. "Inexpensive APPLES, fat CHESTNUTS" trumpeted a dealer (in Spanish) just before Christmas in 1880.[128] Apples from northern Spain came to be loosely connected to the Christmas season in Manila. They were sold in stores and by downtown street vendors. The American consul despaired of breaking into this trade, citing the warm climate and costly delays involved in transshipment through Xiamen or Hong Kong.

The situation changed abruptly following the American conquest. By 1902 the American Northwest was sending carefully chosen and packed apples to "Asiatic ports." Between 1912 and 1914, annual imports of fresh apples topped a million kilos, having doubled in a decade. The taste for American apples grew with remarkable speed. From an average of 9,000 bushels during 1909–13, total annual imports of American apples averaged over 141,000 bushels during 1924–25; they had become a common seasonal luxury in the diets of middle-class consumers. In the 1930s, the October fiesta of La Naval in Intramuros routinely involved apples. Clearly, Manila's more comfortable population had developed a taste for apples, which now arrived during a season lasting from August to February.[129]

Meanwhile, grapes from Spain were still retailed in a few fancy stores in the business district and Intramuros in the 1890s (production of grapes in the Philippines itself was minimal). Grapes, raisins, and wines together had become a critically important part of the Spanish home economy; Jaime Vicens Vives claims, "The vine was the catalyst . . . in the revolution of Spanish agrarian techniques in the nineteenth century."[130] The market for fresh grapes expanded rapidly in the 1920s, but these now came primarily from the American West Coast. The supply exceeded 32,000 containers (kegs) by 1928. A luxury product, the grapes were refrigerated in transit and put in cold storage facilities on arrival in the city. In the 1920s all these fruits—oranges, apples, grapes, and others—were brought into the country by "indent" agents. These

FIGURE 5.9. Tempting American grapes and apples for sale in a public market stall, 1920s. (G. C. Howard, U.S. Bureau of Foreign and Domestic Commerce, USNA II, RG151-FC-85B)

concerns "secured orders from Chinese, Filipino, and other wholesalers, who in turn sell to the small dealers, street vendors, etc." In fact, most of the wholesalers were Chinese. The great majority of the fruit was landed at Manila, and some was then distributed to nearby provinces by truck or boat. Final retailing was through upscale stores, special stalls in the largest public markets, and holiday street vendors (figure 5.9).[131]

In addition to fruits imported more or less fresh, a growing demand for canned fruits was carefully nurtured by American exporters and local wholesale distributors. In the late nineteenth century, exporters had experienced some of the difficulties of shipping fresh fruit to the Philippines or South China, but U.S. consuls foresaw a good market for fruit preserved in cans, which would stand up under tropical conditions. By the 1920s, canned pineapple was popular, despite a potentially adequate fresh local supply, but midlatitude peaches were the big hit. Canned goods kept well, were easy to distribute, and were simple to open and prepare.[132]

~

Within the whole array of *ulam*, we see that some of the fruits and vegetables consumed by Manilans were grown or gathered personally, some were bought from itinerant specialized vendors, and some were purchased at one or another of Manila's markets. Some were widely eaten by the masses of people (or their animals), while others were luxuries that only foreigners and affluent Filipinos could afford. These comestibles had been grown near and far, on large orchards or in small plots, in nearby swamps or faraway mountains, and came to market on the backs of humans and animals, in carts, in canoes, in trucks, and on sailboats and steamships, carried by independent *viajeros* or consigned to major dealers. In the multiplicity of sources and routes and networks, we realize something of the complexity of provisioning Manila.

6

Fishing and Aquaculture

FISH FORMED THE PRINCIPAL *ulam* for all but Manila's more affluent residents, and they often ate fish as well. Families who employed a talented cook enjoyed *relleno* on occasion—a medium-sized fish that was cut open along the backside and cleaned. The meat was scraped out, boiled, and mixed with sautéed garlic, onions, tomatoes, and a little chopped pork. This mixture was placed in the skin and sewn into place. The fish was then marinated in *suka* vinegar together with chopped garlic, peppercorns, salt, and sugar. Finally, it was fried and served hot—haute cuisine for sure.[1] This could be a fish captured in the bay or a *bangus* grown to just the right size in a pond. More often *bangus* were cleaned and cut into sections to form the major ingredient in *sinigang*, a popular dinner soup made with green vegetables and tangy-sour tamarind. Freshly cooked, salted and dried, smoked, marinated raw, and even fermented, fish and other aquatic life forms found their way to even the most humble tables.

The economically comfortable and affluent families of the city consumed considerably more meat protein per capita than did the rest of the population. But not all protein in the diet of Manilans came from fish or other animal life, for ordinary grain foods also contain protein. The percentage of protein in the diet obtained from plants is often a measure of poverty, with elevated numbers reflecting a diet lacking in fish and meat. A dietary survey of more than 100 families of "workingmen" in the city's Paco district in the 1930s concluded that on average only 45 percent of their protein intake was derived from animals, while 55 percent came from plants—38 percent from rice alone. In Philippine society this is a reflection of lowly economic status. Even so, urban workers apparently consumed somewhat more animal protein than did their rural cousins and also more fats.[2] Legumes were also a source of protein, especially in the form of soybean curd—whether ordinary *tofu* (*tokwa*) or *taho*, soft and with syrup and sold by ambulant vendors. Among Filipinos, these products had nowhere near the importance they enjoyed in China and Japan or as *tempe* in Java. Soy sauce (*toyò*) was also manufactured in Manila, still on an artisanal basis in the 1910s.[3]

Among the workers of Paco, the survey found that 80 percent of their animal protein intake came from fish and other aquatic products. Meat, especially pork, accounted for 16 percent. Since pork was readily available and attractively priced in the public markets, the (middle-class) authors wondered why fish consumption was so high, writing, "Many of the families claimed that . . . meat was all right once in a while—on Sundays, holidays, and on festive occasions—but for their daily meals they much preferred fish."[4] Fish in all its various forms remained the cheapest and primary source of animal protein for the great majority of urban residents.

As the city population grew and with it the demand for aquatic products, specialized fish-catching and fish-raising communities expanded and also emerged in new places. In some places older methods of capture were used more intensively, especially in established fisheries. New areas were also exploited and new technologies introduced, especially from or via Japan. Sometimes the introduction of more effective capture methods could quickly deplete fishery resources, especially in restricted water bodies. In addition to capture, great quantities of fish were raised in special ponds for the urban market. Not new, this aquaculture system was very greatly extended during our period—an environmental manipulation by human action on an enormous scale. Where before there was a bayside fringe of mangroves backed by brackish swamps with great numbers of nipa palms, now there emerged mile after mile of open fishpond landscapes.

The prominence of fish in the diet and a coastal location—not unrelated—separate colonial cities in the Philippines from many of those in early Spanish America, which tended to be located inland. In eighteenth-century Mexico City the poor ate the tough meat of bulls and oxen; in Manila they ate fish.[5] Long after considerable effort was invested in developing high-yielding varieties of rice, it began to dawn on postwar policy makers that Southeast Asians eat fish and fish products as a critical central part of the diet. True, Southeast Asians since ancient times have used Carl O. Sauer's chicken-pig-dog complex of domesticated animals, but in any reasonable proximity to a coast, swamp, stream, or pond it was fish they ate most frequently.[6] Fish products are at the center of nongrain protein in the diet. But fresh fish are not always available. In societies where refrigeration is a recent and nonuniversal innovation, various methods of preservation have long been employed. These include salting, drying, smoking, pickling, and fermenting. The last yields fish sauce—widely utilized in both subsistence and elegant cuisines all over the region. Despite a broadening of the diet, most Filipinos still receive half or more of their animal protein from fish. At the start of the last century, the figure was even higher.[7] The majority of fish entered the urban food stream via capture, and this was doubly true during the dry season. The rest came from aquaculture.

Capture Fisheries

For a long time there was a strong proximity advantage to supplying fresh fish to urban consumers from places near at hand. In the days when fish were transported in sail craft without ice, it was often better to preserve them first. Fortunately, Manila is well located in relation to aquatic resources. The site of the city is the delta of the Pasig River where it drains the interior watershed and enters the bay bringing a load of sediment. Smaller rivers and creeks enter carrying materials eroded from the volcanic tuff that forms the broad ridge now occupied by Quezon City. Entering the bay, the sediments were continuously reworked, especially by typhoons throwing up surge tides and great waves. The lowlands of Manila can be appreciated as a series of roughly parallel former beach ridges with gentle swales between, each the site of a sluggish tidal creek, or *estero*. It is this part of the city, with its hundreds of thousands of residents, that may be lost to the rising sea levels brought on by global warming.

Before the massive pollution and construction of recent generations, the creeks, wetlands, river mouths, and bay in and near Manila were rich in aquatic life—a food source available for the collecting. This was also true of the brackish intracoastal waterways and lagoons just north and south of the city. As late as the 1910s large quantities of small crabs were caught in *esteros* during the rainy season and immediately distributed through the public markets. Shrimp and small crabs were caught with a *sakag*, a small "scissors" net on two crossed poles, which was pushed by a person wading in water up to his or her chest.

Many small fish were caught in these places by means of a large square net lowered into the water by a great boom constructed of bamboo and mounted on a long raft. The net and catch were raised by the weight of men climbing a counterbalance pole. The entire apparatus is known as a *salambao* or *salambaw*. An indigenous technology already in use in the sixteenth century, it was widespread in the Manila area in the nineteenth and early twentieth centuries (figure 6.1). *Salambao* were especially effective in catching river mullets, known collectively as *banak*. As long as there were fish to catch, *salambao* operated in the Pasig River, at Malabon, and in the estuaries and shallows of the bay.[8]

There were also many smaller sorts of fishing gear in wide use, including hook and line. With the lowest cost threshold, this simple technology was and is used by a great many people. In some places where the conditions are right the total catch can be astounding, especially when done at night assisted by

FIGURE 6.1. Commercial fishing with a *salambao* on the Pasig River, 1880s. Later the river became too polluted to support fish. (*La Opinion*, Suplemento Ilustrado, April 9, 1888, 8)

lanterns to attract the fish.[9] Besides fish, squid, shrimp, crabs, and mussels were and are widely consumed. One "traditional" use of shellfish in nearby coastal neighborhoods was in a medicinal soup "made from clams boiled in water and ginger" to stimulate lactation in new mothers. Mussels were particularly useful when rainy season storms prevented fishing at sea.[10]

A sickening cargo of human and chemical effluents has left the Pasig River biologically dead. Even in the mid-nineteenth century, when fishing was common, there were occasional events that asphyxiated river fish. By the 1920s, "periodic occurrence" of a phenomenon called *masamang tubig* (bad water) in Laguna de Bay was responsible "for the wholesale death of fish and other animals," including "hardy snails and mollusks" in the affected localities. This mass of polluted water originated in Manila Bay and was made worse by algae decay in the Pasig River. It was pushed through into the lake in the low-water period of the dry season. Local fishermen in the lake eventually learned to set their nets to catch the schools of fish that fled in front of it. Pollution, swamp filling, and construction have steadily reduced opportunities for local capture fishing.[11]

Baklad/Fish Corrals

Offshore, the bay was a rich resource. Writing in 1609, Antonio de Morga mentions the use of corrals for trapping fish. These structures, called *baklad* in

Tagalog, are designed to guide swimming fish into small enclosures. Manila Bay has historically been one of the principal locations in insular Southeast Asia employing this indigenous technology.[12] With variations in design depending on the current, depth of water, and habits of the fish to be caught, *baklad* often feature a row of posts joined by means of vine lashings and set at right angles to the shoreline—a sort of arrow pointing out to sea. This leader ends in a head or heads—often a series of two or three nested heart-shaped structures that confine the bewildered fish in ever-smaller enclosures from which they are removed with nets. In the more elaborate of the shallow-water designs, the arrowhead is so broad that the entire structure resembles an anchor or eagle in shape. There were also narrow V-shaped designs set in rivers and waterways where receding tides in the bay produced a strong current.

One can infer that fish corrals were already well developed in the Manila area in the early 1860s because in one year nearly half a million pieces of *diliman* (*jagnaya*), a fern used for tying parts of the *baklad* together, were landed in the city from the outer zone.[13] By the 1880s and 1890s constructing *baklad* had become a major occupation in half the neighborhoods of nearby Navotas—one of the early specialized fishing communities—and many of the men registered in Tondo as "fishermen" were actually trap owners and workers.[14] The construction and operation of a *baklad* were accompanied by an accretion of supernatural beliefs and practices. In Negros, for example, a floating offering was typically made, which often included a chicken and other consumables. Associated with local parishes, there were also enthusiastic "fluvial processions" of religious images.

The deployment of *baklad* around Manila Bay becomes evident during the dry season of 1882 when more than 430 were registered.[15] Two-thirds were located in the shallows at a depth of one-half to one or even two meters but primarily at wading depth. These sorts of corrals were well designed to catch fish that travel in schools in reasonably shallow waters: sardines, herrings, small mackerels, and anchovies. The greatest concentrations of shallow-water corrals (193) were located along the gentle Tondo-Navotas-Obando shore stretching north of the city center (map 6.1). Along the Bulacan shore, the wide-wing eagle (*aguila*) design was common. The inshore devices were the least expensive to build and the easiest to remove or replace when threatened or damaged by typhoons. They were charged only a nominal tax.

At the opposite extreme, some 52 *baklad* were maintained in water deeper than four meters—ranging up to a maximum of seven or eight. These deepwater traps represented considerable capital investment since they required special water-resistant *palma brava* posts, as well as a lot of labor to construct. They also caught bigger fish—large migratory fish such as bonito, yellowfin tuna, albacore, large mackerel, and scad, including pompano. Judging from

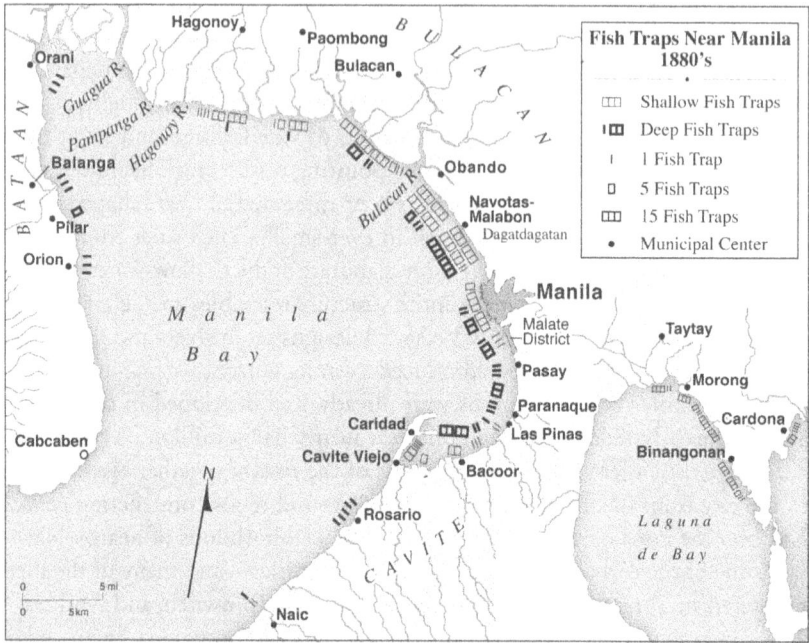

Map 6.1. Fish Traps in Manila Bay, 1880s

their tax ratings, these were the most productive corrals.[16] Half the deep-water devices were located in waters north of the city from Tondo to Bulacan; others were found off the Cavite coast, and along the bayside of Bataan (figure 6.2). There were also 140 registered *baklad* along the shore of the inland Laguna de Bay, especially at Binangonan (67).[17]

As demand from the city population expanded during the 1880s and early 1890s, there was a considerable increase in the number of shallow-water corrals along the Bataan shore across the bay (65). The number in the shallow embayment inside the arc of the Cavite Peninsula also went up sharply. At Caridad on the Cavite Peninsula the shore was now lined with corrals, and some 300 *bancas* were kept busy shuttling the catch to Manila.[18] Many *baklad* were set near the places where rivers and creeks disgorged nutrients and plankton and where fish on their life-cycle migration downriver and into the bay could easily be caught. Because these rivers brought fresh water, the eastern and northern parts of the bay regularly recorded the lowest salinity.[19] Beyond Manila Bay such corrals were also common along the coasts of Batangas and elsewhere by the early twentieth century, if not before.

In the twentieth century the migration of Tagalog fishermen from Manila Bay brought the *baklad* to new locales—from Naic, Cavite, and the Rizal

FIGURE 6.2. A deep-water fish corral in Manila Bay, March 1937. Oblique aerial view off Rosario, Cavite. (USNA II, RG18-AA-185-2)

Province towns to the Ragay Gulf in southern Luzon, for example, and to the expanding commercial fishing center at Catbalogan on the west coast of Samar. Much of the product of these capture devices was delivered for consumption in Manila. Thanks to the investigations carried out by Agustin F. Umali and others, we know a great deal about *baklad* operation in these places.[20] Local topography could easily drive the calendar of *baklad* use, as in Laguna de Bay near the city. In the southwestern reaches of the lake, deep-water corrals were installed and used only during the rainy season of the southwest monsoon. Elsewhere in the lake *baklad* could be left in place all year.

There were more than 200 *baklad* entrepreneurs operating in Manila Bay in 1882. Who were these persons? One was local native Lucio Buzon, whose corral was one of the common half-meter types. It was located within sight of the Tondo church just off the Bankusay shore where he made his home. In 1889 Lucio Buzon was an assistant to the head of a mestizo *cabeceria*, a unit for collecting the capitation tax and monitoring the population. In 1891 he became the *cabeza* himself. It is possible that Buzon took these positions at the bottom of the state administrative hierarchy in order to gain the opportunity to acquire a license to operate a fish corral in deeper water. In any case, by 1892 he had given up the half-meter device for a much more expensive three-meter *baklad*. In each of these years he registered himself as a *pescador*, or "fisherman," but his daughters remember him more specifically as a fish corral owner—a *propetaryo*. The family kept a big *banca* for bringing in the catch. When a typhoon threatened, they called out divers, who dismantled the posts and lashings of the corral and brought them ashore. Of 434 corrals in our Manila Bay database for 1882, only 1 was registered by a Chinese.[21]

At the top of the entrepreneurial spectrum in the *baklad* business was Julian Andres of Navotas. We encountered Andres as the owner of numerous *cascos* used in the rice trade and a substantial storage structure along the intracoastal waterway—this in a town that was central to the daily provisionment of rice to the city. Now we can see that he was also the single largest investor in *baklad* capture fishing in Manila Bay and also a major timber contractor licensed to cut almost 200 beams a year in the forests of Bataan.[22] In the first half of 1882, Julian Andres registered a total of eleven fish corrals divided among Navotas and Bulacan municipalities and Bataan Province. Seven of these were expensive deep-water corrals. Pedro Naval was another large-scale Navotas *casco* owner. In the 1880s he owned two deep-water *baklad*. From 1881 through at least 1892, he served as the *cabeza* of a mestizo *cabeceria*. Other Navals and Andreses of Navotas also owned *baklad* in the 1880s and 1890s, and several served as *cabezas*—presumably two extended family networks. Among those able to organize the resources required to build the deepest devices, most owned only one. Julian Andres and the Navals, it seems, shared a special entrepreneurial vision, and both were based in Navotas, the most concentrated locale of *baklad* ownership.[23]

The continued growth of the urban demand for fish resulted in the ongoing development of this resource and a consequent westward shift in its center of gravity. Whereas in the 1880s–90s there were 400 to 500 corrals in Manila Bay, by 1938 there were 2,900. These comprised three-fourths of the national total and were in the hands of 536 operators. Whereas in the 1880s the corrals were most densely located along the shore from Manila to Bulacan, in the 1930s Pampanga and Bulacan provinces each had 1,200 while Bataan recorded 350. No other province came close to these numbers. Scores of V-shaped corrals crowded the entrance channel of the Bulacan River off Obando and other nearby locales (figure 6.3).[24] Intensively exploited by *baklad* and net, the near shore waters now produced tens of thousands of boxes of shrimp, crab, grouper, and mackerel, as well as great quantities of other fish. After 1901 the politics surrounding the right to operate licensed traps in foreshore waters devolved from the central government and the Port of Manila to the local municipalities.

FIGURE 6.3. A commercial fishing village hugs the shoreline near Malabon, 1926. There are multiple fish corrals and *salambao* in the river and fishponds on the land. (USNA II, RG18-AA-box 185)

The competition for this limited "public" good could be intense.[25] Meanwhile, fishing at sea was an open-access activity.

New Methods

Several new methods of exploiting the fisheries resource were introduced in the twentieth century. Although the use of explosives in Philippine fisheries increased as a result of mining armaments left over from World War II, it began long before. In this low-cost "method," everything in the water within a certain compass was killed. As the dead fish floated up, what was wanted was scooped aboard. This method came into occasional use in Manila Bay in the early 1900s; it was banned by 1906 but not ended. Along with human and chemical effluent, this had the potential to be a major cause of decline in the fishery resource since it killed even the youngest fish. Using explosives primarily appealed to small-scale fishermen.[26]

The *sapyaw* (*sapiao*) was a Tagalog innovation around the end of the nineteenth century. In the 1930s this was a long, round haul seine deployed on dark nights by men using lights and working from two long *bancas*. In the 1930s this method of luring schools of fish into a large seine was put to use in the bay off Parañaque and Las Piñas just south of the city. The system yielded a seasonal catch of herrings, anchovies, and mackerels and quickly replaced most shallow-water corrals in the two municipalities. For a time, it provided an abundance of fish for 30 new fish-drying operations.[27] In the 1920s and 1930s another of the *sapyaw* localities was in the bay off Pilar, Bataan. One family fishing there "would bring the catch to Bankusay in Tondo," where a friend took charge of selling. Unfortunately, fishing around Pilar "did not produce a good income."[28] Also in the 1930s, there were some 80 large gill net rigs, or *bating*, in use in the bay. The nets were used to catch small herrings. Such nets could range up to a kilometer in length when fully deployed, although most were half that. Towed into place by a launch, working these rigs required 20 to 30 men.[29]

Finally, the single major source of captured fish delivered to Manila in the late 1920s through the end of our period came from the use of powerboats and nets. Japanese fishing entrepreneurs introduced bottom-fishing nets to Manila Bay and Southeast Asia more generally around 1920. These were referred to as "beam trawl" rigs because a wooden beam held open the mouth of the long triangular net sock. Lengths of iron chain and stone weights kept the mouth of the net on the bottom, so the rig was only suitable for operation on a bottom that was smooth and flat. Manila Bay was perfect. The beam trawls replaced the much older *palakaya* broad-net system, which had been operating out of Tondo and Malabon-Navotas since the late nineteenth century. By 1928 sail-driven boats had largely given way to gasoline motor trawlers whose operations were bringing in tons of inexpensive fish. The beam trawlers pioneered an effective bottom catch of fish and also supplied Manila with most of its shrimp.[30]

However efficient, investment in this relatively expensive equipment was also motivated by a readily accessible market. Again Manila Bay was perfect. In the early 1930s, 30 beam trawlers were registered there, and 18 based in Dagupan took up operation in Lingayen Gulf. In both places there was initially a terrific negative reaction from fish corral owners, but in both cases local outrage dissipated. It was difficult to compel the government to enforce no-trawling zones. Also connected to the Manila market, other trawlers began operating further afield. In Ragay Gulf to the south, four vessels were each making three trawls a day, returning to their base early in the morning in time to ice the catch for shipment to Manila by railroad—now that it had been extended that far south. In the later 1930s, the catch in Ragay Gulf was said to include larger fish and produce a larger catch per trawl compared to Manila Bay where the fishing pressure was greater. Likewise, three Japanese trawlers began operating in the more restricted waters of San Miguel Bay on the Pacific coast of Bikol in 1936. In this case the catch was sorted, sold to Filipino middlemen, and transported to Naga by truck. The larger fish were then iced in boxes and sent on to the Manila market on the twice-a-day trains. San Miguel Bay was an outstanding place to fish during the southwest monsoon because of its northern exposure and the rich plankton entering from two rivers.[31]

Built in Manila using Japanese designs, the trawlers were operated by trained Japanese crews. Almost all were owned by foreign nationals. However, Faustino

Lichauco, well known in the commerce of feeding the city, was in the forefront of this development as well. His cattle-importing business was concluding, and returning from a world tour in 1929 he stopped in Japan to look into the technology of this new food-related business. Back in Manila, Faustino and his son Tomas immediately turned to trawler construction and operation. In 1932, after Faustino's death, Tomas Lichauco owned 2 of the 70 trawlers registered in the country. In that sense, capital accumulated in supplying beef was converted into capital investment underwriting part of the urban fish supply. Living in Baclaran near his boats, Tomas Lichauco conducted a daily sale of the catch from the open lower floor (*silong*) of his family home.[32] Despite growing concerns about overfishing in Manila Bay, beam trawling continued. The city was growing, and so was its demand for fish, and capture fisheries provided the main dry season supply.

Although the Japanese vessels and crews became famous for their innovations in the later 1920s, they were already helping to supply Manila with fish from 1900 onward.[33] Their arrival had a strong commercial motivation and a military intelligence component. The first is believed to have been represented by Yosobei Yamane of Hiroshima, who arrived in Manila Bay with his boat in 1900. Hayase Shinzo presents evidence suggesting that Yosobei was influenced by persons connected to the Japanese military. Initially, these sailing vessels came from the Inland Sea of Japan, sometimes going back and forth when the southwest monsoon curtailed fishing in the Philippines. Based in Tondo, the Japanese fishing fleet was increasingly built locally. Starting in 1906, one or more Japanese-designed vessels was constructed in the bay area each year, four in 1911. In the early years, changing interpretations of customs regulations made it financially important that foreign fishing vessels be registered to Filipinos or resident Americans.[34] Later most were registered to their Japanese owners. The biggest owners had three or four vessels, but most had only one.

At one point, when 23 boats can be traced, 20 were owned by persons from small islands in Hiroshima Prefecture (Momoshima and Tashima) and 3 by persons from Okayama, located midway between Hiroshima and Osaka. The fishermen came from the same places. Hayase has managed to trace many of these men. Ordinary fishermen and farmers from Okayama Prefecture were important in the early flow. Famous innovators in fish-netting techniques, Okayama fishermen were suddenly disadvantaged by the 1902 Meiji Fisheries Law, which restricted their right to fish in the waters of other prefectures. Instead the Meiji government promoted fishing in foreign waters. According to Hayase, this propelled the Okayama men to fish the Korean Straits, near Taiwan, and in Singaporean waters. The start of Japanese fishing in Manila Bay was a further step. Later the Okayama men increasingly found other opportunities.

Some fishermen can be traced in multiple years going to Manila during the dry season and then returning to Japan. These early Japanese hardly became rich, but they did make approximately double the wages they could have expected as day laborers at home. They used lacquer-coated *utase ami* fishnets from Hiroshima and believed that, although these had a short useful life in tropical waters, they were more flexible than nets in general use in the Philippines and thus were critical to larger hauls.[35]

A late Japanese innovation in capture-fishing technology was the system known as *muro-ami*. It involved a motorized main vessel together with several *bancas* and a large net of cotton twine. The net was shaped like a giant bag with two wings at the mouth. Like the beam trawl, it was designed to capture bottom-dwelling fish, but in this case the target was fish on coral reefs and rocky shoals. While the net wings were pulled open by the *bancas*, swimmers fully deployed the net and swam out to form an arc in front of its mouth. Equipped with goggles and a weighted line, the swimmers advanced toward the opening while jerking the lines, creating a commotion and scaring fish into the bag. This was a very successful capture system, which (perhaps unfortunately) is still in use. As with other capture systems, the availability of ice for preserving the catch and quick access to Manila or smaller market centers were crucial.[36]

Depletion

Sometimes new and more effective methods of capture can deplete fishery resources in short order—especially when the resource territory is restricted, as in a lake or bay. This happened in Laguna de Bay in the early 1930s. This large but shallow lake had long been an important source of fish for the city. In the twentieth century, several varieties of *kanduli*, or catfish, were the main commercial type. Whether delivered fresh or as *daing*, flayed and dried, this was a resource of great livelihood importance to the growing number of fishing families living around the lakeshore. As late as 1928, the value of the annual *kanduli* catch was estimated at a million pesos. Over the next five or six years this value declined by more than half. There were multiple causes, including overfishing by simple hand methods for small fish and other aquatic life for use as feed for Laguna's chickens, hogs, and ducks, and there were occasional outbursts of "bad water" from the Pasig River. But the principal cause was the innovation of enormous drag seines, or *pukot*, operated from motorized boats. The seines were deployed in a circle, which was then steadily reduced in circumference, thereby concentrating and trapping the fish. Because the adult catfish tended to be concentrated and largely immobile on the lake bottom during the portion of the reproductive cycle when the male is incubating and brooding the eggs and larvae in its mouth and because much of the bottom is

smooth and not very deep, they were an easy catch. The resulting high inci-
dental capture rate of eggs and larvae, as well as overfishing the adults, naturally interfered with *kanduli* reproduction. Its numbers and relative prevalence
plummeted.[37]

By the 1930s, a similar decline was readily observable in several other important lakes. In these cases it was usually because local authorities, zealous in their
desire for enhanced revenues, sold licenses to place a *baklad* across the streams
that connected the lake to the sea. These caught many of the fish that migrate
up or down the river and were veritable gold mines in the early years of their
deployment. One example of such migrant fish is the *Chanos chanos,* which
hatches at sea, seeks the shallows and fresh-water inlets as a hatchling, migrates
to fresh inland waters if it can find them, grows for several years, and when
sexually mature seeks to return to the sea for spawning. Called *bangus* in its
smaller sizes, particularly when raised in fishponds, the same fish grown quite
large is called *lumulukso.* Similar migrational behavior also characterized some
species of pompano and mullet. As a result of interference with these migration
streams, the catch declined. On the Pansipit River connecting Lake Taal to
Balayan Bay in Batangas, the commercially important catch of these large fish
declined from 27,000 per year in the late 1880s to 8,500 in the mid-1930s. Something similar happened to the mullet fishery below Lake Naujan in Mindoro.[38]

By the 1930s, a number of fairly distant provincial centers had emerged as
ports that specialized in provisioning Manila with fish. In the cases of the
Ragay Gulf and San Miguel Bay in southern Luzon, the fish were iced in boxes
and sent on by train. From Catbalogan, Samar, and Estancia, Iloilo, fish were
sent by boat. Before ice became available at the point of shipment, both had
difficulty with adequate preservation. This had to do with the relatively low
quality of domestic salt.[39]

AQUACULTURE

In addition to fish brought to the city through capture methods, there were also
fish raised to marketable size from wild fry seeded into special ponds. This was
an elaboration of another long-standing indigenous technology. The fish in
this case began life in the wild but were deliberately raised and harvested for the
urban market in almost the same sense that Manilans might purchase small
chicks to raise for commercial purposes. In that sense, these fish were "livestock." The primary species raised in the ponds until well after World War II
is known locally as *bangus* (*Chanos chanos,* figure 6.4). To Indonesians they are
bandeng, and some English speakers call them milkfish. In the 1920s, and no
doubt for many decades prior, this was "by far the leading fish in Manila markets . . . the daily staple animal diet of tens of thousands of Manilans, and . . .
the only cheap fish available" during the typhoon season. Likewise, the dietary

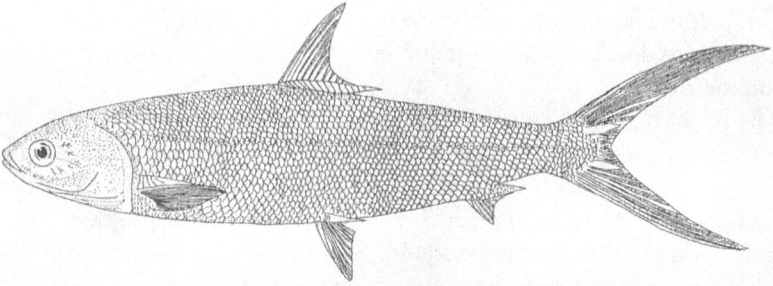

FIGURE 6.4. *Bangus*, the mainstay of brackish pond aquaculture and rainy season market supply. (Ling Shao-Wen, *Aquaculture in Southeast Asia: A Historical Overview* [Seattle: 1977], 44; drawing reproduced by permission of the Washington Sea Grant Program, University of Washington)

survey of more than 100 urban working families during 1936–37 found *bangus* and shrimp to be the most commonly consumed seafood items.[40]

In commercial operations, the *bangus* fry were caught in the wild with fine nets along the shallow margins of the estuaries and bay. April through June was and is the heart of the fry season, with those collected in April considered the best. As demand grew, and the human pollution of spawning waters also increased, the fry were increasingly caught farther afield along west coast shores and brought to Manila in pottery jars. In the city most *bangus* fry were sold near the Tutuban railway station or at the Yangco landing along the Pasig River. In the 1890s, the shallows of Balayan Bay off Lemery in Batangas were a major source of *bangus* fry sent to Manila and Malabon. In the 1930s, Balayan Bay as a whole and the adjacent Batangas Bay provided the greatest quantity of fry coming to the city. Another important supply area was along the western coast of Batangas from Nasugbu to Calatagan. From the outer zone, operators on the Ilocos Coast also sent a substantial number of fry each year, as did several towns in Iloilo and other areas farther south.[41]

Rather than seeding ponds directly with the tiny fry, one could begin with commercially raised three-inch fingerlings (*hatiran*)—an option with higher initial costs but lower subsequent fish mortality rates. Another advantage of stocking with fingerlings is that skillful operators could control their size by manipulating their food supply and thereby keep them available even out of season. This allowed some managers to raise two or even multiple *bangus* crops in a single year. Many fish farmers chose to grow their own fry to fingerling size. By at least the early twentieth century, Malabon had become a major place where nursery ponds were raising fingerlings on a commercial basis.

As with capture fisheries, until there were adequate supplies of ice and bulk transport that was faster than coastal sail craft, there was a substantial competitive advantage to being located in close proximity to the final consumers. *Bangus* fishponds were excavated in low-lying alluvial wetlands. The immediate seaside environs of the city were well suited for this. The sites chosen included mangrove stands—especially those already heavily despoiled for the urban firewood market. In the 1870s, fishponds stretched toward the north in Tondo along the Canal de la Reina. There were many more ponds farther north in the Gagalangin neighborhood and on Balut Island. At the same time, new ponds were being constructed southward in Malate and also in Parañaque and Las Piñas. In the last "there were continual quarrels between the parish and the townspeople regarding the ownership of fishponds."[42] North of Gagalangin, ponds stretched beyond the city along the intracoastal waterway. Hundreds of small enclosures were constructed in the great marsh called Dagatdagatan (map 1.2). By the 1870s, the same was true in Obando and adjacent Bulacan municipality. By the 1920s and 1930s, commercial *bangus* ponds spilled all the way across the great marshes of the Bulacan bay shore and into lower Pampanga. Many Pampangan ponds were owned by the well-to-do and worked on shares by rural folk. Eventually the entire zone from Malabon northwest through southern Bulacan and Pampanga underwent intense aquacultural development.[43]

In addition to the development of more or less regularly surveyed and titled swamplands, ponds were also built in the foreshore shallows by outlining an area with mangroves planted in lines and then waiting for them to grow and trap sediment. Later the mangroves were cut and the dikes filled in. Another method sometimes used by the locally powerful was to appropriate sections of the network of public watercourses by enclosing them for use as fishponds. These misappropriations were under way by the 1890s and were fiercely defended in court and sometimes with the use of strong-arm methods.[44] Some of these structures impeded storm drainage—increasing the likelihood that a storm would cause a flood that would carry away the young *bangus*.

In 1911 there were still more than 18,000 hectares of nipa palm forest along the northern shore of Manila Bay. By 1929 only portions of these nipa stands remained. There were now 18,000 hectares devoted to *bangus* ponds in Bulacan Province and a further 11,600 in Pampanga. The aquaculture production zone now also included the entire northeastern coast of Bataan and a small fringe south of Manila in Las Piñas, Parañaque, and Cavite. At the same time some of the early ponds on the urban margins were filled in, just as the Dagatdagatan was lost a generation ago. Several wealthy Manila families now had substantial holdings in commercial aquaculture as part of their investment portfolios. Some of the biggest were converted from nipa palm tracts formerly used for the production of alcohol such as the Carlos Palanca fishpond estates in Hagonoy and Masantol and the Ayala ponds in Macabebe. Leasing fishponds from absentee owners was also a fairly common practice in Bulacan.[45]

The average size of individual fishpond holdings tended to vary from the southeast near the city to the northwest—farther away and developed later. In Cavite, a minor part of the fishpond zone, operators controlled on average only 1.4 hectares of pond surface. Elsewhere around Manila Bay, the average pond holdings per operator were mostly from 12 to 16 hectares, reaching a high of 18.5 in Pampanga. The averages obscure the really large holdings. Growing fish for urban consumers was good business and nearly recession proof. In 1938 almost 11 million kilos of *bangus* were harvested from the ponds around the bay—75 percent of the national total plus another 6 percent from ponds in Pangasinan. Judging by the reported age of fishponds in 1938, the number of hectares added to the total set a record in every five-year period from 1910 onward—from 3,500 during 1910–14 to 10,200 during 1930–34 and even more during the following three years. Further, almost all the fishpond assets were owned by Filipinos. The result of this remarkable expansion is captured on map 6.2, which depicts one of the more concentrated areas of aquaculture in the world.[46]

The creation of this fish-raising system represents environmental manipulation on an enormous scale. Where before there were swamps with millions of nipa palms and a bayside band of mangroves, now there was an open watery landscape stretching mile after mile (figure 6.3). Increasingly, palm wine tappers and the legion of village women who made nipa palm shingles were replaced with smaller numbers of fishpond managers and workers. In Barangay Santa Cruz of Paombong, for example, 20 of 23 small settlements were said to have been depopulated by the conversion from nipa and mangrove exploitation to fishpond operation.[47] In the conversion process workers cut the vegetation and excavated the ponds, building up the edges. The raised land thus created between the ponds was not used for intensive horticulture nor was the soil there continually enriched with material scraped from the pond bottom as in the Canton Delta in China.

The algae mat on the bottom of the pond is critical to *bangus* nutrition. The growth of the benthic blue-green algae mat can be stimulated by adding chicken manure or fertilizers to the bottom when the pond is empty, but this is a very different system from one that starts with Chinese grass carp fed every day with green leafy materials.[48] In order to reduce snail competition for the nutrient mat, pond managers hired poor rural workers to pick thousands of small snails off the bottom during the dry season when the pond was not in use. By the 1970s, this stoop labor had been largely replaced by liquid poisons, which killed the snails—and quite a lot else in the downstream creek into which the pond drained.

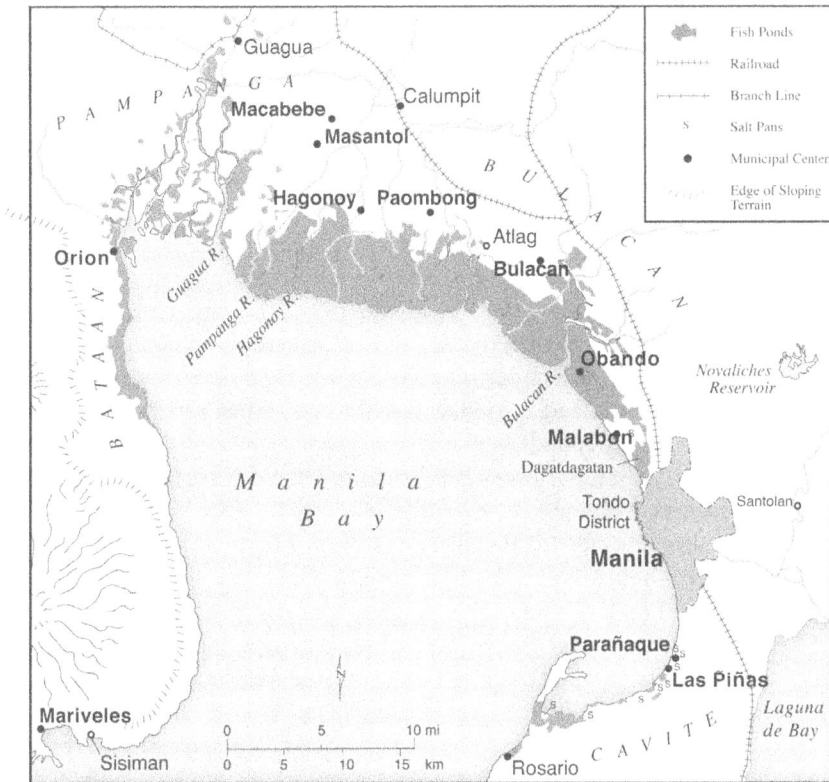

MAP 6.2. The Fishpond Production Zone around Manila Bay in the 1930s

At the end of the nineteenth century this form of controlled production for the Manila market was spreading beyond the environs of the bay to the coast of Tayabas and elsewhere. Not only increasing market demand but impoverishment brought on by the suppression the artisanal distilleries used in the nipa wine trade gave a mean impetus to the conversion of nipa swamps into fishponds.[49]

A critical feature of the *bangus* aquaculture system was its ability to supply large quantities of fish during the rainy season when typhoons and rough waters made it less rewarding and dangerous to fish at sea in small boats. In part this timing was also due to the excellent growth conditions for the algae food mat during the rainy season, which, in turn, led the cultured fish to gain weight rapidly. In this season, the *bangus* reaching the market were only a few months old. The fish could readily be transported to market across Manila Bay in excellent condition. In the ponds around the bay harvesting took place in the evening or at night. Many baskets of fresh *bangus* changed hands in late-night open-air markets along the lower river courses and were delivered to Manila by boat before dawn.[50] Fish coming from the coast of Bataan on the far side of the bay were packed in ice and shipped by small steamer. The combination of local skill and quick market access by boat made the fishponds in Malabon, Navotas, Obando, Bulacan, and Hagonoy the most valuable per hectare in the country.

A perennial problem with the pond system of *bangus* aquaculture was and is the difficulty of maintaining just the right salinity during rains or severe droughts such as those caused by some El Niños. Such drought years included 1874, 1885, 1892, and 1903 when the lack of fresh water flow allowed saltwater to penetrate up the river courses into the fishponds and even into some of the rice lands farther inland as in Hagonoy. At other times, typhoon torrents flooded whole bands of ponds.[51]

With the advent of the railroad and eventually trucks, the product of ponds at greater distances became more competitive in city markets. Truck transport provided greater flexibility in the delivery of fresh *bangus* to Manila and increasingly throughout Central Luzon, a portent of postwar developments. In the early 1930s, the Manila Railroad Company tried competing against trucks by converting a number of freight cars into special cars with louvered sides for the transport of fish and poultry. Having been developed to feed Manila, the same core of fishponds were now also shipping fresh fish daily to interior locales. Well before World War II, areas of expanded commercial *bangus* production had emerged along Lingayen Gulf in Pangasinan, around Iloilo City, and in Capiz Province on Panay.

Brackish water *bangus* long formed the main focus of aquaculture. There were also unsuccessful attempts to culture the native *dalag*, or "mudfish." Still, it was an important urban food source. *Dalag* are air breathers—they must surface regularly to breathe and can be transported some distance and sold alive. In the nineteenth century, *dalag* were usually cleaned and dried before being transported. In the twentieth century, they were often delivered alive in water to the markets of Manila. Indeed, *dalag* have better flavor when killed just prior to cooking. In the 1930s, this was one of the three top fish species consumed by families of workingmen in the city. Many of these were caught by net in the shallow margins of the Laguna de Bay. Greater concentrations, however, were found on the muddy bottom along the eastern shore of the lake. Here considerable numbers of *dalag* were collected with bare hands when they were left stranded toward the end of the dry season.[52]

The Philippines lacked an indigenous fresh-water species suitable for aquaculture. In the 1930s, there was a brief burst of interest in the herbivorous giant *goramy* (gourami) imported from Java. A delicious high-quality fish, it lent itself both to fresh-water aquaculture and to integration with gardening since the adult ate "leaves and tender parts of [*kamotes*], several kinds of land grass, banana, yam, . . . [cassava,] kitchen waste, or a mixture of cooked rice, rice bran, and trash fish."[53] But the problems of adapting the giant *goramy* to local culture included slow growth and a long time horizon—it takes three or four years to reach sexual maturity. The result was that the Philippines remained well behind several other Southeast Asian countries in fresh-water fish production. In the early 1950s, the introduction of the hybrid African tilapia solved this problem.[54]

The cultivation of mollusks is another long-standing form of aquaculture. Oysters were cultured and commercially grown in the major streams of Navotas and nearby Obando at the end of the nineteenth century. They were also raised at Paombong, Cavite, and other coastal municipalities. In all these places the method of spreading old oyster shells on the bottom was used—in hopes that the young oyster "spats" would adhere to them and grow. Increasingly this method was augmented and replaced by the use of large numbers of bamboo stakes driven into the bottom in intertidal areas. Free-floating oyster young tended to attach themselves and grow on these. Held off the bottom on the stakes and submerged racks, the oysters were less likely to be suffocated by high levels of silt or carried away in the waves. Photographs show this method in use at Obando and elsewhere in the 1910s.[55]

Due to the pattern of water circulation in Manila Bay, the Bay of Bacoor is the least salty of any waters along the Cavite coast and a natural oyster bed. In the 1930s, the shallowest cove of Bacoor Bay (Binakayan) was chosen as the site of the government's oyster demonstration farm. Here new and more productive methods of oyster culture were developed, including the practice of using strings of oyster shells hanging on wires to collect the spats. The collector shells were then attached to stakes and set out on the tidal flats. Oyster culture became a substantial success in this environment.[56]

Despite the venerable history of most of these means of protein production, the rise of a concentrated urban market led to a growing role for pond-raised fish. Overfishing and pollution in Philippine waters, and indeed worldwide, have also led to increased reliance on aquaculture. On a national scale, products of aquaculture made up about 10 percent of total fish production by the 1950s and more than one-third by the 1990s.[57]

DISTRIBUTION AND MARKETING

One of the major places for landing fresh fish in the city was the beach in the Bankusay neighborhood of Tondo (now lost due to reclamation). Fresh fish raised or caught around the bay were landed there and bought by brokers at dawn. The fish passed from these brokers to immediate distribution through the public markets and also on the street or door to door. Another landing in the 1890s was at Tangos in Navotas. The informally organized Bankusay beach market was described by the city authorities in 1909 as "clandestinely carried on by many fishermen," and they proposed to build a proper market some distance away at Pretil, Tondo, where the fishermen "will be required to bring their fish and other sea food." The Pretil market was eventually built (1913), but it did not displace the open everyday exchange on Bankusay beach.[58] In the absence of refrigeration, most fresh fish arriving in the city were consumed by the end of the day, and certainly the next. Any attempt to centralize and impede this flow long enough for institutional counting was impractical, and as a result we will likely never have robust quantitative measures of Manila's historical fish supply.

In 1850 fish were abundant in Balayan and Batangas bays and Lake Taal in Batangas Province, and they were a lucrative product. From the small port of Taal, fish were delivered to the city by sailing vessel, for example, more than 1,500 "tuna" (*atun*) arrived in 11 shipments during March and April 1862. Whether caught by *baklad* or other methods or pond raised, fresh fish arrived from all the coastal jurisdictions of Cavite (1870), Bataan, and other places in the estuaries and river deltas around the bay and in the interior Laguna de Bay. In the last case, a special one-week enumeration of traffic on the Pasig River in 1853 recorded 21 large *bancas* carrying fresh fish into the city. Although some of these sources became depleted, the immediate area of Manila Bay continued to play a major role in the fish supply of the city. During the 1930s depression, officials argued that so many people in the coastal areas of Cavite, Rizal, Bulacan, and Bataan were employed in fishing and fish corral building—in feeding Manila—that they "do not feel much the present crisis."[59]

As the provision of nearly fresh fish from greater distances picked up, salt supplies and ice plants became increasingly critical. Ice was also useful in stabilizing the wide swings in the price and supply of fish in localities where bigger catches were not landed every day.[60] Still, ice came at a cost, and most fishermen sold their catch on landing.

Depletion

Interviews can often provide a more multidimensional story than snapshots in time retrieved from archival records. For example, Leoncia Buzon (1866–1949) grew up in Malabon. She became the wife of Lucio Buzon, whom we encountered as a *baklad* owner residing in Bankusay, Tondo. Through the 1940s, she and their eldest son were fish brokers who went to the beach near the family home at 5:00 a.m. and again at 10:00 a.m. to meet the boats arriving with the catch. With the income from the family *baklad* and the fish business, they were gradually able to purchase swampland in the neighborhood, turn it into rental house lots, and provide professional educations for six children. In addition to his business as an early morning fish broker, the eldest son was also a physician.

We met Filoteo Tuason in connection with his father's work as a *viajero*. In 1942 Filoteo became one of the wholesale fish buyers at the Bankusay landing, where he regularly met the afternoon boats bringing a fresh catch from Cavite. He covered the fish with a layer of ice and at dawn the next day brought them to the Divisoria public market. Normally his fish were all sold by the end of the morning, and the cycle began again. The fish business was his fourth line of work.[61]

Many others were also active in the fish-provisioning system. Brigida Alcaire-Pahit's father was a *kargador* in the 1920s and 1930s, helping to unload and carry the fish brought to the Bankusay landing. For this work, he was paid in fish. Brigida and her brothers would then sell his fish to wholesale buyers using the *bulong* (whisper) system. Prospective buyers would whisper an offer in Brigida's ear, and she would sell the lot to the highest bidder.[62] She began this work at about age 10 and stopped at age 18 when her father died.

A native of the Tondo shore, Geraldo Santiago started fishing on his own in the late 1920s at age 14 and made this his livelihood. Using his own *banca*, he sought his quarry two miles out in the bay fishing with special hand gear (a skimming net) for small shrimp during the rainy season and for fish the rest of the time. His wife, Eduarta Gagahasin, was also a native of the Tondo shore. Like the wives of numerous other fishermen, she contributed to the family income by peddling her husband's catch, selling it on her own (*sariling tindera*). This typically yielded a higher price for the fish than selling in a competitive retail marketplace or to a wholesaler, especially one to whom you were in debt. So she carried her husband's catch to various parts of the city by bus, horse cart, and foot, leaving the house by 6:00 a.m. each day and even buying fish to sell if her husband's catch was too small. When the fish sold well, she might be back by 8:00 or 9:00 a.m.—on a poor day it might be noon. Often she would go out again to sell fish door to door during the late afternoon. Some men also worked as ambulant fish vendors—displaying the fish in shallow basket-trays slung on a balance pole.[63]

All this highlights the roles of urban Filipinos in the business of buying and selling fish. Nevertheless, the surviving business tax license records of persons active in this business in the Manila area in the 1890s are for Hokkien Chinese—two licensed to the beach at Bankusay and eight to the Tangos landing in Navotas. Both places were well situated to handle fish entering the commercial system. At the same time, Isabelo de los Reyes tells us, "Like the Chinese, [mestizos] go to the fishponds and the fish pens of Navotas, to buy fish that they then resell in Manila or in the markets of Malabon."[64] Chinese were involved at the commercial level in the trade in fresh fish, but they did not dominate the system.

PRESERVED FISH PRODUCTS

Considerable fish protein arrived in the city in various preserved forms. *Bagoong*, tiny fish, including anchovies and sardines pickled, fermented, and aged in rich brine, arrived during the 1860s and 1870s in large earthen jars from ports in Ilocos, Pangasinan, and Iloilo—and no doubt from nearby places as well. It was used in small quantities to add protein and flavoring to rice and also, on occasion, to compliment slices of tart green mango. A tablespoon or two of *bagoong* could be the *ulam* in a simple meal of boiled rice. A bit up the economic scale, it often served as the sauce on some other *ulam*, whether vegetables or meat. In Ilocos *bagoong* was made from September to February from tiny gobies (*Gobiidae*) known in Ilocano as *ipon*. These fish hatched at sea and gathered at the mouth of the Abra and other rivers in a brief monthly run to what would become their adult upriver habitat. In the early twentieth century, these were caught in huge fine nets set in the river mouth. By the 1930s, this flow to Manila had reversed and *bagoong* was going to the now fish-impoverished markets of Ilocos from Cavite, Pangasinan, and elsewhere.[65]

In the Manila area the minnows used for *bagoong* were anchovies (*dilis*) or young herrings. Another variant, *bagoong alamang*, was made with tiny shrimp—an outstanding source of dietary calcium and phosphorus.[66] In San Miguel Bay in Bikol and Balayan Bay in Batangas, *bagoong* was made in the same way as in Manila, but by the late 1930s overfishing had ruined the industry at Balayan. By the same time, year-round anchovy fisheries in the vicinity of Catbalogan, Samar, had helped to turn that municipality into one of the chief suppliers of *bagoong*. In 1934 several Chinese and one Filipino manufacturer in Catbalogan were marketing their product in Manila, the Ilocos provinces, and even Hawaii.[67] Almost never advertised but consumed by many, *bagoong* was and is a significant source of protein in the diets of the least affluent urban residents.[68]

Small fish in salt were also fermented and strained to produce a liquid, *patis* in Tagalog, for flavoring rice and vegetables. A more refined product was the result of further fermentation in liquid form for more than a year. As Doreen Fernandez tells us, "The best *patis* is not fishy in smell at all, but amber-colored and aromatic, with only a faint suggestion of its source."[69] This condiment is generally important in Philippine cuisine for carrying flavor and salt and is a significant source of both vitamin B_{12} and protein. It is often used for sautéing vegetables. Important in Philippine cuisine, fermented fish sauce (*nuoc-mam*) is also at the heart of the cuisine and cultural identity of Vietnam.[70]

For generations *patis* was manufactured in Malabon-Navotas on a small- to medium-scale artisanal basis. Some of these operations eventually evolved into modern companies—including at least 18 *patis* and *bagoong* producers operating just after World War II. Processing was concentrated in Malabon-Navotas because the fresh *dilis* that served as the principal raw material were prodigiously abundant in the local waterways and salt was produced there as well. Later the river became heavily polluted and the Dagatdagatan paved over. Ready proximity to the most concentrated body of consumers was another locational factor.[71]

Fish also arrived in Manila in dried, or "jerked," form, usually because distance, transport difficulties, or weather interfered with fresh delivery. This was especially true in the general absence of refrigeration. Known as *Daing* to nearly everyone, dried fish were nevertheless recorded as *pescados secos* by the Spanish port authorities, meaning fish that had been flayed lengthwise, salted, and exposed to the sun. Included in the same category was *tuyo*, preserved in the same way but whole rather than split. Some dried fish were fresh-water mudfish, *dalag*, known for going into suspended animation as ponds and sloughs dried up. These were harvested—*mined* is the word Brian Fegan uses—at the Mangabol swamp in Pangasinan, the Candaba swamp, and elsewhere (map 1.3). In 1862 dried *dalag* arrived in Manila from Dagupan and Calasiao and, no doubt, in larger quantities from Candaba. A decade later they came especially from Balayan. In 1870 Laguna sent out more than 6,000 bundles of dried fish—quite possibly *dalag*. Other dried fish came from the sea. In three towns on the Bataan coast from the 1880s to at least 1910 people were salting and drying fish for sale in Manila and nearby provinces.[72]

Briefly reviewing the management of the Mangabol swamp and "lake" in southwestern Pangasinan, local authorities concluded that it had produced little of commercial note in Spanish times and instead was a "forbidding wilderness" inhabited by wildlife. During the early American occupation the fisheries there were leased to Chinese fishermen who marketed their product in Manila. With the establishment of civil government in 1901, "ownership of the fisheries was transferred to the municipality of Bayambang . . . and leased out to private individuals, mostly Filipinos. . . . About 1912 the municipality declared the lake 'public fisheries' [and] divided [it] into lots leased at public auction."[73] Thereafter a large portion of the catch was sent to Manila, as well as Tarlac and Pampanga, by rail.

In so-called dry salting "the catch is washed first in sea water, then in fresh water, then immersed in brining tanks" for several hours. The fish are then washed again in fresh water and "dried thoroughly in the sun." In 1862, such dried fish entered Manila during May–September from the southern portion of the outer supply zone: Capiz, Masbate, Pasacao, and Zamboanga, all of them places from which Manila could be reached despite the contrary winds of the southwest monsoon. The modest quantity then arriving from these places was dwarfed by the flow from nearby provinces. Over time, however, the locales sending dried fish to the metropolis proliferated. By the 1930s, the fisheries of southwestern Samar and Catbalogan had become a major source of dried fish shipped in sacks and boxes to Manila and elsewhere. Estancia in eastern Iloilo Province would also become a major supplier.[74]

Another popular preserved form is known as *tinapa*. Millions of medium-small fish were smoked and turned into *tinapa* in the city itself, especially in the seaside Bankusay neighborhood of Tondo. These fish were briefly soaked and cooked in brine, laid out on open air racks to partially dry in the sun, and then smoked over charcoal and sawdust fires set in a long smoking furnace. In the early years of the twentieth century there were 5 of these fish-smoking facilities, known as *umbuyan*, close to the Bankusay fish landing. By 1911 their number had grown to 36 and then remained around 30 through the 1920s and 1930s. This industry was a great consumer of coarse salt, particularly that produced by the indigenous solar leaching and crystallization method used in Rizal and Cavite. These *umbuyan* gave a special character to the neighborhood, both for the imposing physical presence of great nipa-roofed smokehouses and the extensive stretches of open air drying racks and also for the welcome employment they offered to neighborhood women. This provisioning activity had been going on in the city for a long time. Processing by smoking was a local craft associated with the former Omboy section of Binondo in the 1820–50 era and later with Bankusay. In the 1880s, the press sometimes referred to Bankusay as the "barrio de Umbuyan."[75]

In the city these *umbuyan* were organized and managed by Chinese entrepreneurs, although some were actually owned (or the lots on which they stood were owned) by local Filipinos. In Cavite, by contrast, smokehouse operations were usually in the hands of Filipinos. In essence, fish with significant consumer value were immediately sold through the public markets or door to door, but small herrings and shad, caught in great numbers in Manila Bay during the rainy season, were more often smoked. In the 1930s, fresh herrings for smoking were supplied from the local fish landings and also from the catch of fishermen using large cast nets. These were supplemented with partially processed fish arriving from Rosario on the Cavite coast, Ragay Gulf, and other locales.[76] *Tinapa* remains a much-appreciated part of Manila cuisine.

In the 1930s, Malabon and Navotas and the new operations at Las Piñas and Parañaque were also centers for smoking fish. At that time, the fish to be smoked in Malabon and Navotas were said to come mainly from shallow-water fish corrals in the bay. By contrast, the smokehouses at Las Piñas and Parañaque processed fish caught locally by means of the round haul seine and light (*sapyaw*) system. Several locales on the Bataan bay shore regularly sent dried and smoked fish to Manila in the 1930s. In Tondo, the Chinese operators of the smoking sheds were simultaneously dealers in smoked fish in the Divisoria public market, and retailers around the city were supplied from Divisoria. From Tarlac and Nueva Ecija provinces in the north to Laguna and sometimes Tayabas in the south, wholesale dealers in smoked fish were supplied from the same source. By contrast, in most of adjacent Cavite Province, smoked fish came from small-scale smokehouses in Rosario. In Manila markets, the volume of *tuyo* and *daing* together was about four times the volume of *tinapa*—reflecting greater demand for the less expensive product.[77]

In the nineteenth century, there were occasional reports of other maritime products arriving in the city, some for local Chinese restaurants and celebrations and more for export to China. This was part of an important and long-standing trade that formerly passed in Chinese junks from the Sulu Sea to the ports of South China and later through Singapore. After the Spanish established a stronger presence in the Sulu Sea, some of this trade proceeded through Manila. The products included valuable shark's fins from Zamboanga and dried *trepang* (*balate*, or sea slug) from Jolo and the central Visayas. For generations these products were collected by kidnapped and impressed laborers, Visayans, Bajau, Samals, and others.[78] Earlier *balate* was collected around the island of Alabat in Tayabas and traded to Manila through several localities in the same province, including Mauban.[79] *Trepang* was in common use not only among the Chinese of Manila but also among the Europeans. In "soup or stew, it has a taste between the green fat of a turtle and the soft gristle of boiled beef." According to Edmund Roberts, about three-quarters of a million pounds were shipped to Canton annually in the 1830s. In the 1920s and 1930s, the Philippines was still exporting hundreds of thousands of kilos annually to China, Hong Kong, and Singapore.[80] Similarly, the nests of swiftlets used in making bird's nest soup also sometimes arrived in the city, principally from northeastern Borneo via Sulu and Zamboanga in the nineteenth century, Calamianes, and later Palawan.[81]

In the domestic coastal trade, Manila was a net distributor of 40,000 kilos of salted fish in 1870. At the same time, it was also the premier center for importing fish. Most of these came from the China coast and also from Hong Kong and Singapore. After a trickle in the 1850s and 1860s, imports of large quantities of dried, salted, and smoked fish and also shellfish entered Manila from Hong Kong each year, usually ranging from 200,000 to 300,000 kilograms. A lot of this would ultimately have come from Vietnam. Dried cod for use by the affluent in Lenten recipes formed a persistent subset of imported fish products. Inexpensive dried cod in packages came from Japan in the 1930s.[82]

Salt

Salt for direct human consumption, as well as for these several forms of preservation, was produced in solar evaporation beds during the dry season. Found in several parts of the archipelago, Manila's bay shore collectively was the single major domestic salt production center throughout our period. The process started with channeling seawater into shallow ponds, using evaporation to increase the saline concentration, and moving the resulting brine into smaller beds—sometimes in liquid form and sometimes in dried clay squares cut from the bed or surface scrapings transferred to a leaching vat. The final stage took place in a flat crystallization pond. Many of the starter beds doubled as fishponds during the rainy season. In the immediate vicinity of the urban area, salt beds were concentrated at Parañaque, Las Piñas, and the three closest towns of Cavite, including Bacoor. Salt from Bacoor was a common item of mass consumption in Binondo in the 1880s. Salt was also made in Tondo and at Malabon just north of the city at Tinajeros and Dampalit. With local variation, this is an ancient technology in Southeast Asia, and many production facilities were in operation long before 1850 (map 6.2).[83]

In the 1910s, more than half the national production of solar salt took place in Rizal Province—especially at Parañaque and Malabon—and a quarter in Cavite. By the 1920s, salt had become an important seasonal product of ponds from Malabon-Navotas and Obando north to coastal Malolos in Bulacan. In the immediate post–World War II period, the five coastal municipalities of Bulacan emerged as the leading salt production area—driven by urban displacement and the escalating demand of the *patis* industry and various fish preservation processes.[84]

Salt also arrived in Manila from the outer zone—in 1862 from Vigan and the Ilocos Coast, Zambales, and Pangasinan, especially the first. The Pangasinan coast along Lingayen Gulf was a traditional source of salt supply to inland rice farmers. The very name Pangasinan is derived from *asin*, "salt," and denotes a region of salt making. In all these areas, the process began conventionally with seawater and solar evaporation, but it was finished by heating the concentrated brine in large iron pans or *kawas*. In the 1910s almost 30 percent of the national production of salt took place by means of this method, mostly along the Ilocos and Pangasinan coasts. In one analysis of samples, "Ilocano salt" tested at 94.2 percent pure salt and contained appreciable amounts of calcium while "Manila salt," produced solely by means of solar evaporation, was found to be 91.2 percent pure with a significant magnesium content.[85] Salt making then was the focus of seasonal specialization in numerous locales, especially, though not solely, in the vicinity of the metropolitan market.[86] Why wasn't the Pangasinan-Ilocos method of deriving a purer salt by means of heating used in areas closer to the city? Deforestation and the removal of coastal mangrove had proceeded faster and farther in the Manila area, and as a result fuelwood there had to bear an added cost of transportation.[87] Salt making near the metropolis was economical but only when done using "free" solar evaporation.

Outside the Philippines in East and Southeast Asia, the universal human need for salt often made it an important revenue source for administrative authorities. A salt monopoly produced a critical revenue stream for the Chinese imperial state for centuries and also for the indigenous rulers of Javanese port capitals in the seventeenth and eighteenth centuries. In Vietnam, the French assertion of an imperial salt monopoly generated revenue but also undercut indigenous fish sauce manufacturers and traders.[88] In the Spanish Philippines of the mid-nineteenth century, however, the state monopolies were for tobacco and alcoholic beverages rather than salt.[89]

As demand in the fish preservation industries expanded, local salt production proceeded apace. But by the late nineteenth century domestic production had fallen behind rising demand—part of the increasing commercialization and decreasing subsistence trend in the economy. The shortfall was made up from China. In the later 1930s, millions of kilos of salt were imported annually.

Canned Fish

Preserved fish were also increasingly imported in cans. For decades, sardines packed in oil in little square tins came from Spain and Europe more generally. By the 1870s and 1880s, these often came packed in tomato sauce and oil. Price limited their consumption. Following the American conquest and a tariff reduction, canned salmon from the United States became widely popular. Starting in 1903 large quantities were imported. By 1912–14 canned salmon was arriving in amounts of five to ten million pounds a year. Hayase uses this as an index to the unmet demand for fish from local sources—thus revealing the comparative vacuum into which Japanese fishermen sailed in the first years of the twentieth century.[90] Canned salmon was finally surpassed in 1917 by the introduction of cheap California sardines packed in tomato sauce. Canned salmon continued to enjoy some following until 1928 when it was briefly replaced in the Philippine market by the less expensive "salmon-style" mackerel. Thereafter, depression conditions limited sales. In good times, the Philippines had become a very important market for American exports of processed fish. Ironically, many of these fish were packed on the Pacific coast from Monterey to Alaska by poor Ilocano men and other Filipino seasonal-labor immigrants to the western United States. In 1935 the first full year of the Commonwealth, the United States and Japan shared the Philippine market for sardines. Imported canned fish had long since come within the economic range of a majority of urban consumers.[91] During the long siege of Bataan in 1942, it was canned salmon and a little rice that (barely) sustained the Filipino defenders.

Summary

In retrospect, the technology of capture fishing underwent several permutations. One of the more important for Manila was the arrival of Japanese trawlers in the early twentieth century, and their subsequent adoption of the "beam trawl" rig to scour the floor of Manila Bay. The use of very large purse seines by Filipino fishermen also contributed to an increased catch. By the end of our period important supplies of fish were also coming to Manila from distant ports not accessible by train or truck. These included especially the port towns of Catbalogan in Samar and Estancia in eastern Iloilo. At the same time some other aspects of capture technology hardly changed at all. Individual fishermen continued to use hand lines and scoop nets; *salambao* continued to operate in the less polluted estuaries and coastal waters of the bay. The design of many *baklad* corrals may have changed little even as the geography of their use expanded with the outward migration of fishermen from the Manila Bay communities. In these cases rising urban demand led to intensified use of long-standing capture forms.

Against this background of successful provisioning, overexploitation in various bodies of water began to be a notable problem, especially in confined waters. By the 1920s, some coastal fishing grounds, such as Balayan Bay in Batangas, were on the road to depletion. Manila Bay and other shallow embayments were also among these places, as were small inland lakes and swamps. In the second half of the twentieth century, this would become a more widespread phenomenon. The management (or nonmanagement) of "public" resources is often highly contentious and notoriously difficult to police in a way that long-term depletion is avoided.[92] By the 1930s, new motorized capture techniques were leading to major changes in the fish population of Laguna de Bay. Such capture operations are better labeled "mining" in the sense of appropriating a resource in such a way as to make it nonrenewable. Here the actions of some impacted the many.[93]

Aquaculture provided a critical alternative to capture fishing. In this case, raising *bangus* (milkfish) in brackish ponds was a survival and escalation of a pre-Hispanic technology. Known to some as "blue deserts" for their lack of integral gardens, these ponds nevertheless came to produce an important and growing rainy season fish supply for the metropolis.

7

"Generations of Hustlers"

Fowl and Swine in Manila

ON A MASS BASIS, fish constitute the primary *ulam*, the principal item of cuisine, along with rice. But chicken and pork are also ancient and important in regional culture. The major exception has to do with Islamized peoples, who avoid pork. Given the ability to afford such things, the two provide families with a diversification of the protein diet. In our era provisionment with fowl and pork also provided a seasonal compliment to the fish supply. The tonnage of captured fish routinely declined during the rainy season, and utilitarian chickens took on an enhanced role. Along with aquacultured *bangus*, many ordinary consumers turned to fowl as a substitute.

In our period many Manila families of a certain modest economic standing enjoyed chicken meat and both chicken and duck eggs as a matter of course. Given the way most country chickens were kept, boiling was the optimal mode of preparation. Chickens were almost always available, live, in Manila's public markets either to consume or to raise at home. Managing large flocks of ducks for their egg production was already a highly specialized undertaking in select locales before 1850. Only very slowly, however, and toward the end of our period, did raising chickens begin to emerge as a set of specialized operations on a similar scale. For the most part, chicken raising in a more or less "industrial" form of organization came after World War II.

Pork consumption did not initially fit the rainy season profile; rather, the festivities of May produced a marked peak of consumption. This peak was accentuated by the Catholic Lenten observance of an extra meatless day per week, which produced a notable downturn in hog slaughter and pork consumption during some portion of February, March, and April. This calendar of consumer demand changed during the twentieth century, particularly during the 1920s and 1930s. The former peak of demand in May was replaced by a less episodic profile in which pork consumption was lower during Lent but then rose to a higher level, which was sustained during the entire period of July–October plus the December holidays. In this manner, the seasonal volume of pork consumption came increasingly to resemble that for *bangus* and chickens. As the securely employed urban middle class grew rapidly during the 1910s and early 1920s, this led to a general rise in pork consumption.[1] And as railroad and then truck transport came into general use, there were less likely to be occasional rainy season supply shortages. Urban pork prices and consumption tended to even out during the second half of the year—a sign of modern commerce and consumption.

Like chickens, hogs were raised all over the archipelago, except in the Muslim parts of the south. For a long time, most animals that ended up in the urban food supply stream were raised singly or in twos and threes, not on specialized livestock farms. These were purchased, aggregated into small shipments, and conveyed to the city alive. Here they were quickly slaughtered in the single central city abattoir and the meat distributed by wholesalers to the retail stallholders in the public markets. Veterinary postmortem inspection in the slaughterhouse began early in our period and was intensified in the 1880s when members of the Spanish government became concerned about the health implications of trichinosis. Very few Filipino urbanites chose to live as vegetarians for ideological reasons.

Fowl

In terms of tradition, cultural familiarity, and general frequency of consumption, fowl and eggs constituted an important occasional protein source for the Filipino population of the city. A central dimension of change is the slow rise of organized commercial poultry production using hybrid stock. Chicken egg production and marketing forms were even slower to change, and for a long time inexpensive eggs from South China dominated the Manila market. Preserved duck eggs were a different and much more localized matter.

Chickens were domesticated in the region and have been widely kept for millennia—part of Carl O. Sauer's chicken-pig-dog complex known from ancient times in Southeast Asia, including what later became South China.[2] Recognized in the sixteenth century as "the principal sustenance of this land," chickens loomed large in the perceptions of the new authorities as they grappled with questions concerning how to feed an urban population foisted on a society that was largely unaccustomed to such demands outside of "personal alliance and clientage networks."[3] Accordingly the Spanish *audiencia* (tribunal) in the sixteenth and seventeenth centuries promulgated a series of legal codes that required each inhabitant of the country to raise a small number of fowls—six hens and one cock—and bring chickens to the city from time to time to sell for the official price. At the same time, local Chinese market gardeners were required to keep a dozen hens. Likewise, the new authorities ordered a rotation among nearby provinces as to which was responsible for the urban supply in various months.

At the same time (ca. 1668), Father Alcina reports, "In [the] matter of raising fowl, the Chinese without doubt excel all the people of the world and they provide those where they live by contract, as in Manila, with hens, young pullets, eggs, both of chickens and ducks, in the greatest abundance." Edgar Wickberg also mentions that some Chinese held contracts "for supplying meat to Manila."[4] How the contractor (*abasto*) system was organized in Manila begs investigation, but it continued into the early nineteenth century. In addition, provincial governors came to see it as their right to dominate the flow of supplies to the city. For both animals and rice the compulsory food supply system of the early colonial state, the contract system, and domination by provincial governors eventually gave way to an increasingly commercial system of supply. The steps by which an open commercial market arose have not been thoroughly explored.[5]

As relatively free-ranging animals foraging on their own and roosting in trees, chickens were and are part of the national dietary. Often their foraging was supplemented with varying quantities of leftovers and grain feed—in part to keep them localized. In general looking after chickens was a family responsibility accorded to women and children. However, some roosters were and are lavishly cared for by men (*sabongeros*) for their value as fighting cocks, a major pastime and passion. But compared to the quantities likely consumed, the nineteenth-century reports of maritime arrivals of live fowl from the outer zone are modest in the extreme.

In 1862 hens and gelding cocks (*gallos capones*) were reported as minor components of 9 cargoes arriving from places as diverse as Ilocos Norte, Marinduque, Batangas, and Zambales. There was also a consignment of 25 geese from Dagupan. The capons were destined for affluent dinner tables. Turning cocks into capons was widely practiced in South China—carried out by specialists who traveled from village to village. There is little comment on such a practice in the Philippines, although Antonio de Morga in 1609 mentions that some large chickens bred in the Philippines made excellent capons.[6] In 1872 four shipments of chickens were recorded from the Ilocos coast, three from Marinduque, and one from Batangas. Spanish agricultural authorities believed that the average Manila European ate a chicken per day. So the supply of chickens was an important colonial concern.[7]

Despite the sparse documentation of maritime arrivals, live chickens were readily available in the markets of the city in the nineteenth century for immediate consumption or rearing. Most of the supply came from the inner zone.

During 1870 more than 50,000 chickens came from Laguna Province alone. Frederick H. Sawyer, in particular, points to Laguna for its supply of poultry to the public markets of Manila—brought on shallow draft steamers or other small craft making the daily shuttle.[8]

In the late nineteenth century, a decade after the collapse of coffee, it is clear that the microhinterlands of various Batangas ports were increasingly specialized in the supply of nonrice provisions to Manila, with poultry and eggs shipped from Bauan and Batangas municipalities on a weekly basis. No doubt other nearby places were doing the same. The overall distribution of chickens was diffuse, with the highest per capita rates in the five rice-growing provinces of the Central Plain.[9]

Raising Chickens for a Rapidly Growing Population

Between 1900 and the mid-1930s the population of the metropolis roughly tripled, from something over 200,000 to approximately 700,000.[10] Demand for all classes of animal protein grew apace. In the 1890s one of the foreign consuls said that people had no interest in raising fowls.[11] Nevertheless, the Bureau of Agriculture responded to the need for poultry by encouraging commercial-scale operations. But even in the 1910s there were still few chicken growers of any significant scale. The bureau recommended that persons thinking of starting such a business first develop a hybrid, higher-productivity, disease-resistant stock. Later it actually introduced improved breeding stock for distribution.[12] In the 1920s, the poultry station at Pandacan was producing a well-acclimated hybrid fowl by crossing Cantonese hens with Rhode Island Red roosters. By then two moderate-sized poultry businesses had emerged, the Baclaran Farm in nearby Parañaque and another in Los Baños, Laguna. Each of these had 500 birds, and there were now at least 200 commercial poultry producers operating on a backyard scale with an average stock of about 60. Unfortunately, all these flocks were devastated by a new disease known as avian pest; apparently the hybrid stock either was not in mass use or proved to have little innate resistance to this disease. The epizootic of September 1927 killed more than 80 percent of the chickens in the capital region.[13] Poultry yards then were in six urban districts and three suburban towns, as well as several municipalities in the portion of Laguna closest to the city.

Despite these tentative moves toward a greater commercial scale, the Bureau of Animal Industry estimated that three-fourths of the national production of chickens and eggs was still carried out by small town and country people who ate relatively few of the birds themselves, preferring instead to sell them for cash. Felice Prudente Sta. Maria wrote, "Not much money is spent on feeding or caring for them." Their chickens "come from generations of hustlers and find most of their own living" in great contrast to those raised in poultry yards.

Such sinewy free-range chickens were more likely to end up boiled in a dinner soup than served as a stand-alone viand. She adds, "Slow cooking could soften the most flavorful but toughest of chickens," even "a muscular fighting cock that had lost or managed to survive past his prime." No wonder *tinola* soup was so popular.[14]

More numerous and larger scale commercial operators emerged during the 1930s. Shortly before his death in 1930, former cattle importer and trawler owner Faustino Lichauco purchased a small farm near Antipolo where he planned to start a poultry business.[15] By 1931 a local Japanese company was advertising live fowls from its breeding farm in suburban San Juan, as well as chicken feeds and incubators. A few years later there were 11 Japanese poultry farms in suburban Rizal.[16] In the same era Susana Madrigal, the wife of industrialist Vicente Madrigal, arranged to start a poultry and piggery operation on a large property she purchased in Alabang-Muntinglupa south of the city. Larger scale poultry farms were now on the rise. The 1939 national census recorded 80 with an average of almost 800 birds each. Entrepreneurs were seeing the potential of chickens as a commercial meat source, and Bureau of Education poultry clubs were giving young people some experience with quality breeding and care.[17]

Beyond the more immediate environs of the metropolis Batangas was the leading supplier of poultry in 1933–36, continuing its important role in the nineteenth century. The According to the Commerce Department, chickens from there, and also from provinces bordering Manila, were cheaper than those coming from more distant places because the cost of shipping to market was less. By 1938, however, both Pangasinan and Nueva Ecija had surged far ahead, with Pangasinan alone supplying 1.5 million birds, about 27 percent of the total recorded. Tarlac and Camarines Sur also sent more birds than Batangas in 1938. The total reported that year, 5.6 million, would have provided more than 7 chickens per capita. Apparently, the main production area was shifting outward to the Central Luzon Plain, although Cebu and even Sulu were also active in this trade. Reflecting this dominance, almost half the arriving chickens entered the city by train at the Tutuban Station.

According to the Commerce Department, large restaurants and institutions now maintained standing contracts with Manila-based dealers for the timely supply of chickens. The names given were all Filipino. Mass contract supply of dressed chickens for the coolers at supermarkets and restaurants such as Max's lay ahead in the 1960s and megafarms with 100,000 birds in the 1970s.[18]

CHICKEN EGGS

The record of shipments of eggs to the metropolis is scanty. Laguna supplied Manila and its environs with a million eggs in 1870, and Montero y Vidal points to the Marikina Valley and "towns close to the capital" as further sources of eggs in the 1880s. Of fresh domestic chicken eggs there is practically no other trace: a single shipment from Vigan in November 1862. Unless preserved, eggs could not be kept for more than a week. Manila was also supplied with eggs from China. From small quantities in the 1870s, deliveries rose to exceed 800,000 annually in the 1890s. These were generally listed as fresh chicken eggs, but such shipments also included the famous preserved "century eggs."

In the city chicken eggs were frequently consumed by those who could afford them. When visiting Manila in the 1890s, the Umalis, an affluent family from Tayabas, would take a "full Filipino breakfast" consisting of hot chocolate, fried rice, *tinapa* or *daing* (smoked or dried fish), sliced *tapa* of venison or boar, and so on, and "always there would be scrambled or soft-boiled eggs."[19] Eggs had a place in many light meals taken at home or on shipboard and were also used in various dishes of celebratory meals. Soft-boiled eggs were standard hotel breakfast fare in the late nineteenth century, and hard-boiled eggs were frequently carried on travels. Likewise, salted eggs might be eaten with *bibingka*, a sweet rice cake. The total urban demand was substantial even though eggs remained a modest part of proletarian diets.[20]

While raising chickens for meat was advancing, Philippine production of chicken eggs remained grossly inadequate to meet urban demand, and whereas the supply of live chickens was from domestic sources, a great many eggs were imported. In the early twentieth century, "thousands of dozens [of eggs] were shipped by each Tuesday's steamer from Hong Kong," a continuation of the pattern of the late nineteenth century. Probably these came from various locales in the Canton Delta, but the agricultural district centered on the port of Swatow was exporting tens of millions of eggs.[21]

Importing fresh eggs would not necessarily have been a bad thing if it had allowed the population to produce and export something else in which it had a comparative advantage. After all, British cities received hundreds of millions of fresh eggs by sea at this time from nearby places with lower feed costs.[22] Still, in Manila the situation was little changed at the start of the 1930s. A report from the Bureau of Animal Industry for 1930 remarks, "It is a rather anomalous situation that in the last twenty years chickens for table purposes have almost doubled in value, while the price of eggs has not advanced proportionately." At this time, large quantities of imported eggs were retailing "at from 2½ to 5 centavos per egg, depending on the season. . . . It is very doubtful if [domestic] commercial units can produce them for sale at less than 5 centavos per egg." This left local producers at a decided disadvantage. "The very cheapness of the great bulk of the imported eggs . . . [was] a factor militating against increased egg production in the small towns and barrios." The Bureau of Animal Industry faulted domestic chicken growers for keeping too many non-egg-laying roosters and not replacing them with hens.[23] After all, a fowl population with few males would produce eggs at a much higher rate than one with many males. The record is largely silent on whether there was a seasonal variation in egg consumption or in the position of domestic eggs in the total supply that paralleled the seasonality of the chicken business.

There followed campaigns to raise productivity and enact a more protective tariff. On the production side, one of the problems was a lack of feed variety. Rice and maize were the main feeds. Some experts recommended six parts *palay* and five parts maize supplemented with one part mongo beans, but copra meal left over from the manufacture of coconut oil (as in Laguna-Tayabas) was the only significant domestic supplement. In Batangas, sun-dried prawns were sometimes used to supplement poultry feed.[24]

Imports of eggs numbered about 4.5 million dozen per year during 1911–15, much lower during 1916–18 due to the shortages and high shipping costs of World War I, and then higher in the 1920s at 5 to 7 million dozen. Almost all these came from South China, and there were numerous complaints that Chinese import merchants had inordinate power in the marketplace. During the depression of the 1930s, various interests urged the adoption of tariffs to favor local producers. A protectionist tariff passed the legislature in 1931 but was vetoed by the American governor-general, who cited harm to consumers and the skullduggery of lobbying interests. Still, from highs of 7.0 to 7.2 million dozen in 1929 and 1930, imports of eggs declined, becoming precipitous in 1933 and 1934. Japan suddenly became a factor in the local supply of hen's eggs in the early 1930s, averaging about 150,000 dozen supplied during 1931–33, but its initiative was lost in the general decline of imports attendant on the arrangements for the new Commonwealth. The following year the United States replaced Japan as a supplier at about the same level. The import decline was nearly matched by increased Philippine domestic production. Thereafter, the total flow of imports declined further to 2.0 and 1.2 million dozen in 1937 and 1938, respectively.[25] The question of enactment of another tariff proposal and whether that was related to the New Deal political thrust from Washington or to the establishment of Commonwealth autonomy remains unanswered.

Following a drive to achieve a more comprehensive record, 2.0 to 2.3 million dozen domestic eggs were recorded as arriving in the markets of Manila in the years 1935–37. This total reportedly doubled to 4.9 million dozen in 1938. During these same four years, figures on the origins of these domestic eggs are inconsistent but suggestive. Cebu, a center of maize production, was now the leading supplier at 1.3 to 1.7 million dozen eggs per year. Most of its weekly shipments came from backyard producers. In the next rank of provincial suppliers were Iloilo, Pangasinan, and Batangas. Suburban Rizal was reported as unimportant in this regard until 1938, when it was suddenly credited with 836,000 dozen coming from Caloocan, Quezon City, and other nearby communities. Both

Nueva Ecija and Pampanga became important suppliers of chicken eggs in 1938 and 1939. Twenty-eight other provinces also sent eggs to the urban market in the 1930s. As with chickens, almost half the domestic eggs arriving in Manila and captured by the recording system arrived via the Tutuban railroad station. About a quarter was recorded as arriving at the Pasig riverside.

Some of the production increase was due to the emergence of increased-scale commercial operations such as those of the Llenado brothers in Malinta, Bulacan, and the Yuson family in Jaen, Nueva Ecija, which won the contract to supply the army's Clark Field with 2,000 eggs per day. But many eggs were also now coming from clusters of second-income poultry ventures run by the woman of the house at the scale of a few hundred laying hens kept in cages.[26] The rise of Cebu and Iloilo as important sources of poultry products for the metropolis and the general broadening of the supply zone were major developments reflecting greater effective demand, improved transportation, and deepening commercialization. Only in the postwar era were very large, integrated, layer breeder farms created, farms that required professional veterinary involvement, a pharmaceutical industry, and substantial capital outlay.

A persistent problem with Manila's egg supply was the lack of entrepreneurship in organizing the collection of domestic eggs. *Viajeros* certainly helped, but an everyday gathering and delivery organization was missing. In the penumbra of several large cities in China, by contrast, a system of commercial collecting had emerged in the 1890s and later. This marshaled eggs every day from small farms that each produced a small number of fresh eggs from a few hens running at large. Collectors made their rounds on foot with two large baskets and a balance pole. With baskets full, they returned to substations where the eggs were repacked in baskets of 800 and sent on to the main stations located on the principal water and rail transport routes. In the 1930s, hundreds of such baskets arrived at each main station each day and were sent on to the big urban egg dealers.[27] In South China these dealers bought the eggs of a large number of Cantonese small farmers, gathered them in Hong Kong by boat and rail, and quickly sent them on to Manila by ship.

What was missing in the Tagalog region was effective commercial organizations that could collect and deliver domestic eggs to the mass urban market in a timely way. There were Filipino *viajeros*, including some who purchased eggs, but apparently no one found it worthwhile—at the compensation rates of the urban penumbra—to go door to door every day to collect and concentrate the fresh eggs for shipment to the city.[28] Given an economic choice, Filipino consumers preferred domestic eggs to Chinese eggs because they were fresher and the Chinese product was said to taste fishy. But the price of domestic eggs was higher, and many could not or chose not to pay the difference.

We may never know with a great degree of assurance how many eggs were consumed in the city or exactly where they came from. However, oral testimony provides some evidence. For several years during the 1930s, Maria Balboa and her husband operated a modest wholesale egg business from their home in the Kapampangan neighborhood on Antonio Rivera Street in Tondo. Periodically, they traveled to Sta. Maria, Bulacan, in order to purchase native chicken eggs. *Itlog na Tagalog*, she called them. Those who wanted to order 100 or 200 eggs at a time just came to their house in Tondo. It was no accident that Balboa and other buyers went to Sta. Maria, for that municipality was emerging as the "egg basket of Bulacan" and in the immediate postwar era would be host to large-scale operators.[29] An agglomeration advantage developed: as more producers emerged in Sta. Maria, more buyers like Maria Balboa made their way there, confident that they would find an available supply at competitive prices. At the end of our period, eggs from Bulacan commanded the highest prices in Manila, followed closely by those from Batangas. Both sold for more than eggs from Cebu. Balboa and her husband operated at one scale. At the same time, a number of Philippine Chinese were engaged in supplying eggs on a much larger scale—operating from shops clustered about the largest public markets. Thousands of eggs were destroyed in a fire in one of these shops next to the Divisoria Market in 1939.[30] Aside from the period of the Japanese occupation, this ethnic bifurcation would not change materially until the emergence of very large commercial egg and chicken operations that contracted directly with upscale grocers.

Duck Keeping and *Balut* Production

In contrast to the trade in chickens, the production and supply of ducks and duck eggs attracted considerable attention. The great flocks of Pateros, just east of Pasig, were mostly composed of young female domestic ducks kept as egg layers, so Pateros was also the center of the early duck egg industry. There were a lot of ducks. Charles Wilkes, visiting in 1842, reports, "The number of ducks of all ages may be computed at millions."[31] Perhaps a majority of these eggs were processed as *balut*, a delicious duck embryo food product boiled in its own broth. Clean in its eggshell, *balut* was and is a godsend for ordinary travelers and snack seekers in a situation with less than ideal sanitation and refrigeration. In *balut* making, the fertilized egg is 17 or 18 days old at boiling. The duck and *balut* production facilities at Pateros were amply described from the middle of the nineteenth century onward when they attracted considerable foreign comment on their way to becoming one of the first Manila area tourist sites. By that time women from Pateros were already delivering and selling clean rice and *balut* in the metropolis.

The population of Manila formed the critical mass customer base for the *balut* industry. Two large *bancas* loaded with eggs were reported traveling to the city on the Pasig River during a sample week in 1853.[32] Along with Pateros, nearby Taguig and Pasig also became duck egg production centers by 1880— all three using the several Pasig River channels near the Laguna de Bay for gathering snails as part of the duck nutritional system (map 1.1). The ducks were also fed shellfish and small shrimps caught in large numbers in Laguna de Bay. Elsewhere, in East Asia, flocks of domesticated ducks often fed themselves by gleaning recently harvested rice fields.[33] No similar *balut* production locales are known outside of the immediate hinterland of the city in the mid-nineteenth century. Not all the duck eggs were converted into *balut*. In 1906 the governor of Rizal claimed that the duck farmers of Pateros supplied "the confectioners and bakers" of the city with up to 3,000 fresh duck eggs a day.[34] In addition to eggs, ducks themselves were provided to the markets and Chinese restaurants of the city from the same three municipalities. De la Cavada invites us to conclude that Manila was not only consuming but also supplying quantities of both ducks and eggs to other coastal localities by 1870.[35]

By the 1920s, the duck husbandry system had expanded to five lakeside towns in Laguna, new locales in Rizal, Hagonoy and Paombong in Bulacan, and Masantol and Macabebe in Pampanga. Increasingly, duck eggs came to be transported into the city by truck. Bulacan and Laguna provinces also now sent hundreds of thousands of duck eggs, many arriving in *balut* form.[36] Most *balut* were retailed by ambulant street vendors, but there were also shops, often Chinese, on side streets next to the largest markets.[37]

～

Various public markets of Manila were also supplied with wildfowl. "The markets are well supplied with chickens, pigeons, young partridges, which are brought in alive, and turkeys" according to Wilkes in the 1840s. Later wild ducks were captured or shot over the lakes and swamps of Central Luzon and Zambales. Pigeons and endemic quail were also found in the Quinta Market in the 1890s–1910s coming from Parañaque and other points in Rizal or farther afield. Sawyer mentions snipe, quail, and wild ducks as welcome seasonal diversifications of the diet. More than occasionally such game birds were cooked and partially preserved in a hunter's stew that included garlic, vinegar, *patis*, and cayenne—a variation of adobo. The British consul mentions that, despite the commercial dislocations associated with the Revolution and war in 1899, the availability of game in the markets was not seriously affected.[38]

Doves or "pigeons" were also available. These were raised on *palay*. They could be consumed at home or they could easily be carried alive in baskets and on arrival cleaned and cooked in *suka* (nipa vinegar) and garlic adobo style. Since adobo keeps relatively well, jars or pots of birds in adobo were sent from provinces such as Tayabas to students and other relatives living in the city.[39] In Rizal the municipality of Taytay was already known for raising domesticated birds in the 1850s, presumably doves. In the 1930s, Batangas and Rizal each provided Manila with 10,000 doves a year. Though prized by some of the affluent, in general these comprised a minor part of the urban supply of fowl.[40]

Turkeys were also part of the mix. Presumably the turkey was introduced from Mexico, its place of domestication. Much less common than domestic ducks, turkeys nevertheless had a provincial distribution much like that of ducks. In the 1870s, turkeys were especially raised in the then suburban town of Pandacan for consumption in the city. Sawyer says that they were sometimes served along with *lechon* at wedding feasts. Gonzalez notes, "Since poultry roamed freely and were not tender, the American custom of roasting was still not practiced." By 1920 Hagonoy and Binangonan had become the leaders in turkey raising.[41] Still, turkey meat hardly entered the mass urban dietary.

HOGS AND THE "RED MEAT" SUPPLY

In addition to fish and fowl, meat animals sent to Manila were primarily swine and cattle. This was true in all ten decades except during war-related famines. *Karne*, from Spanish, became the general term used for such animal meat in Manila. The formerly common Tagalog term, *lamán*, has narrowed in meaning to represent a specific lean piece of meat without bones. At this point, one is discussing the food supply of the somewhat more affluent portion of Manila households, since many residents could afford only *tuyo* or other inexpensive fish and some not even that. Affluent households have long been heavily concentrated in the metropolitan area, where in the 1930s average per capita pork consumption was five or six times that of the surrounding provinces. In the nineteenth and early twentieth centuries, a great many Filipino households would likely have been unable to afford such a diet apart from a few celebratory occasions such as baptisms or feast days. This was true even if they raised a pig at home as a way to recover the nutritional value of household garbage and rough plant materials, including chopped and boiled banana stalks or rice bran.

Beef consumption was also concentrated in the city, and its consumers were socially located even higher on the pyramid of affluence. The resident European population and cosmopolitan mestizos were the most frequent consumers of cattle flesh. Unlike pork, most urban Filipinos of modest means would have been quite unfamiliar with the taste of beef. Roughly half as much beef relative to pork was sold through the Divisoria Market in the early twentieth century. On a dressed weight basis, 65 percent more pork than beef left the abattoir. While pork was sold in public markets throughout the city, the sale of beef was skewed toward the Quinta Market, which catered to the better off.[42]

The Supply of Hogs

Swine travel reasonably well, and in the days of poor land transportation they represented a relatively efficient way to get some of the agricultural product to market (i.e., feed it to pigs and then move the pigs to the urban market). The daily record of vessels arriving in Manila during 1872 tracks 6,898 hogs arriving by sea from an astonishing 68 ports in the outer zone (map 7.1).[43] Nearly all these animals would have gone quickly to slaughter for the consumption of the urban population. This is an extraordinary breadth of supply—50 percent greater than the number of ports sending rice to the city in the same year. Within this pattern, a few ports accounted for an outsized share, Dagupan (1,122) and Vigan (969) in particular. San Narciso and San Antonio in the Ilocano settler area of southern Zambales each sent more than 400, as did Bolinao, then part of the same province. These five accounted for about half the coastal movement from the outer zone, but still the breadth of flow is impressive. As in the rice trade, the ports sending the largest numbers were drawing on interior riverine hinterlands. Bikol in southern Luzon is notably absent in this record, and almost nothing came from the Cebuano Visayan zone at that time.[44] This was Manila's integrated food supply territory in the late eotechnic age.

As it happened, the geography of hog supply changed radically with the seasons. The broadest extent was registered in February and March in the dry season and its narrowest in the rainy season, the *tagulan*, July through November, when many provincial sail craft were taken out of service (graph 7.1). The swine producers of the inner zone provided almost all the domestic urban supply during the rainy season but tended to hold back their hogs during the dry season when competition from outer zone locales was at its maximum. In the mid-1890s it was precisely during the rainy season that the largest of the monthly flows from nearby Laguna were received in the city abattoir and, at a lower level, from Bulacan, Pampanga, and Manila as well. From Batangas this ranged from July to January.[45]

There was also a distinct seasonality in the consumption of pork in the city. The slaughter of swine for the supply of public markets was at its lowest point during February and March, the same months in which the flow of hogs from the outer zone was at its annual maximum. This was the case in 1872 and for most other years for which there is a monthly record, including the 1930s. The seasonal decline was real. Ethical and seasonal concerns about the consumption of certain types of animal flesh are a near-universal characteristic of religious systems of meaning. In this case, the dip concerns the Catholic fasting proscriptions connected to the observance of Lent—the 46 days before Easter. Fr. John Schumacher explains, "Depending on the lunar calendar, Lent could

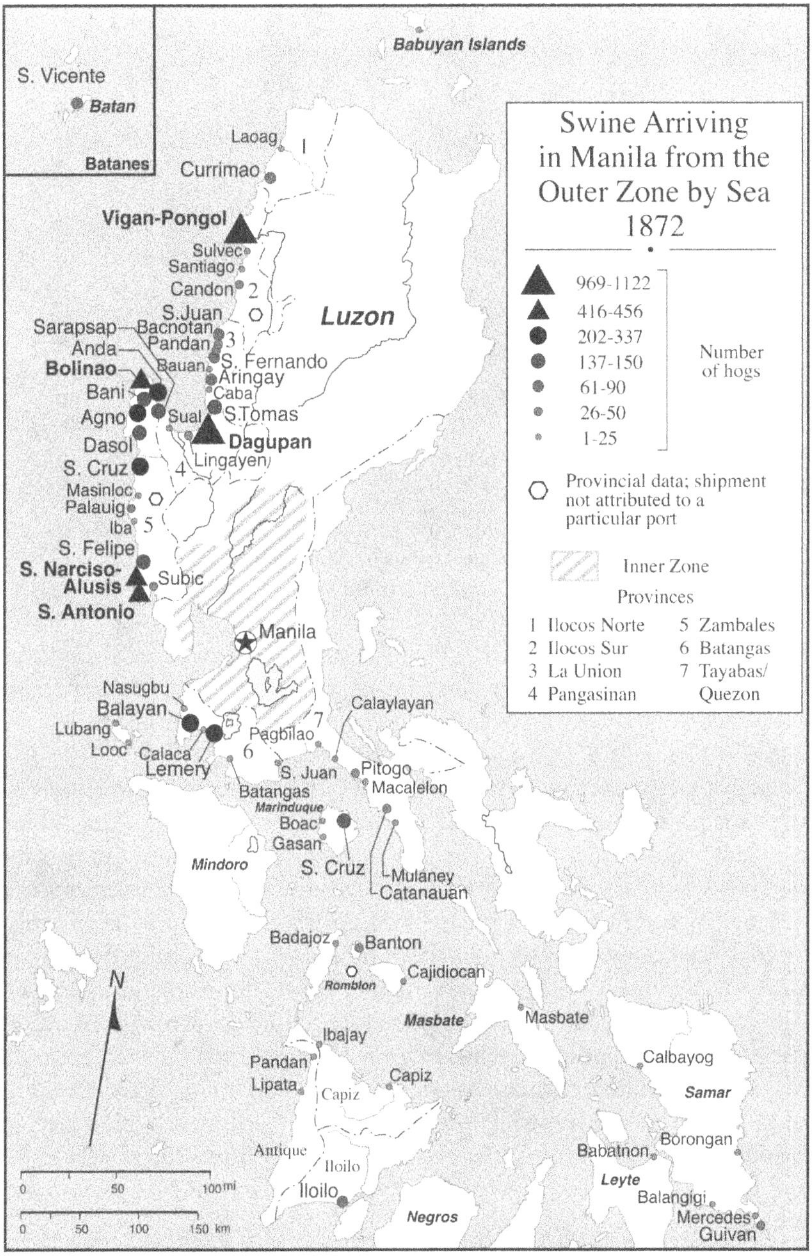

Swine Arriving
in Manila from the
Outer Zone by Sea
1872

Symbol	Number
▲	969-1122
▲	416-456
●	202-337
●	137-150
●	61-90
●	26-50
●	1-25

Number of hogs

⬡ Provincial data; shipment not attributed to a particular port

▨ Inner Zone Provinces

1 Ilocos Norte   5 Zambales
2 Ilocos Sur   6 Batangas
3 La Union   7 Tayabas/
4 Pangasinan   Quezon

Map labels: Babuyan Islands, S. Vicente, Batan, Batanes, Laoag, Currimao, Vigan-Pongol, Sulvec, Santiago, Candon, S.Juan, Luzon, Sarapsap—Bacnotan, Anda—Pandan, Bolinao, Bauan, S. Fernando, Aringay, Bani, Caba, Agno, Sual, S.Tomas, Dasol, Dagupan, S. Cruz, Lingayen, Masinloc, Palauig, Iba, S. Felipe, S. Narciso-Alusis, Subic, S. Antonio, Manila, Nasugbu, Balayan, Calaylayan, Lubang, Looc, Calaca, Pagbilao, Lemery, S. Juan, Pitogo, Batangas, Macalelon, Marinduque, Boac, Gasan, Mindoro, S. Cruz, Mulaney, Catanauan, Badajoz, Banton, Romblon, Cajidiocan, N, Masbate, Masbate, Ibajay, Calbayog, Pandan, Capiz, Lipata, Capiz, Samar, Antique, Iloilo, Borongan, Babatnon, Leyte, Iloilo, Balangigi, Negros, Mercedes, Guivan

MAP 7.1. The Origins of Hogs Arriving in Manila from the Outer Zone by Sea, 1872

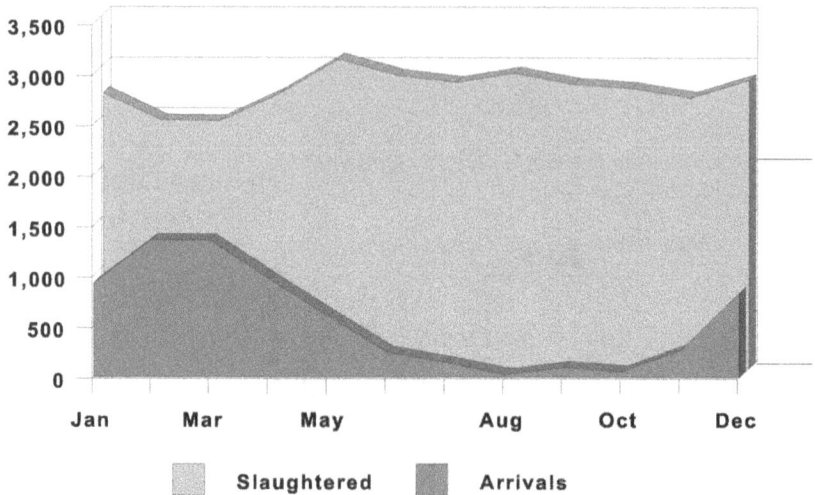

GRAPH 7.1. The Manila Pork Supply, by Month, 1872: Hogs Slaughtered for the Supply of Public Markets and Swine Arrivals by Sea from the Outer Zone. (Compiled from *Gaceta de Manila*, various issues.)

be the last half of February, almost all of March, and sometimes a good part of April. The old church laws called for 40 days of fasting (i.e., except Sundays). . . . In the Philippines, as derivative from the Spanish Bula de la Cruzada, the rule was no meat on the seven Fridays of Lent."[46]

One can see Lenten practice already in operation in the Philippines in 1598, when the governor-general and *audiencia* ordered the nearby provinces—those required to send provisions to the new Spanish city—to substitute eggs for shipments of meat animals and fowls during Lent.[47] Three centuries later this stricture applied in church law to both Spaniards and Filipinos but in actual practice only to those who could afford meat. Clearly this rule affected and guided operations at the city slaughterhouse. It also affected popular behavior. The result was the dip in consumption implied in the monthly record of slaughter. At the opposite extreme, May celebrations—parish fiestas, Flores de Mayo, and wedding celebrations—often featured *lechon*, a clean, medium-small pig turned into a special delicacy by careful roasting on a spit. Pieces of well-basted skin and the crunchy ears were special delights. During the 1870s and early 1880s, more hogs were slaughtered in the city in May than in any other month.

The historical evidence for Lenten abstention and fasting strikes a strong chord with older Filipino Catholics today. One lay testimony can stand for many.

During Lent . . . all Catholics are expected to fast every Friday. Fasting means avoiding meat of any kind. I remember how strict it was before the war when I was growing up. . . . On Good Friday, fasting means eating only one full meal. One is entitled to two light meals. . . . At noon, or early afternoon, we eat chocolate porridge and dried fish. Supper may be a bit heavier [or vice versa]. . . . [Our cook] prepares a Spanish dish of codfish in olive oil, which we eat with rice; this is the heavier meal. This codfish dish is a Lenten recipe for many traditional houses, mainly among the middle class. It is called *bakalao*, referring to the codfish. The lower classes, however, cannot afford *bakalao*. The chocolate porridge, or *champorado*, eaten with dried fish is a more affordable traditional Lenten meal.[48]

The Spanish practice of eating bakalao (or *bakalaw*) was widespread in Manila, and it led to the substantial import of this special fish—as opposed to designating one from Philippine waters. From 1854 through the 1890s, trade statistics allow us to see this flow—from Liverpool and Hamburg with a trickle from Spain in the 1860s and from China, Singapore, England, Germany, and even Scandinavia in the 1890s. Likewise one sees pickled tuna advertised as "a good dish during the fast."[49] These deeply held Lenten dietary habits were strongly reflected in the seasonality of the meat supply in the city.

Animal Buyers in the Late Nineteenth Century

There were apparently no open swine markets in the provinces in the nineteenth century.[50] Rather, the calendar of urban market demand was communicated to the producer households by custom and more directly through buyers operating in their locale. In the 1890s, when they stand out in the provincial records of the *contribución industrial*, we can see these buyers as licensed "speculators and dealers in swine." Others in the food system were licensed to deal in cattle and carabao, and still others worked as dealers in meat or as *tablajeros cortadores* (retail meat cutters and vendors) in particular localities.

Who were the hog buyers for the city? At the level of buyer-speculators in hogs, they were ethnically mixed in the 1890s. Ten were licensed in Batangas in that decade—4 Chinese and 6 Filipinos. The 4 Chinese were based in the interior at Lipa and Tanauan. All 6 Filipinos were on the intensely commercialized coast, especially in Taal and Batangas towns. In Tarlac Province in the Central Plain, the tax licenses of 14 hog dealers have survived, all apparently Filipinos. Seven were based in the important rice town of Camiling. Three each were located in Moncada and Anao and 1 in Victorias. This last was a woman, one Segunda Marbel, specifically licensed as a *viajera*, or traveling buyer. Three of the men in Camiling were also recognized as traveling buyers. In Candaba, Pampanga, the 4 hog buyers were all Filipinos. Nearer to Manila, the buyer in

Antipolo was Chinese. Likewise, all 3 apparent hog buyers in Cebu City were Chinese. So at the level of wholesale purchasing, the ethnic pattern was mixed and locally variable. All but 1 of the identified swine buyers were men.

A somewhat different mix of ethnic participation emerges at the level of the local fresh meat systems. In Tarlac at the retail meat level, all the *tablajeros* and meat dealers in Camiling were Filipinos and almost half were women, but in the rest of the province Chinese outnumbered the locals. In Pampanga (excluding Candaba) the split was 46 Chinese and 10 to 13 non-Chinese. It was the same in Nueva Ecija, 14 to 7. In Cavite Province, it was 61 Chinese to 14 indigenous *tablajeros*, and 5 Filipino to 3 Chinese meat dealers. At a distance, in urban Cebu, 9 of 12 pork vendors in 1891 were Chinese.[51] At the end of the nineteenth century, Chinese participation in the everyday fresh pork delivery and retail system was substantial.

In the twentieth century, many of the hog buyers would be called *viajeros* in the sense of being traveling buyers. Primarily, this was a special occupation, a matter of basing oneself in a locality for a time and putting out the word concerning one's willingness to buy. Often the buyer visited the seller, assessed the size of the pig, and paid the negotiated price. In more recent decades, the *viajero* would truss up and weigh the pig on the spot—the animal protesting vigorously. In more remote places, the broker may have moved with the local system of periodic markets and bought what was offered for sale on market days. Purchased stock needed to be fed, cared for, and protected from theft or loss, so it was essential to move them fairly rapidly through the system. In the days before railroads and trucks, that meant getting them to a landing and off to the city in lots moving by flat-bottomed *casco* in the inner zone or coastal sail craft from the outer zone. During 1872 swine arriving by coastal sailing vessel came in 436 lots averaging 16 animals each. Only a few shipments (14) arrived with 50 hogs or more. The largest shipment (with 75) came from San Vicente in Batanes, a small place with infrequent shipping services. Batanes was a special case. Until the mid-nineteenth century, hogs were raised there for sale to whaling vessels. Thereafter local coasting vessels (*pontines*) "of about 20 tons," made two or three trips a year to Manila and Aparri. After 1900 in ordinary years, these vessels returned with "petroleum, ironware, cheap fabrics, matches, and in bad years rice."[52]

Inner zone families evidently produced more than three-quarters of the hogs slaughtered for consumption in the city in 1872. But where did this domestic supply come from more precisely? In a survey of livestock in 1886–87, Batangas officials reported that there were more than 53,000 hogs in that province—the second-highest provincial total—and that approximately 11 percent of these, or 6,000 hogs, were sent to the metropolis each year. Likewise, the reporting officer in the interior Ilocos province of Abra mentions some 2,000 being sent.

These would have been shipped via Vigan downstream. Other provinces listing Manila among the destinations of 100 hogs or more included Ilocos Norte (though more of its hogs were destined for Cagayan), La Union and Tarlac (both with some of the hogs going to Pangasinan), Cavite, Morong (now eastern Rizal), and Leyte.[53]

The census of 1903 provides a more comprehensive record. It confirms that swine rearing was extremely widespread and that pigs were by far the most common red meat animals raised at 17 per 100 people enumerated. Cattle, their numbers catastrophically depleted by rinderpest, and goats were next at less than 2 per 100. In the more "normal" times that followed, hogs easily held the lead over the larger meat animals. In 1939 hogs were reported at 27 per 100 of the national human population, while cattle had recovered to 8.4 per 100. Using the per capita incidence of swine in 1903 as a rough basis for comparison, I have calculated the same index using 1870 data.[54] In both years, swine were most common in the Cagayan Valley at more than twice the national average. This was followed by the Central Plain (except Pangasinan), Ilocos Norte and Abra, Batangas and Laguna, and Negros. In neither year was there any particular concentration in the outer environs of the city, while Bikol, Samar-Leyte, Panay, and the city of Manila itself were all below average. The two lists reveal some change over three decades: hog raising in Cavite surged to match that of Laguna and Batangas, Cebu regained its position, and Negros Oriental joined it to become a notable swine producer.[55] But mostly one is struck by the rough correspondence between these lists (see table 7.1). As expected, the notably low production areas match up in a general way with the areas in the outer zone sending the fewest swine to Manila during 1872.

De la Cavada, the statistician, partially lifts the curtain on production and supply in the inner zone. He reports that Laguna Province sent almost 4,000 hogs to market in Manila in 1870. In other words, this one nearby province sent a number of animals that exceeded half the total outer zone flow! Clearly commercial stock raising reached much greater intensity in some areas with easy access to the urban market. Laguna was already a major center for raising coconuts and pressing the pulp of the nuts in order to extract the coconut oil used for lighting and cooking, as we have seen. This activity produced coconut meal as a by-product, an excellent animal feed or supplement. Sawyer says that use of this by-product was divided between pig feed and a fertilizer for coffee plants. Still, there is no sign of livestock raising in concentrated piggeries using coconut meal or the more widely available rice bran and broken rice as feedstock. Writing in the 1880s, Montero y Vidal points to both Batangas and Laguna provinces as the places that provided Manila with quality animals. In the 1890s, the port of Batangas was said to be sending out 3,000 to 4,000 hogs per year. The Batangas municipality of Taal, as well as Balayan and Lipa, also sent "beasts" (*reses*) to the Manila abattoir. Calaca and Rosario were also important centers for hog raising. When peace finally returned after the Philippine-American War both shipped animals to Manila every week. Calaca is located in the same coastal zone that includes Balayan and Taal-Lemery, important regional ports for hog shipments in the 1870s.[56]

So the overall geography of supply was broad, but with a notable concentration in two close by, but not suburban, southern Tagalog provinces. In the days before modern forms of transportation, many hogs came from beyond the relatively nearby places supplying chickens, ducks, and eggs.

More detailed patterns of swine ownership in parts of the city and suburban zone emerge from the ill-fated Spanish census of 1896. Although this never-completed census was not a proper house-to-house enumeration of the human population, statistics were nevertheless presented on economic animals, apparently compiled by survey. The conclusions for 1896 fit well with those derived from the counts cited earlier. The available returns from four districts of the built-up urban area enumerate 374 swine plus 845 more in parts of the ring of surrounding municipalities. The largest owner thus revealed was one Mateo de Vega y Oliveros in Navotas, with 15 hogs, and there were 7 other owners scattered in Sampaloc with 6 or 8, but most who kept pigs had 1 or 2. In many districts, it was unusual to encounter anyone with as many as 5. In sprawling Sampaloc, then on the edge of the city, 120 owner households kept 235 pigs. In Santa Ana, 44 families owned 54 pigs. Many others owned none. Only 2 owners in the entire set were Chinese, both on a tiny scale.[57] We may conclude that, other than Mateo de Vega, there were probably few notable concentrations of swine raising in or near the city in 1896 and that most urban and suburban pigs were owned by indigenous Filipinos who fattened them for market or fiesta.

Live hogs imported from other countries never became a significant part of the urban diet. Between 1854 and 1890, fewer than ten hogs per year were specifically listed among the import flows. In the 1890s and early 1900s a few hundred came from China and Japan. Saigon was exporting thousands of live hogs per year during 1900–1902, but there is no comment on the destination of these animals. Preserved pork was a different matter. Between 16,000 and 20,000 Chinese hams were received annually in the 1850s and 1860s, and European-style cured hams also arrived in some quantity from England and China and in lesser amounts from Spain and the United States in the same years. The British provisioner H. J. Andrews was advertising hams from York in the 1890s, while in 1902 the Pacific Oriental Trading Co. was offering hams from the huge American meat packer, Swift and Co.[58] Imports of canned meats from the United States expanded smartly following a tariff reduction in 1902. In addition to hams, pork sausage products from Germany were regularly advertised for the foreign merchant community and other upscale consumers in the 1890s. These imports were followed in the early twentieth century by the establishment of Max Druseidt's German Sausage Factory in the city itself.[59] *Tosino* (bacon) and other salted and smoked meats tended to be retailed by special provisioning shops. Hams continued to be a regular item of import—amounting to more than 600,000 kilos in 1935—and China continued to be a major source of pork imports, with the United States and Australia as secondary providers.[60]

Arriving in the City

In the early twentieth century, the city landing for hogs coming from the provinces was the point called Murallon. This was a site at the mouth of the Pasig River in San Nicolas district. A preliminary screening was done there by an official of the Bureau of Agriculture in order to eliminate from the supply stream some of the downed hogs (too injured or sick to stand up) and those with obvious signs of hog cholera—usually reddish hemorrhages beneath the skin of the abdomen.[61] Hog cholera was a serious menace. It took 4 to 14 days to incubate and in its more severe form usually killed the animal in a day or less after becoming apparent. In an outbreak more than 80 percent mortality among exposed animals was not unusual. Many cases were contracted directly during movement to market—another reason for moving the purchased hogs quickly. In other cases it was transmitted by contaminated feed, on shoes, from large animals, or in wallows. Hogs passing the initial inspection at Murallon were then moved to a corral near the slaughterhouse.

Already by the mid-nineteenth century, city authorities had established an official abattoir together with a requirement that the slaughter of all major animals for food in the city must be done there. It appears, however, that many hogs were slaughtered privately before 1872 and some even after. Operations were shifted to a new slaughterhouse in 1893. This was the Azcarraga Matadero on the *paseo* of the same name (now C. M. Recto), and its task was to supply fresh meat to the public food markets. This was done, ultimately, by private pork dealers who brought their animals to the abattoir for slaughter and cleaning and then distributed the fresh meat. This Matadero remained in operation throughout the prewar period. In the late nineteenth century and early twentieth, many of the butcher workmen were Chinese (figure 7.1).[62]

Both urban and rural Filipinos fattened a pig for market on household wastes. These often included human wastes. Allowing swine access to human waste could be a problem. Sawyer explicitly warns Europeans to avoid Philippine pork precisely because of this. Some late Spanish authorities were concerned about trichinosis as a major cause of human mortality—if it was that—so the workload of the veterinary inspectors at the municipal abattoir was doubled in 1882 by a royal order mandating a post-mortem inspection for the disease.[63] At

FIGURE 7.1. A Chinese butcher workman demonstrates his technique in the Manila abattoir around 1900. (USNA II, RG350-P-E-20-2, box 21)

the Manila abattoir in 1907, some 300 carcasses were condemned as unfit for human consumption because of tapeworm cyst infections. Upon investigation by the Bureau of Health, the authorities reported that "as there are no modern sanitary sewage systems in the municipalities of the provinces, swine are looked upon as the natural scavengers and are allowed free access to human excreta, from which source they become infected." In 1908 both Victor Heiser, director of the Bureau of Health, and Dean C. Worcester, secretary of the interior, acknowledged the public health problems inherent in hogs feeding on excreta, but both felt this risk paled in comparison to their priority work on clean water systems and against diseases such as smallpox, cholera, and leprosy.[64]

Many of the cyst-infected hogs found in Manila were from Batangas, but the practice was widespread. At the Manila abattoir during 1926–35, cycsticero-sis was the leading cause of whole carcass swine condemnation at 1.3 percent of more than 1 million animals. Again, of the provinces providing more than 1,000 hogs for slaughter in Manila in 1935, only Batangas had an above-average rate of cyst infection. This is a paradox because Batangeños led the archipelago in the ownership of purebred hogs and in commitment to careful husbandry practices for cattle. In addition to cysticercus, at least seven other Helminth parasites (worms) were common in the digestive systems and lungs of Philippine hogs.[65]

Feeding on wastes from infected persons was also the cause of tuberculosis in hogs. More than 100 whole hog carcasses were condemned for this in both 1916 and 1935, and more than 1,000 tubercular carcass parts were condemned in 1917 and 1927–30. With increasing scientific awareness that they were dealing with widespread actual tuberculosis, the heads of almost 20,000 hogs were condemned in 1934–36, about 4.5 percent of the total. Some of these animals were raised in Manila itself. To the extent that it had a rational utilitarian basis—many such preferences and avoidances do not—the preference for beef among Europeans living in Manila may have been related to the fear of trichinosis, tuberculosis, and other diseases in pork.[66]

Most hogs were hardy and usually successful in foraging for themselves, but allowing village animals to run at large also made it easy for hog cholera to spread. In 1908 the Bureau of Agriculture reported that this disease had "practically ruined the swine industry. . . . [But it] has not received a great deal of attention, owing to the fact that rinderpest [in cattle and carabao] has been considered the most important disease [of] domestic animals, and has claimed almost the entire attention of our small force." With a small technical staff and considering the economic devastation of rinderpest, it is difficult to fault this decision, but it was also typical of foreign authorities to place a higher priority on beef than pork. As a result of the cholera, the supply of pork to the public markets in 1908 and 1909 declined notably, accompanied by a partially offsetting increase in beef. There were further outbreaks of hog cholera in 1916 and 1920, in each instance accompanied by substantial swine mortality.[67] In the following year, 1921, there was again a notable decline in the supply of fresh pork to the public markets, once more partially offset by an increase in beef (graph 7.2). It was left to the government's Veterinary Research Laboratory to produce a serum to immunize hogs against this form of cholera. Given the way most swine were raised, it was slow going. Following the end of bovine rinderpest devastation of the 1930s, hog cholera came to be seen as the number-one livestock disease.

～

There was little change in swine-raising practices in the early twentieth century. In 1901–2, an American was trying to raise hogs in San Juan del Monte, "feeding them with refuse collected from the hotels" of the city.[68] But there were still no large-scale piggeries on the eve of World War I. Rather, as before, some families kept one or two "brood sows." The indigenous domestic hogs, though shaped like wild boars, were pronounced "wonderfully prolific and hardy" by the new animal science authorities. Like chickens, "Very little attention is given to them, and they are allowed to run at large . . . compelled to forage for the greater part of their subsistence."[69] In urban areas, young pigs were often kept about by knotting a tether cord through a slit in an ear. In the Manila abattoir during 1928–33 slaughter hogs averaged 48 to 51 kilos dressed

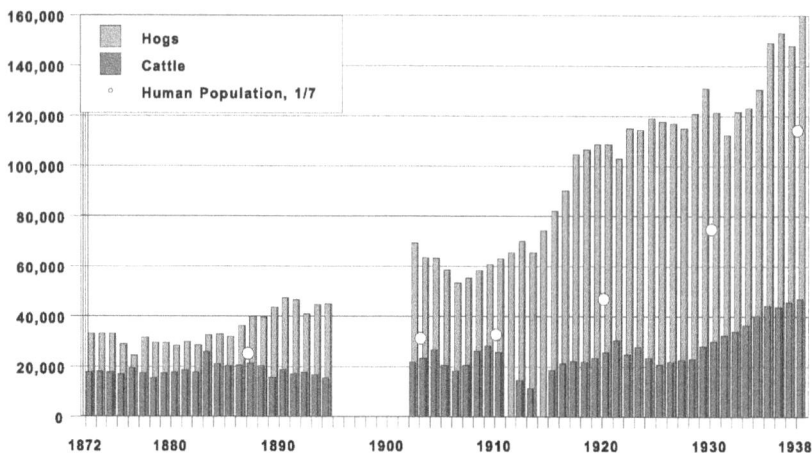

GRAPH 7.2. Hogs and Cattle Slaughtered for Consumption in Manila in Relation to the Metropolitan Population, 1872–1938. Data are missing for both animals for 1896–1901 and for cattle for 1911 and 1914. Symbols approximating one-seventh of the human metropolitan population are shown. Consumption of both animals failed to keep up with population growth in the late 1920s and 1930s. Missing data are estimated for one month in 1872, 1893, and 1901–2 and for two months in 1890. The data for cattle also include carabao for 1928–38. (Compiled from *Gaceta de Manila,* 1872; *El Comercio,* "Matanza de reses," 11Jan1881 (for 1876–80), 13Jan1885, 1Feb1886, 15Jan1887, 9Jan1888, and monthly reports for 1890–94; Report of the Municipal Board of the City of Manila, 1901–2, III; RPC 1902, pt. 1, 137; RPC, 1903, pt. 1, 650; ms. Report of the Municipal Board of the City of Manila, in ms. RPC, 1910, 547–48 [for 1906–10]; PAgR 5:12 [Dec12]: lxvi–ii, and 6:12 [Dec13], 650–55; ms. annual reports of the Collector of Internal Revenue, 1917, 28, and 1919, 27; ms. annual reports of the Bureau of Agriculture, 1916–29 [esp. 1919, 39]; ms. annual reports of the Bureau of Animal Industry, 1930–33 [all reports in ms. ARGGPI for the years indicated]; SBPI, 1928, 59 [for 1924–28]; and BPS, 1939, no. 4, 109–22 [for 1929–38].

weight, which dropped to 42 to 43 kilos in 1935–38. In truth there was no standard size. Every day saw a wide variety. Many hogs were sold at lower weights since good prices were paid for animals (*lechones*) slaughtered at 11 to 18 kilos (25 to 40 pounds). Others were sold early because of a pressing need for cash. But when pigs were allowed to reach adult size, they could easily weigh 68 kilos (150 pounds) or more dressed weight.[70]

If the husbandry systems were changing but slowly, some progress was being made with the statistical system. Annual animal census numbers still came from special forms filled out by municipal officials, and apparent year-to-year population changes were as often due to the vagaries of this system as to any real change. The fantastic rates of increase in swine numbers reported during the early 1920s turned out to be fictional, and, on a reconsideration, the official national total was abruptly lowered from 10 million animals in 1927 to 2.3 million for 1928, explained as a "decrease . . . that did not really exist." Thereafter, annual changes in swine numbers have greater consistency, growing haltingly to a reported 3.1 million in 1936. The national census in 1939 returned hog numbers of 4.3 million. The 1928 disjunct in hog numbers has a parallel in 1872. The monthly slaughter of hogs in the city appears to have been underreported by half until mid-1872 when the authorities admitted, "[F]rankly, we don't understand" the discrepancy.[71] Possibly more were slaughtered but not reported so that some persons in the system could pocket the fees. The number of hogs reported at slaughter rose sharply in the late 1880s–90s. In this decade the treasury received its slaughterhouse payments in lump sums directly from the tax farmer/contractor—leaving less reason to conceal the larger numbers. Clearly, learning the statistical system is critical. Finally, in 1933, the Bureau of Animal Industry began to interest itself in tracking the larger supply of both hogs and poultry to the city. Armed with a new form, "[L]ivestock inspectors [were] assigned at the river front along the Pasig River, piers, Tutuban Railroad Station, Velasquez Corral, and in the Pritil, Quinta, and Paco Markets." Although still incomplete, the urban statistical system was now extended for a time to the places "where most animals are usually unloaded."[72]

Several insights come from this effort. First, with regard to the seasonality of consumption, the traditional Lenten downturn persisted but became more complex. May as a peak of consumption was replaced with most of the rainy season—July through October—plus the December holidays as the highest hog slaughter months of the year. Increasingly, increased pork consumption tended to match the season of low fish supply and to approximate that of chickens. One interpretation is that as the securely employed urban middle classes grew rapidly during the 1910s and 1920s and as railroad and truck transport came into general use, urban pork consumption tended to even out and become less episodic. With the departure of much of the orthodox military and some of the lay Spanish population of the city as a result of the change of regime in 1899, one might expect to see a decline in Lenten meat avoidance. But that was not the case. With improved forms of overland transportation and a broadening in the seasonality of higher consumption demand, prices tended to flatten out and become more constant throughout the year.[73]

Commercial Networks in the Twentieth Century

The main buyers of hogs in the city in the mid-1930s were "Chinese pork vendors in the different public markets of Manila, who operate either through middlemen or by themselves" and who purchased during 1933 through the La Loma stockyards. The Department of Commerce, however, advised the public that "it would be more advantageous for you to sell your pigs in . . . [the corral] on Velasquez Street, Tondo, Manila, where buyers and sellers meet every day to do business."[74] Left unstated was the fact that the Velasquez corral was a smaller, specialized stockyard owned and operated by Filipinos. In 1937 it was the only city stockyard that specialized in hogs. Initially on the edge of the built-up portion of Tondo, the Velasquez corral came to be surrounded by the homes of those involved in supplying hogs to the city. A whole network of traveling buyers/*viajeros*—some called them commission agents—worked out of this neighborhood, operating in many cases on credit advanced by Tagalog merchant capitalists who also lived in the neighborhood. Commercial capital was accumulated and lent by the *ventadores*, the businessmen hog dealers. The successful *ventador* gradually built a network of scores of *viajeros* who went out to buy hogs. The *viajeros* and the *ventadores* that backed them were competing with Chinese networks, but there were areas (unnamed) where Chinese buyers feared robbery, effectively leaving the territory to the Filipino competition.

The *ventador* handled the wholesaling of the livestock. There was no public auction, just a one-on-one transaction between the prospective buyer and the *ventador*—not unlike the whisper system used at the Tondo fish landing. If the offer was accepted, then the buyer would transport that batch of hogs to the public slaughterhouse by truck and later return there to pick up the meat. Usually, the buyers coming to the corral were Chinese, and they were often public market stallholders. Others were supplying a restaurant, hospital, or school.

The *viajeros* associated with the Velasquez corral mainly traveled to places in Luzon, including Central Luzon and Bikol before World War II, although some also went on to Negros and other parts of the Visayas. In several cases, the *ventador*'s network of hog buyers was eventually passed on to a son and continued in operation over two or more generations. The Velasquez corral was built about 1932 on marshland along the Canal de la Reina. In the 1930s, hogs were frequently delivered there by *batele*, a small sailing vessel, for example, from Zambales. They also arrived by river from Laguna. By the mid-1930s, trucks had become a form of major transport—often traveling at night when the roads were unclogged and the heat stress on the animals was less. After the war it became quite common for the *viajero* to rent a truck and use it to send back loads of hogs from, say, the Cagayan Valley in the northeast or from Central Luzon. The buyers now also traveled by airplane to Mindanao, sending back animals on the open decks of interisland freighters.

Still based in the Velasquez-Nepomuceno neighborhood, today one finds whole families of sons and some daughters who have followed their fathers into this occupation. Other *viajeros* were and are *provincianos*. For a very long time, the actual slaughtering was done at the city's Azcarraga abattoir, as we shall see. Finally, after World War II, this facility was replaced with a new slaughterhouse built on reclaimed land farther north in Vitas, Tondo.[75]

The Geography of Supply

A proper statistical profile of the origins of the hog supply of the capital can be constructed for the 1930s. In 1934 and 1935, some 86 percent of all hogs sent out by the 50 provinces (and captured in the statistical system) came to Manila. In the latter year, 150,800 hogs were slaughtered in the Azcarraga Matadero. Almost 9,200 of these (6 percent) came from Manila itself and suburban Rizal. The rest were supplied by the provinces. Among these, Batangas led with 41,000 (31 percent). From Lipa, hogs were "brought into Manila in truckloads . . . as fat as Berkshires and mostly of that blood." In the 1920s and beyond, Batangas was a notable corn surplus province, and no doubt the density of hog production was in part related to that.[76] Batangas was followed by Pangasinan and Nueva Ecija at 19 and 12 percent of the total, respectively. Both of these were provinces where rice bran and broken rice were routinely available for use as the major components of hog feed.[77] By 1939 Pangasinan led the nation in the commitment of its farmers to raising imported and interbred hogs. Almost 10 percent of the total supply now came from the Cagayan Valley, most notably Isabela (11,800). Laguna and Cavite together supplied another 6 percent, while 3.5 percent came from Ilocos Sur–La Union. In the 1930s the hogs supplied from Ilocos were ordinarily native hogs as opposed to the increasingly common admixture of imported breeds typical of slaughter animals from Batangas and Central Luzon. Beyond Luzon, Cebu was now a major domestic exporter of swine at more than 17,500 head per year, followed at a distance by Iloilo and Negros Occidental. However, none of these then supplied any significant number to the metropolis. Bikol, the Visayas, and Palawan were essentially outside the hog supply shed of the city, and altogether they sent fewer than 800. Sulu and the Muslim portions of Mindanao of course sent none because of their avoidance of hogs and pork. Beyond the top provincial suppliers there is little evidence of high commercial concentration; rather, a great many places now forwarded hogs to Manila, which in turn increasingly organized the market for hogs on most of Luzon.[78]

The principal supply zone, then, included almost the entire territory stretching from south of the city in Batangas all the way north to Vigan in Ilocos Sur. The major change since the 1870s was the addition of parts of the Cagayan Valley. But with by far the highest per capita swine production over 70 years, the flow from the provinces of Cagayan and Nueva Vizcaya was grossly underrepresented in the urban supply. Something similar could be said of Negros Oriental after 1900 (table 7.1). Relative proximity, cost-distance as it is called in geography, continued to count for a lot. Improved infrastructure, especially roads, also helped.

Most hogs were still being raised individually—an average of 2.7 per farm grower nationally in 1939 with only modest provincial variations. Further, the type of farm made little difference for the average numbers of swine. Almost 16,000 hogs were counted in urban Manila itself that year. Indeed, a third of all pigs in the archipelago were being raised off farm in aggregations that averaged 2 animals in Manila, as in the provinces.[79]

Taken together, this forms a remarkable record, a measure of the increasing, though still subnational, commercial integration proceeding in the archipelago, and of the adaptive persistence of long-standing household swine husbandry practices. In this system of pork provisionment on the eve of war, one sees a rising prevalence of imported and hybrid stock in the domestic stream supplying the city, but there is as yet little hint of the rise of huge integrated operations four decades later in which feeds are produced and hogs raised and fattened in large numbers, slaughtered, and the meat processed and delivered directly to grocery stores—all under one corporate management team.

Other Animals

Swine and bovines provided a great proportion of the red meat consumed in the city, but they do not exhaust the list. Throughout the second half of the nineteenth century small numbers of sheep were slaughtered every week in the Manila abattoir for the consumption of affluent Spaniards and other residents. Most of these animals were imported from or via Hong Kong or Shanghai, for they tended not to do well in the Philippine lowlands. Eventually the cause was shown to be an intestinal parasite. Almost all the live imports were for slaughter. Few sheep came to the city from within the country. Although many of the consumers of mutton were foreigners, the mestizo elite also developed a taste for it, occasionally incorporating roast mutton as part of the social display at wedding or other feasts.[80] The price for mutton per kilo was approximately 25 percent higher than for beef in the Quinta Market in the 1910s. Manila also had a small number of residents from the Middle East who would have been consumers of this meat. There was even a ritual slaughterer operating in the city—"a Baghdadi Jew" whose occupational niche also included small-scale "dairy farming and raising sheep, the latter often sold to Middle Easterners."[81]

Horsemeat was also occasionally consumed, but unlike sheep, horses were not ordinarily butchered in the public slaughterhouse. Occasionally, however, the clandestine untaxed slaughter of horses, as well as other animals, came to the attention of city authorities. Without comment, the 1903 census reports that 60 horses were slaughtered in the city of Manila during 1902—versus almost 2,000 in Batangas, Tayabas, and Laguna and 4,270 in the entire area of the archipelago covered by the enumeration. Many of these were horses that had become incapacitated and then were slaughtered for their food value.[82]

Eating goats or dogs was more common. Among Filipinos, less affluent urban men occasionally roast and consume a dog or goat as a *pulutan*, something to go with alcohol when a serious bout of male bonding is in prospect.

In that sense, both are feast animals; rarely are dogs raised or treated as "pets." They may also be consumed in family settings, but neither is ordinarily an important component of anyone's ongoing diet in the city, although one version or another of goat stew (*kaldereta*) is common enough among those who can afford it. A few Muslims lived in Manila in the nineteenth century, and their numbers expanded thereafter to 100 in 1903 and (almost unbelievably) to more than 14,000 in 1918 according to the census. Together with the tiny Jewish community in the city, they would have been likely consumers of both sheep and goats. Sawyer reports that both meats were occasionally served at Christian wedding feasts along with *lechon*. The diminutive goats, at least, could readily have been slaughtered at home. The city abattoir reported 1,052 "sheep" slaughtered for consumption in the city during the fiscal year 1910–11. However, in the same year, fees were collected for the transport of 1,033 goat carcasses from the slaughterhouse to the public markets but only 132 sheep. "Sheep," or *ganado lanar* in Spanish records, seems to be a category that often included goats.[83] In any case the totals are small.

Occasional consumption of dogs is a practice of great historical depth in the region, but in the twentieth century it is doubtful that a majority of Filipino urbanites participated in it even occasionally. The practice is said to vary by language group in the lowlands while remaining most common in the uplands. In Manila society, members of certain language groups are stereotyped by the Tagalogs as likely dog consumers—though dog consumption is known in the Tagalog provinces as well. Since dogmeat is a *pulutan*, a sizable percentage of those who consume it in Metro Manila are male.[84]

The sale of both goats and dogs in Manila takes place outside the context of public markets. In the case of dogs, it is semiclandestine as it violates western norms, although the Philippine Catholic Church does not condemn it as such. The dogs are ordinarily sold alive and whole and can readily be seen to be in reasonable health. Eating an already slaughtered or dead dog risks a gruesome death from rabies. There is nothing like the public marketing of portions of butchered dog carcasses as in, say, Canton.[85] During the mass starvation of World War II, most dogs in the city were eaten.

Goats were occasionally encountered in the sixteenth- and seventeenth-century Philippines even beyond the principal Muslim contact areas in the south of the archipelago. Writing in 1609, Antonio de Morga says that goat meat in the Philippines was not very savory and the animals did not do well because they ate "certain poisonous herbs." Commenting on Morga in 1890, Jose Rizal said that goats were held in low esteem. Even though breeding stock was "often brought" from Mexico, the animals failed to multiply. In 1862 two small shipments were received in the city from the island of Lubang off Mindoro and from Zambales.[86] In 1870 goats were quite spectacularly overconcentrated in the tiny Batanes Islands, a rocky place where crops are regularly damaged by typhoons. They were also moderately common in Pampanga and Cebu. Elsewhere they were broadly found in small numbers. The livestock survey of 1886–87 omitted goats and sheep, but 1,000 of each was reported by Misamis Province. Some goats and sheep were recorded in Batangas municipality in the 1890s. In 1919 and 1920, some 1,276 and 1,476 goats, respectively, were slaughtered in the Manila abattoir versus only 80 and 158 sheep.[87] At 2 percent or less of the recorded national slaughter, it seems that the city of Manila was not a major center of consumption.

In the 1930s, to the limited extent that goats were picked up by the statistical system, one finds them entering interprovincial commerce from Cebu, Ilocos Sur, Pangasinan, and Nueva Ecija, each sending out 1,000 or more per year. The 1939 census reveals that the highest concentrations and highest per capita raising of goats were in the Ilocano provinces of La Union and Ilocos Sur at 10 and 8 per 100 residents, respectively, and also in Cebu—on more than a quarter of all farms—and Negros Oriental, both also at 8. The hills of Cebu and Ilocos were rapidly becoming the most environmentally denuded places in the nation. (The people of Batanes were raising goats at similar rates but not sending them to Manila.) After the war, anthropologists observed of an Ilocano village, "Most families have at least one or two goats, which are slaughtered by the owner for goat feasts or sold locally for this purpose," and "Although dogs are occasionally raised for meat, few . . . families do so—they are considered expensive to feed."[88] Goats were inexpensive to feed since they had a broad diet and could be raised on grasses alone.

There were other occasional food items. In Tayabas and elsewhere at the end of the nineteenth century, little rice-eating birds called *mayas* (or *munias*) were eaten in *adobo*, especially, though not solely, by poor rural folk. Sta. Maria mentions "mole crickets" and "locusts called *durun* in Pampanga, considered seasonal delicacies, [and] cooked like a meat *adobo* or *guisado* but allowed to crisp."[89] Under the starvation conditions of late World War II, rats and mice were consumed by the desperate in the city—a reasonable protein source if properly cleaned and cooked.[90]

~

Nothing at this point prefigures the powerful position various business organizations would eventually develop in supplying chickens to both the supermarket and fast food sectors. There was a small "supermarket" sector before the war in the sense that certain purveyors specialized in high-quality foodstuffs, especially imported foods, and they also made abundant use of print advertising to develop and inform their clienteles. There were also cold storage companies that sold imported cuts of meat. But price made such businesses relatively limited in their appeal.

TABLE 7.1. Number of Swine per 100 Persons by Province, 1870, 1903, and 1939

Province	1870	1903	1939
Isabela	18	39	59
Batanes / Batan	56	39	54
Cagayan	19	32	47
Nueva Vizcaya	18	31	44
Nueva Ecija	17	29	44
Zambales	9	27	35
Pangasinan	4	16	32
Pampanga-Tarlac	12	29	29
Bulacan	9	29	25
Bataan	6	15	19
Abra	11	21	54
Ilocos Sur	7	14	34
Ilocos Norte	10	28	32
La Union	5	20	25
Batangas	9	23	30
Cavite	5	23	23
Laguna	8	22	15
Morong / Rizal	3	14	14
Manila	2	3	3
Albay-Sorsogon	2	7	27
Mindoro	11	8	25
Ambos Camarines	5	6	23
Masbate y Ticao	2	13	23
Tayabas	6	17	20
Negros Oriental	11	28	52
Bohol	7	17	39
Cebu	6	23	37
Leyte	8	9	37
Samar	11	8	23
Antique	4	7	19
Negros Occidental	11	13	16
Capiz	3	6	16
Iloilo	5	10	15
Romblon	13	15	12
Misamis	n/r	18	46
Surigao	n/r	13	37
Luzon	7	19	—
Visayas	7	—	—
Philippines	—	17	27

SOURCES: Calculated from Montero y Vidal, *Archipiélago Filipino*, 329, 342, 361; de la Cavada, *Historia Geográfica*, 1:372, 2:337; *Census 1903*, 4:236–43; and *Census 1939*, 2:1124–36.

NOTE: Provinces are grouped by geographical region and are listed from north to south. Figures in italics are well above average for that year. Morong / Rizal and Batanes / Batan are not exactly equivalent units. Not all provinces are shown. La Union 1870 includes Benguet, Tayabas includes Marinduque, and Negros 1870 is for the entire island. Underreporting may be a serious shortcoming of the earlier data.

8

Beef, Cattle Husbandry, and Rinderpest

BEEF, THOUGH NEVER a major item in the ordinary Philippine dietary, is important to this study in two ways. As an item of elite consumption, it was better documented than many other foodstuffs, while the importation of beef cattle precipitated epizootic crises that affected the entire economy.[1] Eating beef has become relatively common in contemporary Manila, following the advent of mass-market hamburgers skillfully retailed by the Jollibee and McDonald's corporations.[2] In colonial Manila, however, beef was primarily consumed by Spaniards and other foreigners, as well as some of the more affluent Filipinos, for whom it would have been a status marker. Ordinary Filipinos would scarcely have tasted it except on the occasion of a feast back in their home villages. Europeans, on the other hand, tended to feel deprived if they could not obtain beef regularly, and some advocated a meat diet as a protection for European constitutions in the tropics. Later the idea that red meat consumption was protective was well known (though not universally accepted) among American military physicians in the Philippine campaign.[3]

Northern Europeans found the beef available in nineteenth-century Manila less than tasty. Before refrigeration, the practice of developing the flavor and tenderness of beef by aging it was an impossibility. Further, except as jerky, beef does not cure into an acceptable product, in contrast to hams and other pork products. As a result, cattle were not particularly valuable; a cow cost less than a good-sized hog in Pangasinan in the 1850s, and fresh beef was less expensive than pork. It was the same in Bikol a decade later.[4] This changed as cattle became more important as draft animals in the sugar industry and as the number of potential beef consumers rose with the engorgement of the Spanish establishment during the last three decades of the century. By 1871 the price of beef had doubled in Manila and continued to rise.[5] The Spanish required that all cattle, carabao, hogs, and sheep intended for Manila's public markets be slaughtered in the city abattoir. By 1869 large numbers of cattle were being slaughtered; annual recorded totals tended to oscillate between 17,000 and 20,000 (50 or so per day), although in 1878 the reported number declined abruptly to 15,500. British consul George Mackenzie attributed this downturn to the fact that "the natives have been excessively poor, and they have been obliged to take cheaper food in consequence." This was due to a depression at least partly traceable to the effect of the 1877–78 El Niño on rice production.[6] The count rebounded to 17,000 to 18,000 during 1879–80, rose to 20,000 to 21,000 during 1884–86, and then declined to 15,400 by 1894 following the devastation of Philippine bovine populations by rinderpest.

CATTLE FROM THE PROVINCES

Unlike swine or fowl, domestic cattle are an introduced species—not of deep antiquity—in most of insular Southeast Asia. The major exception is Indonesia, where Bali cattle (and the original Java cattle) were domesticated from the native *banteng*.[7] By contrast, the Philippine archipelago has relatively few native ungulate species; cattle were introduced under Spanish aegis in the sixteenth century from China, Mexico, and Spain.[8] The nomenclature reflects this: in Tagalog and the major Visayan languages, the common term meaning "cow" is *baka* or *vaca*, from Spanish, in contrast to the widely shared indigenous cognates for carabao.

European cultures lived in close symbiosis with cattle, sheep, and horses for millennia and prized their ability to judge, manage, and care for such stock. Austronesian Malay peoples had exactly this kind of long-standing relationship with the carabao, or water buffalo, an animal that thrives on poor-quality browse and is well suited for work in muddy or swampy environments. William Henry Scott and others assert that the agricultural use of carabao in the Philippines postdates the Spanish conquest, as the plow was not yet in use at contact, but wet rice cultivation was well established in several areas of Luzon and Panay when the Spanish arrived, and it is difficult to believe that water buffalo were not used to prepare paddy soil by trampling.[9] Fr. Pedro Chirino, writing in 1604, notes the presence of "a tame and domestic breed" of carabao, as well as many "mountain buffaloes," which were hunted. Francesca Bray notes that the use of bovines to trample the soil of rice paddies remained common for a long time in low population areas of Indonesia, Malaysia, and the Philippines. Filipino rice farmers were past masters at managing water buffalo.[10]

Carabao meat had been a prominent item in the competitive feasting of larger pre-Hispanic Philippine chiefdoms, but by the nineteenth century many Manilans were prejudiced against eating it. At various times there were also official regulations aimed at preserving the carabao population for agricultural and draft purposes.[11] Still, carabao meat—or "carabeef," as it is sometimes called today—was eaten on occasion, though during 1926–33 an average of fewer than 650 carabao were slaughtered annually in the city for human consumption versus 28,000 cattle.

◇

In the 1860s and 1870s only a small portion of Manila's cattle supply was brought by coastal shipping: 2,058 head in 1862 and about 1,000 in 1872, as against total consumption of around 18,000 animals. The most important months for domestic arrivals by sea were December through February; arrivals were uncommon during the rainy season, when shipping was impaired and there was a (partial) Lenten proscription on meat. In 1862 more than one-third of the cattle arriving by sea from the outer zone came from Ilocos, especially Vigan, followed by a scattering from numerous localities in Zambales and the Bolinao peninsula, Iloilo, Masbate, Lubang Island (off Mindoro), and a group of sites in Bikol that had recently begun commercial cattle raising.[12] This pattern is remarkable for its relative dispersion. Cattle coming by sea in 1862 entered the city in small lots—less than 13 head per boat on average—arriving as secondary items of deck cargo. The rest of the slaughter cattle came primarily from the inner zone around Manila Bay and Laguna de Bay, although "large numbers of bullocks" were walked overland to the city from Pangasinan and the Bolinao peninsula.[13]

Laguna sent out more than 5,300 head of cattle in 1870; presumably most of these went to the city, completely dwarfing the number arriving by coastal vessel. Many of these animals would have come from the friar haciendas in the western part of the province. Tayabas (now Quezon) sent 1,000. Nueva Ecija, in the then lightly populated inner Central Plain, was also a major supplier of range cattle—inferior stock, according to Montero y Vidal. The industry there later declined, as land was converted to wet rice production, but in 1870 the ratio of cattle (raised mainly for meat) to carabao (raised for work) was still more than double the ratios of the other provinces of the Central Plain. By 1886–87 this had changed, and carabao now outnumbered cattle.[14] By the mid-1880s only about 10 percent of Manila's slaughtered cattle came from the inner zone, with 75 percent coming from the outer zone of northern and Central Luzon and 14 percent from farther away in Bikol and the Visayas. Batangas, Laguna, and Tayabas to the south accounted for 43 percent of the total supply; to the north, Pangasinan, Ilocos, and Zambales sent another 34 percent of the slaughter cattle. More than 1,000 head came annually from Nueva Ecija, while the island of Masbate was emerging as a major source, accounting for almost 2,000 head a year.[15]

Not all domestic beef arrived in the city on the hoof. Lightly salted sun-dried strips of flesh, called *tapa* (jerky) were already an item of interprovincial trade to Manila in the 1810s.[16] De Bosch encountered dried and salted beef and dried venison, as well as fresh beef and pork, for sale in Pangasinan in 1857. Modest quantities of dried meat were brought to Manila from both Luzon and the Visayas; some of this was then exported to China, where it was reported to be on sale in Amoy in 1861. De la Cavada mentions some 16,900 kilos of *tapa* arriving in Manila via coastal shipments circa 1870. Venison *tapa* (*tapang usa*) was relished as a delicacy among urban Filipinos as late as the 1940s, but it was tough and had to be pounded in order to make it palatable.[17] Horsemeat was also occasionally made into *tapa*.

Tapa was primarily an outback product, from places where overland drives of cattle were not feasible or where deer could still be hunted or trapped (such as fallow plots in areas of swidden production). More than a few rustled carabao and cattle also ended up as *tapa*. There was even a special word in the Tagalog of the day for the clandestine slaughter of stolen animals: *patani*.[18] Beef and carabao *tapa* came especially from the Bolinao area in 1862 and 1872, from the islands of Masbate, Marinduque, and Mindoro in 1862, and later from Marinduque, Tayabas, and Dagupan. Venison *tapa*, when it can be traced (as it was in 1862), arrived from Leyte, Masbate, and southern Tayabas.[19] In the early twentieth century, "Syrians" traded supplies for *tapang usa* at Butuan in northern Mindanao.[20] Another semipreserved animal product was the spinal cords of slaughtered bovines, which arrived in nineteenth-century Manila as minor items of domestic cargo. These were part of Chinese cuisine—eaten "sliced and stir-fried with vegetables"—and were mostly exported.[21]

Commercial Dealers

Early in the nineteenth century, itinerant horseback merchants from San Carlos, Pangasinan, were known for traveling throughout Zambales, Ilocos, and even the Cagayan Valley to purchase cattle to be sold in Manila. They also delivered many packloads of dried venison and hides.[22] Even by the third quarter of the nineteenth century, there was little sign of commercial concentration in the maritime cattle trade. Numerous consignees, most operating on a modest scale, took charge of animals arriving in Manila; the largest operators handled only about 150 head. Live cattle also arrived in Manila via illegitimate commerce, as rustling was a persistent feature of rural life, especially in nearby Cavite.[23]

In the 1890s, stock buyers and dealers were among those required to pay the new business tax (*contribución industrial*); thousands of the annual one-page tax forms survive. For this inquiry, my colleagues and I searched the provinces of Central Luzon and southern Tagalog plus Ilocos Sur.[24] Only one or two cattle speculators or dealers were encountered in Nueva Ecija and Tarlac in the 1890s and none in Ilocos Sur, Bataan, Morong (now eastern Rizal), or Pangasinan, even though Pangasinan natives had once been notably active as seasonal cattle buyers. Five cattle dealers each are listed for Pampanga, Bulacan, and Cavite, some taxed as traveling buyers; two, charged a triple tax, were explicitly licensed to deal in cattle in Manila and the provinces. There were more buyer-speculators (17) in Laguna. Most were authorized to purchase cattle and carabao throughout the province, though some were limited to particular municipal clusters; two, who paid the highest tax, were licensed to trade in a broader region. Tayabas Province was home to 41 large livestock buyers, half of them in Lukban municipality, but the tax categories in this province were generally not limited to cattle. Some records note particular dealers as "ambulatory" but fail to list their commercial territories.

Batangas, however, was the major regional center for the Manila cattle trade. Some 84 persons in this province were licensed for various periods as speculators and dealers in cattle. The commercial port of Taal was their principal home base. A premium on informal training and experienced judgment meant that whole families were active in this specialized commerce.[25] In all, at least 37 different persons were licensed in Taal, 5 in Bauan, and 1 to 3 each in eight other Batangas towns (others failed to list a home community). A number of Batangas cattle dealers were explicitly free to trade in most provinces in the archipelago; others could purchase livestock in the Bikol provinces and sometimes in Tayabas and Laguna as well. At 45 pesos (in 1895), these licenses were moderately expensive—hog dealers paid only 8.80. All the Batangas cattle dealers were male, mostly indigenous Filipinos and mestizos with some few Spaniards. Only 4 dealers in all the provinces considered here were Chinese, although Chinese of the 1890s had attained a prominent position in the movement of hogs to market.

Local Conditions and Systems of Husbandry

Three local cattle systems can be identified in the nineteenth- and early-twentieth-century Philippines: free-range grazing, the Batangas practice of considerable individual bovine care, and, finally, the emergence in the Bukidnon highlands of imported zebu cattle using upland pastures. The most widespread system involved such little intervention that it amounted to rounding up near-feral animals. To Matthew Turner, a leading expert on Sahelian cattle, this suggests "limited and sporadic marketing channels combined with pressing needs for cash."[26] The practice of raising cattle in largely uncared-for herds was long-standing, dating from the first and second generations of Spanish control. Although relatively few animals were brought to the Philippines from Spain or its colonies—cattle from China being more accessible—the early Spanish managers of the Philippine open range system surely came from the milieu that had developed in Andalusia after the Reconquest and was transferred via the Canary Islands and Santo Domingo to central Mexico, the Acapulco coast, and elsewhere.[27]

Cattle proliferated in the absence of tigers, leopards, and other major predators.[28] De la Costa and Cushner cite the existence of more than a score of cattle ranches in the environs of Manila by 1606, some with thousands of head. In an age of relatively scarce labor, most haciendas were cattle ranches at first. The ranches were unfenced, and wandering animals could become a danger to crops. This was more beef than was needed, given the size of Manila at the time, but it is possible that Spaniards in that day ate prodigious quantities of meat—as did the entire population of Mexico City in the sixteenth century.[29] By 1700, rustling from the estates had become common, and cattle raising was being "replaced by large-scale farming." The damage inflicted by the British military occupation of the 1760s "dealt a final death blow" to the remaining estate cattle ranching near Manila.[30]

Cebu also had a range cattle industry on church lands in the seventeenth and eighteenth centuries, but its market was small, and the number of livestock had declined considerably by 1630. A reason for this was that people as well as unruly native dogs began eating the calves. A Jesuit ranch in Iloilo with thousands of cattle circa 1610 also declined. In the 1810s members of the new Cebuano mestizo *principalía* (municipal elites) began grazing cattle on the slopes along the southeast coast of Cebu without fences or close herding, which led to considerable conflict over trampled crops.[31] Masbate, an island province with a modest human population only recently made secure from "Moro" slave raids, was less conflicted. In the 1860s it became a leading source of firewood and timber for Manila and was rapidly deforested, becoming increasingly well suited to cattle grazing. More than half the island—and smaller islands nearby such as Burias and Ticao—was covered with grass.[32] Jagor describes sailing to Bikol in the 1860s via Masbate, where the vessel anchored next to a volcanic islet with excellent pasture: "Nearly a thousand head of half-wild cattle were grazing on it. . . . They are badly tended . . . [and] could scarcely be said to have any real owners."[33] In 1870 Masbate stood out from the Visayas and most of Luzon in that cattle there outnumbered carabao 2 to 1. A livestock survey in 1886–87 reported more than 50,000 bovines on the three islands, nearly 40,000 on Masbate alone. Its ratio of 21 cattle to 1 carabao was matched only on a tiny scale in frontier Davao; even in Batangas it was only 9 to 1.[34]

At about the same time, Spanish ranching was expanding on the eastern side of the Central Plain of Luzon. Marshall McLennan finds little in the record that directly reveals management practices, except that stock raising flourished on the open frontier and on so-called common lands then existing around established settlements, writing that "besides the use of burning to extend grasslands and to renew grazings, little management of pasturage occurred." Widespread grazing in the Nueva Ecija area came to a sudden end with the first wave of rinderpest in the late 1880s. After a decade or more of relative disuse, during which *parang*-savanna landscapes apparently expanded at the expense of *cogonales*, the entire area was settled and developed as a rice production zone. Already in 1886–87, the number of carabao used for rice farming had grown until the ratio between cattle and carabao was close to parity.[35]

Grazing is one thing, careful animal husbandry another. At the Manila abattoir in the 1920s and 1930s, cattle from Masbate consistently weighed less than 60 percent of the mean for Batangas cattle and included more than 40 percent females, both marks of indifferent stock management. During the first two-thirds of the nineteenth century Nueva Ecija was similar. Here many bovines were simply turned loose to forage for themselves on the hill slopes and forest edges when they were not immediately needed as plow animals, facilitated by the practice of regularly burning the vegetation cover in order to prevent the recovery of woody forest.[36]

Such casual cattle husbandry, reflecting abundant land resources in a lightly populated area, was also employed by successive French owner-managers of the famous Jalajala Estate along the northern shore of Laguna de Bay. A foreign naval officer visiting in the 1850s described it as a rice and sugar estate of 10,000 acres, of which only about one-tenth was cultivated. The rest was hill slope and mountain land, "run over by large herds of horses and cattle in a wild state, excepting inasmuch as . . . they are driven between the harvests from the hills to pasture on the cultivated land. Horses, at this time, fetched a high price at Manila, and cattle, too, would have sold well, being of excellent breed, but partly from the difficulty of getting the [local people] to catch and bring them in, and partly from a liking [the owner] had to keep them scampering about, . . . [and] he seldom sent any to market."[37]

In part, the cattle and horses at Jalajala were being used to gather nutrients over a broad area and concentrate them on the cultivated fields during the off-season, a practice common in the management of native horses in several parts of the archipelago.[38]

Although most cattle in the Philippines in the first half of the twentieth century were not raised in large herds, this seemed to make little difference; most stockmen practiced neither castration nor deliberate culling of most males. Those that did practice castration are alleged to have used methods that were extremely painful to the animal and left it unable to function normally for a long period.[39] Uncastrated male cattle are notoriously unmanageable, and there were numerous reports of difficulty in rounding up and controlling animals being sent to Manila. Peninsular Spaniards would have had only limited experience with husbandry in the humid tropics, so advanced bovine management techniques were not widely practiced in the Philippines. The result was that herds were made up of nearly equal numbers of females and scrub bulls. With no attempt to control breeding, little progress in livestock quality could be expected; on the contrary, selection was for ability to survive on poor browse. Further, mayhem and trampling by the males reduced the survival rates of calves and thus depressed the growth of livestock numbers. "The cattle of these large herds are very wild," one observer wrote, "and as they are never worked, and as no fences are used, it is a difficult undertaking to capture them for shipment to market."[40] Because of these persistent practices in the Philippines, 50 to 70 percent of free-range female cattle sent to market were pregnant, reducing the herd growth multiplier even more. Such range cattle also tended to lose weight quickly in transit because they were unused to stall feeding.

Despite the availability of imported barbed wire in the Philippines, its use in fencing—common in North America by the 1890s—remained rare. This inhibited segregation of stock by ranch and made it exceedingly difficult to impede the transmission of bovine disease or develop more productive, disease-resistant animals. In the 1930s, six of the eight provinces sending annually more than 1,000 cattle to slaughter in Manila were still using the open range pattern, at least in part (table 8.1). The widespread survival of this system of

TABLE 8.1. Indicators of Cattle Herd Management by Province, 1930s

Sending Province	Average number slaughtered in Manila	Percentage female of those slaughtered	Average dressed weight (kilos)	Zebu and mixed as a percentage of herd
Batangas	4,263	7	152	<1
Bukidnon	4,970	28	133	66
Cotabato	2,152	42	110	53
Zamboanga	2,597	42	104	30
Palawan	1,612	35	93	12
Masbate	8,532	43	92	28
Sulu	2,559	52	99	2
Tayabas	1,121	37	84	2
Mean	—	36	110	—

SOURCES: Ms. ARBAI, 1932, 220; Ms. ARBAI, 1933, 153–54; PSR, 1st–2nd qtrs., 1937, 4:1–2, 197.
NOTE: Provinces providing more than 1,000 head per year are shown. Average number and dressed weight are for 1931–33 and 1935. Percentage female is from 1933. Zebu are humped Indian cattle.

husbandry speaks to the lack of a long-standing tradition of commercialization in these places.[41] At the same time, the rising percentage of zebu or part-zebu animals implies some control of breeding such as would come with fencing.

Could what might be called "leisure preference" have led livestock owners to reject more labor-intensive husbandry activities once their minimum subsistence needs had been met?[42] Did owners simply decide not to recruit, train, and pay the skilled labor required to manage the animals to a more productive standard? They may have done so in part because of the greater risk created by all too common epizootics. Although there was little feed cost in the remote grassy places where herding cattle became a local specialization, transportation charges (when calculated per kilo of dressed weight) could be considerable for underweight animals and the profits small. In the 1930s prices in Manila for poor quality beef were about half of those paid for Batangas beef. Ranchers blamed both *ganaderos* (cowboys) and middlemen, but in an editorial entitled "Driving Skeletons to the Meat Market," Dr. Victor Buencamino, director of the Bureau of Animal Industry, placed the blame on absentee owners, indifferent to quality breeding and unmotivated to initiate other good husbandry practices.[43]

≈

For generations, the best beef cattle came to the city from Batangas. Even before 1850, Taal (and Batangas Province more generally) sent many cattle— "highly esteemed," according to Buzeta and Bravo—to the Manila market.[44] In addition to the cattle trade, Batangas also developed a leading role in the commercial supply of swine, chickens, and horses; clearly Batangueños saw that raising animals for sale in Manila could be a profitable enterprise. The cattle husbandry system used in Batangas was the opposite of "free range." Many cattle were used for work and raised in small groups with considerable individual attention. The producers of Batangas enjoyed easy commercial connections to the Manila market, creating opportunities for more sophisticated and labor intensive husbandry.[45] Several characteristics set Batangas apart. Much of the province is hilly or sloping with volcanic soils, not well suited to a wet rice monocrop. The prime use for water buffalo is in rice cultivation on deep muddy soils, but in Batangas bullocks, rather than carabao, were widely used for fieldwork; cattle would have been kept by many farmers in small numbers. Westernmost Batangas was committed to sugar and cattle *estancias*, while the rest of the province had extensive areas devoted to coffee or wheat and certain locales focused on raising upland dry rice, onions, garlic, mongo beans, or citrus fruit.[46] Further, the several coastal port communities enjoyed easy year-round access to Manila via small and inexpensive sail craft, operated by a relatively sophisticated and educated local elite with a strong tradition of commercial participation. Finally, in the nineteenth century most Batangueños were not restricted to landed estates held by conservative religious orders but had some latitude for experimentation.

By the middle of the nineteenth century the Batangas economy was already partly specialized in the production of cattle and horses, along with coffee. Buzeta and Bravo note the excellent pastures of San Jose, the many cattle of Batangas municipality, and the fact that families in Bauang and Taal were routinely selling cattle in Manila. In 1870 almost a quarter of all cattle in Luzon were in Batangas, the only province where cattle significantly outnumbered carabao. The all-Luzon average was something like 50 or 60 cattle per 100 carabao; in Batangas the ratio was reversed at 3.5 cattle for each carabao. A decade later Montero y Vidal said that Batangas possessed an "immense wealth of cattle and hogs" and its inhabitants were "excellent stockmen." The ratio of cattle to carabao was now 9:1. From the mid-nineteenth century onward, Batangas was the site of an annual fair in which there was keen competition for the prizes awarded for excellent livestock: horses, cattle, and bullock teams. Many of the animals were raised and bred individually, not as undifferentiated members of a herd; controlled breeding allowed selection for useful characteristics.[47]

In the twentieth century, Batangas stockmen were credited with hand-feeding their cattle, understanding the general health benefit of removing ticks (thus lowering the animals' exposure to parasites), and routinely culling surplus males. At the Manila slaughterhouse, cattle entering from Batangas included the lowest percentage of females and, partly as a result, the highest average dressed weight. Batangas stockmen were rewarded for their care with the highest price per kilo in the Manila market. In physical type, Batangas cattle were compact and made "splendid draft animals." Although they differed in proportion, both Batangas and Ilocos cattle were said to have been developed from Chinese stock (figure 8.1).[48]

～

A third form of stock raising emerged early in the twentieth century in the pioneer uplands of Bukidnon in Mindanao. Under American rule, large blocks of grasslands were leased at minimal rates for grazing in an attempt to meet Manila's demand for beef. As part of the Bureau of Agriculture's stock-raising experiments, zebu cattle (*Bos indicus*) of the Madras Nellore variety were found to prosper in the Bukidnon uplands. Further, because of natural selection over a long period in a disease-rich environment in India, the imported zebu proved less susceptible to rinderpest than most local cattle. Secretary of the Interior Dean C. Worcester, former Philippines Constabulary officer Manuel Fortich—soon to become the political hegemon of Bukidnon—and agricultural college graduate and stock specialist Florentino F. Cruz were all centrally involved in the development of high-quality Nellore and mixed-blood herds in the province.

Figure 8.1. A prize Batangas bull draws an appreciative crowd, 1928. (Bureau of Agriculture photo, USNA II, RG350-P-Am-4-10-1, box 4)

Bukidnon cattle raising became a substantial success for rancher-entrepreneurs and led to increased beef production, although it was not so profitable for indigenous (prior) inhabitants of the province. In the mid-1920s regular shipments of steers to Manila commenced, and by the 1930s this system enabled Bukidnon to become one of the three most significant provinces in the supply of live beef to the city (table 8.1). Stockmen specializing in zebu and mixed-blood cattle were rewarded in the Manila marketplace with a value per kilo well above that for average grade "native" beef. Animal-breeding and care practices advocated by the University of the Philippines College of Agriculture became increasingly influential. By the late 1930s zebu-derived animals comprised over 50 percent of the herd in Cotabato and about 30 percent in Zamboanga and Masbate. Other practices employed in Bukidnon included castrating and later culling many male animals and even spaying subpar heifers.[49]

CATTLE IN MANILA

Once in Manila, some cattle were kept briefly in the corrals of urban dealers. There were four approved corrals in the city in 1933; that same year four suburban corrals were closed because they were "not in condition to guard against the leakage of infection."[50] Once the animals were presented for processing into fresh meat, they entered a system overseen by the state and designed to produce both revenue and an adequate supply of fresh meat free of obvious disease. The meat was retailed through the system of public markets, along with fish, vegetables, fruit, and many other things. By far the most important of these were the great Divisoria public market and food wholesale center in Tondo-San Nicolas and the Quinta Market on the river in Quiapo (figure 5.8).[51] Eighty percent of the city's market fee collections in 1908 came from these two; a decade later it was still above 70 percent.[52]

The slaughterhouse, or *matadero*, was also owned by the city. This was an important "power center" in the provisionment system.[53] The Matadero de Dulumbayan—literally "slaughterhouse at the edge of the city"—on Cervantes (today's Avenida Rizal) in Santa Cruz was built in the 1870s. Here the daily slaughter began by torchlight at midnight. Postmortem inspection followed at dawn, when retailers came personally to select meat to sell in the public markets. They also arranged to have it transferred in large baskets (*batulanes*) carried on balance poles or in a cart. A substantial abattoir on Azcarraga was built in 1892–93 to replace and improve on this 20-year-old *matadero*.[54] The abattoirs of Manila were entire neighborhoods of butcher workmen, where hundreds of thousands of animals raised in the provinces met their collective end and were unceremoniously transformed into food and nutrition for the more comfortable families of the city or for life-cycle feasts. Whether the slaughter took place by torchlight, as it did in the 1850s, or in the early afternoon, as in the 1920s, only the profoundly deaf could have been unaware of it.

All medium-sized and large animals destined for use as food were required to be slaughtered here. A fee was charged for each animal killed and cleaned: 50 centavos for each bullock or cow and 25 for each hog in the 1870s. In addition, a 1-centavo tax was collected on each pound of beef sold in the public markets. The slaughterhouse and beef-vending fees were important sources of municipal revenue; such fees produced 67 percent of the municipal income of Cebu City in 1895.[55] In the 1880s, fees in Manila were increased to help finance the new public water system. The right of collection was auctioned off in advance ("farmed out") in two-year contracts, a typical premodern weak state arrangement. Only those with very substantial resources could hope to win the bid and take over the daily collection; in the 1890s, these were Chinese merchants. Naturally some attempted to avoid the *matadero* system and its fees; now and then the government would try to enforce the monopoly rights of the contractor by going after these malefactors.[56]

The American regime continued the practice of requiring that slaughter of medium-sized and large meat animals be restricted to the public abattoir. Even hogs destined for the local German sausage factory were slaughtered in the *matadero*. At the same time, however, the new regime ended the practice of auctioning the right to collect the slaughtering fees in favor of direct adminis-tration.[57] As before, the abattoir monopoly was surreptitiously challenged and had to be defended by continuing police action against back-alley slaughtering.

The requirement that all animals be slaughtered at the single city abattoir, not in the streets or dispersed sites, was also intended to provide a modicum of meat inspection. By the 1830s an inspector known as a *veedor* was attached to the slaughterhouse. A royal order of 1882 mandated a postmortem inspec-tion of hog carcasses for trichinosis, a requirement said to have effectively doubled the workload of the veterinary inspectors.[58] The Americans rebuilt or retrofitted the slaughterhouse and major public markets, in each case increas-ing the provisions for sanitation and meat inspection. Specially designed city meat wagons now carried the product from the *matadero* to the markets. As a result, in the early twentieth century the Azcarraga Matadero became noted for its sanitary facilities, especially an overhead trolley of hooks that got the carcasses off the floor.

Four decades later, with the urban population roughly tripled, the *matadero* had become grossly inadequate to handle the volume of daily work. Improve-ments were slow in coming because the slaughterhouse was caught between municipal authorities trying to run it with revenue in mind and insular vet-erinary health officers bent on protecting the health of the consuming public. In 1933 this institutional conflict reached the Supreme Court, which ruled that the local authorities did not have exclusive power over slaughterhouses vis-à-vis the Bureau of Animal Industry. Finally, the city proposed to build a new abattoir on a new site with its own stockyard.[59]

Although large numbers of animals were slaughtered in Manila, the city remained almost entirely a center of meat consumption, not distribution. Most of the flesh was sold fresh in the city itself; relatively little was packed and preserved. Nor were sides of beef or cleaned hog carcasses widely shipped elsewhere in the archipelago. Considering the modest level of demand for beef outside the city and the widespread availability of hogs, provincial towns could easily slaughter their own animals locally. Once the Manila-Dagupan Railroad was in operation in the 1890s, meat might have been shipped relatively quickly along the main line, but inexpensive refrigeration was lacking.[60]

In the 1930s urban cattle buyers divided arrivals into three general catego-ries. These included average-sized native stock called *partida* cattle (because they were sold in lots); the heavier *nonpartida* animals (usually coming from Batan-gas or Ilocos), which were often sold individually; and "inferior" smaller cattle. The larger animals were prized as a source of quality beef and often became the objects of competitive bidding, while the smaller animals suffered consider-able discrimination in the market. By this time big ranchers had often elimi-nated the middlemen, as had large-scale importers of live animals. Increasingly big cattle raisers and importers oversaw the slaughter of their animals in the abattoir and then distributed the sides of beef directly to their market stall operators. In the 1920s importer Victor Buencamino "organized a small crew of *matanzeros* who slaughtered the cattle around noontime and distributed the individual carcasses by 5 p.m. Each carcass was then prepared for sale in the market stall by next morning and was usually sold out before noon." Middlemen bought up and handled the selling of animals coming from small-scale cattle raisers.[61]

In the last several decades of our period, most cattle arrived in the city by ship, commercial sail craft, or railroad rather than being driven to the city overland. These animals were then sold on the hoof to bidders who came onboard the vessels or waited in the railway yard. There were many middlemen, who often bought sizable lots of cattle in concert and subsequently allocated the animals among themselves according to the capital each had put up, effectively replacing competitive bidding with collusion.[62] In 1933 eleven persons each handled more than 1,000 slaughter cattle for the year, approximately 83 percent of the total. The top three each handled 4,000 to 5,000, totaling about 36 percent of the whole. At least one of the Manila buyers was a woman. Despite the fact that animals were "bought" on shipboard, owners were often paid only after the cattle had been slaughtered, the cleaned carcasses weighed, and the meat sold. Retail butcher-vendors (*tenderos*) remained financially obligated to the middleman or large-scale owner. Ordinary fresh meat needed to be moved quickly, but dressed and chilled or frozen beef arriving from overseas was delivered to one of the cold storage companies that in turn sold individual cuts directly to affluent consumers.[63]

The Changing Geography of Supply

The changing geographical pattern of domestic beef supply in the 1880s and the 1930s is depicted on map 8.1.[64] The contrast between the two patterns is clear. Except for Batangas, which retained and further developed its culture of high-quality bovine husbandry, all the other major beef production zones in Central and northern Luzon declined as sources of supply for the city; from 12,500 cattle per year in 1885–86, they sent fewer than 1,000 in the 1930s. By the 1930s the central and southern islands of the country were providing 78 percent of the domestic supply to the Manila market. Annual shipments from Masbate increased from fewer than 2,000 to almost 9,000 head per year; Bukidnon was now sending nearly 4,500 good-sized animals; and Palawan, Zamboanga, Cotabato, and Sulu had also become important suppliers. This represents a massive displacement from Luzon to more distant places and a change in mode of transport, mostly to steamers, although nearby Batangas shipped cattle more cheaply by truck.[65]

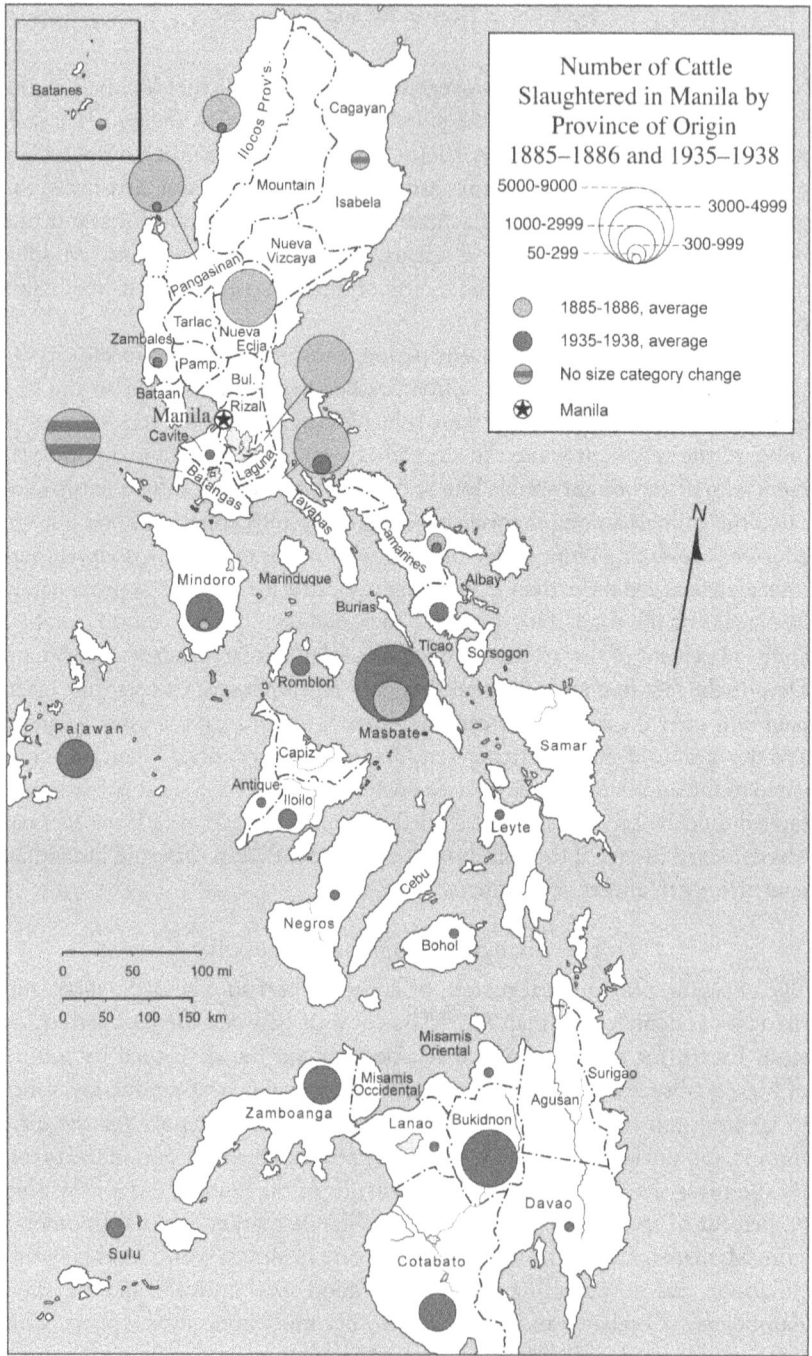

MAP 8.1. The Provincial Origins of Domestic Cattle Slaughtered in Manila, 1885–86 and 1935–38

IMPORTED ANIMALS AND EPIZOOTIC DISEASE

In June 2011, ten years after the last discovery of a living case (in Kenya), the United Nation's Food and Agriculture Organization (FAO) declared that the rinderpest virus had followed smallpox into the dustbin of history. It did not go quietly.[66] The Philippines experienced three waves of rinderpest in the late colonial era. They were tragic enough in their effect on ordinary cattle (*Bos taurus*), but their destructiveness was much more serious because the death of large numbers of carabao (*Bubalus bubalis*) seriously undercut domestic rice production. "The importation of the rinderpest virus," claims Ken De Bevoise, was "arguably the single greatest catastrophe in the nineteenth-century Philippines." Without carabao, farmers "could not cultivate as much land as before. The consequent reduction in food supplies, in turn, aggravated malnutrition and debt. Untilled land that returned to scrub . . . provided favorable conditions for both locusts and anopheline mosquitos."[67] Only in the 1920s were biological intervention and quarantine regulations finally effective in stopping the disease.

~

Caused by a virus (technically a morbillivirus), rinderpest attacked the mucous membranes of the bovine body and especially the digestive tract. High fever, ulcers of the membranes, dysentery, and death in a week or less were typical. Rinderpest was primarily transmitted by means of close association with an infected animal, through contact with nasal and other discharges, dung, or urine or through food. It was unlikely to spread by insect transmission or through the air, and even direct discharges were believed to lose their virulence following two days of sunlight.[68]

Initial outbreaks with 85 to 90 percent mortality were typical of this "cattle plague" (the old English term). It was a familiar disease in Europe, with major seventeenth- and eighteenth-century outbreaks associated with the continental oxen trade. The notion that it was the job of the state to take action against epizootics dates from at least this time, as do ideas of quarantine, animal health certificates, the mass slaughter of infected animals (and those likely to have been exposed), and even policies aimed at restocking affected areas.[69] The broader diffusion of rinderpest and other bovine diseases was a consequence of the quickening international commerce in animals during the late nineteenth and early twentieth centuries—part of the general escalation of world trade—leading to what some have called a "rinderpest panzootic" in the late nineteenth century.[70]

Within Southeast Asia, Thailand annually exported thousands of bullocks to Singapore, Sumatra (Aceh), and British Burma in the 1880s and 1890s, a trade that crashed in 1897 when rinderpest "ravaged the whole of central Siam, attacking both buffalo and oxen with such severity that the [rice] harvest prospects [were] seriously threatened."[71] There was a major rinderpest event in West Java starting in 1879 and possibly limited outbreaks even earlier.[72] In the Malay Peninsula adult water buffaloes imported from India, where rinderpest was already well known, proved more resistant than local stock precisely because they had grown up in a disease-rich environment.[73]

In the Philippines, rinderpest became critical in 1886, but its initial arrival has not been precisely documented and its source may never be known. It might have been lurking unrecognized for years, bursting forth at this time for unknown reasons, or it might have been brought by other animals such as sheep. Rinderpest can be carried by small ruminants such as sheep or goats, which are less likely than cattle to die from it. In many years from 1854 to 1881, the Philippines simply recorded the importation of "live animals." In those years, with some breakdown of import statistics, they were generally classed as "sheep": *carneros* or *ganado lanar*. The numbers range from scores annually in the 1850s to hundreds in the 1860s. Significantly, 2,398 *animales lanar* were imported to Manila during 1884–86, just before the first rinderpest outbreak attained epizootic status.[74]

In all likelihood, however, rinderpest arrived with live cattle imports. To meet the rising demand for slaughter animals in Manila, a trickle began to arrive from the Asian mainland—principally from South China at first and then (in the early twentieth century) in greater numbers from Cambodia and Australia. Not until 1884 did imports of "bovines" reach a significant number, with 60 arriving from Hong Kong and 8 from Australia. The first wave of rinderpest almost certainly may be traced to animals imported from Hong Kong, since by then the disease was enzootic (endemic) in the hinterlands of all the major ports of the China coast. Although most of these cattle were intended for slaughter, they might well have been held in private corrals in the city, where animals ordinarily mixed without aggressive quarantine. A few would likely have been sold to buyers from nearby provinces. It takes only one infected animal to start a chain of infection and just a few imported animals to introduce a world of pain.

The First Wave

As the first rinderpest outbreak reached crisis proportions, Dutch consul Hens in Manila wrote of nearby provinces, "Bands of hundreds or even thousands of starving and miserable people are found in several provinces pillaging and murdering on occasion only to disperse and hide in the forests and mountains at the approach of a military force. At the highest point of these problems . . . an epizootic broke out for six months that killed two-thirds of the farmers' beasts, especially buffaloes and cattle and the government couldn't do anything to stop it. The cadavers infested the air and the rivers; we bless providence that the epidemic didn't attack our species." Hens reports that the epizootic was first noted to the east and southeast of Manila and then seemed "to follow the wind of the southeast monsoon, to stop, we hope, in the northwestern provinces of Luzon on the China Sea."[75] His account of the early track of the epizootic is almost exactly confirmed by the dramatic decline in shipments of slaughter cattle to Manila, first from Laguna, to the southeast, and subsequently from Pangasinan and Ilocos to the northwest.

Military veterinarian Gines Geis Gotzens states that the epizootic predictably tended to follow the lines of commerce and communications.[76] These lines ordinarily ran through Bulacan and up the west side of the Central Plain. By the end of 1887 what was subsequently understood to have been rinderpest had spread from Laguna and the Marikina Valley (just east of the city) and Bulacan (just northwest), blanketed central Pangasinan, and entered the Ilocos coast and Nueva Vizcaya in the far north.[77] Cattle shipments from Pangasinan arriving in the Manila abattoir declined steadily from 4,500 in 1885 to fewer than 1,900 in 1887. As this diminished flow of animals moved overland, it further spread the disease. From central Pangasinan, rinderpest was transmitted west into the hilly Bolinao peninsula and then south down the Zambales coast. From Ilocos the numbers of slaughter animals sent to Manila actually increased before declining by half with the arrival of the disease in 1887. During the following year Ilocos was further devastated; in Ilocos Norte the greatest mortality hit in August 1888.[78] By 1889 the worst of the first wave was playing itself out in the Cagayan Valley in northeastern Luzon.

In parts of Central and northern Luzon, this wave was a disaster of the first order. Foreman reports a stockowner in Bulacan who lost 85 percent of his animals. De Bevoise cites the loss of at least 84 percent of the carabao and cattle in Pangasinan. Rodell writes that this epizootic wave "almost completely destroyed the Zambales cattle industry," with reported numbers collapsing from more than 23,000 in 1886 to fewer than 2,900 in 1892, an 87 percent decrease. In the heavily deforested northwest peninsula—the main source of charcoal for the city a generation earlier—the important stock-raising communities of Bolinao and Anda fared particularly badly; cattle declined from a combined 9,660 to approximately 385. In nearby Alaminos, with losses of 8,000 cattle and 5,000 carabao, it was the end of a stock-raising era. Carabao numbers also declined in Zambales, though less drastically than those of cattle, probably because they were more isolated on local farmsteads than cattle, which were often kept in herds.[79]

Laguna, one of the first provinces hit, sent 1,711 cattle to be slaughtered in Manila in 1885, the last year before the outbreak; this declined to 328 in 1886 and 28 in 1887, eliminating the province as an important supplier of beef to the city.[80] Rinderpest spread to Batangas, Tayabas, and Cavite provinces much later, however. There had been no immediate decline in shipments; indeed, cattle coming to the city from Batangas and Tayabas actually increased during 1886–87, picking up the slack in supply caused by the decline from other places.[81] Some 2,200 had been sent from Tayabas in 1885; shipments doubled in 1886 and were up to 6,000 by the end of the following year. Likewise, slaughter cattle sent to Manila from the island of Masbate were up by 125 percent in two years, and even more thereafter. In Batangas, where cattle were not generally allowed to run free, disease diffusion was slow, although rinderpest was recorded in Rosario in 1887. More than 700 cows died in this wave in Santa Cruz (now called Tanza) in neighboring Cavite.[82] Rinderpest also spread by sea when infected animals were carried to Iloilo and Capiz provinces on Panay, whether as domestic shipments or as direct imports to the port of Iloilo. Otherwise it did not appear in most of the Visayan Islands or even much of southern Luzon. Years later veterinary Vicente Ferriols—clearly unaware of rinderpest's impact on places like Bolinao and Pangasinan—wrote of the first wave, "[T]he disease must have run a sporadic and mild course after the first severe onset."[83]

By mid-1888 the domestic supply of beef in Manila was insufficient to maintain the normal level of provisionment. From circa 21,000 in 1886 and 1887, the annual slaughter of beef cattle in Manila fell by a quarter, to 15,700. The slaughter continued well below normal into the mid-1890s. There were places, said Consul P. K. A. Meerkamp, where "hardly any carabao or cattle were left alive."[84] Local ethnoveterinary practices for treating observable conditions in carabao, including the use of a wide variety of herbal medicines, did not prove effective.[85]

The response of the colonial government was slow and uncertain. A major circular of regulations, aimed at both combating its spread and educating the public, was issued by the Inspección de Beneficencia y Sanidad in October 1888 and renewed at the end of 1890.[86] The regulations quite sensibly attempted to impede the transport of diseased animals and required "scrupulous vigilance" by the public meat inspectors at the slaughterhouses to keep such animals out of the human food supply. They provided for a 15-day quarantine of suspicious animals in areas where the disease had broken out and recommended keeping those animals away from goats, dogs, pigs, and other animals. They recommended disinfection procedures and the cremation of animals that died from the disease. These regulations came too late to seriously impede rinderpest diffusion in Central Luzon, but they may have helped to protect the livestock industry of places less integrated with Manila.

What the regulations did not mention was the proactive slaughter of diseased animals or those likely to have been exposed in the vicinity of outbreaks, a major weapon in the arsenal of animal disease control. This and tough quarantine measures were "strong state" approaches that had been used extensively in the Netherlands and United Kingdom in their successful efforts to eradicate rinderpest two decades earlier and by Dutch officials combating rinderpest in Java during 1879–83, but neither Spain nor its Pacific dependency was a strong state. This may have been what Consul Hens had in mind in writing that "the government couldn't do anything to stop it." The slaughter of "sick and exposed animals, with a certain amount of indemnity" was tried briefly in 1911 in the aftermath of the second wave, but it was discontinued because of strong farmer opposition and attempts to hide sick animals. Subsequent American and Filipino authorities largely eschewed this weapon. All three regimes worried about the popular response.[87]

The Second Wave

The disease was now at large in the archipelago, and the incidence of new cases did not decline to zero, although the numbers of carabao and cattle gradually recovered. Work animals do not seem to have been in notably short supply in 1895 and 1896, just prior to the Philippine Revolution. A little more than a decade after the first wave, sufficient animals had been born to sustain another. Although animals born to rinderpest survivors would initially have had some immunity to the disease, this immunity generally lasted less than a year.[88] Moreover, extremely cheap range cattle from Queensland were imported in the late 1890s, likely a further source of infection.

The second wave began in 1898 or 1899 in the southern Tagalog areas most affected by the Revolution and spread outward, devastating Central Luzon in 1899 and 1900, with reports of gruesome river-edge tangles of rotting carabao corpses. In 1900 and 1901, very large numbers of bovines died in the extremities of the island, to the far north in the Cagayan Valley and to the far south. Owen flatly states that "the rinderpest epidemic of 1900 . . . virtually destroyed the local cattle industry" in Bikol.[89] Now the disease claimed many victims outside Luzon in island provinces such as Marinduque, Masbate, and Leyte and was carried south to Zamboanga. A retrospective census question on cattle and carabao mortality during 1902 reported 629,000 animals having died that year, as against only 80,000 slaughtered for meat. The greatest mortality was concentrated in the central and western Visayas, with Negros, Bohol, Cebu, Iloilo, and Leyte each reported more than 50,000 large animals dead.[90] With the exception of Iloilo, the bovine populations of these places had largely been spared during the first wave and thus might have been expected to suffer high mortality rates. Only a scattering of isolated locales escaped into 1903; judging from cattle-to-human population ratios, these included Palawan, Mindoro, Cagayan, and parts of Mindanao and the Mountain Province. The little northern islands of Batanes escaped entirely.[91]

The second wave was made incomparably worse by the general disruption of the country due to the collapse of stable administrative authority during the Revolution and then the invasion of American forces. American commanders were intent on asserting imperial control in every province, and they moved troops and draft animals around the archipelago with scant regard for the disease. We cannot be sure exactly how much of the severity of this wave can be ascribed to warfare, but clearly it exacerbated the crisis. Reynaldo Ileto describes the disruption in western Batangas when local farmers took their carabao with them to Cavite to participate in the revolt against Spanish rule; few returned. Ken De Bevoise suggests that rinderpest transmission increased as a result of refugees taking surviving carabao into concentrated and unsettled conditions and of the American army requisitioning carabao for military transport.[92] Later veterinary authorities thought that a "continual intermingling of animals" in the inner lowland plain of Central Luzon had amplified its impact. Considerable rice land remained out of production there in 1908 and later for lack of plow animals.[93]

Even in the provinces hit hard before, the toll in the second wave was enormous. In Batangas, where armed resistance and subsequent American "pacification" measures were strongest, mortality was close to 90 percent. In Cavite about two-thirds of all cattle died. At the northern end of the Central Plain, there were "great tracts of land in Pangasinan which were idle for lack of laborers and by reason of disease which had recently destroyed the live stock." Notables from Dagupan, testifying in February 1901, reported that it "would be several years before the lands could be properly worked; that when land is allowed to lie fallow for a year it virtually grows into a forest [of *Imperata*] and requires much time and great expense to clear it." In Vigan, farther north, by August 1901 the cattle and carabao "had nearly all died" of rinderpest, and glanders had taken a toll of the horses as well.[94] A majority of Philippine horses, in fact, died of disease during the Revolution and Philippine-American War.[95] It has also been suggested that the epizootic contributed significantly to human mortality, as malaria-transmitting mosquitoes, deprived of their favored large-animal targets, turned increasingly to prey on people.[96]

Having imposed themselves on the Philippines, the Americans now tried to induce the Filipinos to accept their rule. With the economy of rice-producing areas in collapse by 1902, many families had resorted to roots and tubers for subsistence. One initiative of the new government was an attempt to speed the replacement of work animals by purchasing and importing them from abroad. During 1903 an estimated 35,000 carabao were purchased in China and brought in, with more following in 1904 and 1905.[97] Imports were wide open. The mail and passenger steamers *Rubi* and *Zafiro* always carried cattle on the crossing from Hong Kong to Manila in 1904. More than 23,000 live bovines arrived from Hong Kong during fiscal year 1907–8, with lesser numbers coming from Xiamen (Amoy), Hainan, and Taiwan, surely bringing disease with them.[98] Many died en route or upon arrival. Without more effective disease control, acquiring animals and distributing them in good health were impossible.[99]

The other major bovine supply zone in this period was French Indochina, especially Cambodia (through the river port of Phnom Penh) and Vietnam (Saigon and Vinh). Although "Indochina" was the source of small numbers of "live animals" in 1877, 1880, and 1891, there is little direct evidence in the *Balanza* that it was a major supplier to the Manila market until just before the turn of the century. The total exports of Cambodian cattle peaked in 1898 and 1899, coinciding perfectly with the start of the second rinderpest wave in the Philippines; plummeted by half during a period coinciding with the Philippine-American War; and then recovered sharply during 1911–1913, surpassing the peak of 1899 in the last year. The destination of the exported animals is not precisely identified, but the Philippines was the major market for Cambodia.[100]

Some 16,600 bovines arrived in the Philippines from Indochina in fiscal 1907–8. Such animals were essentially wild and so numerous that they had little value at home. Pierre Brocheux notes, "The centers of greatest cattle breeding [there] are in the hills of southern Annam [central Vietnam] and in Cambodia, where the animals run at large, unattended." The internal commerce in livestock in these regions was usually in the hands of Chams; neither Vietnamese nor French were much involved.[101] A report of 1908 describes this trade.

> The people engaged in [the business of exporting livestock from Indochina] . . . must surmount many difficulties. The cattle centers are some distance from the ports, so the buyers must proceed inland buying up cattle in small bunches, two here and three there, until a herd has been gathered of sufficient size to warrant the chartering of a steamer. The natives will not accept bank notes, so silver coin must be carried inland by pack horses. There are no facilities for transportation in those mountainous districts from which cattle are obtained. . . . The animals must be driven overland to the ports and fires have to be built at night to protect them from tigers and leopards, which abound.[102]

Although the animals were held in corrals in Phnom Penh and Saigon pending shipment, or in small lots along the river awaiting transfer to a steamer, they were not routinely subject to inspection by veterinarians. Even when quarantined, resistant animals from areas where rinderpest was endemic might well display only subtle symptoms. In one 1907 shipment 100 out of 375 animals were found to be suffering from rinderpest by the time the ship arrived in Manila.[103]

The major buyer-importers were mostly Filipinos. The best known was "cattle king" Faustino Lichauco. Lichauco was a grandson of Cornelia Laochanco (vda. de Lichauco), one of the premier Manila rice dealers of the 1860s, who had gone on to make a fortune in export sugar processing, gold buying, and urban property management. He was also the godson of Alejandro Nable Jose, a leading Manila rice merchant of the 1880s, and he was well connected politically as a result of his services to the Revolution. Drawing on family wealth, he invested in the lightering business in Manila harbor (successfully) and the rice-milling business in Dagupan (unsuccessfully) before his attention was drawn to opportunities in the Indochina cattle trade. He arranged to visit Saigon and set up a local office to facilitate purchase and shipment of the animals. Subsequently his cattle-importing interests were extended to Australia. He soon became rich and was known for maintaining a large and stylish household (across from Malacañang Palace), lavish social and political entertaining, and the extended residences of his wife and children in Europe and the United States.[104] Lichauco's first cousin (another grandson of Cornelia Laochanco) and business rival Ramon Soriano was also a prominent importer of Indochinese cattle, as was Smith Bell, the British trading and management company (figure 8.2).[105]

Meanwhile, the livestock division of the Philippine Bureau of Agriculture, having made little progress in the biological control of rinderpest, renewed its efforts to seize geographical control. "Cattle plague" had been effectively combated in Western Europe by the aggressive use of quarantine, segregation, and slaughter, so the bureau decided to attempt such internal restrictions in the Philippines. Lacking personnel and resources for a national program, it imposed a strict quarantine at the northern end of the Central Plain beginning in 1911. Pangasinan was selected because it was a critical rice-producing province, because rinderpest was again spreading there, and because it lay astride strategic choke points on the routes south into the Central Plain from the Ilocos coast and the Cagayan Valley. These routes were used by migrating Ilocano rice farmers seeking to settle along the railway lines and by dealers chasing the high prices for work animals in the reviving sugar industry of Pampanga.[106]

East-central Pangasinan—the prime "intermingling" area—was a well-chosen place to start. The army agreed to cooperate in the quarantine effort by providing five veterinarians, some cavalrymen, and 1,200 Filipino scouts, the last essential because farmer hostility to "government quarantine measures had built up to a point of violence," as Buencamino later said. "The battle against rinderpest had to be carried out by force."[107] The Customs Bureau also cooperated by banning the transport of carabao and cattle in small boats and requiring health certificates for movement on larger boats. Quarantine stations were established along the major land routes from the Ilocos coast, Benguet, and

FIGURE 8.2. Faustino Lichauco in an undated family photograph from "Indochina." (Reprinted by permission of Jessie Coe Lichauco and her daughter Cornelia Lichauco Fung)

Nueva Vizcaya, and a barbed-wire *cordon sanitaire* was constructed along the narrow Pangasinan–La Union border. As a result of this well-focused effort, rinderpest was gradually eradicated from the eastern two-thirds of Pangasinan, impeding further diffusion southward. In the general absence of fencing, however, one infected animal entering after the military quarantine was withdrawn could undo it all. At the end of fiscal year 1911, at least 81 municipalities across the archipelago still reported active cases of rinderpest.[108]

The Philippine authorities were in a bind. On the one hand, the shortages due to disease mortality created an immediate need to import cattle for agricultural work and slaughter. On the other hand, such importation risked the reintroduction of lethal disease vectors.[109] Relatively affluent and influential urbanites, to say nothing of the U.S. Army, wanted beef, and well-connected importers like Lichauco and Soriano wanted the continued opportunity to make a profit in this commerce.[110] So the Bureau of Agriculture compromised.

It would have been happy to ban live imports altogether, but given these coun-
tervailing pressures all that was possible was to continue the work of public
education and to institute a quarantine system based at first on the Sanitary
Code of the City of Manila rather than on national legislation. This would at
least intercept diseased animals coming from abroad, though even this was
vigorously opposed by some cattle dealers.[111]

As part of this effort, a quarantine station was constructed in the upriver
Pandacan district, accessible to international vessels via lighters. Meat animals
passing final inspection there were walked through the city to the newly reconstructed Azcarraga Matadero for slaughter. Carabao for work were released to
dealers. This facility promised some protection from the transmission of bovine
disease, more than the private corrals scattered about the city. Pandacan was
where animals from Cambodia and Vietnam were landed, quarantined, and
inspected; eventually animals were slaughtered there as well. A similar quarantine station was set up and maintained at Iloilo, though not at Cebu, which
was now closed to live animal imports.

Cattle continued to arrive in Manila from the ports of China during the first
months of 1911 but were held on lighters in the bay for ten days before being
certified as disease free and allowed to land. Not surprisingly, both rinderpest
and foot-and-mouth disease appeared among the stock thus quarantined. The
Lichaucos report (without supplying a definite date) that disease once forced
Faustino to dump an entire shipload of live cattle into Manila Bay. The
authorities feared that the workers tending the animals on the lighters would
spread the disease in the city and thence to the country at large, as cattle coo-
lies had spread the disease inadvertently among scattered dairies in Shanghai.[112]
Accordingly, after a decade of battling the importers, the flow from South
China was effectively stopped by imposing an uneconomic three-month quarantine requirement.

Shipments from China were quickly replaced by major arrivals from Wynd-
ham, on the north coast of western Australia. The importers of Australian
animals were required to build holding pens and an abattoir across the bay at
Sisiman cove on the Mariveles military reservation in Bataan, 30 miles from
Manila (map 6.2). Wild animals and domestic stock were kept well away from
potential disease contact by effective fencing. Starting in 1911, Australian arrivals in their thousands were landed and slaughtered at Sisiman, with the sides
of beef delivered to Manila daily by steamer. The new quarantine system was
not ideal, but it was considerably better than the chaos of prior practice.[113]

The new Sisiman system was successful in reducing the threat from Australian cattle, but the stations at Pandacan and Iloilo both released live animals to
dealers and were less successful at stopping further disease introduction. Rinderpest and foot-and-mouth disease were discovered in animals under quarantine,

and 1912 outbreaks of rinderpest in Laguna, Rizal, Iloilo, and Capiz were all traced to cattle recently imported from Indochina. "These animals had been passed" by veterinary officials and "had undergone ten days quarantine in the Philippines."[114] As a result, the 90-day quarantine rule was extended to animals arriving from this region, and for the next two years, few cattle were imported from Indochina. Cattle importers were offered simultaneous inoculation of their stock in Hong Kong or Phnom Penh at their own expense as an alternative to a three-month quarantine, but this was not adopted because of high death rates following the inoculation.[115]

The hiatus in continual reintroduction from abroad, together with provincial quarantine efforts in Pangasinan and elsewhere, were the major reasons why the period from mid-1911 through 1915 was the low point between the second and third waves of the Philippine rinderpest epizootic. Rinderpest was also becoming enzootic in places like Pangasinan, Panay, and the provinces around Manila, so mortality in these areas of chronic outbreaks could be much less than elsewhere, even as low as 20 percent. This period saw considerable recovery in domestic livestock numbers.[116]

Why were imports from Indochina resumed? Veterinarian and sometime cattle importer Victor Buencamino relates in his memoirs a story that seems to float in time, unconfirmed and undated. In the process of trying to assure that imports were disease free, colonial authorities cut off live imports from French Indochina, where animal sanitation measures were known to be lax. The French retaliated, Buencamino alleges, by banning the import of American kerosene, hitting the powerful Standard Oil Company in the pocketbook. "Whereupon," he wrote, "Standard pressured the State Department into ordering the American authorities in Manila to find means of resuming the cattle importation. Manila's reply was that the Cambodian cattle might come in provided the French guaranteed that shipments shall hereafter consist of only immunized cattle."[117] Unfortunately, it was some time before an effective vaccine was developed.

The Third Wave

Despite the clear public interest in controlling animal disease, bovine imports from Indochina and China soon resumed in earnest—following rises in beef prices (up 100 percent in the last few months of 1913) and ongoing agitation by affluent traders and beef eaters—amounting in 1915 to perhaps 16,000 head. Predictably, a third wave of the rinderpest epizootic surged early in 1916, affecting 18 provinces by the end of the year (graph 8.1).[118] Because of more effective intervention, and perhaps more regular disease exposure, this wave did not develop the same intensity as the first two. Annual mortality peaked at 27,000 animals in 1917 and again at 35,000 in 1921 and 1922. It took eleven years to

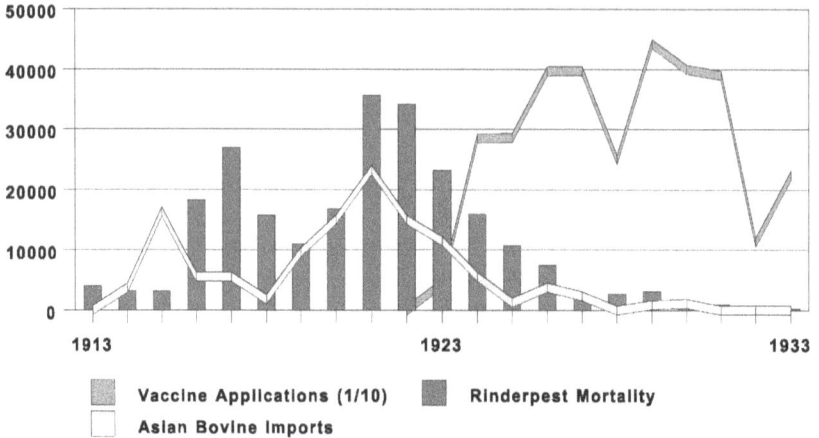

GRAPH 8.1. Bovine Rinderpest Mortality in the Third Epizootic Wave in Relation to the Number of Cattle Imported from China and Indochina and the Application of Improved Vaccines. Only 10 percent of vaccine applications is shown to avoid distorting the graph.

bring the figure below 10,000 again, but during this entire period fewer than half as many animals died as had perished in 1902 alone.

The Veterinary Division of the Bureau of Agriculture attempted to respond vigorously to the new epizootic but with resources grossly unequal to the task. Medical interventions that had worked moderately well in India and Shanghai either did not work or caused high mortality among less resistant animals in the Philippines. There now existed an immunization that often conferred long-lasting immunity and was not lethal to animals in good condition, but many animals arriving from abroad were not in good condition. Despite the bureau's efforts, during 1918 only 3 provinces in northern Luzon had been (temporarily) cleared. The epizootic stayed active in 27 other provinces, plus Manila, and had spread to 5 more, from Davao to Bikol to Ilocos Norte. The largest numbers of deaths were recorded in 4 provinces of the inner zone around the city.

The bureau attempted to launch a major campaign in Masbate during 1918, where "smugglers" pursuing private interests were obviously avoiding the restrictions on movement. The bureau suspected that this particular outbreak had occurred when a local dealer tried to carry a few cattle to Leyte for sale; unable to make a profit, he returned the now diseased animals to Masbate. The bureau responded with a handful of veterinarians, 30 livestock inspectors, and 50 Philippines Constabulary troops as quarantine guards. On this occasion, the disease was stopped just beyond the municipality of the first outbreak—no mean feat on an island with an open-range cattle economy and few fences. In 1921 the greatest mortality was recorded in the western and central Visayas. In these peak years of the third wave, the bureau calculated that the annual death rates were just less than 3 percent—significant but a far cry from the devastation of the previous waves.

On the biological front, an effective vaccine suitable to Philippine conditions was developed in stages by the scientists of the Bureau of Agriculture. In 1923 William H. Boynton, a pathologist, developed a tissue vaccine incorporating finely ground material from the organs of infected animals.[119] In careful application by well-trained personnel, it represented a breakthrough that helped greatly to lower rinderpest mortality and eradicate the disease from some enzootic areas. In 1927 the bureau came up with a similar vaccine, treated with chloroform, which could be more readily prepared and kept under refrigeration for extended periods.[120] A practical drawback to both vaccines was that they required three injections over a period of weeks, but in 1928 the single injection format was adopted. Finally, in 1934, veterinarians M. M. Robles and J. D. Generoso developed a dried vaccine requiring only a single injection, which could be kept a month at room temperature and more than two years under refrigeration.

Meanwhile, rinderpest again invaded the Ilocos coast and Mountain Province in 1925. Starting at the southern tip of La Union and running northward in a chain of infection among animals grazing on the hillsides, it was finally stopped at Tagudin in southern Ilocos Sur by targeting a mass vaccination campaign just ahead of the disease. Again the Philippines Constabulary, 300 strong, established an effective quarantine cordon running from the coast to the hills. Owners of semiwild cattle grazing in the hills were warned that animals found running loose would be shot. Vaccine was aggressively administered in mass campaigns to 200,000 to 300,000 animals a year from 1924 through 1931. Photographs show mass vaccinations in various communities of Ilocos Sur, with each event treating about 600 carabao and cattle (figure 8.3). By 1927 losses from rinderpest were below 3,000 per year.[121] Progress against the disease was now such that "scouting parties composed of veterinarians and livestock inspectors" could be sent to scour the outback for hidden cases.[122] Finally, inspectors found the last case in the wilds of southern Negros in 1938. After half a century of intermittent devastation, the combination of vaccine, quarantine, and near zero imports worked—a very substantial public health and economic achievement.

Through all this, some urban consumers continued to demand beef. Imported cattle continued to meet most of the demand for fresh beef in Manila until a shortage of shipping developed in World War I. Imports

FIGURE 8.3. Front line action in the campaign against rinderpest, 1927. A team of skilled vaccinators injects hundreds of cattle and carabao passing through a stock chute at San Esteban, Ilocos Sur. (Bureau of Agriculture photo, USNA II, RG350-P-Am-9-6, box 5)

from Australia ceased during 1916–21, and the Sisiman facility in Bataan was temporarily closed. Overall arrivals dropped to quite low levels, though not quite to zero.[123] The war also brought rapid price inflation, so local owners rushed to bring their animals to market. For the first time in the twentieth century, domestic animals provided the large majority of the fresh beef supply of the city. The number of native cattle slaughtered rose from fewer than 4,000 in 1915 to almost 20,000 in 1918. With the end of the war the previous high import pattern was reestablished. The number of domestic cattle slaughtered in the city declined temporarily, to only 5,000 in 1921 (graph 8.2). Presumably the imported animals were cheaper and yielded better quality meat than run-of-the-mill Philippine range stock.

But 1921 proved to be the peak year for live bovine imports, including more than 9,000 cattle from Phnom Penh alone.[124] During the rest of the 1920s the flow of animals from Asian locales declined. Effective August 1, 1922, the secretary of agriculture ordered a prohibition on the importation of cattle, carabao, and pigs from places and ports designated as having dangerous communicable animal diseases—initially identified as Hong Kong, French Indochina, and British India. A month later the department partially

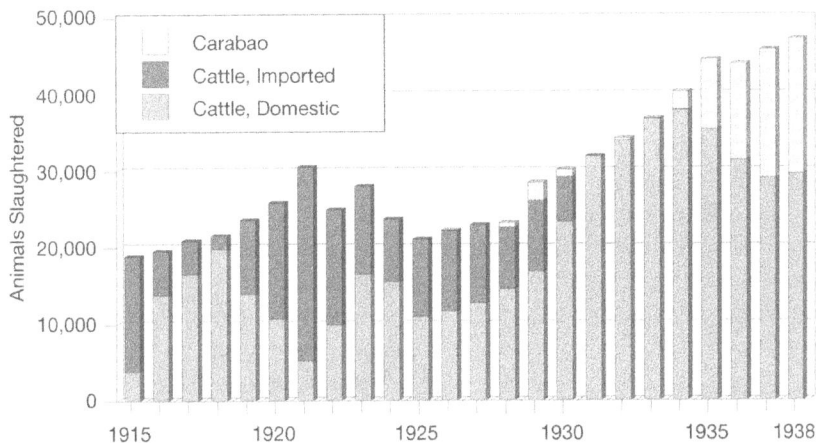

GRAPH 8.2. Bovines Slaughtered for the Fresh Meat Supply of Manila, 1915–1938, Domestic and Imported Cattle and Carabao. The number of carabao slaughtered is for 1919–20, 1925–26, and 1928–38. Slaughter of foreign cattle passed for consumption is estimated at 90 percent of those imported to the city in 1915–18. (Compiled from ms. annual reports of the Bureau of Agriculture, 1916–29, and Bureau of Animal Industry, 1930–33; Antonio Peña, "Agricultural Conditions in the Philippines," PAgR 14.2 [1921]: 153–57, and 15.2 [1922]: 139–47; PSR, 1st–2nd qtrs., 1936, 197; PSR, 4th qtr., 1937, 615; BPS, 1st–2nd qtrs., 1939, 133–37; and BPS, 4th qtr., 109–16.)

backtracked and offered to grant an exception, on application, for imports of animals that had been effectively immunized and then held in quarantine abroad for ten days. Faustino Lichauco challenged the order but lost before the Supreme Court.[125]

Lichauco was increasingly displaced as the leading importer by a partnership consisting of his cousin Ramon Soriano and Victor Buencamino. The latter credits the partnership's success to both Soriano's business experience and capital and his own expertise as a veterinarian. At one point, Buencamino traveled to Phnom Penh "to set up a system of large-scale immunization before loading the cattle for shipment to Manila." Their return on investment was almost 100 percent in 1920, the first year of the partnership. Major cattle importers announced their intention to cease bringing animals from Indochina, but 1,100 cattle from Phnom Penh were consumed in the city in 1930. Meanwhile, imports from northern Australia resumed in 1922 and were suspended briefly in 1924 (due to an anthrax outbreak) but averaged 7,000 to 8,000 cattle for the rest of the 1920s, when the supply from Indochina began to dry up.[126]

Despite prolonged lobbying, live cattle imports came to an end after 1930. It was in this protectionist environment that the ranches in Bukidnon specializing in hybrid zebu animals boomed. No longer needed, the facility at Sisiman was closed. Faustino Lichauco closed his import business and turned to trawling and commercial poultry production, Victor Buencamino became the director of the new Bureau of Animal Industry, and Ramon Soriano got involved in the domestic beef business.[127] But the domestic cattle supply system could not meet the volume of consumer demand for beef in Manila indefinitely at prevailing prices. By 1933 the average weight of slaughter animals was declining, leading to speculation that some own-ers were forced into marketing young stock "in view of the money crisis." The number of cattle slaughtered in Manila peaked at 38,000 in 1934 and by 1937 had declined to 29,000.

The importation of chilled and frozen beef, meanwhile, continued to expand.[128] Decades earlier, at the turn of the century, sending dressed beef very long distances was uncommon. The United States had resorted to innovation to supply its troops, shipping a quarter of a million pounds of beef and an equal amount of vegetables to accompany the fleet from the American East Coast to the Philippines. In ordinary commerce, however, such beef was more likely to come from relatively nearby Australia. By 1898–99 a commercial steamship trade in frozen meat from Queensland had started.[129] Shunned by many in the early decades of the century as tasteless, its quality improved, and imported frozen and chilled beef was more widely embraced in the 1930s.

At the same time, the slaughter of carabao in Manila increased rapidly to meet the demand for bovine meat, from an average of fewer than 100 head per year during the early 1920s, and still less than 1,000 during the early 1930s, to 17,000 per year in 1937–39 (graph 8.1). The rapid expansion of carabao slaughter reflected both small farmers' need for cash and urban consumers' desire for cheaper meat. In 1937 the municipal board passed an ordinance allowing beef and carabao meat to be sold in the same market stalls, an open invitation to subterfuge. "Carabeef" suddenly became a commonplace—as it was during the first years of the Japanese occupation and is again today.[130]

～

The first two waves of rinderpest in the Philippines were plagues of biblical proportions. In the late 1880s and early 1900s, respectively, they completely devastated lowland wet rice production and drastically changed the volume and geographical pattern of flow of beef animals to the city. The epizootics also helped lead to the creation of a professional veterinary school associ-ated with the University of the Philippines.[131] In each case, the numbers of

TABLE 8.2. Changing Sources of Domestic Cattle Slaughtered in Manila by Province, 1885–1886 versus 1935–1938

Province Average	1885–86	1935–38
Inner zone		
Nueva Ecija	1,031	35
Laguna	1,020	24
Morong / Rizal	9	232
Bulacan	2	46
Cavite	0	76
Pampanga-Bataan	0	4
(Subtotals)	(2,062) 10.3%	(417) 1.3%
Outer zone: Central and northern Luzon		
Batangas	4,001	3,691
Pangasinan	3,790	131
Tayabas	3,590	483
Ilocos	2,423	187
Zambales	670	75
Cagayan-Isabela	441	629
Batanes	151	80
Mindoro	77	1,010
Tarlac	0	43
N Vizcaya-Mtn.	0	19
(Subtotals)	(15,143) 75.4%	(6,351) 20.3%
Outer Zone: Bikol, Visayas, and Mindanao		
Masbate	1,960	8,893
Camarines	896	265
Albay-Sorsogon	3	548
Leyte	18	202
Iloilo	0	864
Romblon	0	432
Antique	0	230
Negros	0	124
Cebu	0	46
Bohol	0	231
Samar	0	30
Bukidnon	0	4,461
Palawan	0	2,547
Cotabato	0	2,415
Zamboanga	0	2,317
Sulu	0	717
Misamis-Lanao	0	174
Surigao	0	78
(Subtotals)	(2,877) 14.3%	(24,577) 78.4%
Totals	(20,082) 100%	(31,345) 100%

SOURCES: *El Comercio*, "Matanza de Reses," 1Feb1886 and 15Jan1887; PSR, 1st–2nd qtrs., 1937, 197; PSR, 4th qtr., 1937; BPS, 1st–2nd qtrs., 1939, 136–37.
NOTE: Percent refers to the percentage of the Philippine total for that average year. Masbate 1885 includes Burias (54). The Bolinao area was in Zambales in 1885 and Pangasinan in 1935.

carabao and cattle were greatly reduced, but the reductions were not pro-portional. Between 1870 and 1903, epizootics reduced the recorded num-bers of carabao by approximately 40 percent and the number of cattle by 77 percent, changing the ratio between the two species from 51 cattle per 100 water buffalo to only 20. Carabao were needed for muddy-field preparation and even for transportation of produce into the city on the quagmires that passed for roads, so the number of carabao in the country recovered first, despite their slower reproduction rate. Even on the eve of World War II, the ratio between the two stood at just 46 cattle per 100 carabao—not quite back to the level of 1870—despite the large number of cattle now being raised for beef. The Philippines had a special dedication to the carabao.[132]

PART III

Fluids and Fashions

9

Fluids of Life

Water and Milk

WATER AND MILK may seem like ordinary matters, but without safe sanitary supplies of both the modern megacity is impossible. The city would be a death trap, especially for the infants and children who rely on milk. The death rates would be horrendous for all age groups, and the rapid replacement of individuals would affect the urban economy by impeding the buildup of skill and experience in the work force. Having a sanitary public water supply is often a matter of political will to prioritize and commit the resources necessary to build and operate the necessary infrastructure.[1] Beverages are an essential and ordinary part of daily existence. They provide hydration for the normal operation of bodily systems as well as essential nutrition for infants. At the same time, in their variety, beverages offer some of the marvelous diversity of life, and as delivery mechanisms for psychoactive compounds such as caffeine and alcohol they can provide a stimulus, tension reduction, a context for social bonding, and/or escape from daily reality. Here, as with other aspects of the urban dietary, one can see long-standing indigenous patterns of production and consumption and their alteration by choices made in the face of prestige structures stemming from foreign practices, industrial production, and mass advertising.

Changing beverage production, delivery, and consumption in Manila and the Philippines is a subject of great diversity, meaning, and economic magnitude. It is also one that becomes an arena for playing out clashes of social interest. Further, it has been too little addressed by scholars. Works on water infrastructure by Heutz de Lemps and cocoa by Clarence-Smith are significant recent exceptions. Unfortunately there is as yet nothing like Erik Swyngedouw's pathbreaking monograph on social power and water in Guayaquil.[2]

THE MANILA WATER SYSTEM

Despite the development of a more or less workable infrastructure for water delivery in a number of early Hispanic cities in the Americas—often incorporating aqueducts, tanks, masonry conduits, and public fountains—there seems to have been little like that in early Manila. An exception was the tiny canal dug in 1690 to carry springwater from the then rural Dominican sanctuary in San Juan del Monte to a tributary of the San Juan River, where it could be accessed by urban water carriers using small boats.[3] During most of the nineteenth century fresh water for urban use came from shallow wells and also cisterns that collected rainwater from the tile or galvanized iron roofs of the more substantial dwellings. Considerable water for ordinary folks was taken directly from the Pasig River between Guadalupe in Makati and Sta. Ana and from various *esteros*. The water was transported in large earthen jars known as *tinajas* or *tapayan*, carried in *cascos*, and retailed along the river and *esteros* throughout the city. Early in his business career one of the persons distributing drinking water by *banca aguadora* to ships and residents was Luis R. Yangco, later a *casco* and steamboat magnate, as we have seen.[4] These practices became increasingly problematic as human density increased and both well pollution and waterborne diseases increased concomitantly. Sandy seashore places such as Malabon-Navotas ran out of fresh groundwater before 1850 and had to be supplied with fresh water brought from springs to the east at Tinajeros (appropriately named) and Malinta. Some of the wells of other coastal districts also provided only brackish water, including parts of Tondo.[5]

At intervals in this environment, cholera, a disease transferred from person to person via contaminated water, became a major scourge, and a majority of the population was also preyed on by intestinal parasites. Persons of poor nutritional status, that is, subsistence urbanites, were more likely than others to become infected and experience serious or lethal illness.[6] This was not limited to the city by any means, but because of density, sanitation problems were generally more acute there, resulting in higher death rates in the city than in the average countryside—the so-called urban penalty. Increasingly Manila required a stream of new migrants—even when it was not growing appreciably—to replace those who died of waterborne diseases. Many foreigners and the well-to-do in general learned to drink distilled or boiled water, but for ordinary impoverished urbanites and provincial migrants, the combination of lack of knowledge and the cost of urban firewood made this an unaccustomed behavior.[7]

The water problem was long-standing. Recognizing this, the Spanish *cabildo* (municipal board) member and *alcalde* (mayor) Francisco Carriedo left a substantial legacy in 1743 to be used for the construction of a properly engineered water system for the city. After many vicissitudes and a delay of 140 years it was used for exactly that purpose. A Manila resident for 20 years, Carriedo became wealthy in the galleon trade. In his lifetime, he distrusted local sources of water and "was accustomed to have his water conveyed in *cascos*, either from Laguna de Bay or from old Cavite." In those days "water for sale was conveyed through the streets in carts drawn by belled carabaos or sold from licensed *bancas* at Santa Ana and San Pedro Makati." Not sufficient to finance a water-works initially, Carriedo's money was left to accumulate by investment in trade and much later by making institutional loans for the construction of the big public markets in the city. Part of this fund was looted by the British and Indian forces that occupied Manila in 1762–64. Thereafter, the remainder grew rapidly. Frederick Sawyer maintains that long after the amount had reached a critical threshold, the trustees declined to take action. One reason may have been their use of the fund as a source of loans to themselves. Another, perhaps, involved the financial interests of those involved in water sales and delivery. Finally, in 1878, after an extended mano a mano between the central authorities and the city council, the remarkable governor-general Domingo Moriones forced them to use the funds to install the city's first piped water system. In the event, the fund contributed more than 365,000 pesos to this critical project. This amount was almost matched by the proceeds of a spe-cially enacted tax of 1 percent on beef and pork cleaned in the city abattoir. Since ordinary families ate mainly fish, the slaughterhouse tax fell on those best able to pay. Xavier Heutz de Lemps points out that along with a growing scientific appreciation of the mechanisms of disease transmission, the later Spanish administration was concerned with heading off a repeat of the indig-enous riots that were occasioned by the first cholera epidemic in 1820. For the colonial authorities the possibility of another such mass event becoming con-nected to rising Filipino self-confidence and "national" feeling was something to avoid.[8]

In 1884 numerous public water taps and the Carriedo fountain became important infrastructural fixtures of the city—with a population then of about a quarter million. With pride the newspapers of the day pointed to the new waterworks as one of the great accomplishments of Spanish rule. And indeed many other important colonial cities in the region lagged well behind Manila in this regard. The intake for the new system was located in the valley to the east on the Marikina River opposite Santolan. Here large coal-fired steam engines pumped river water up to large subterranean reservoirs newly exca-vated within the volcanic ridge in suburban San Juan del Monte (map 1.1). From this *depósito* the water moved by gravity in a large pipe across the river next to the San Juan Bridge and then through the Sta. Mesa district into the city. The cast-iron water main ran above ground to Calle Alix (now Recto) in

Tanduay, Quiapo, and smaller-diameter distribution pipes branched out underground from there. The small size of the typical distribution pipes greatly limited the water pressure outward from Quiapo-Sampaloc. Far from immediately becoming the "in" thing, some very affluent households were in no hurry to use this service and waited almost a decade to bring piped water into their homes. Almost 20 years later, in 1902, only 1,825 residences were connected directly, perhaps 5 percent of the total.[9]

Carriedo had specified that the system must benefit the poorer classes, and in its realization there were hydrants from which the general public could draw water for free. There was also clandestine tapping into the water main, which began almost at once. Still, as Victor Heiser (director of the Bureau of Health from 1905 through 1914) later said, the Carriedo "system did not reach all the people in Manila. As a result, the poorer classes, among whom the danger of cholera was greatest, were [still] accustomed to take water from shallow wells, ponds, *esteros*, or other questionable sources . . . in many instances for drinking purposes."[10] Shallow neighborhood wells were almost made to order for spreading cholera. The Carriedo system was a substantial gain for the city, but it was not designed to serve high-quality water to a majority of ordinary residents.

Given the relatively small "footprint" of the urban area in those days and the general slope of the city's land surface in the direction of the bay, almost no surface runoff from the urban area reached the Santolan intake site (map 6.2)—a very good thing. Unfortunately, however, this was not pristine water. Santolan was downstream from the then small towns of Marikina, San Mateo, and Montalban and the agricultural valley surrounding them. Both people and carabao bathed in the river. In 1902 an expedition traversing the riverside found "five dead animals in varying stages of putrefaction" floating between Montalban and Santolan, as well as several villages using the stream as an open sewer. This left the water subject to pollution, and, while some solids would have settled out in the *depósito*, no effective method of filtration was included in the system. Investigation of the water supply yielded counts in the thousands of bacteria per cubic centimeter versus a reasonable standard of less than 100, as well as evidence of amoebic contamination.[11]

In this environment, imported home water filters and drinking water brought from Sibul Springs in Central Luzon and Mount Makiling were widely advertised and marketed. In the 1890s, locally manufactured carbonated beverages— *aguas gaseosas*—became the rage. Commercial interests even tried importing springwater from Japan. Companies vied to advertise their product lines, establish their "brands," and make it known that they operated regular home delivery routes. At the same time they attempted to cast doubt on the purity of the water used by their competitors. Soon the new authorities warned that impure bottled waters were appearing on the market.

To keep up with consumer demand for water, in 1897 the Santolan pumps were in use during the dry season six days a week. Two years later, after the American takeover, two additional steam-operated pumps were installed, doubling the capacity of the Santolan pumping station, and a footbridge was authorized to keep some of the people and carabao out of the Marikina River. But by this time the total use of piped water in the city was such that the *depósito* supply was only sufficient for one day's use, leaving the city instantly vulnerable to any mechanical failure. Although experiments were carried out treating the water with copper sulfate crystals in an effort to remove amoebic dysentery, these were not immediately successful. Little wonder that resident engineer Frederick Sawyer advocated taking wine at meals rather than water in order to avoid "dyspepsia." Of course some people, including some Americans, pronounced the water of the Carriedo system "exceptionally pure and agreeable to taste."[12]

Still, many foreigners and members of relatively affluent families learned to drink only distilled or boiled water. This was especially true following the onset of a terrifying cholera epidemic in March 1902.[13] Distilled water was delivered to their doors by the new government ice plant. Several private companies continued to advertise their water products in the years before an expanded system was functioning. "How to Avoid Cholera" trumpeted the Distilled Water Company. "Isuan" brand mineral water from Los Baños was promoted with very large advertisements as "the peerless table water," "agua esterilizada." Likewise, retailers offered pumps for household water systems. At least in part, the local Chinese were protected by the customary dietary practice of boiling drinking water and consuming hot tea. They also took readily to distilled water. But for provincial migrants and other poor inhabitants of the city, boiling drinking water or obtaining distilled water were not routine behaviors, and this was reflected in the morbidity and mortality rates of the city. Ulcerated colons and in severe cases liver abscesses were all too common results of amoebic dysentery in drinking water supplies.[14]

The new American colonial regime was anxious to tackle big issues and set infrastructure, sanitation, and public health problems right—as it soon began to do in dealing with yellow fever and other diseases in the construction of the Panama Canal.[15] Further, the cholera experience of 1902 left Americans and others in the city with an acute sense of their biological vulnerability. With the use of bond issues, a greatly augmented water system and a new sewer system were placed near the head of the list of projects—in both cases because the health implications of not doing so were understood by both the military and the civilian bureaucracies.[16] As a start, a number of deep wells of generally 500 to 1,000 feet were dug and all the shallow private wells ordered closed. The new wells—labeled "artesian" whether or not there was enough pressure to force the water to the surface without pumping—were deep enough to get under the brackish water stratum in the deep alluvium under the city. These wells were completely enclosed to prevent contamination by surface runoff. In the eastern parts of the metropolis such as Sta. Mesa, the porous volcanic tuff that was the chief medium for fresh water was near the surface, so the wells there did not need to be so deep. Where deep water was available and the wells properly constructed, there was little need for it to be distilled. These early wells were heavily augmented in the 1920s.[17]

In addition to the deep wells, a dam was constructed above Montalban at Barangay Wawa on the Marikina River where it cuts through the limestone ridge (map 1.1).[18] This created a new supply of surface water for urban consumers. The San Juan *depósito* was now put in reserve as an emergency reservoir. The head of Government Laboratories warned that the new system must add a filtration facility to remove "animal parasites [amoebas] which are present in the Mariquina River," but as in the Carriedo system no filtration plant was included. On the plus side the watershed of the new reservoir was demarcated and cleared in order to reduce the surface sources of pollution entering the municipal water system, and former soldiers were hired to guard it.[19] Nancy Peluso reports that in Java the Dutch created "reserves" to protect urban watersheds in which local individuals and families continued to assert ownership of individual fruit trees. If local people in Luzon took the opportunity to claim or plant scattered fruit trees on the Manila watershed, it was not recorded.[20] Everyone concerned admitted that the river was subject to some contamination. This was a surface water system, so after a heavy rain the bacterial count would greatly increase, and without a filtering system, amoebas were always present. Even so, after 1908 the reservoir at the Wawa Dam provided cleaner water than before for the enlarged piped water system, and its elevation was sufficient to provide water by gravity flow rather than having to rely on lift pumps. Waterborne disease and overall death rates in the city went down smartly as this improved system came on line and the clean new wells were drilled and opened for free public use.[21]

The new piped water system opened without sufficient water pressure in the downtown area—prompting some acerbic commentary.[22] And it did not deliver water of ideal quality. The authorities added chlorine—reducing bacterial counts by 70 percent—helpful but insufficient. Without effective filtration, amoebas, ciliates, and flagellates were routinely present in the piped water system. Health authorities found it prudent to compel market food stall managers to "sterilize the city water by thoroughly boiling it." Despite these shortcomings, a further substantial apparent decline in the city death rate in 1913 was attributed primarily to the "decrease in water-borne diseases, owing to the radical improvement in the water supply and to the increasing use of distilled and artesian waters for drinking purposes." At least typhoid was not endemic, and 1912–13 passed without a single reported case of cholera in the city.[23]

∽

By 1915 rapid urbanization had produced a growing demand that outstripped the capacity of the Wawa Dam to deliver sufficient water to get the city through the dry season.[24] Water from the upriver reservoir now had to be supplemented with the downriver pumps at Santolan. During the 1920s, with the metropolitan population jumping from more than 300,000 to more than 500,000, the piped water system could no longer maintain consistent pressure during the peak months of the hot season. Not only did this produce a general water shortage, but it also meant there was too little flow to flush the mains and wash out the accumulated debris, which acted as a medium for the growth of amoebas and other contaminants. From 1921 through 1929, the hot-dry season became a public health officer's nightmare. The managers coped with the shortfall by using the Marikina River pumps to feed water into the system. Initially this was added mainly at night. In the 1920s, however, the pumping often went on around the clock. The worst conditions came during the great drought of 1926 when the old Santolan pumps and *depósito* system had to be used for 83 days. Thereafter, stopgap improvements to enlarge the Montalban Dam reduced the use of polluted water from the lower river.[25]

In the meantime, the Metropolitan Water District was established in 1921 as an autonomous entity run by a professional manager under an appointed board of directors.[26] The first manager, former city engineer Abraham Gideon, and the appointed board wrestled with the renewed and growing threat to public health posed by the dry season water crises.[27] At the direction of the board and with the help of geologists from the Bureau of Science, Gideon prepared a report containing elaborately considered options. These included (1) building dams and settlement basins in eastern Rizal Province on two upper tributaries of the Marikina River known as Bosoboso and San Isidro and (2) building a dam at one of several alternate sites on the Angat River in eastern Bulacan. Both had acceptable and comparable water quality, but the Angat solution promised to provide much greater supplies over the long run. Throughout the report there runs an open disagreement between Gideon and an unnamed assistant manager plus others. Both sides favored some sites on the Angat River in eastern Bulacan—all but one of which were quickly rejected by the geologists—but the assistant wanted to piggyback a hydroelectric facility onto the project, requiring a different kind of dam and much higher construction costs. Gideon, trying to solve the water supply problem, found these additional costs far beyond any reasonable projection of return from the sale of electricity.

Going to the root of this expensive alternate proposal, Gideon traced the legal transfer of the "concession for power rights on the Angat River" from the

late-nineteenth-century company La Electricista to Salvador Farré and the last
Spanish waterworks manager, Col. Carlos de las Heras, then to "Mr. Swift,
President of the Manila Electric Railroad" (Meralco) "about 1912." He notes
that de las Heras and Farré "tried on several occasions to get capital interested
in this power project, and that the engineers of the Manila Electric and the
J. G. White Co. of New York investigated this proposition but apparently did
not deem it advantageous." Finally, his professional patience at an end, Gideon
concluded, "We cannot possibly afford to construct this power plant [at triple
the cost of simply supplying water, as] it would never prove an asset even [if]
we were able to carry such a staggering loss."[28] He was joined by the geologists,
who were alarmed by the prospect of any dam that was not a low-rise earthen
structure because of the extensive folding and faulting of the entire archipelago, the high likelihood of further seismic shocks, and the very rapid rate of
discharge by the Angat River during typhoons—in one case rising ten meters
in a single afternoon! To his credit, Gideon repeatedly built in the costs of adding "sand filters" to the proposed systems, even preparing elaborate mortality
tables and costs of ill health due to unfiltered water supplies. Evidently no
positive action was taken on his report in 1922. Was action blocked by persons
doing the bidding of the private electric utility—the only possible commercial
buyer of the proposed electricity—or by some other crisis such as the problem
of sorting out the public losses due to unwise and corrupt government lend-
ing policies at the end of World War I? It remains a question. For its part, the
interest group known as the American Chamber of Commerce soon recommended immediate construction of the Angat water project.[29]

Three or four years later, perhaps as a result of public outrage over the
effects of the great drought of 1926, a giant new earthen dam was constructed
at Novaliches (map 6.2). The new reservoir was filled and ready to take over
supply by the dry season of 1930. The water of this reservoir system was consistently of better sanitary quality than that from the Wawa Dam above Montalban, but water quality really began to achieve universally acceptable standards
with the opening of the filters at Balara in 1935 in what is now Quezon City.[30]
In all, this was a huge achievement first of the colonial engineers and later of
Filipino professional leaders.

It quickly became evident that the new dam and filtration system could
consistently deliver safe potable water. The system was also robust—able to
quickly handle both a major leak and an attack with explosives.[31] However, in
drier years the Novaliches Reservoir, by itself, could not handle the rapidly
escalating demand. In 1936 almost 24 percent of the 44 million cubic meters of
water delivered to the city population came from the old Wawa Dam. By then
construction of the new Ipo Dam on the upper Angat River in eastern Bulacan

was under way.[32] Both 1938 and 1939 were comparatively dry years in Manila. By mid-March of the latter year the great Novaliches Reservoir was nearly empty—just as work was completed on the essential parts of the new system, including a siphon aqueduct. A public health disaster of great magnitude was narrowly averted. With the Angat River supply on line, the Wawa Dam was retired. Best of all, 99 percent of the water supplied was said to meet public health standards for drinking water.[33]

Despite the expanded system of piped water, the deep wells continued as "the main source" of drinking water. From the free hydrant pumps of these wells, much of the city population had its water delivered by *aguadores* or family members who waited to fill up in the ever-present lines. Whether in Francisco Carriedo's time or in the nineteenth and twentieth centuries, a significant number of persons made a bare living as water carriers. In the 1870s, many of these were Chinese, especially in the area of Quiapo-Sampaloc. After the change of colonial regime in 1899 brought strong prohibitions against the continued immigration of Chinese as laborers, the *aguadores* were Filipinos. At the Plaza Miranda well in Quiapo in 1931 water carriers began lining up between 4:00 and 5:00 a.m. to fill five-gallon water cans packed 15 to a pushcart. The lines continued until late at night.[34]

In Manila, as in many other colonial and Third World cities, water carriers were essential in the absence of high-standard utility infrastructure. Even in quintessentially middle- and upper-middle-class San Juan district in the 1920s and 1930s, suburban residents got their drinking water in this fashion. For example, accountant Estevan Munarriz and his family moved to the San Juan Heights subdivision in 1922. The piped water supply in their new home was useful for washing but not good to drink. As with the neighbors, their drinking water was brought by a Filipino *aguador* who operated a stable long-term business supplying a string of regular customers. He came on Wednesdays and Saturdays carrying two large cans suspended from a balance pole. This was "artesian" well water, and "he emptied it into our *banga* in the kitchen. In pouring the water, he used a very clean cloth as a filter. For years he brought drinking water to our house."[35]

Across the city in Tondo in the 1930s in the mostly working-class Antonio Rivera Street neighborhood, one of the *aguadores* was Antonio Sumbillo. Mang Antonio carried water from a deep city well to homes in the immediate area. The going rate was two centavos per can. If asked by a customer, he would sometimes deliver water late in the day as well. He had earlier worked as a construction laborer and souvenir maker. This was his third livelihood, and by itself in a poor territory it was not sufficient, so he combined carrying water in the mornings with vending bread door to door in the afternoons. Both were self-organized, informal sector occupations. In 1940 he gave up this dual liveli-hood to become a *kargador* in the great Divisoria Market.[36]

What is all too often missing from the public record on water availability is reports of social conflict, including careful analysis of who was largely left out of the piped water systems and what it cost them to purchase water from door-to-door vendors as part of their household expenses for basic provisionment. In some cities at the same time, water carriers were part of large social organi-zations. In Beijing they were powerful hometown associations. Groups of car-riers came from the same provincial locality or had something else strongly in common. These combines were well known for their social and political "conspiracies" to monopolize this poorly paid work—sometimes by threat of violence—whether to put themselves in position to charge higher delivery prices or simply to avoid being undercut by a surfeit of carriers or by changing water technology.[37] Such carriers could hardly be expected to lend support to an expanded infrastructure. Evidence for these sorts of dynamics connected to water distribution in Manila has so far eluded me. (Actually, it was the *cocheros* and cart owners who cohesively defended themselves against demands that they vacate narrow business district streets choked with automobile traffic.)[38] One wonders in particular how the transition from Chinese to Filipino water carriers played out. Organizational and infrastructural differences between Swyngedouw's observations and Manila in our period include the dispersed system of free hydrants in Manila versus the monopolized water source and central organization of motorized carriers in Guayaquil.[39] Finally, despite the fact that the residents of the large impoverished district of Tondo were fre-quently able to elect one or more of its residents to the city council, it was poorly served for fire protection by the small size of the water distribution pipes. This lack of effective "clout" came at great cost when 10,000 or even 20,000 residents were suddenly made homeless in one of the great dry season conflagrations.

Periodic additions were made to both water and sewer systems. Individual water services that tapped into the piped water system expanded from fewer than 2,000 in 1902 to 11,000 in 1920 and 46,000 by 1939. Of these last, 36,000 were in the city of Manila alone—servicing about 32 percent of all households. The rest were in the suburbs, including 1,000 or more in Pasay, San Juan, and Caloocan. Overall, this was a significant supply and public health achievement. At the same time, however, only 10,600 homes and businesses were connected to the sewer system. Further, there were rapidly growing parts of the metropoli-tan area, especially less affluent parts, where the water mains were chronically too small and the water pressure too low. After each great fire in the nipa hous-ing zone of Tondo, the city council vowed to rectify the situation.[40] Follow-through was painfully slow.

"Soft Drinks"

So-called soft drinks were the rage of late-nineteenth-century Manila. This was due to their chilled and refreshing appeal aided by modern mass advertising (figure 9.1). And it was spurred on by the health risks associated with problematic domestic and public water supplies. *Soft drink* refers to a lack of alcoholic content—as opposed to *hard liquor*. It is a problematic term because from its formulation in the 1880s until 1903 Coca-Cola was fortified with cocaine—as the brand name conveys—and through 1911 at least with very high levels of caffeine.[41]

In Manila various soft drinks were imported initially. Increasingly, however, local pharmacies and *aguas gaseosas* (carbonated water) companies vied with each other to promote their diverse product lines and even operated regular horse-drawn vehicle routes for home delivery. At the same time, there was considerable advertising for ice and *bebidas heladas* (cold drinks). Only medium amounts startup capital were required to enter the soft drink business, and this quickly became a substantial arena of small industrial entrepreneurship, attracting risk takers of several nationalities. After a time, the imported machinery for "soda water factories" could be purchased right in the city. Given all the sales promotion and consumption of bottled mineral and distilled water, it is difficult to draw a firm boundary between these products and our contemporary concept of soft drinks. In the 1890s the flavored products advertised included "nectar-soda . . ., tamarindo, piña gaseosas, zarzaparrilla extra, agua carbonica, limonada, tonica, [and] aperitiva gaseosas."[42] *Zarzaparrilla* ("sarsaparilla" in English), better known to later twentieth-century Manilans as *sarsi*, became a particular favorite. Derived from a tropical tuber and similar to root beer (though not identical), *zarzaparrilla* had been prescribed for some time

FIGURE 9.1. This ad suggests that nearly everyone likes soft drinks, with a skillful mingling of bandwagon and snob appeal advertising techniques. But note the lack of ordinary workers in line. The ad ran frequently in Manila's newspapers in the 1890s. (*El Comercio*, October 1, 1894)

by medical practitioners as a drink that would help restore health and help the body fight off sores.[43] These industrial beverages quickly entered the diets of those able to afford them.

Well-trained persons from the German lands pioneered western-style pharmacies in the Philippines starting in the 1830s. By 1850 at least, they had introduced the retail soda fountain and western pharmaceutical preparations. According to Wigan Salazar, all the German pharmacies in Manila were using soda machines from Europe and relying on "soda water and lemonade" sales as a major source of profit. By the 1890s, they were actively bottling and distributing these soft drinks. Early soft drink advertising aimed for brand recognition and generally omitted any mention of ice. But ice manufactured locally was available for household iceboxes from the 1870s on, and by the 1880s it was strongly associated with *boticas,* suggesting its commercial use with iced drinks.[44] One of the largest bottlers of the 1890s was A. G. Sibrand Siegert, based in Quiapo. A wholesale druggist and distiller of *ilang-ilang* essence (for use in perfumes), Siegert's advertisements emphasized modern production equipment and the capacity to produce 10,000 bottles a day. He was a manufacturer and wholesale distributor of the full range of soft drinks.[45]

The profit potential of this consumer industry was also apparent to others. If it was begun by pharmacies, hucksters and others were quick to seize on its presumed health benefits. In 1881 promoters claimed that "La Zarzaparrilla de Bristol" was an infallible cure for nearly all that might ail you, including tumors, syphilis, and malignant eruptions—afflictions that were said to begin with impure blood and bodily humors. Others were still advertising this product for similar ills in the 1890s.[46] In Manila, however, the commercial pitch soon switched from quasi medicinal to refreshment. At the same time the Manila market was being eyed by successful industrial bottlers based in the China treaty ports. In 1881 A. S. Watson & Co. was importing a mix of soda, lemon, tonic water, and *zarzaparrilla* from its bottling plants in Hong Kong and Shanghai. Soon Watson was manufacturing its several lines locally and in the process making use of the largest steam engine in the industry—a clue to its capital intensity. But it was on the defensive concerning rumors about the quality of the water it used. To counter this, the company installed its own deep well and took out ads in the new English-language press to bolster the reputation of the "filtered" water used in the aerated products manufactured at its downtown factory.[47]

A number of Spanish and Filipino entrepreneurs also entered this industry in the 1890s. One of the smaller *fábricas* producing soft drinks over the long term was Adelante, based in the Angustia section of Tondo, which made use of two on-site deep wells and a small steam engine. This was the business of

Silvestre Bautista from Santa Cruz, Laguna, a mestizo with a Chinese father and Filipina mother who had been educated in Spanish. Adelante was a going concern from at least 1903 until the plant was destroyed in the great Tondo fire of 1941. Sales were sufficient for Bautista to send some of his children and grandchildren to college and to build homes for them.[48] At least 21 aerated water factories still operated in the city in the 1930s.

Manila paved the way as usual, but in the 1890s local soft drink plants also sprang up outside the city. In Barasoain, Bulacan, the leaders of the Philippine Revolution had the chance to purchase soft drinks from the Fábrica de Limonadas K K K.[49]

Large-scale production came with the entry of a capital-rich and advertising-savvy beer company: the San Miguel Corporation. In 1919, San Miguel bought out its erstwhile competitor, the Oriental Brewery, and took over its physical plant. This facility was soon turned into the manufacturing center for San Miguel's already successful line of soft drinks, including Tru-Orange and Royal Soda. By the mid-1920s Royal Soft Drinks was claiming annual sales of 6 million bottles. Soon it was offering 23 flavors. Advertising in several languages, San Miguel aggressively cast doubt on the water purity of its competitors. The San Miguel Corporation and its Royal soft drinks plant also became the Philippine franchise bottler of Coca-Cola. By 1927 Royal Soda was producing Coca-Cola for the city and nation, but in the early years it spent far more advertising its Royal line of drinks. Actually M. A. Clarke was the first agent for Coca-Cola in Manila and offered it at his soda fountain, restaurant, and chocolate shop on the Escolta, a favorite gathering place for Americans in the city in the early twentieth century. He went on to bottle it as well.[50]

In the long run, much larger bottling companies beat out the initial brands by employing higher advertising budgets as well as more expensive and productive manufacturing facilities. Even so, several small volume brands continued.

A SANITARY MILK SUPPLY

Among the range of potables needed in a tropical climate, early-twentieth-century Manila had large, lively industries to service adult demands that included beer, other forms of alcohol, distilled water, and carbonated drinks, but not, to any significant extent, did they supply milk-based beverages. Yet in a city and metropolitan area with a youthful age profile and 23,000 children less than five years old in 1903, and more than 100,000 in 1939, milk was a major source of nutrition and thus became a focus of official and public concern. But public health work in the early twentieth century initially addressed other priorities: expanding and improving the water supply, instituting a proper sewage system, and enhancing sanitation in public markets and homes. Further, the great epizootic of rinderpest that broke out in the first years of the new century carried off the large majority of bovines. This affected not only farm work and the meat supply but also the urban milk supply.

Milk was used, especially for infants, but it was not an item of adult mass consumption in the Philippines or insular Southeast Asia more generally. Milk and cream were used in Spanish-derived pastries and custard deserts, especially *leche flan*, but it was not a particularly common element in adult cuisine, nor were milk- and cream-based sauces widely used. A dietary survey of more than 100 working families in the city in 1936–37 found a mean expenditure on dairy products equal to only 2.6 percent of the total food budget—a tiny allocation compared to that in western countries. Further, the families with the least to spend on food spent nothing at all on dairy products.[51]

As it happens, many adult Filipinos are lactose intolerant. They do not share in the genetic change common in northern European populations that allows most adults to digest milk without gastric distress. The geographer Frederick J. Simoons and others have shown that the incidence of the ability to digest milk among contemporary adults—"lactose tolerance"—varies widely among human populations and can be shown to have very deep cultural and ecological roots. Nearly all infants produce lactase, the critical digestion enzyme, but by the teenage years or earlier, lactase production has typically declined and, as a result, a gastric encounter with milk will produce gas, abdominal cramping, and diarrhea. This corresponds to long-entrenched culinary habits.[52] Still, the rich milk of water buffaloes is not a new element. The everyday word for milk in Tagalog is *gatas*, presumably from an indigenous Austronesian root, rather than *leche* from Spanish. Setting aside for the moment questions related to ice cream and cheese, fluid milk is and was primarily consumed by the young, especially infants, for whom it can be essential in the absence of an adequate human milk source.

Milking, as an ecological adaptation, was not an early feature of most Southeast Asian culture groups. However, Paul Wheatley has gathered inscriptional evidence that seems to show that the practice of milking cows eventually spread across portions of the region during the first millennium CE in connection with tiny numbers of Indian migrants and a more general (if quite incomplete) process of Indianization. The inscriptional evidence is particularly dense in Cambodia. Milk or melted butter was associated with Hindu and Mahayana Buddhist ritual practice. There is little to suggest that the peoples of the Philippine archipelago or eastern Indonesia were included within the zone of this innovation. In any case, milking as a local specialization declined as these ritual practices faded in the early second millennium CE. This left the geographical pattern as before, with milking and milk use among adults largely confined within a broad Old World territory that included India and what became western China and had its eastern margins in Bengal and western Yunnan. Antonio de Morga, writing in the early seventeenth century, points to water buffaloes "brought tame from China" and "used only for milking."[53]

Dairying and milk use among adult Filipinos and Southeast Asians today principally derive from recent influences. For a long time, first Spaniards and then northern Europeans and Americans have resided in Manila in some numbers. In the twentieth century a small population of South Asians also developed. The per capita consumption of milk by these groups has always been much greater than among Filipinos or Philippine Chinese in general. In Spain dairying is highly regionalized—relatively uncommon on the dry Mediterranean coasts and Meseta but widely practiced in Galicia and among Basques along the grassy northern coast. Here both bovine and goat's milk drinking is more or less common. So the demand for milk by resident Spaniards was likely variable.

Supplying uncontaminated milk to the urban population long remained an intractable problem and one of the prime causes of high infant death rates. There is almost no record of milk arriving in Manila from the outer zone in the nineteenth century and for good reason. Without refrigeration in warm weather, natural milk has a very short shelf life. Supply in this case was not just an inner zone product, but almost entirely a suburban phenomenon (as von Thünen predicts). The soft white cheese made from carabao milk, *kesong puti*, might come from Cavite or Laguna in small quantities, but fluid milk came from the margins and interstices of the built-up area.[54]

Throughout the nineteenth century and into the twentieth, milk was delivered by vendors walking their rounds. In the 1850s *lecheros* carried fresh milk in large pitchers and delivered it to the kitchens of a regular set of customers (figure 9.2). An essay by an anonymous Spanish resident captures the nature of this work. Addressing female *lecheras*, he observes that they tended to be fit and energetic because every day they made urban deliveries in two round trips on foot from the suburban dairying neighborhoods where they lived. These were in Caloocan and Makati, most distantly, as well as Gagalangin and Lecheros in Tondo, and Sampaloc.[55] Arriving in the city between four and five in the morning, they made their rounds—in all weathers and with great punctuality. The milkmaid was not particularly concerned about the quality of the milk— to which she might have added adulterants, he thought—but she certainly lived up to her agreement to deliver it every day without fail. As a consequence, the *lechera* habitually moved with dispatch, and her rapid, almost ritually repeated movements struck the essayist as akin to the military manual of arms: "She enters a house and quickly goes to the place for delivering milk, stops like a soldier, lowers the earthen jar and bamboo *chupa* or half-*chupa*, measures out the daily ration, returns the jar and *chupa* to their place, and departs like a shooting star," repeating this process in the homes of all her customers. With the deliveries completed around eight, the *lecheros*, including male and female young people, tended to gather on Real Street in Intramuros to wait for companions who came from the same neighborhood. The Escolta was another meeting place. All met up, they set off for home in groups. There was a second round of home deliveries in the late afternoon except for those coming from distant Makati.[56] Fifty years on, there were still multiple "Barrios Lecheros," in Tondo and Sampaloc and others now farther east in Sta. Ana and San Juan del Monte. And *lecheros* still made a livelihood delivering milk in open vessels. In some parts of Java in the 1880s, by contrast, the milkman on foot delivered his product to Europeans in bottles with sealed paper covers. Still, adulteration remained a significant problem.[57]

Product quality was a chronic problem. Spanish concern in 1889 led to a study of milk samples obtained from ambulant vendors in Binondo. This

FIGURE 9.2. *Lecheros* making home milk deliveries in the 1850s. Sometimes river water was added. Similar work routines and equipment were still used in 1900. (*Illustrated London News*, December 26, 1857)

revealed that in some cases the milk had been diluted with various starches, rice dust, coconut milk, and unnamed other ingredients "more or less harmful to the health of the consumers." In other cases, the vendor's milk was stretched using contaminated river water with potentially fatal results for infants.[58] The typical door-to-door milk supply at that time was rich in fats and slightly bluish in color because it came from carabao. There were only a few European dairy cows in the city. These were not well suited to make good use of the available dry season browse and fodder, and they were especially vulnerable to the tropical disease environment. By contrast, water buffaloes were well adjusted to these conditions as long as they could avoid the major epizootics.

In the early twentieth century public health work tended to focus on priorities other than milk. Further, as we have seen, the second mass rinderpest epizootic had just carried off the large majority of water buffaloes and cattle. This directly affected the urban fresh milk supply. Finally, starting in 1907, the new American authorities made a major effort to clean up the milk being sold in the city. In particular, this effort attempted to reduce the use of adulterant contaminants. For a time at least, the use of coconut oil, rice dust, and some other additives declined.

Still, Manila continued to record an infant mortality rate that was horrendous by modern standards. The health authorities believed that this was heavily due to intestinal disorders and malnutrition often caused by contaminated food. Scientists with the Bureau of Science said at the time, "We cannot recommend the use of unboiled fresh milk in Manila obtainable in the open market, and until conditions are much improved, its use as food for infants seems almost criminal." This was as true for carabao and goat's milk as for that from cows. W. E. Musgrave of the University of the Philippines' College of Medicine summarized, "[T]he milk sold on the streets of Manila . . . is from 26 to 30 hours old; has been diluted with tap water, or worse; has been collected and transported in dirty receptacles," and so on. The state of the milk supply was clearly a major cause of the high urban infant death rate.[59]

In the absence of an abundant, affordable, and reliably sanitary fresh milk supply, Manila and the Philippines more generally became a prime market for imported industrially packaged dairy products—an important nutritional development. Nestlé, a European brand, was advertising in the city by 1883, but a few years later the American consul reported, "As a rule everybody, except the very few who own cows, uses American condensed milk ([Borden's] Eagle brand)."[60] By the first decade of the twentieth century, canned condensed milk had become "the most available source of supply" for feeding middle-class infants. Bear/Oso Brand sterilized fresh canned milk and Milkmaid sterilized milk, both from Switzerland, and Dragon Brand from Milan had also become major brands. In the subsequent deluge of advertising, these and others

were placed before the literate public at weekly or even more frequent intervals. In this, Royal Brand condensed milk with the Swiss eagle trademark
was the exception in its concentration on the Tagalog print media.[61] By 1911
Nestlé had begun direct distribution through its own subsidiaries rather than
local sales agents—creating an important commercial advantage.[62] By 1913
there were at least 50 different brands on the Manila market. In some weeks,
the density of newspaper ads for imported milk products rivaled those for
beer. As with the mass advertising campaign for beer, there were fewer later.
In their day, these products may have been essential in making up a great
shortfall between local production and demand, although it is also possible
that the increasingly affordable product stimulated the demand. Unfortunately, such tinned condensed milk was often mixed in ways that were dangerous to infant health.[63] In a major study of the spread of milk consumption in
Indonesia during this period, Adel P. den Hartog credits canned condensed
milk with eventually piercing the price barrier to reach indigenous families
for use in infant feeding. He also points to the emergence of a major nutritional problem for infants with the wide use of cheap sweetened condensed
skim milk.[64]

The Philippine government in general and Victor Heiser as director of the
Bureau of Health promoted a sanitary milk supply. Initially, this focused on
sterilized, canned milk. A clean local dairy product was also advocated through
the work of several bureaus and experimental dairy farms. Unfortunately,
almost all of this proved too expensive for the least affluent component of
urban society. One finds in the technical literature analyses of the biochemical
content of mother's milk and discussion of cases in which the baby was doing
poorly following the mother's loss of milk. Further, in the early 1910s infantile
beriberi was rife, and it appeared that mortality among breast-fed infants was
greater than among the artificially fed. A lack of thiamin (vitamin B_1) in the
diet of poor mothers due to the nearly exclusive consumption of polished rice
was being passed on to infants via the mothers' milk. During this time the
recommended treatment for beriberi in infants involved cessation of breast-
feeding for a month or more. But poor families could not afford sanitary artificial feeding much less a wet nurse. An effective solution to this nightmare
was found to be the administration of *tikitiki*—rice bran polishings.

At the same time, we may be looking at the wrong period. Hartog reports
that Dutch women in the East Indies in the eighteenth century only rarely
"breast-fed their infants but made use of a wet-nurse, who was a slave." The
use of a wet-nurse apparently remained the norm among European women
in Indonesia in the nineteenth century. Hartog believes that "most likely they
were following the habits of the sophisticated upper-class women of Europe,
where breast-feeding was declining."[65]

I do not find in all this a campaign against breast-feeding embedded in the official crusade for a sanitary milk supply in Manila. In this period most of the abundant ads placed by the imported canned milk industry were in media chiefly consumed by the educated, although there were also large downtown signs. The trend against breast-feeding over time, if it was that in the prewar period, seems to have been the consequence of the general communication of an affluent American lifestyle in movies, cigarette advertisements aimed at women, and the ceaseless bombardment of other consumer advertising.[66] By contrast, the profit imperative could sometimes lead to advertising that tended to acknowledge and promote breast-feeding. The San Miguel Brewery originally developed and marketed a bland dark beer as a health tonic for dyspepsia and general ills. Three decades later the company mounted an elegant advertising campaign to market this same beer to comfortable and affluent nursing mothers (figure 9.3). The changing incidence of breast-feeding is a broad question deserving further study in its own right.[67]

A committee formed by the Bureau of Health in 1912 to look into the causes of the continued high infant death rate quickly zeroed in on the milk supply. While the more comfortably situated urban residents tended to use imported sterilized milk for their children, that was not a practice the poor could afford to emulate. Taking their lead from the committee, a press story on problems with the milk supply observed that "the principal source of supply of milk for the poorer classes [is] the town of Bocaue, on the railroad. The milk is from carabaos, which are milked in the afternoon of the day previous to that on which the product is sold. The milk is sent to Manila on the evening train, without ice, and remains in the Tondo market over night. It is sold to the peddlers during the morning, part of it not reaching the consumer until four o'clock in the afternoon, still without ice. The condition of the milk can be imagined, and it is this that is fed to the helpless infants."[68] In fact, the sort of milk in question came from several places, but the point was made.

Victor Heiser noted that, despite many fines imposed on vendors of such products in the past, the improved quality of the milk supply lasted only as long as the authorities were giving the matter special attention. In 1912 the authorities mounted a crackdown carried out by the police and especially the several district boards of health. Early in this campaign, 20 vendors of contaminated milk were apprehended in Tondo. Then the Paco regional Board of Health brought 5 offenders to "municipal court. . . . The most flagrant case was that of Mariano . . . who was fined ₱25." This man was caught in the act of adding polluted Pasig River water to his container of milk. Two sick children were found on his list of customers. All the *lecheros* brought to court that day "were from Santa Ana, the headquarters for the milk dealers in that part of the city, as it is nearer to the *zacate* fields, and rents are cheaper. . . . Filthy

The Food - Tonic That Appeals is labeled "Negra"

Mothers of young children depend on this refreshing beverage — it renews strength, and promotes health under strenuous conditions—

You will do well to take it regularly each day—Order a case and begin at once—Ask for

Negra

made by

San Miguel Brewery

FIGURE 9.3. Dark beer for nursing mothers, part of San Miguel's campaign to attract female consumers. This high-end advertising image employs both maternal resonance and snob appeal. (PFP, January 11, 1930, 6)

stables and still worse facilities for cleaning bottles were found."[69] Heiser now came to believe that universal pasteurization offered the only practical solution. Not coincidentally, 1912 was also the year in which both New York City and Chicago finally imposed the same requirement on all milk supplies—an entire generation after Pasteur's discoveries. In American cities, this public health innovation almost immediately eliminated the annual hot weather surge in infant deaths.[70]

On a tiny scale, Manila's Gota de Leche was already trying to provide healthful milk for needy infants. This was a voluntary organization of elite local women dedicated to helping the infants of poor mothers and thereby lowering the infant death rate. Under the auspices of this humane organization, one sees photos of poor Manila babes being suckled directly by a goat—avoiding all of the problems of adulterants, contamination, and lack of refrigeration.[71] The work of this organization came to the attention of Nathan Straus, the foremost advocate of pasteurization in New York, and he in turn donated milk sterilization equipment for use in Manila. The Philippine Legislature was persuaded to fund construction of a building for this activity. As Straus and the Gota de Leche evidently hoped, this effort eventually evolved into public pasteurization stations for milk brought into the city by vendors. By 1935 there were three of these facilities: one in Intramuros handling 80,000 liters and others in Meisic (Binondo) and Tayuman (Tondo) handling about 25,000 liters each. Independent milk vendors were required to use them. In 1937 an estimated 75 percent of the carabao milk brought in by peddlers every morning was boiled in these facilities prior to distribution.[72]

This sterilization was required by health service regulations, but it was avoided by a persistent minority of vendors. One of those quietly avoiding sterilization in the 1930s was a midwife living in a then grassy area near Tayuman Street in Tondo. Neighbors brought bottles to be filled with fresh milk from the family's carabao. The proceeds from the sale of milk helped the family buy a modest house. Oscar Evangelista recalls that his mother always boiled the carabao milk before serving it to her children. To the extent that institutional pasteurization and home boiling were effective, they greatly lowered the exposure of infants and children to lethal disease agents, including tuberculosis of the bowel.[73]

In addition to the public sterilization stations in the city, there were also now a few local companies buying and pasteurizing carabao milk from producers on the metropolitan fringe. The largest of these was Ramon Arce's Selecta Sterilized Carabao Milk plant on the Novaliches Road in the hills to the north. In the mid-1930s, Arce was selling 200 liters a day to Manila ice cream parlors, as well as colleges and hotels. Two smaller operations sprang up in Sta. Mesa using milk collected from Alabang south of the city and in Tondo using milk from Caloocan.[74]

In the end, there is no way to determine the full composition of the milk supply of metropolitan Manila, but the Census Bureau estimated that local production in Manila and Rizal in late 1938 was 81.0 percent carabao milk, 18.8 percent cow's milk, and 0.25 goat's milk. This is at considerable variance from the national figure of only 8.0 percent cow's milk, a reflection of the very different consumption patterns in Manila. There were a few provincial places where cows were said to be the chief source of milk: the Baguio hill station, Batanes, Sulu, Antique, and a few others. But only in these few did milk cattle appear on more than one out of 1,000 farms.[75] This is our best estimate of production, but the affluent and merely comfortable in Manila, including Americans, northern Europeans, and Indians, consumed a lot more milk per capita than other social groups, and most of it was cow's milk—with the difference between consumption and production made up by imports.

~

The milk distributed by the ordinary *lecheros* came from carabao, but there were a few small dairies in Manila in the 1890s producing fresh milk from Australian cows. According to the American consul, "the quality" of milk was "not nearly so good as that from cows in a colder climate, but we are glad to get it." At the same time there were at least five licensed *tiendas* selling milk and/or cheese. Another was the Vaqueria Australiana, which operated in Malate in the late 1890s and then relocated to Sampaloc where it advertised *leche pura* available at all hours.[76] A photograph taken in front of this dairy depicts a deliveryman and colorful horse cart just returned with empty recycled bottles. The Bureau of Insular Affairs labeled this scene "milk vendors, new style," as opposed to the ambulant *lecheros* delivering milk from pitchers (figure 9.4).[77]

Meanwhile, students of the nineteenth-century milk supply of London recognize a major transition there from reliance on tiny, embedded, urban dairy herds kept in congested and less than sanitary conditions to a system in which these urban dairy herds were increasingly replaced with on-farm production at a distance with the milk transported into the city by railroad.[78]

Circa 1900 in Manila a few other microdairies offered their products to Manila's consumers. San Juan was briefly the site of a dairy operated by the Bureau of Agriculture, but it was moved out to a former friar estate in Alabang. At the same time imported Australian milk cows could be purchased from one William Van Buskirk, who also distributed crushed feeds from Australia and India. By the early 1910s there were three Manila dairies managed by Americans with a total of about 75 cows, three tiny dairies managed by Spaniards with 14 cows, and one Filipino dairy with 12 cows. All seem to have been using European cattle breeds brought from Australia.[79] There were also tiny dairies employing native cattle and carabao, especially in Taguig and Meycauayan.[80]

FIGURE 9.4. Milk cart, deliverymen, and the premises of the Vaqueria Australiana in Sampaloc, Manila, early twentieth century. The name and advertising picture suggest that the milk came from European dairy breeds rather than carabao. (USNA II, RG350-P-E-10-2)

Coincident with the clean milk campaign of 1912, a new dairy opened to the enthusiastic approval of the Bureau of Health. The Australian Machine Dairy Co., Inc., apparently not the former *vaqueria* of similar name, began operations in Manila with 60 Australian cows on hand and more coming. Initial production was said to be 1,000 bottles a day.[81] The company was a branch of a large dairy in Australia, and its managers had extensive prior experience in the industry, though perhaps not in the tropics. Starting out in Tondo, the dairy soon moved farther out to La Loma beyond the city limits. It continued in business at this site through at least 1921. Thereafter the name is lost, although a new La Loma Dairy Farm—possibly the same operation under new management—had appeared by 1926 and operated through 1941. One Abboudi Perez owned La Loma Dairy in 1937 when it had 30 Holstein and Ayrshire dairy cows—all descendants of stock brought from the United States and Australia. The farm included a pasteurization plant.[82]

Most of the dairies concentrating on cow's milk were trying to use western midlatitude dairy breeds. The San Miguel Dairy (named after the district rather than the beer company) was using Holsteins to produce 450 liters a day in 1929.[83] The tiny Santol and Manila City Dairies were using Holsteins and a few Ayrshires. The Santol Dairy's Holsteins came from Japan, as did the owners. Vicente Araneta's Hacienda Carmelita Dairy was using about 100 Holstein and Jersey cows.[84]

Few of these operations were intensively managed with an eye to reproduction. One owner was cited as saying that if a Holstein could be kept alive and producing for three years, she would have earned enough to satisfy her owner. In fact most of the Holsteins were judged to be in poor condition. There were constant problems with both feed and disease. Directly imported dairy animals were expensive and required special nutrition, often in the form of imported feeds. Such animals, if pastured, could do fairly well on tender grasses during the rainy season, but as these grasses dried out in the hot season, the proportion of protein declined and the grass became less digestible. Well-cured leguminous hay would have helped, but very little was produced. Instead, some combination of by-products from agricultural processing was employed, including rice bran, copra meal, corn bran, and molasses. In fact most of these dairies were side interests of their owners, and few, perhaps none, were producing milk of the high sanitary and nutritional quality that the newly separate Bureau of Animal Industry thought the population should have. The general result was that locally produced fresh milk was not high quality and was so expensive that it was "only within reach of the moneyed class" or the modest number of poor families aided by the Gota de Leche.[85]

Long ignored by the dairy entrepreneurs was the obvious possibility of employing well-acclimated milk cattle already used to tropical feeds. In 1911 one of the "agricultural explorers" employed by the U.S. Department of Agriculture had written to officials in the Philippines recommending the Sindhi breed of Indian cattle (*Bos indicus*) for its milk production. He assumed that the consumers of this milk would be "white people."[86] Despite publication of this report, commercial dairies were not interested—quite possibly because many of their customers were indeed Europeans or Americans who wanted milk from cows they recognized. But 20 years later the government stock farm at Alabang was milking almost 60 Sindhi cattle with good results. But the initial aim of the Red Sindhi milk cow project was not realized. This was to provide milk animals to small farms, for example, in Batangas, thereby requiring the impoverished farmers to grow legume forage crops that would benefit the soil in addition to producing milk for sale. By the 1930s, the government was also experimenting with crosses between Nellore zebu cattle imported as meat animals and both Ayrshire and Sussex midlatitude dairy breeds. There were also experiments with milk goats in the 1930s, which came to little. Cultural blindness to the possibility of using Indian cattle for milk production was hardly unique to the Philippines and was widely replicated near the big cities of Java, Shanghai, Hong Kong, and elsewhere—except in the small Indian dairies.[87]

Still, just before World War II, dairy animals interbred among Ayrshire and Nellore—between dairying *Bos taurus* breeds and tropically acclimated *Bos indicus*/zebu animals—were beginning to win a following. Few of these animals survived the war, and early postwar dairies started by foreigners began again with Holsteins.[88] As it turns out, the initial enthusiasm concerning the interbred animals was probably due to some success with first-generation offspring. Subsequent generations were less productive—a phenomenon known as "hybrid breakdown." Today one can see this was not unique to the Philippines. After an entire century of attempts in numerous countries to breed a stable race of dairy cattle that is both tropically acclimated and productive, no such breed exists.[89]

~

Eventually reconstituted cow's milk using imported concentrate became the standard liquid milk product. In 1926 the San Miguel Corporation expanded beyond its central interest in beer to purchase the existing Magnolia Ice Cream plant. Retaining the Magnolia label, San Miguel invested in new facilities, including a 600-foot well. Three years later it added a full dairy plant and began distributing "fresh, pasteurized and reconstituted milk, ice cream, sherbets, table and pastry cream, cottage cheese and buttermilk." Quickly Magnolia ice cream began to feature a kaleidoscope of flavors. "French vanilla" was introduced in 1931 to be followed later by the famous lavender "ube." Just as it had hired an experienced German *braumeister* to oversee technical brewing processes, so the company now hired an experienced Scandinavian American dairyman to manage Magnolia and train a generation of indigenous technicians. By 1940 the technical director was Dr. Felipe T. Adriano, who had been hired away from the Bureau of Science. In the meantime, a great deal of advertising and education was done to acquaint the public with the nutritional value and sanitary standards of the reconstituted product. Looking for a broad customer base, there was now extensive advertising for milk in the Tagalog print media.[90]

By 1935 San Miguel had swept the field with high-standard Magnolia reconstituted milk accounting for more than 60 percent of all milk sold in the city, including both carabao and cow's milk. Despite the fact that the concentrate was imported from the midlatitudes, Magnolia reconstituted milk was substantially underselling locally produced fresh cow's milk. In 1940 Magnolia was producing daily 1,300 gallons of milk and cream, 1,000 gallons of Chocolait brand chocolate milk, and 1,200 gallons of ice cream. Its products were distributed in Manila and over a growing territory beyond. At the same time,

Magnolia was also distributing a limited amount of fresh milk supplied under contract by Araneta's Hacienda Carmelita Dairy in Novaliches. Magnolia's milk and ice cream business was helped by the installation of improved refrigeration equipment by retailers and of the appearance of refrigerators in affluent homes.[91] Magnolia's prime position in both milk and ice cream continues to the present day (although it is now owned by Nestlé rather than San Miguel).

～

Water and milk are the critical beverages for public health, and both presented major problems in Manila life. All over the world at various times, investments in the expensive infrastructure of piped water have been institutional responses aimed at providing a sanitary supply and lowering urban mortality rates. Until nearly the start of the twentieth century, worldwide, urban death rates routinely exceeded birth rates—more people died in cities than were ever born in them. In metropolitan Manila the urban penalty traceable to poor water supplies was increasingly brought under control for a while by aggressive public health measures and a commitment to building the needed infrastructure. Closing open wells, installing numerous deep "artesian" wells, and creating a piped water system were all part of this effort. The policy objective was to reach the entire urban population with a dependable potable water supply.

Infants and small children require milk or a nutritionally adequate substitute. Developing a mass milk supply that did not also infect babies with lethal bacteria proved to be a problem all over the world in the nineteenth and early twentieth centuries—in Chicago in summer as in metropolitan Manila. Once the problem was understood, the technology of pasteurization was easy. Requiring and actually enforcing sanitary standards proved more difficult. The problem of milk production in the tropics remains.

10

Foreign Fashions

Flour and Coffee versus Cocoa

WHEAT FLOUR AND COFFEE are food products that increasingly came in for mass everyday use in Manila relatively late in the colonial period. Both were comestibles introduced by outsiders—actually the introduction of wheat is obscure. As a beverage, coffee increasingly replaced cocoa, or *chocolate*, another originally foreign food but one accepted into the diet at an earlier time. Wheat also was not a native product (although it had long ago become so in North China and India) and was not a significant part of what one might call a "traditional" Filipino/Manila diet. In the rise of wheat-flour-based products and coffee, one can see long-standing patterns of production and consumption being altered by choices made in the context of foreign dietary practices, foreign trade, industrial production, and mass advertising. Wheat for making flour (rather than porridge) and coffee beans were both grown in Luzon in the nineteenth century. But in both cases local mass consumption was little related to this but rather to models of consumption and prestige structures set by Europeans and Americans locally or picked up by cosmopolitan Filipinos traveling and living abroad.[1] Both came to appeal to large segments of the population.

Modest quantities of excellent wheat flour were produced in several Philippine localities in the early to mid-nineteenth century. But blight attacked standing wheat and, together with other demands on the land, led to its domestic demise. Consumption of flour-based products became a mass phenomenon only after relatively inexpensive high-quality flour began to arrive from California. Just when the consumption of baked goods began to spread through Manila society beyond the resident Spaniards, flour itself came to be almost wholly supplied from abroad. Without an adequate internal wheat supply and with little possibility of gaining local control of the Philippine tariff structure, only a modest modern local milling industry emerged during the period. Foods baked from flour added a tasty diversity to the diet but nothing essential.

Coffee was grown in much greater quantities than wheat, reflecting its importance as an export commodity in the second half of the nineteenth century. But Filipino consumers at that time remained far more dedicated to hot liquid chocolate than to coffee. In this they were behaving in a culturally conservative way, matching Spaniards, Mexicans, and Columbians, who also held tenaciously to *chocolate* while people in many other parts of the world switched to coffee. Beverage choice had become a "global" cultural and commercial event. The demonstration effect of locally resident Americans after 1899, to say nothing of coffee's greater psychoactive effect and ease of preparation, made a difference. Residence abroad, particularly in the United States, may also have contributed. The paradox is that mass coffee drinking began to become popular at about the time Philippine production collapsed—again, initially due to blight. A number of other novel imported items, including apples, grapes, and canned peaches, among others, were also taken up in the diets of more comfortable Filipinos as a result of western preferences and western exports, but few came to be as important as flour-based products and coffee.

THE REMARKABLE RISE OF FLOUR IN FILIPINO DIETARY PRACTICE

Wheat flour products entered the country through two distinct channels, Chinese and European. The Chinese have a long tradition of both wheat (and other) noodles and steamed foods. Both were sold from temporary stands on the streets of Manila in the mid-nineteenth century and likely much earlier. At the same time, the Spaniards introduced the European tradition of baked goods, especially breads and cakes. Both noodles and bread became available in the Manila marketplace early on, but aside from occasional artistic renderings it is hard to track who among the indigenous population ate them and in what settings. Of the two, the European tradition of baked goods is somewhat easier to document, "both because the Europeans were doing the documenting and because the adoption seems to have been strongest among the Europeanized elite."[2] By the end of our period both wheat flour cuisine traditions were well established in the metropolis and, in provincial cities and towns as well, but still, then and now, calories in the diet come primarily from rice or a rice substitute such as maize.

Baked goods made from wheat flour were not a prominent characteristic of "traditional" Southeast Asian cookery. Spaniards were raised on bread as a quite ordinary and everyday component of the diet. Naturally they (and other Europeans) wanted bread and other wheat flour concoctions when they resided in other parts of the world. Thus wheat was introduced to the Americas—an important element of the Columbian Exchange of biologic forms between continents and cultural traditions.[3] As an Old World crop common in both northern China and South Asia, it is possible that wheat preceded the Spaniards to the Philippines. But in Tagalog, wheat and wheat flour are still called by their Spanish names, *trigo* and *harina*, and for a long time the consumption of bread was part of the cuisine of foreigners, not of indigenous Filipinos. Certainly there were a few Chinese bakers in early colonial Manila, and bread was said to be available in the everyday marketplace in the 1580s. But in general breadstuffs did not fit well with rice in the lowland dietary and were not a hit with local people.[4]

Still, in the Philippines, Japan, and other Asian countries, the consumption of wheat based baked goods rose markedly during the nineteenth and twentieth centuries. And bread acquired an indigenous name, *tinapay* in Tagalog, from *tapay*, a name for dough, for "something that is kneaded," although in Cebuano it is *pan*, the common Spanish term.[5] Bread is still no threat to rice as the central source of starch in the diet, but it has become a notable part of urban cuisine—expanding in recent decades as buns for mass-market hamburgers. Between the late 1930s and the early 1960s, the national annual per capita consumption of wheat flour doubled. Surveys circa 1960 show a continuing urban bias with per capita consumption in metropolitan Manila at three times the national average.[6]

Many Filipinos were introduced to baked wheat flour through the communion host in Catholic Eucharist services—since the clergy was adamant that the wafers be produced from wheat flour—and in the city by Chinese venders hawking bread. Flour for use in baking the host and consumption as food in the Spanish colonial towns was imported early on from Mexico. But even before the seventeenth century it came from China and Japan until access to the latter was ended. Also, for 200 years, from the mid-seventeenth century onward, modest quantities of wheat were grown domestically, especially in the Batangas and Laguna uplands. Diaz-Trechuelo mentions Taal and Balayan in Batangas, as well as Ilocos, in the eighteenth century (map 1.4).[7]

In the early nineteenth century, de la Gironière points to Batangas as the most notable of several upland locations where wheat was grown. In Ilocos Norte a French visitor reported excellent bread made from local flour in 1807. Small quantities were shipped to Manila from there in the 1810s. Díaz Arenas reports wheat grown in Ilocos, Tayabas, and Laguna in the 1830s.[8] In the periodic provincial reports of 1861–62 prices for wheat were quoted for Tayabas pueblo and the market town of Sta. Cruz, Laguna, most commonly and sometimes for San Pablo and both Vigan and Ilocos Norte. The locality of Badoc in Ilocos Norte was said to produce the best wheat. In Tayabas the price rose from an extended low during April–July escalating in September through a

notable peak in March (1862) a month before the new harvest.[9] Wheat was not
common, but it was mentioned more often than one might expect. J. E. Spencer concludes that, unlike introduced maize, "[R]ural Filipinos in the Spanish
period did not grow wheat for their own home consumption, but produced
it for Spanish consumption and for use in the religious rituals of the Catholic
Church."[10]

From a Manila perspective at midcentury the main domestic wheat production area was an irregular strip of municipalities running southward from
Bay in Laguna and San Pablo through Tanauan, Lipa, and San Jose to Batangas
town and Taal, all in Batangas Province. Buzeta and Bravo report wheat grown
in abundance in San Jose and list it first among the crops grown at Tanauan,
Taal, and Batangas. It was a dry season crop sown at the end of December and
harvested in late April. Wheat was also grown at Marigondon, Cavite, and at
various localities in central Tayabas for use as pottage. Buzeta and Bravo also
report wheat in abundance at Cabatuan, Iloilo, in modest amounts at Janiuay
and Sta. Barbara in the same province, and in scattered other places. A summary of coastal trade for 1853 lists 550 *picos* (approximately 34,000 kilograms)
delivered to Manila, and a one-week survey of the trade carried on the Pasig
River in the same year reported four large *cascos* carrying wheat—likely coming from or via Laguna.[11] In Spain by the end of the seventeenth century
barley bread and gruel and eventually even maize bread and gruel had become
notable foods of the poor but not in the Philippines. Although the production of wheat was slowly coming on in the Philippines, it was not a factor in
the location of Spanish settlement, as in Spanish America, and bakers did not
become such pivotal figures in the food supply system as they were in Mexico
City, Lima, and Paris.[12]

Wheat and flour production subsequently collapsed in Batangas Province,
probably in the 1870s, reportedly as the result of blight. In Spain itself, where
a singular focus on wheat was now seen as backward, more than a million hectares were taken out of its production during the 1860s and 1870s.[13] De la
Cavada lists only 1,500 liters as the Philippine production circa 1876, and the
1886–87 survey reported only 200 *cavans* produced in Batangas. In the 1890s,
Foreman wrote that wheat had been grown in "large quantities" in the vicinity
of Lipa in earlier generations, but its cultivation had "quite fallen into disuse."
In Tanauan (as in California later) it was replaced by citrus production.[14]

Blight was blamed, but the quickening of world commodities trade was central to the decline of wheat growing in the Philippines and ultimately to the
rise in flour consumption. In January 1862, a large packet boat arrived in Manila
harbor on a voyage from San Francisco, California. It carried 500 barrels of
flour consigned to Peele Hubbell. By the end of that year, seven transoceanic
sailing vessels had unloaded wheat flour in Manila, five from San Francisco

and one each from Sydney and Newcastle in Australia. One of the American vessels was the sleek wooden clipper ship *Fearless*. From its launch in Boston in 1853, *Fearless* made eleven voyages from the American East Coast carrying men and equipment around Cape Horn to the California gold fields. This was a lucrative run in the decade before transcontinental railways, but it presented the practical problem of where to get a profitable cargo for the return to the Atlantic coast. This led to a rapidly expanding flour supply for Manila. In eleven voyages, the owners of *Fearless* chose to return via Manila and the Cape of Good Hope nine times. On its first seven visits to Manila, *Fearless* had arrived in ballast and left with a cargo of abaca and sugar.[15] But in 1862 exports of California flour were getting under way, and this was the first time *Fearless* had arrived with a paying cargo. It helped pioneer California wheat growers and millers that Philippine-bound vessels found it "convenient to take flour in lieu of ballast." As the regional economy of California grew and the demand for flour in Manila increased, a few sailing vessels began traveling to Manila and back again in a closed loop across the Pacific. They returned to the United States with sugar, hemp, and some coffee.[16] Supplying Manila with American flour became a long-lasting business.

When such international vessels landed flour in Manila, the consignees were usually Anglo-American trading houses: Russell & Sturgis, Peele Hubbell, or Findlay Richardson. The shipments of flour (and coal) from Australia were handled by Aguirre, a Spanish Basque firm engaged in the rice trade and in carrying tobacco for the monopoly. Overall, imports of flour were small in the 1850s, but they expanded five- or tenfold in the early 1860s and ran over two million kilos per year by the middle 1860s. By 1864 the duties on imported wheat flour had been reduced and in 1869 were zero.[17] To the Spanish authorities in Manila, flour had become a necessity—just as the Spanish bureaucracy and military in the archipelago was about to enter a prolonged expansion.

Within the country the modest flow of flour was reversed, and in 1870 Manila sent out by coastal shipping a net of more than 275,000 kilos. From 1873 through 1882, flour imports were usually in the 3 to 4 million kilo range, exceeding those in the late 1880s, and they expanded smartly to 7.4 and 8.7 million kilos in 1892 and 1894, respectively. California was a leading recorded port of departure early on and again in 1894, but China and/or the British "possessions"—presumably Hong Kong—were the main registered points of origin most of the time. In 1894 6 million kilos of flour arrived from Chinese ports and 2 million directly from the United States. Actually, almost all the imported flour was coming from the West Coast of the United States. Hong Kong was simply functioning as the regional distribution center.[18] Daniel Meissner, the leading authority on this trade, writes, "As a duty-free transport hub [and centralized transshipment port] . . . Hong Kong was second to none.

Moreover, the city's inexpensive warehouse charges and labor costs were attractive to western exporters." In 1872, one can see that Chinese merchants in Manila were among the recipients of flour from Hong Kong. In the 1880s, the flour trade was "in the hands of the Chinese, who import it from Hong Kong."[19]

The business organizations advertising flour in Manila near the end of the nineteenth century included Smith Bell, Warner Barnes, two stores on the Escolta in the central business district, and the local Molino de San Miguel. The *molino* was then milling flour and animal feed from imported California wheat.[20] Still, most of the flour supply of the city was imported in a finished state. One store advertised the receipt of 4,000 sacks of "superior fresh flour . . . direct from San Francisco." All of it carried the trademark of "XXX Señorita," which was produced in California by Sperry Flour Mills. In Cebu in 1894, Sperry's XXX brand was "practically the only kind" sold. Sperry had captured the high end of the Philippine market and then broadened its appeal. It did this by consistent attention to quality and distinctive brand promotion. Immediately following the American takeover of the archipelago, the Philippine flour trade became a near monopoly of American milling companies.[21]

American flour arriving in Manila came from California. Wheat had been grown in California well before the Gold Rush, but only in the 1850s, as cultivation mushroomed to feed the miners and new urban populations, did a surplus emerge. Much of the Central Valley was put to wheat growing, and new flour-milling enterprises proliferated. Local production quickly exceeded regional demand. Austin Sperry, with a commercial background in the Boston dry goods trade, was one of the "49ers," as participants in the California Gold Rush were called. His first flour-milling venture was opened at Stockton in 1852. By the 1860s, Sperry was developing a knack for sales promotion and buying out other mills. By the 1880s, he was a regional leader in both production technology and mill acquisition, building a system that spanned the West Coast. The urge to realize economies of scale and simultaneously reduce "overcompetition" led to waves of consolidation in the milling industry. In 1892 a new corporate entity was created, which absorbed eleven formerly independent enterprises. Sperry Milling was part of this, and its name and brand were retained by the new organization. As in Manila and Cebu, the consolidated company arranged for the careful marketing of its quality brand-name flour along the Pacific Rim.

Meanwhile, in California yields declined as prolonged wheat growing depleted the initial soil fertility. Increasingly the land was put to irrigated commercial fruit and vegetable culture. California's total wheat and flour export peak came in the early 1880s.[22] Over time the West Coast wheat industry shifted to interior production areas in the Northwest. The flow of flour to Asian destinations from northwestern U.S. ports grew rapidly. From 500,000 barrels of flour in 1898, the Puget Sound ports sent out 2 million barrels in 1903. In 1905 Sperry contributed to this trend, opening a big mill in Tacoma, Washington, specifically for the Asia trade. In some years flour was funneled through Tacoma from the Great Plains and even Minnesota.[23]

During the 1870s 1.8 million acres of new wheat lands had been brought into production in South Australia, and the same was happening in New South Wales and Victoria on a smaller scale—part of the expansion of commercial production serviced by railways. But El Niño–related droughts kept Australian wheat exports nearly level for at least two decades starting in 1884. Low-priced Australian flour became increasingly important in Manila during 1905–14, and this competition held American exporters to two-thirds of the local market. With lower prices, one distributor of Australian flour sponsored ads addressed to Filipino workers in the nationalist newspapers *El Renacimiento* and *Razón*. Yangco's Bazar Siglo XX was now advertising "harina de Australia." For a few years in the twentieth century Australia benefited from its relative proximity to the Philippines and lower transport costs. During 1915, however, Australian flour exports were affected by a disastrous drought of the year before and practically disappeared from the Manila market.[24]

The same years witnessed the near collapse of American flour exports to China after a great expansion during the 1890s, an event that bears on the Philippine experience. Initially the problems centered on "the Boxer movement, fluctuating exchange rates, [and] rising shipping costs" from the West Coast to Asian ports, and then the Russo-Japanese War. But as these were solved or diminished, a great nationalist boycott led to a sharp decline in sales of American flour in China. The Chinese boycott of 1905 protested the latest in a series of American coolie labor exclusion acts and treaties. This had the effect of greatly restricting American flour sales and enlarging the market space for new, state-of-the-art, native-owned flour mills. American exports to China recovered briefly during 1907 due to the floods and famines of the previous year and the resulting shortage of Chinese wheat, but thereafter the new Chinese mills met local demand for flour in nearly all quality categories.[25]

The Sperry Flour Company was not taking the assault on its Pacific market share quietly. In 1906, according to the rabid *Manila American*, Sperry was reported to be building an alliance with other mills to establish a "trust" in the spirit of the Standard Oil Company. Having seen Standard Oil drive Russian kerosene off the China market, Sperry was reported to be forming a group that would employ "the usual tactics" to reverse its loss of market share. It would try to break the Chinese boycott by lowering the price of flour, underselling the competition, and accepting short-term losses in order to drive the others out of the business in China. Then it would recover its losses "when the trust has full sway." The newspaper reported approvingly, "It is believed that" other

American mills "will align with [Sperry] in this scheme for administering pun-
ishment to the Chinese for the boycott sustained."[26]

These tactics were well known in the American flour industry because they
were used successfully during the 1880s and 1890s by Theodore Wilcox, head
of the Portland Flouring Mills Company, to crush some of his competitors in
the Pacific Northwest with the goal of dominating American participation in
the China flour trade. At the end of the nineteenth century, Wilcox used them
again in China. If these tactics were used in the Manila marketplace, no record
has come to light.[27]

Production of wheat in the tropical Philippines had already declined and
was unlikely to expand again given the environmental and land base and the
relative ease of international trade. The medium amount of capital needed
to purchase and install a modern roller mill could likely have been raised in
Manila business circles, and certainly the acquisition of needed technical skills
would not have proved a barrier—witness the pioneer establishment of a competitive brewing industry in these same years. Indeed, Tornow mentions a flour
mill established on Luzon just before the American takeover—no doubt referring to the Molino de San Miguel.[28] But the Philippines would not control its
tariff policy for many years to come and would have been unlikely to win a
dispute with American millers and wheat farmers in Washington, D.C. Most
flour continued to arrive in its final form.

Although Hong Kong had long been used as a base for flour distribution, a
new tariff policy starting in 1909 linking the Philippines and the United States
gave U.S. ports the financial advantage for direct shipments to Manila: no
tariff at all. The Connell Brothers Company, a new Seattle-based distributor in
the Philippines, expected that this would allow it to beat out Australian flour
and dominate the Philippine market vis-à-vis the product coming from other
countries. Smith Bell and Warner Barnes, encountered earlier in the context
of the rice trade, were also persistently involved in the Manila flour market—
the former as agent for the Portland Flour Mills Company and Warner Barnes
as agent for Sperry. A number of other participants in the flour trade came
and went. Clearly this was a market worth having, especially as the China
market shrank. To this end, colonial market protectionism aided American
flour producers.[29] In Manila local Chinese wholesale distributors increasingly
dealt directly with the several western commercial agencies.

As with imported fruit, the change of colonial regime led eventually to a
substantial alteration in trade patterns. Except for the years around World War
I (1917–20), more than 60 percent of Philippine flour imports came from the
United States. In 1922–24, American flour exceeded 75 percent of the total.
Whereas wheat flour imports ran between 7 and 9 million kilos annually in
the 1890s and 10 to 15 million annually during 1901–5, by 1912 they exceeded

40 million. Following the end of the shipping shortage during World War I, this level was reestablished. Imports exceeded 50 million kilos for the first time in 1923 and passed 75 million in 1928 before settling back to 70 million during the more difficult economy of the mid-1930s. The United States was supplying 86 percent of the total in the late 1920s through 1931 (figure 10.1). This was considerable growth in trade and consumption by anyone's standard. Filipino tastes were changing rapidly.

In 1929 Sperry was acquired in the general industry consolidation that resulted in the formation of General Mills, Inc., a giant new company based in Minneapolis. The Sperry unit continued to service the Pacific trade for this megacompany. Still, Australian flour did not disappear. Macondray and Company continued to import "Gillespie's Roller Flour," a.k.a. Anchor (Ancla) brand flour, from that country and to promote it sometimes with full-page ads. Macondray also handled American flour and in the 1930s became the biggest importer of flour from both countries. After many years of American dominance in the Philippine market, Australia reemerged as the number-one supplier during 1935 and 1936. Flour from Canada and Japan was also important

FIGURE 10.1. The arrival of a shipment of wheat flour from abroad provides a burst of dockside employment, Cebu, 1928. (G. C. Howard, U.S. Bureau of Foreign and Domestic Commerce, USNA II, RG151-FC-85D, box 85)

during these two years. American prices were apparently much higher than the others until June of the latter year. The American trade commissioner reported that the "use of Canadian flour is mainly confined to Manila . . ., as it requires more kneading . . . and only the largest bakeries have machines for kneading." He also believed that Japanese flour was "used only by Japanese bakers."[30] Despite considerable consolidation in the industry, at least 21 brands were still being imported in bulk in 1934.

Manila and Philippine flour consumption escalated so rapidly in the twentieth century that it cannot possibly be accounted for by the food needs of foreigners alone. Filipinos were patronizing the bakeries of the city and inventing their own variations. Middle-class Filipinos were routinely eating and relishing *pan de sal* (literally "bread of salt") hot from the neighborhood bakery. It had become a major cuisine item of breakfast and *merienda* (light food eaten in late afternoon) and was hawked in the streets at dawn. Likewise, there were delicious cinnamon buns and other bakery confections at parties. This had been going on for a long time among the more comfortable Filipino families.[31]

The expanding acceptance of flour-based baked goods and noodles in the diet accelerated in the nineteenth century and really took off in the twentieth. The 1850 map of Intramuros displayed by Buzeta and Bravo locates a single *panaderia* (bakery) and bread shop on a central block—the better to serve the Spanish population of the city. There were also four bakeries located in the commercial section north of the river, three run by Chinese. To this point, according to Wickberg, Chinese had served as the bakers and confectioners of the city since the sixteenth century.[32] By 1895 there were at least 22 bakeries paying the *contribución industrial* tax in the city, a few more outside the urban center, and perhaps others in clandestine operation. These were fairly well distributed about the urban area. Most of the bakeries made considerable use of firewood, as did the *pansiterias,* large specialty noodle restaurants.[33] Of the 22 bakeries in the city in 1895, 8 were registered by Chinese. The other proprietors had Hispanic or European names. Felice Prudente Sta. Maria points out that once the travel time from Spain had been reduced by steamships and the Suez Canal a new generation of "Spanish pastry and sweet-makers opened new cafés and take-home outlets around residential and business centers."[34] By 1904 there were more than 50 bakeries in Manila; in the 1930s there were hundreds. Nor was this phenomenon confined to the capital. Bakeries also began to flourish in the larger and more commercialized provincial towns. In the 1890s Dagupan, Malolos, Vigan, Lingayen, Bautista, and Lipa each had multiple bakeries, as did the string of Bulacan towns on the waterways just north of Manila.

Consumption of baked goods was diffusing geographically and socially. Consul Stigand may have been right when he reported in 1893 that Philippine "laborers never use bread," and Sta. Maria is equally certain that the "principal consumers" of the nineteenth-century breadstuffs were "Caucasians."[35] But things were changing. Fifteen years later, Buzon and Company was addressing its ads to "Filipino Workers" in media read almost exclusively by the indigenous population and touting its various brands of flour along with a broad range of local and imported comestibles. And a shop on Gandara Street was manufacturing baking powder.[36] By the 1930s a dietary study of working families in Paco district reported that 64 percent had bread for breakfast. This was certainly a dietary change, but given the impoverishment of industrially milled white flour it may have represented a net nutritional loss.[37]

One of the successes of this period was the bakery business of the Paglinawan brothers. Felix and younger brother Mamerto began in the printing business, but business was slow, and they decided to try another. They began what became a highly successful coffee- and cocoa-processing operation in Sampaloc in 1914. They were looking for a business for which they could afford the capital equipment needed to be competitive. Subsequently, they added baking, and this became their main enterprise. The Spanish trade name of their products, La Patria, which means "Native Land," suggests that one should patronize Filipino stores as opposed to those owned by Chinese or others. The company's advertising stressed that its products were made using "scientific methods" and baked in "modern gas ovens." By the early 1930s its baked goods were being made and sold by 100 employees at five locations in the city. Ultimately, much of the retail distribution took place through a network of Filipino neighborhood *sarisari* stores. The brothers began with horse-drawn delivery vehicles with advertising painted on the sides, but as their business grew they switched to a fleet of trucks used to deliver to outlets in the city and surrounding provinces.[38]

Noodles, or vermicelli noodles (*fideos* in Spanish records), were already a significant trade item in the 1860s, with 100,000 to 200,000 kilograms imported each year. By 1894 the annual total had grown to 475,000 kilograms. Noodles were manufactured in Xiamen in great quantities, and some portion of this arrived in Manila with the regular steamship traffic from that port.[39] Mobile open-air food stands along the streets of Manila became the subject of artistic representation by the 1850s. At least some of these served *pansit*—one or another of the thin Chinese-style noodles. By the 1890s, and probably well before, the large specialty restaurants called *pansiterias* had made their appearance downtown, often run by Cantonese and patronized by the ordinary public. By the 1890s, networks of hundreds of small stores were licensed to sell foodstuffs from Europe and China. These, together with large *sarisari* stores, were retailing both imported wheat flour and Chinese noodles in every large neighborhood of the city and many provincial towns. Of course, not all noodles were made from wheat. Noodles were now a part of the national cuisine. Despite their importance, however, they were only equal to about 5 percent of the imports of flour in bulk.

By the late nineteenth century, a proliferation of specialty flour-based products had become a routine part of affluent cuisine. Small meat pies known as *empanadas,* also sometimes filled with oysters or fish, were on the menu of an Escolta restaurant in the 1870s.[40] At a great luncheon for revolutionary leaders organized by Pedro Paterno in Malolos in 1898, several courses included flour products, including a filled puff pastry and a *pastel* of chicken wrapped in a pastry crust. Ambeth Ocampo points out that mass baking was possible in Malolos because of the large number of beehive masonry ovens used for cooking *ensaimada,* or "sweet roll," a locally famous puff pastry. Growing up in Manila in the 1890s in the household of his wealthy grandmother, Victor Buencamino later recalled that for breakfast they had "all the *ensaimadas* you could eat, washed down with thick chocolate."[41] From 1887 through 1941, Intramuros residents bought their *ensaimadas* at La Palma de Mallorca on Solana Street, sometimes advertised as a *"panaderia y pasteleria Europea,"* a famous bakery, confectionary, and more. And in the 1930s children of affluent families who did well in school were sometimes treated to sandwiches at the air-conditioned Botica Boie on the Escolta.[42] All these were flour-based products.

Although we know that beginning in the mid-nineteenth century the consumption of baked goods increased dramatically, what is lacking is a contextualized description of the myriad ways in which wheat flour came to penetrate the household dietary. How did people think about such items and when it was appropriate to eat them? Did they distinguish things you sometimes ate as snacks, or *merienda,* from what was considered "real food" suitable for meals? Such questions are important in their own right and also because the answers provide a bridge to the mass-market sandwich buns of recent decades. Fortunately, historian Ruby Paredes has provided a memoir on the changing cuisine of one mildly affluent household in postwar Manila that addresses these questions. It turns on her mother's strongly held food values in the prewar period set against the changing views of a new generation.

Flour-based food items were a staple in my mother's household—wheat noodles (*mami, miki, pancit canton, soba*) were around as much as rice (*bihon*) and bean thread (*sotanghon*) noodles. Mother's favorite dinner recipe that used flour was *pastel*—a meat pie with chicken, chorizo, ham, carrots, potatoes, and *champignons* [mushrooms]. Just quickly recalling other flour-based recipes that she liked, I note that these were mostly for *merienda* or dessert (*postre*): fresh *lumpia ubod* (spring roll with heart of palm), *empanadas,* bread pudding, chiffon cake, brownies, pineapple upside down cake, angel food cake, and fruit tarts or *empanaditas* and fruit pies. Gradually, mother began having spaghetti and macaroni dishes served, but not too often. She considered these as *merienda* food because they do not complement the real dinner dishes, that is you just don't serve pasta with rice. . . . We did pizza from scratch occasionally, but again for *merienda*. As kids, my sisters and I considered bread a snack food, and not something to have at a real lunch at home, or for dinner. A flour-based item that was considered "real food" . . . was *pancit*—whether wheat-or rice-or bean-noodles. Fried *lumpia* was another, because you could serve it with rice and the other dishes.

In addition to the universally popular *pan de sal*, my mother liked certain regional specialty breads and pastries like *monay, pan de limon, ensaimada*, and *hojaldres* from Cebu. I don't know how long these things have been around, but I would guess since the latter half of the nineteenth century. Speaking as a lay person with no claims of expertise in the matter of food history, I would say that bread and other flour-based recipes in Filipino food habits are generally from Chinese and Spanish and later European cuisine—I think this is clear from the names of the items themselves. . . .

As Doreen [Fernandez] has often pointed out, in Filipino society food carries enormous status implications. Imported food . . . is considered high status.[43]

The dietary role of breadstuffs enlarged over time. By the 1960s informal parties often included sandwiches, macaroni salad, pastries, and brownies. Soft drinks and sandwiches became "the inevitable combination in young people's diets." These were joined by hot dogs and "the great thick juicy hamburgers at the Tropical Hut. . . .Young people and more and more of the older ones were becoming used to the idea of a bread-based menu for more serious eating." By the end of the 1970s the fast food chains had appeared. Increasingly the consumption of their mass-marketed products cut across class lines: "Despite the snob-and-mob-appeal of these new arrivals, however, traditional snack items like *siopao*, the steamed meat- or vegetable-filled bun, and *mami* (*soba* noodle soup) remain Filipino *merienda* favorites."[44]

In this way, incrementally over more than a century, foods based on wheat flour became a significant part of urban Filipino cuisine. In a number of cases the new foods and food practices started out fairly high in the status hierarchy and gradually diffused down through the ranks of society and the financial ability to consume. But we have also seen that in the 1930s some two-thirds of urban working families were eating bread for breakfast, the Paglinawan brothers were operating five bakeries and retailing through a network of Filipino neighborhood sarisari stores, and an *aguador* in Tondo was vending bread door to door. Diffusion of new food items could be slowed by conservative versus practical change-oriented assessments about what constituted food suitable for meals as opposed to snacks and in what settings.[45]

COFFEE REPLACES *Chocolate*

One day I was suffering from a bad cold, and a friend introduced me to *sala-bat*, a hot tea made with fresh ginger and sugar. It had a marvelous effect. *Salabat* and some herbal medicinal teas are representative of a not particularly elaborate early tradition of hot beverages.[46] Given the usually warm tropical environment, this is not surprising. Eventually, however, resident foreigners introduced all three of the mildly stimulating hot beverages whose consumption spread around the world over the generations and centuries—especially after 1492. These were tea, cocoa, and coffee. In the cultural geography of hot caffeine-rich drinks, coffee and tea came to dominate world consumption patterns with some areas of South America left over for *maté* and *guaraná* and with coffee gaining over tea in recent generations.[47] Historical patterns of exploration and trade, imperial systems of integration and diffusion, and religious preferences and taboos all factor in to produce the changing geography of consumption. In the Philippine lowlands, except on the coolest nights and dawns around January, there is little need for something warm. So consumption of these beverages comes down to chemical stimulation and taste and to what such consumption might symbolize about the associations and place of the consumer in society.

Tea was in mass use in China and other parts of East Asia. It was later adopted in Inner Asia, Russia, Portugal, and the United Kingdom. Surely introduced to Manila by Chinese, Filipinos came to drink tea when they were sick. But given the commercial interaction and habit of tea drinking among the Chinese resident in Manila before the 1750s, it seems remarkable that Filipinos did not embrace tea more broadly.[48] But it was the male Spanish and creole clergy that set much of the tone for early culture transfers and promoted and gave status to chocolate. In Philippine life tea was primarily for Chinese and other foreigners, especially British employees of the merchant houses. In 1851 the value of tea imports was one-fifth that of cocoa. Later, in 1926–31, it was about 15 percent of coffee imports. For much of our period it was almost never advertised in the general press.[49]

Chocolate, *Chocolate*

Hot liquid chocolate is often ignored when discussion turns to caffeinated beverages. But, as William Clarence-Smith has shown, a consideration of the changing geography of mildly psychoactive beverages is woefully incomplete without chocolate, even if its active alkaloid is theobromine rather than caffeine. Theobromine "makes chocolate mildly stimulating and slightly addictive," according to Clarence-Smith.[50] More specifically, recent research has shown that the complex chemistry of the cacao bean contains certain amino acids that are "precursors of *adrenalin* . . . and *dopamine*, a neurotransmitter that relays signals between nerve cells in the brain. Scientists postulate that dopamine induces feelings of pleasure; if so, the passionate craving of the true chocoholic may have a neurochemical basis."[51] These same qualities may trigger migraine headaches in others.

In my experience in Manila, it is coffee people drink when they want a hot drink and a mild caffeine kick. This coffee product is generally made on the spot from hot water and a spoonful of dehydrated crystals. This made sense to me in the present context because in the nineteenth century a great deal of coffee was raised in the hills of Batangas and some other places.[52] Coffee beans were an important Philippine export during much of the nineteenth century. More than a million kilos were exported each year from 1856 to 1892. During 1861 and 1874–91, coffee beans constituted 5 to 9 percent of all Philippine exports by value, occasionally outpacing tobacco products. Then came coffee leaf blight and insect infestations and a rapid production collapse.

~

The cacao tree comes from Mesoamerica, as does the name and the evolution of the Spanish mode of consumption from its native roots.[53] Despite the difficulties involved, seedlings of the Central American Columbian type known as Criollo were introduced to the Philippines via the Acapulco-Manila galleon in the seventeenth century.[54] Meanwhile, the Dutch were introducing Venezuelan varieties to Ceylon and later Java. Both introduction streams were successful, and by 1810 chocolate was "being greatly extended among the natives of easy circumstances" in the Spanish Philippines and Sulu sultanate.[55] In subsequent decades, when the chocolate beverage was being replaced by coffee in Cuba and numerous other places, it was still favored in the Philippines. In Manila by the mid-nineteenth century, if not well before, the consumption of hot chocolate had broadened well beyond the Spanish foreigners and local affluent. The form of consumption varied by class and ethnicity. Spaniards— in the Philippines as in Spain—and the better-off mestizos and others drank hot chocolate made with water from roasted and ground cocoa beans sweetened with considerable amounts of sugar. Within this practice there were status distinctions. In a famous passage, Jose Rizal speaks of an abstemious foreign Franciscan offering an ordinary visitor "*chocolate-ah*," a watery version (*ah* for *agua*), versus offering an honored guest "*chocolate-eh*," the rich, thick version.[56] In Dagupan, the well-to-do spoke of *chocolate padre,* meaning a very thick hot chocolate served in a small cup and fine enough to serve to the local priest should he visit. Cinnamon and vanilla were common flavorings. By contrast, ordinary Filipinos took chocolate in a hot rice porridge rather than as a drink. Known as *tsamporado*, from the Mexican Spanish *champorado*, this chocolate-flavored porridge contained roasted *pinipig* rice and frequently roasted pili nuts as well.[57] In these various forms, cocoa came into wide use in Manila and more broadly in the archipelago.

Cacao is a tropical plant, and since it was transferred to the archipelago it has remained in modest production for more than 300 years, but its cultivation is and was risky business. Cacao is easily damaged by the winds of tropical storms—to say nothing of typhoons—is subject to a variety of diseases, and must be provided with appropriate shade. As a result, throughout our period cacao trees were grown mostly by smallholders as an adjunct crop mixed in gardens near the house rather than as a plantation specialty.[58] Circa 1770 Franciscan revenues included cocoa from nine of its parish-missions in Bikol, three in Tayabas, and one at Polo, Bulacan. About the same time and in line with the Bourbon agricultural diversification policy, some 20,000 cacao and 20,000 coconut trees were planted on the San Pedro Tunasan Estate in Laguna, although what happened to them after the confiscation of Jesuit lands is unknown. In the 1810s, Cebu was said to produce high-quality cocoa but in very small quantities.[59] Nevertheless, cocoa from Cebu, as well as Tayabas, was traded in Manila. In the 1860s, tiny quantities also arrived from Polloc on the Cotabato coast of Mindanao. According to Buzeta and Bravo, Batangas and Cebu were the most notable provinces for quality production circa 1850, but they also report it being grown in more than 160 localities scattered in many parts of the archipelago. Among these, Argao, Cebu was singled out for the volume and superior quality of its cocoa. Cocoa was a leading product of Bauan in Batangas and was notable in several other coastal communities in the same province, including Balayan and Calaca.[60]

Circa 1870 cocoa production was tiny and scattered says Montero y Vidal. Albay, the Camarines, Bohol, and Batangas were then among the leading provinces, although in truth no province was producing very much. Further, both Bikol and Batangas are subject to regular typhoon damage, making them chancy locales. An 1873 ad identifies cacao coming to the city from the Bondoc Peninsula of Tayabas. In the 1880s, advertisements for superior grade cocoa from Davao in southern Mindanao appeared among the small number of relevant commercial notices. Further, Davao was the only source of cocoa listed in the regular biweekly commodity price reports for 1888–91.[61] This was promising in the sense that Davao is only rarely visited by typhoons. The provincial agricultural reports gathered during 1886–87 list Basilan Island, near Zamboanga, as the only major producer.[62]

The annual price cycle for 1891 suggests both a cool season and a holiday consumption tendency—low during February–August and quickly rising to a peak during November–December. Tax records for Laguna in the 1890s allow us to see several buyers of rice and other commodities also buying cocoa and coffee and forwarding these to larger merchants in Manila. All four of these buyers were Chinese. Advertisements for cocoa from San Pablo (Laguna) at this time stressed its genuine unadulterated character—suggesting that adulteration was a common problem.[63]

The 1903 census reveals Cebu as the clear production leader, but the pattern of production remained dispersed—the farther south, the less exposed to typhoon damage. Annual domestic production trended upward according to Clarence-Smith, from 200 tons in the 1860s to 350 in the 1900s to perhaps 600 a decade later and 3,000 in 1960. Even in small stands, cacao could be a remunerative crop. As Foreman says, growing cacao "pays handsomely in fortunate seasons" and not at all in others. A good year was like a windfall, but you could never count on it. A lot of the gardens in and around the larger towns included cacao trees.[64]

Despite producing good-quality cocoa, the Philippines became an important net importer. Small quantities of cocoa from Guayaquil in Ecuador, and sometimes from Caracas, crossed the wharves of Manila in almost every year of which there is a record in the 1860s, and were still being sold on the Escolta in the 1880s and 1890s. In Manila it undersold local cocoa, but many consumers preferred the Criollo variety raised in the archipelago. These imports represent resumption of a trade that had flourished in the late eighteenth century before the independence movements in Spanish America.

William Clarence-Smith points out that from about 1728 Basques controlled the cocoa trade between Caracas and Spain. Soon after, "Basque merchants . . . mastered the burgeoning cocoa trade of Guayaquil . . . " and from that point became important in the Acapulco galleon trade. "The Caracas Company lost its monopoly . . . and was dissolved in 1784, but its Basque promoters formed the Real Compañia de Filipinas, recognized as a royal monopoly trading company, the following year. This "multi-colonial octopus . . . had interests around the globe," in particular in the trade between Cadiz and Manila via the Cape of Good Hope. Among other things, "[I]t obtained a monopoly on exports from South America to the Philippines."[65]

The surnames of businessmen who became prominent in Manila's commodities trade during the second half of the nineteenth century strongly overlap with those seen in the South American cocoa trade. However, Basque traders who had done relatively well in the eighteenth century were increasingly edged out by Catalonians, whose wholesale trade, financial capital, and wheat production made their home region relatively wealthy. By contrast, the multiprovince Basque homeland became an increasingly poor place—offering an insight into the motivation of Basques to migrate to the Philippine archipelago.[66] Joaquin Ma. Elizalde and Valentin Teus Yrissary, then teenagers, were brought to Manila circa 1846 by their uncle, Juan Bautista Yrissary. Once in the city, they went to work for Jose Joaquin Inchausti, who was already well

established.[67] In a roundabout way, the cocoa trade may have had something
to do with acquainting these future Spanish Basque immigrants with opportunities in Philippine export commerce, but I agree with Benito Legarda Jr.,
who points out informally that the strongest link to the Philippines among
these people was service as maritime officers.

According to Clarence-Smith, it was the trans-Pacific trade hiatus following
the independence of many former Latin American colonies that led Philippine-
based traders to search for other sources of cocoa in the sultanates and nearby
Dutch colonies. From the 1850s, at least, Hokkien traders brought cocoa to
Manila from Ternate and other Moluccan islands in the Dutch East Indies.
Using the ensign of Sulu as a cover and with the connivance of local Dutch
officials, this illicit trade was able to avoid the regulation requiring that exports pass through Java (thus involving two of Eric Tagliacozzo's contrabander
categories, foreign Asians/Chinese and Europeans with personal rather than
state-directed agendas).[68] As the rules evolved, so did the pattern of trade.
Through the late 1870s, at least, cocoa continued to enter the Philippines from
the Moluccas and Sulawesi, including Menado. Almost all of it was landed in
Manila, but the carrying vessels often stopped in Zamboanga on the way. It
was the best-developed port of the Spanish Philippines near this cocoa-trading
zone and was located on the trade route between the Strait of Makassar and
the Mindoro Strait leading to Manila. At least some of the import trade was
organized from there.[69]

The magnitude of the reported import trade was highly variable, a reflection of inconsistent local production. Jagor reports that much of the local 1856
crop was destroyed by a typhoon. In response the government authorized
duty-free importation, thus allowing Guayaquil cocoa to undersell the local
product by half. From annual reported import totals of circa 150,000 kilograms in 1859 and 70,000 during 1861–63, imports of cocoa rose to a high of
350,000 kilograms in 1873 and then retreated to oscillate between 60,000 and
180,000 through 1894. The pattern of trade also changed, so that from the
early 1880s onward most cocoa entered from Singapore: 93 percent at the
end of this time series in 1894. These were actually reexports of cocoa coming
from the Moluccas and Sulawesi. Given the production estimates offered by
Clarence-Smith above, imports might have accounted for roughly one-quarter
to one-half of consumption. Imports increased further in the early twentieth
century.[70]

∽

Processing techniques shaped the product for consumption. Until the late
nineteenth century, "excessive fat in chocolate posed problems of solubility
and digestibility, [and these problems were] not overcome until cocoa butter
was regularly and effectively pressed out." Chocolate imported from Germany

advertised exactly this quality in 1891. In the last decades of the nineteenth century, Swiss technology "revolutionized the quality of eating chocolate and created milk chocolate."[71] This introduced milk chocolate as a popular form and helped to set off a great expansion in consumption and commerce. Now fancy chocolate entered Manila as part of the confectionary trade. One could also see foreign merchants consuming cocoa made with hot condensed milk in the European way. At the same time a lot of cocoa was processed informally. Farm families that grew cacao typically made their own by "roasting the beans over a slow fire," separating the husks, and then pounding the beans together "with wet sugar, etc. into a paste, using a kind of rolling pin on a concave block of wood." In the city, itinerants, including Chinese *chocolateros,* went door to door, grinding the beans and creating a mass they "kneaded to chocolate-dough with sugar, peppers and other favorite spices." For use, balls of this "dough" were placed in a cup with boiling water.[72]

Hand methods remained in use, but the tasks of grinding and mixing became more mechanized for those who could afford it. In the 1850s there were two Manila *fábricas* making chocolate with mills powered by oxen.[73] In the 1870s and later, the Café Oriental advertised the product of its *fábrica de chocolate* made with both steam-powered machinery and handwork, with and without cinnamon (figure 10.2). Some 25 makers and sellers of chocolate are identified in the Manila tax records of 1895, each working from a fixed address. Twenty-three were Chinese with their workshops concentrated in Intramuros and Sta. Cruz. Two others were using small steam engines to crush and blend the beans and sugar. One of the steam-powered chocolate *fábricas* was La Bilbaina, which suggested a Basque connection (Bilbao). The other steam-powered *fábrica de chocolate* was the well-advertised bakery and confectionary La Palma de Mallorca. New circa 1887, it was still in business in the 1930s. Its winning formula was to operate like a German *konditorei,* or confectioner's shop, combining beverage, bakery, pastry, and confectionary functions with light meals and long hours.[74]

At the start of the twentieth century, city directories identify 14 urban chocolate makers and dealers. Ten were Chinese shops; the others included La Palma de Mallorca and La Bilbaina. Finally, from 1914 through 1928 the Paglinawan brothers operated a coffee- and cocoa-processing plant in Sampaloc. The brothers had the idea that coffee and cocoa were items of "first necessity" and would sell well. Further, they chose this business because of the relatively low capital requirements.[75] They were also successful in the bakery business, as we have seen.

Even after modern processing removed the excessive fats, hot chocolate re-mained time consuming to make. Depending on the day and the occasion, this might be fine for people with household help. Again technology was helpful,

FIGURE 10.2. A steam-driven *fábrica de chocolate*, as well as meals and billiards, were
hallmarks of the Café Oriental during the 1870s–90s. Under the same roof, La
Malagueña sold imported foodstuffs and beer to an upscale clientele. Rosario Street,
Binondo, facing Plaza Moraga. (E[benezer] Hannaford, *History and Description of the
Picturesque Philippines* [Springfield, Ohio: 1900], 78)

creating modern powdered and nearly instant cocoa. But, while the Swiss led
the early-twentieth-century world in the development and export of eating
chocolate, it was the Dutch who created the alkalization process that made
powdered cocoa palatable. The Manila ads for Van Houten's Cocoa in 1907 are
almost certainly for the powdered form, since it was a leading Dutch firm in
the export of cocoa powder.[76] But it wasn't the same in status or consistency.
At the same time, San Miguel's Magnolia division found an abundant market
for chocolate milk among comfortable middle-class youngsters.

For comfortable Manilans in general, Ruby Paredes reports that her mother
would occasionally request "*chocolate puro*" for breakfast, a "thick heavy morning chocolate made from *tableas* that had to be beaten frothy with the *bati-
dor.*" This became an occasion on which to have "toasted bread slices or *pan
de sal* with lots of butter that we dunked into our chocolate." As noted, ordinary Manilans consumed chocolate primarily as *tsamporado*. Celebrated author

Bienvenido Santos recalled that while living with his brother's family in Tondo in the 1930s, his sister-in-law "had a good, very early morning business serving *tsamporado*." Their dwelling on Antonio Rivera Street was close to the wall that surrounded the railroad yard, and her customers included "those who go over the wall to steal."[77] Across class lines, *tsamporado* remains in wide use for breakfast.

Coffee

Among the suite of hot beverages, it is coffee that is psychoactively the strongest—at least stronger than tea and chocolate. This might provide an explanation for the enormous gains made by coffee in the twentieth century, but by itself this is not enough. We get somewhat closer with Clarence-Smith's meditation on chocolate in Europe as a drink associated with the absolutist regimes of the baroque era and a mass drink primarily in Spain and southern Italy. By contrast, "[C]offee and tea were . . . perceived as incarnating the values of the rising bourgeoisie. . . . Tea and coffee represented 'sobriety, serious purpose, trustworthiness and respectability'"—and less expense.[78]

Legarda describes hot coffee being delivered to American troops on the firing line in Sampaloc on the first day of the Philippine-American War. These soldiers were given bitter black coffee with almost every meal throughout their training and service. No doubt it was meant to keep them alert. The stepped-up pace of life in Manila after 1899 lent itself to the diffusion of coffee drinking—particularly among the expanding Filipino middle class. Coffee consumption was not new in Manila. At least two chocolate *fábricas* were also grinding and selling coffee in the 1870s and 1880s, and many small kitchen hand grinders were imported at this time, but the magnitude of urban use certainly picked up.[79] It is one of those economically unfortunate paradoxes that this happened mainly after Philippine coffee growing had declined.

Like cacao, coffee arrived in the Philippines well before our period—in the late eighteenth century—and became a commercial product by the early nineteenth. Still, it is not listed among provincial products landed in Manila in 1818. This is quite late by contrast with Java, where the Vereenigde Oost-Indische Compagnie (a.k.a., the Dutch East India Company) introduced coffee in the late 1600s, leading to enormous crops being harvested by 1725.[80] In the Philippines, coffee was first grown on a plantation scale on the famous La Gironière estate known as Jalajala—part of what is now eastern Rizal Province. In 1837 its owner was awarded a prize for being the first to exhibit more than 60,000 coffee trees in a second harvest.[81]

The main geographical thrust came in the southern Tagalog provinces. Like wheat, it was in and around Batangas that coffee flourished. Coffee was introduced to Lipa, Batangas, in the 1810s. Buzeta and Bravo give it no special emphasis there circa 1850, although they list coffee in most Batangas towns at one level of importance or another. The traditions of the town indicate desultory development until 1859, followed by increasingly large-scale production.[82]

This timing is typical. William Clarence-Smith and Steven Topik describe an "explosion of [world] coffee cultivation beginning in the mid-nineteenth century reflect[ing] sharp tax reductions in the West, as free trade spread, and towns, industry, and population grew rapidly. This helped to make coffee a staple part of the diet of North America and much of Europe, especially France and the Germanic lands."[83] Indeed, many Americans had learned to drink Philippine coffee between 1860 and 1872 when more than half a million pounds were exported to the United States nearly every year, sometimes more than a million pounds.[84] By the 1880s, about two-thirds of the municipal territory of Lipa was planted to coffee monocrop—yielding a local harvest of 70,000 *picos* in the banner year of 1887. It was primarily grown on locally owned estates, and those who did well in Lipa soon bought land in the adjoining municipalities for planting more coffee. Perhaps 95 percent of the crop grown in Batangas entered the commercial system and was sold in Manila. Of sixty-two substantial cargoes of coffee received in Manila by coastal shipping in 1881, all but one was from Batangas Province. Lacking modern land transport, strings of "coffee-laden ponies" carried the crop from Lipa and elsewhere to the little ports.[85] Coffee in its nineteenth-century Philippine heyday was an unusual commodity in the sense of the lack of involvement of Chinese commercial networks in its collection and wholesaling. Batangas was a special province—with coffee, vegetables, citrus, an important livestock economy, medium-scale shipping, and considerable medium-scale entrepreneurship.

Coffee plants require nurturing for four or five years before they begin to bear in commercial quantities. Planting in the late nineteenth century involved growing strips of durable shade trees first and setting out coffee the following year. In 1903 Governor Simeon Luz described a process of reducing shade by removing limbs of the sheltering trees as the coffee plants grew larger. Once under way, the plants were said to go through three flowering and fruiting cycles per year. Picking the beans was generally done by workers on a one-fifth share basis.[86]

Domestic coffee arrived in Manila every month of the year, although the greater part came during the dry season. In addition to Batangas, low-grade coffee came to be grown in northern Mindanao and Cotabato. Basilan coffee shipped from Zamboanga was routinely listed in current Manila price reports during the 1880s and also sometimes appeared among the arrivals in the 1860s, as did shipments from the Cotabato port of Polloc. Jagor points to Laguna, Batangas, and Cavite as producing the best coffee circa 1870. A decade later

Montero y Vidal locates coffee as highly concentrated in Cavite and Batangas.[87] Small boats connected tiny ports of the Cavite coast to Manila, but the shipments were typically not reported. Still, by about 1891 coffee had become the leading crop in the upland municipalities there from Silang to Alfonso. There were many complaints from Cavite growers about the high cost of transport, but in several ways coffee was well suited to this locale. It was environmentally sound, could withstand the jostling and delay of rough transport, and its bulk selling price was 20 times that of *palay* and at least 10 times that of crude sugar, so it might still pay something to the grower despite the cost of moving it to the coast for shipment to Manila.[88]

The local economies of all these places were devastated by the sudden collapse of coffee production in the 1890s. The impact of this on the incomes and lifestyles of wealthy landowners has drawn attention, but we lack studies of the effects of the collapse on household incomes and prospects of the legion of coffee pickers, small-scale shippers, and retail shopkeepers. Like the rinderpest and cholera outbreaks, this was another calamity of the 1880s–90s that tended to undermine the position of ordinary people. Once the Revolution and the stubborn resistance to the American military had ended, many survivors voted with their feet. By the early 1910s, the commercialized core of Batangas Province had become an area of high outmigration.[89]

The leaf blight that laid waste to the arabica coffee plants of Batangas and Cavite was caused by the fungus *Hemileia vastatrix*. It apparently originated in Sri Lanka in the 1860s and spread, engulfing many stands in the Philippines and parts of Malaya (1894), among others, and affecting a huge territory from West Africa to Samoa in the South Pacific by 1914. Under some conditions, *Hemileia* was survivable with some loss of yield, but in most Philippine growing areas it was the simultaneous attacks of *Hemileia* and an insect stem borer that led to the sudden collapse.[90]

Several crops were tried as replacements for coffee in the former production areas. According to Clarence-Smith, "[C]ereals were the annuals more likely to replace coffee at higher altitudes, especially in densely populated islands moving into overall food deficit and experiencing rising food prices. Luzon . . . was one such example, with lands formerly devoted to coffee not only turned over to sugar, but also to maize and rice."[91] At the same time, there were attempts to defeat the blight by planting what were hoped would be resistant varieties. The common arabica variety was hit hardest, so others were tried. Robusta was widely planted in upland Java and was increasingly adopted for southern Tagalog plantings between 450 and 700 meters of elevation, where it proved to produce well if the soil was friable and fertile—no small requirement—but its berries required artificial drying. Below 350 meters, the liberica variety also worked reasonably well with proper preparation, but with lower productivity. This last plant was imported from West Africa, in part with promotion by the Genato grocery family in the 1890s, but it took some time to become established.[92]

Batangas, which had led in coffee production during the export boom, regained and held on to the number-one position, in part by planting these and other more disease resistant varieties. Still, its days as a producer of significant export commodity earnings were over. In the generation after the crash, most of the former coffee estates were broken up into smaller parcels and let on a share arrangement commonly known as the *kasama* system. Few share tenants could afford the multiyear costs associated with planting a coffee specialty farm. The result was that the more disease-resistant coffee trees were set out and managed in patches rather than over extensive areas. Soon neighboring Cavite, which had led or been number two in coffee production during the 1920s, fell out of the top group, overtaken by the efforts of growers in Mindanao. Together, Batangas, Lanao, and Davao accounted for half of the national coffee production in 1938.[93]

Whereas the established coffee-growing areas were devastated in the early 1890s, *Hemileia* was slow to invade the northern uplands. Coffee was introduced there by a military governor in the 1870s. In 1877 it was being tried out above 4,000 feet. By 1881 a new governor was attempting to force local people to plant coffee, but they resisted. Between 1896 and 1900 the German entrepreneur Otto Scheerer, in the uplands for his health, established a small coffee estate in the vicinity of Baguio.[94] In the northeastern portion of Benguet at Kabayan, a second upland group accepted coffee sufficiently for it to be grown on a small scale. As the price rose, so did the enthusiasm of local cultivators. By 1900 occasional traders were taking small loads of coffee to sell in Manila. Two years later the total annual harvest was still less than 1,000 *cavans*. Much of this was purchased by "the Tabacalera Company year after year, and . . . shipped to Spain, and there disposed of at fabulous prices," bypassing the Manila marketplace. North of Kabayan in Bugias, Martin Lewis indicates that coffee spread as a garden planting among the local elite in the same period. They took up coffee drinking and also found the beans a "valuable trade item. . . . Having planted sizable orchards of *arabica* trees, wealthy [uplanders] soon lost their inclination to [relocate] their homes periodically" as had been the custom.[95] But about 1906 the "blight struck, damaging especially those orchards located on clay soils." Thereafter, coffee growing in Bugias was limited to the "gardens of a few wealthy households situated on rich loam." By 1911 high-quality coffee was being grown "in isolated patches all through the Subprovince of Ifugao, and at greatly varying elevations."[96]

Despite the decline of domestic production, coffee consumption was selectively replacing chocolate in Manila and elsewhere by the early twentieth century.[97] Imported brand advertising reinforced this trend. At the start of the twentieth century, cocoa was an important part of both breakfast cuisine and late afternoon *merienda*. According to Clarence-Smith, coffee quickly became established as a Philippine breakfast rival. Mamerto Paglinawan, the entrepreneurial coffee and cocoa processor, implied that his family was already drinking coffee every morning in 1914.[98] Over time thick morning chocolate remained deeply appreciated among those who could afford it, but increasingly it was a special treat rather than a routine beverage.

Coffee consumption gained rapidly enough to attract fast buck artists, unscrupulous dealers who adulterated the product. In 1912 the American Pure Food and Drug Act was used to prosecute several Chinese grocers. Prosecutors alleged that the coffee in question was "imported from Singapore in the bean, [then] ground and mixed with Chinese peas and beans for the local trade in adulterating establishments on Calle Nueva."[99] One measure of change in consumption is that the number of Manila coffee dealers, never as many as 10 before 1926, expanded to almost 20 by the late 1930s. Another measure is that the relative production of coffee and cocoa reversed between 1902 and 1938. In 1902, a decade after the blight dramatically shrank production, coffee made up only 21 percent by volume of the total Philippine production of cocoa and coffee. In the 1920s, coffee pulled slightly ahead and by 1938 accounted for 78 percent of the combined total. Domestic coffee production expanded by half between the mid-1920s and 1938. Cocoa production plummeted by the same fraction in these years.

Philippine growers were not meeting national demand with either beverage crop. In the 1920s, while the domestic harvest tonnage of the two remained close, the balance of imports switched back and forth. Imports are recorded by value. Between 1922 and 1928, a greater value of cocoa imports was reported four times, coffee twice, and they were virtually tied once. The pattern of origin of cocoa imports remained much as during the 1880s–90s. The British East Indies (presumably Singapore) remained by far the leading source during 1922–28, handling cocoa grown in Indonesia and Ceylon. The value of cocoa arriving via the United States exceeded that coming directly from the Dutch East Indies starting in 1926. Imported coffee came primarily from Java, rather than the Moluccas, and was credited to the Dutch East Indies rather than Singapore. A distant second was coffee coming from Latin America via the United States. In a dietary survey of more than 100 Manila working families in 1936–37, coffee was found to be "a very common beverage."[100] This mirrored the long connection between coffee in American society and manual labor.

By 1941 La Tondeña, the leading domestic distillery, was offering its own nicely packaged Jai-Alai brand coffee advertised in the Tagalog media. And there were other reputable local coffee packers, such as Ah Gong Sons' A & G Issue Coffee. Possibly both firms were distributing Philippine coffee even if they refrained from saying so in their advertisements. At the same time, fancy imported grades of both coffee and cocoa products were disproportionately consumed in metropolitan Manila. According to Nick Joaquin, that was exactly the problem. While typical Philippine coffees were likely to contain somewhat more caffeine than the imports, the quality of local coffee in the market was uneven while imported brands in sealed cans maintained freshness, were of more consistent quality, and were backed by "snob appeal advertising."[101] Coffee reigned supreme—though not necessarily locally grown coffee—leading all forms of chocolate by better than five to one by the 1960s.

〜

Two hot beverages introduced and promoted by Europeans and Americans were adopted by Filipinos, one largely replacing the other. Tea, introduced by Chinese, apparently remained identified with Chinese tastes and did not become well embedded in Manila dietary culture. The inherent qualities of the products and the social attachments associated with each provide plenty of reasons to make sense of this. At the same time, Chinese noodles of all kinds (especially *pansit*) and steamed buns (*siopao*) were widely and enthusiastically adopted. Eventually, so were European baked goods, though not at the expense of rice. Of course food transfers in this era were not just in one direction. As Kristin Hoganson tells us, "the curry-eating British were not the only ones" to develop an "imperial cuisine."[102]

PART IV

Wartime Provisioning and
Mass Starvation

11

Subsistence and Starvation in World War II, 1941–45

MORE THAN THREE YEARS of Japanese occupation took its toll, even before the Battle of Manila in February 1945 resulted in massacre and destruction on a horrific scale.[1] To chronicle the decline of the provisionment system in this era entails the use of oral history from survivors across the full spectrum of affluence and from different parts of the city. More than 100 interviews were conducted in 1985 with Filipinos who were working and forming families before and during the war. Most of these interviews were conducted in Tondo and San Juan districts, one renowned as the largest poor area of the city and the other well known for having been developed by a subdivision company catering to the securely employed and affluent. This is supplemented by interviews with persons then living in other districts and by several diary-based accounts of the war. The resulting body of testimony reveals the remembered experience of mostly ordinary people and provides the primary basis for what follows.

The interviews that constitute this oral history were open ended, and most took place in the home of the interviewee. The great majority of encounters were spontaneous; fewer than a dozen involved a prior appointment. Most were carried out by Loreto Seguido and the author working as a team.[2]

THE COMING OF WAR

Emergency Measures

The Japanese invasion of the Philippines began on December 8, 1941. The initial air attacks and the troop landings in Lingayen and Lamon Bay (Tayabas) were a terrible shock, as Gen. Douglas MacArthur had predicted that there would be no attack before April 1942.[3] But war itself was not unexpected. In the city, some buildings were partially protected by great walls of sandbags, model air raid shelters had been erected, and evacuation planning was under way.

Although there was commercial gardening in the suburbs, home vegetable growing had not been common in prewar Manila. A survey of families of manual workers in Paco found that "only a negligible quantity" of household food came from either "home gardens" or "home poultries."[4] So it was a major departure when a new entity called the Food Administration Office of the Civilian Emergency Administration (CEA) tried to interest the citizenry in starting private "victory gardens." Little popular enthusiasm met Manuel Quezon's order that "idle lands" (often vacant lots) in the city be planted to food crops. During the December invasion period, the government strongly recommended, along with conserving food stocks, planting *kamote* (sweet potatoes) and other root crops and suggested that the poor be allowed to start gardens on public school grounds.[5]

The government took more direct action on behalf of its civil service. Luciano Salanga of the Bureau of Commerce recalled, 'Before the invasion, the Commonwealth had encouraged us [government employees] to lay up canned goods, like corned beef, against the possibility of war, and we did. Also, as war approached, the government gave each civil servant three months advance pay. It seemed like a lot of money then. With our canned goods and advance pay, my wife stopped teaching, and we opened a *sarisari* store in San Juan. It had been a Chinese store, but the proprietor ran away. So we took over and displayed our accumulation of canned goods.'[6] Both manual and clerical city workers were included in the three-month advance salary plan, though not everyone laid in food stocks. Meanwhile, a pictorial ad by one of the smaller distilleries suggested that people stock up on its alcoholic products for drinking while confined in bomb shelters, and San Miguel Brewery ran a graphic public service ad giving advice on what to do in a downtown air raid.[7]

Manila under Attack

Military violence initially came to Manila with air raids on Nichols Field (now Ninoy Aquino International Airport), shipping in Manila Bay, and the Cavite Naval Station. These aerial attacks were followed by extraordinary lapses in public order. Looting was concentrated initially on Japanese stores, but surging crowds also looted various Chinese stores downtown; reportedly, some Chinese grocers just gave away their stock on the assumption that the invaders would take it anyway. At the end of December, believing that goods owned by the U.S. forces would otherwise fall into the hands of the invaders, the quartermaster general threw open the warehouse doors. The frenzied stripping of the great South Harbor warehouses continued until the entry of Japanese forces brought it to a halt. American trading companies were in the same predicament, and their warehouses were also looted. The Red Cross bodegas in Santa Ana and Santa Mesa were broken into, and more than 200,000 bags of cracked wheat, as well as medical supplies, were taken. Mobs threatened the NARIC rice warehouse on January 2 but were turned back. On top of this, there were rumors—fortunately untrue—that the metropolitan water supply had been poisoned.[8]

More than forty years later, residents of San Juan recalled these events. A printer observed, '[T]here was plenty of looting during December 1941, especially of Japanese shops and bazaars on Avenida Rizal [in the central business district] and in Sampaloc and Quiapo.' A housewife agreed: 'Early in the war everyone was well fed. There was looting of Japanese stores . . ., so many had a lot in their pantries.'[9] In San Nicolas, a housewife recalled participating in looting the "Pacific Oil Company," and taking a 'can of sardines and some *bigas*.'[10] An American wife from one of the Mountain Province mines (soon to become an internee) recalled these weeks in Manila: "We decided that three of the [now husbandless] mining families could live in one apartment. . . . We bought some canned food and acquired a case of Eagle Brand milk and a drum of salad oil from the bay area, which was being looted of its millions of dollars' worth of cargo. It was a sight, that. Everything on earth was in those warehouses . . . and [it] was all looted with the army's approval. . . . Much of it was later rounded up by the Japanese and stored in great piles. I remember passing one huge 'dump' that had nothing but flour and cigarettes in it."[11]

Despite everyone's worst fears, however, the rape of Nanjing was not repeated, and the initial takeover of Manila went peacefully. It had been officially declared an "open city" on December 26, 1941, when General MacArthur directed the army's withdrawal to Bataan. This declaration—and its acceptance—was one of the more humane moves of the war.[12] Armored vehicles entered Manila late on January 2, 1942, and the Japanese army took control. A month later a Filipino intelligence officer on reconnaissance was "impressed at the calm indifference of Manilans to [the Japanese] Occupation" and attributed this to the expectation of early liberation.[13]

The Early Occupation
Evacuation

The invasion brought a military call-up; thousands of Filipinos left the city to join their units and go into action. Their departure was followed by a civilian evacuation, simply to get out of the way of marauding soldiers. There were directives to evacuate neighborhoods near the airfields and terrible stories and rumors concerning the treatment of women in the areas of Central Luzon occupied by the invading troops.[14]

By December 20 the *Philippines Free Press* estimated that about 150,000 souls had been evacuated, perhaps a sixth of the metropolitan population. The Tutuban and Paco railway stations were jammed with a "sudden, feverish exodus of bewildered city residents." Sensing opportunity, *cocheros* began charging seven pesos "for a trip to Rizal Avenue extension" on the northern edge of the city, which made "speedy evacuation of the poor . . . almost impossible."[15] Most of the evacuees came from the inner districts of the city; rates were much higher in Tondo than in San Juan. The great majority of the evacuees went to the nearest towns of the inner suburban ring or to the adjacent provinces of Bulacan and Laguna.[16] Among those who left was a carpenter from San Juan who evacuated to Lukban in Tayabas. His wife's father 'had a coconut grove there, and we grew maize. [So] we were not hungry. Also, we had a small *sari-sari*, and there were no Japanese.' The family returned after liberation.[17] Another family operated a transportation business with 20 horses and 15 drivers. They were located on the outer edge of Sampaloc because "there were few houses then, and *zacate* was all around." But "we left . . . on December 8 and evacuated to Laguna for the entire war. We took the *kalesas* and horses to Biñan [their home place.]"[18] With many differences in detail, thousands did the same, and for a time metropolitan Manila's civilian population declined substantially from its prewar high of 800,000 to 900,000.

Rice Shortages

During the decades before the war the commercial movement of rice to Manila had become increasingly concentrated in fewer hands, many of them Chinese. In 1936 the NARIC was established as a vehicle for putting Filipinos and the political administration in charge, though by the end of the 1930s the NARIC still handled less than 1 percent of the trade.[19] The Japanese attempt to establish a controlled economy was later characterized by rice economist Leon Mears as "a bleak example of the futility of an extensive control organization—even supported by the sword—to bring even a semblance of order to the marketing of rice in the Philippines at a time of extreme scarcity when those governing did not have the support of the people." The occupiers attempted to impose rural rice cooperatives "to facilitate the collection of rice for military and civilian use," and to restrict the legal movement of rice to official channels.[20] In Manila a system of neighborhood associations was formed to monitor and control the population but also to carry out food distribution. The NARIC bureaucracy was assigned the role of buying from the producer cooperatives and rationing rice through the associations.

To buy rice one now needed to possess a residence certificate and endure long waits in line. The official policy was that persons could purchase a certain amount of rice per day from a designated neighborhood store at the controlled price. In July 1942 the legal quota was set at 1,200 grams, but this declined to 300 in October 1943, then to none as rations stopped (coincident with a November typhoon). It rose to 120 grams in January 1944, then fell to

60 plus an equal amount of sweet potato. At the same time the price escalated exponentially. In what became Indonesia, the occupying Japanese were able to achieve some food production improvements, but in the Philippines their much-trumpeted attempt to introduce *horai* early-ripening rice from Taiwan was not a general success.[21]

The rice shortage in the city began immediately. The main harvest was still under way in the Central Plain when the invasion began. Some standing rice was destroyed, and the harvest was interrupted and left incomplete. As time went on, shortfalls of rice production resulted from the escalating slaughter of carabao for meat, disruption of truck transportation, lack of maintenance on irrigation facilities, and further damage to crops by efforts to root out guerrillas, as well as from corruption, bureaucratic inefficiency, and unrealistically low prices paid to producers. Another factor was the creation of virtual autonomous zones in some of the principal rice-growing areas of Central Luzon where the Japanese system held minimal sway and landlords increasingly feared to go. The most notable were organized by the Huks, a legacy of the deteriorating agrarian conditions before the war. Martin Tinio Jr., reports that his father "closed the farm [in Talavera, Nueva Ecija] in the latter part of 1942 [because] the Huks began threatening the *hacenderos* . . . and most of them . . . chose to abandon their farms and stay in the large towns or Manila." Two cousins who did not leave their estates in Talavera were killed.[22]

Imported Foods Cut Off

Since the late nineteenth century, severe shortfalls of rice supplies had been met with imports from the riverine deltas of mainland Southeast Asia. Despite the improving domestic rice situation before the war, metropolitan Manila still relied on such imports to one degree or another as the war began. During the war, there were reports—unconfirmed in local sources—of continued cargoes of rice arriving from Saigon, which was also under Japanese control, through 1942–43. Such shipments must have become rare as the American submarine campaign got going.

Routine imports of foodstuffs from the United States, Australia, and Western Europe were cut off for the duration. Imported coffee, Sunkist oranges, apples, grapes, and other foreign fruits quickly disappeared, but canned goods and other provisions with a long shelf life, including preserved milk products, remained commercially available well into the first year of the occupation. Yet resupply was slow or nonexistent under wartime conditions, and prices escalated with astonishing speed. Margaret Sams reported, "I bought a case of Carnation milk. Before the war a case . . . had cost a little more than seven pesos. Already, after only three months of war, I felt very fortunate in getting it for sixty pesos."[23] Stores owned by foreigners from Allied nations were forced out of business, and their managers—except for the Chinese, who were far too numerous to lock up—were incarcerated as enemy aliens. Since Switzerland remained neutral, one Swiss store was able to continue in business, but as the months wore on without resupply, there was less and less to sell.

Much of Manila's prewar supply of eggs, garlic, onions, and various leafy vegetables had arrived in weekly shipments from the Canton Delta, now under Japanese military control. What flow may have continued for a time cannot be ascertained, although we know Hong Kong experienced major food supply problems of its own. Garlic, at least, remained available in Manila markets.

Flour, a prominent item among looted commodities, began to disappear within a few months. In August 1942 an elite baptism buffet featured a cake only because the family had sifted and resifted its store of flour in order to keep it fresh for the occasion. In September 1942 there were frequent ads for domestic cornmeal, fewer for whole wheat flour. By 1943 "cakes" were made of cassava flour. In the near absence of domestically produced wheat flour, the military administration's five-year plan encouraged the production of corn and cassava flour instead.[24] Most residents would have to wait until liberation to taste bread.

Sugar also became increasingly hard to obtain; by the start of 1943, a significant shortage had developed.[25] Filipino and American military forces had managed to hold on in much of the Visayas into April 1942 and had gone about "blowing up [sugar] refineries" in the main production areas of Negros and its environs. They also burned the "remaining stocks of sugar in the warehouses." As a result, "[T]he shortage of alcohol [derived from sugar] in Manila has become more and more acute and the authorities are taking drastic measures to conserve their fast diminishing supply."[26] In mid-1943 an Allied intelligence assessment reported, "Sugar has practically gone out of existence. . . . It is estimated that for the next crop year [1943–44] practically no centrifugal sugar will be produced. Generally, the fields previously planted to sugarcane have been planted with Taiwan or upland rice and cotton." This was part of the occupation government's ill-fated cotton initiative.[27]

Fish and Fishing

Manila Bay was a saving resource for many families that otherwise would have gone hungry. Commerce in fish continued during the war, although the Japanese attempted to force Filipino fishermen into a supervised cooperative. Many fish traps and some fishponds remained in operation. Fishing at sea in powerboats, which required imported petroleum supplies and ran the risk of encounters with Japanese naval vessels, was more problematic, so there was a significant decline of beam trawling. In the prewar period this activity had largely been controlled by Japanese entrepreneurs and fishermen, but because of the fuel shortage and the call-up of men for the imperial armed services, they found it difficult to continue. As a result, during the first rainy season of the occupation fresh fish became "rich man's fare."[28] Still, an ordinary commerce in fish persisted. Filoteo Tuason, with a long family background in public market provisioning, began a new livelihood in fish supply in 1942, primarily fish harvested from the *baklad* of Cavite and landed at Bankusay, Tondo. He would buy the fresh fish in the early afternoon, ice them down, and convey them at dawn the next day to the Divisoria Market. By the end of the morning, he would have sold the fish and prepared to start the cycle again.[29]

Artisanal fishermen continued their work on the bayside, joined by a growing number of Tondo people looking for subsistence. Small-scale fishing also had risks. Local fisherman Adolfo Jose 'was caught once and beaten by the Japanese. It's a bad memory. One night during the Battle of Bataan we were out on the bay using flashlights to fish for crabs. The Japanese took me for a guerrilla. They thought we were making signals. I was arrested and beaten up. That is why I am weak [now]. I was held underwater in the river. One of my friends tried to escape, but I held him back. I feared we would be shot. [Despite this experience,] I continued to fish during the war. I had a *banca* and used a light and . . . an improvised spear gun.'[30] Another fisherman said that when he caught shrimp from his small *banca* during the war, he would travel upriver to Malolos and exchange the shrimp for rice, later returning to his home in Tondo. In Gagalangin, a plumber reported, '[B]efore the war, this area was in fishponds. During the war, these fishponds were not [kept up or stocked with *bangus* fry]. Still you could get crabs and catch some fish, but the water was dirty.'[31]

Viajeros

Working as a *viajero* (a traveling buyer of vegetables, fruits, hogs, eggs, and the like) during the war was not for the faint of heart. In ordinary times a sizable proportion of Manila's nonrice food supply was commercially organized and brought to the city by *viajeros*. In a controlled economy, however, they could easily be regarded as dealers in contraband, depending on the interpretation of changing regulations. Thus many Filipino *viajeros* feared losing their goods or even their lives at the hands of the Japanese sentries arrayed around the urban area and at checkpoints in the provinces. For their part, the Japanese authorities believed they were confronting dangerous guerrilla groups that might use the steady passage of *viajeros* to convey military intelligence, arms, and other contraband.[32]

In northern Bulacan, normally an area oriented toward raising, milling, and shipping rice to Manila, the occupation authorities established restrictive blockades. In the city, old *casco* routes took on renewed importance. After the Japanese arrived, carpenter Arturo Bautista recalls:

"*Mahirap ang buhay noon.*" [Life was difficult then.] 'Thereafter, my livelihood was buy and sell clothing, furniture, and other things. I bought here in the city and bartered it for *palay* in the provinces, especially Bulacan. I transported the *palay* to Manila by *casco* and sold it for Japanese money, but it was very hard to transport foodstuffs then. Once I encountered a Japanese soldier at PMC.[33] I was the captain of the *casco* that belonged to my brother. The soldier stopped us and asked if we were carrying coconut oil. I said we were not, but he did not believe me, came onboard, and found some of the oil. I was ordered off the *casco* and told to stretch my arms. The soldier started to bayonet me, but instead I was pushed into the dirty waterway. To appease the man, I gave him cigarettes, whereupon he bowed and saluted me. But I was terrified.

'I had evacuated to Bulacan in December 1941 but made trips into the city throughout the war. I met my future wife doing this. She was a *viajera* and became a passenger on my *casco*. We were married in December 1944. These were difficult times. [By then] the Americans were already conducting air raids, and bombs were falling. The Japanese prohibited the transport of rice from the province to Manila, so we covered the rice in the *casco* with stacks of sugarcane until we were past the sentry post at PMC. The sentries did only an "ocular inspection," so they were unaware. We were afraid of being caught, but if you do not have courage then, you would die of starvation.'[34]

Marcial Lichauco reports that in December 1943 Japanese sentries were shooting farmers who were smuggling a bag or two of rice into the city in dugout *bancas* at night. In rice-producing towns like Baliuag, the *kempeitai* (Japanese military police) tracked clandestine rice merchants. One group from Baliuag was caught in Navotas attempting to smuggle a whole truckload of rice into the city. Such harassment was constant. But the returns were high, and the transport of contraband rice persisted.[35]

~

More and more people bought secondhand clothes and took them to Bulacan to exchange for *palay*. "Almost everyone was doing it," said one *cochero*. Another evacuated to Bocaue, Bulacan, said, 'During almost all of 1943, I bought rice in Bocaue and sold it in Valenzuela [Polo]—one-half sack at a time carried by bicycle. I did that together with my brother-in-law and nephew.' This was not necessarily clandestine, for the Japanese occupation police sometimes authorized half-*cavan* "export" passes even while they deliberately restricted the commercial flow.[36] Not everyone dealt in rice; it was safer to carry noncontraband items. A Tondo mechanic 'sold fresh watermelon and *singkamas*' during the war: 'We [also] got sugarcane in Calumpit and brought it to Manila in a big *banca* via Vitas to the Canal de la Reina and on by water to Divisoria to sell. [En route,] we were usually stopped by a Japanese guard at the PMC.'[37]

In 1943 a new *viajero* began. 'I was a . . . student when the war broke out, but [soon I was] engaged in buy and sell—used clothing as well as hogs and pigs. Father helped me with the capital. My first trip to the provinces was to buy hogs. I bought them in places like Bikol and Pangasinan [the war-damaged rail line to Bikol having been put back into limited operation].' Not only did many of the *viajeros* continue to circulate in Luzon, but *ventadors* in Tondo also remained active, helping to finance purchases and selling the live hogs the *viajeros* brought back. This part of the prewar supply system remained in operation, though under increasing stress and now mainly restricted to Luzon. Lichauco mentions an acquaintance who bought small numbers of slaughter cattle in Leyte in late 1942 and brought them to Manila in a sailing vessel, but the loss of major interisland steamers in the first days of the war choked off much of the routine commercial interaction with the Visayas.[38]

Most petroleum supplies in the city were destroyed in advance of the Japanese takeover. As the war continued, the flow of imported energy was severely attenuated.[39] The Japanese initially permitted some businesses to operate trucks, especially those engaged in transporting vegetables to the city. Martin Tinio Jr. recalls that his father, based in Nueva Ecija, used several trucks in 1942 to transport rice from the family haciendas in Talavera and Luar to the city: "The . . . trucks were fueled with gasoline or alcohol that he bought at the black market from Japanese Army officers." A year later he fueled the trucks with coconut oil purchased on the market.[40] By late 1942, charcoal-based power systems, which produced carbon monoxide fuel, were increasingly common[41]

Seeking Foodstuffs in the Provinces

Many ordinary citizens, especially from Tondo, also went directly to farmers and provincial markets to purchase rice for their families, contributing not only to urban provision but also to the intense circulation of goods, especially dry goods, suddenly made scarce by the virtual end of foreign trade. *Umbuyan* worker Felicisimo Soldaña recalled, 'From 1942 on we bought and sold anything—clothes, furniture, anything with a little profit. Sometimes I went home to Bataan and got rice, but transportation was very hard, and if a sentry catches you with rice, he will take half or even all.' Another said he sometimes journeyed to Nueva Ecija to buy rice for his family. 'Once the charcoal-powered bus I was riding in was stopped by highway robbers in San Miguel, Bulacan. Fortunately, they took only my pocket watch, leaving the three *gantas* of rice I was able to buy.'[42] Others went all the way to Pangasinan and Nueva Ecija to exchange goods for rice. Some attempted to smuggle rice into Manila by crowding onto the train in Bulacan and then risking life, limb, and robbery by jumping off in northern Tondo before the train arrived at the Tutuban Station.[43]

One way or another, enough rice came into the city for some hoarders and speculators, Filipinos and Chinese, to make a killing on the black market, selling at many times the officially allowed prices. In February 1943 more than 200 were arrested, but it made little difference. Historian Teodoro Agoncillo, who lived through this, wrote, "It was not uncommon for some City Hall employees and even officials connected with the market administration to make purchases or to accompany their wives and friends to make purchases in city markets for the purpose of buying goods and foodstuffs at low prices" that is, at the officially allowed prices, when the public was not able to do the same.[44]

Public Markets and Small Stores

Wholesale suppliers, retail vendors, and managers endeavored to keep the public markets of the city going even as inflation escalated and stocks of supplies were curtailed; at least these vendors and their families generally escaped having to eat starvation food. What continued to flow most reliably were supplies from the inner zone, including the immediate environs of the city, such as Gagalangin and other swampy areas where *petchay*, *kangkong*, and even a tiny amount of *palay* remained in production through 1945. As the war dragged on, *kamote* came to be widely grown by city families; it was nutritious, relatively fast growing, and able to produce both tubers and edible vine leaves on restricted areas of soil. Although much was grown for home consumption, significant quantities also moved through the public market system.[45]

~

Both within and beyond the public markets there was a plethora of individual stores. The Japanese disrupted this arena, targeting the Chinese and trying to force them to withdraw from retail activities, part of their reaction to vigorous public anti-Japanese campaigns in China and among the sizable overseas Chinese communities in Manila and elsewhere.[46] Many Chinese were forced into hiding—some spent the occupation living quietly in out-of-the-way Philippine villages, and more than a few organized an armed resistance—but a few Chinese restaurants, bakeries (using cassava flour), and other businesses, including black market traders, continued to operate, though always subject to Japanese financial demands.[47]

~

This situation provided some opportunities for Filipinos. Luciano Salanga, who used his advance pay as a civil servant to begin a *sarisari* store, continues his story.

> The Japanese took over the NARIC three or four months into the occupation. Now the NARIC with its Filipino staff was selling rice and looking for reliable distributors. NARIC appointed us to handle 200 sacks of rice a day in this area.

People queued up every morning. Our compensation was the leftover rice, about one and a half sacks out of 200. It was a very good opportunity. We had to pay a *kargador* plus two women who measured out the rice. All three were paid in rice. We built up the store, adding coconuts, *gulay*/vegetables, sugar, and "wine." The Japanese liked the wine, and they came to buy. They were friendly then, except for the Koreans, who were rough. The Koreans were mad, and they took it out on the Filipinos. Sometimes, the Japanese would steal rice from their own supplies and bring it to my store at night for wine. This was difficult for us, since the Japanese took part of our stock.

Another difficulty was supply. I brought eggs and vegetables from Marikina, two baskets full. I did this by walking, leaving San Juan at 4:00 a.m. and returning at seven or eight. There was no other way. There were only a few *karretelas*. Even the rich had to walk, like former Manila mayor [and San Miguel Brewery chairman] Ramon Fernandez, who had a house in San Juan. He walked to my store for eggs—200 a week. He said, "No problem with money. I have plenty of Japanese money." The problem was in getting the eggs.[48]

The urban neighborhood associations, besides collecting information and reporting daily on personnel movements, served as channels for the rationing and distribution of commodities such as rice, sugar, matches, soap, salt, and cigarettes. Lack of experience in food commerce was not unusual among those chosen as neighborhood outlets.[49] Grim as the occupation became, it is also remembered as providing a climate for an expanded Filipino role in food commerce.[50] Outside the official network new *sarisari* stores and makeshift food counters also appeared. One interviewee's husband had been recalled to the U.S. Navy and was away for the entire war. 'I started a *sarisari* store in the house with merchandise and food from Divisoria,' she said, 'including coffee, soft drinks, biscuits, and some other things. Ultimately, I operated that little store from 1941 to about 1952.'[51]

Early Food and Health Crises

By far the worst initial conditions of diet and health were faced by the Filipino and American soldiers who survived the fighting in Bataan, the notorious Death March, and the terrible conditions in the prisoner of war camp at Capas, Tarlac. More than half of those held in the camp died there (28,500 Filipinos and 1,650 Americans). Thousands of emaciated and sick Filipino survivors were eventually released, starting in June and July 1942. This quickly became a city problem, since many of those released came to Manila, as they could not get back to their provincial homes even if they were physically able to travel.[52]

The first strictly urban subsistence crisis of the new order emerged among the civilian Allied nationals interned in the city campus of the University of

Santo Tomas (UST) in Sampaloc. One of the internees was Margaret Sams, who later recalled the minimal diet available to those who had neither accessible resources nor good outside contacts: "I remember paying $1.50 for a box of cornflakes. For weeks I had been having such an upset stomach every morning after my breakfast of cracked wheat that I thought I would go mad. . . . But after stretching my box of cornflakes into breakfasts for a month, I no longer had such pains . . . and even got so that I craved cracked wheat. The more the better, no matter how many little pink worms were floating upside down in it. The worms had revolted us all at first, but after we convinced ourselves that they were really dead, from boiling, we were quite calm about them."[53]

Also in the UST camp were members of comparatively well-off business families, some with long-standing relationships with local Filipino family members or other Filipino Manilans. One, "Benny" del Carmen, said, 'My boss at Manila Trading was William Douglas McDonald, vice president. Along with other American nationals, he was interned at the University of Santo Tomas. . . . The food made available at the camp was inadequate, so Mr. McDonald set up a small "restaurant." Our arrangement was that I would visit them and bring food. [Mr. del Carmen cries with emotion.] I did visit them nearly every day until I was stopped. In return I lived throughout the war on funds given me by Mr. McDonald.'[54] The internees lost weight—as did nearly the entire city population—and eventually emerged gaunt, but they did not starve to death.

Starvation was, however, an immediate threat to those who were arrested. Assistant solicitor general and future justice J. B. L. Reyes recalled, "I didn't experience hardship until I was arrested on suspicion of having connections with the guerrillas. A Filipino spy for the Japanese denounced me. We were kept in Ft. Santiago from February until May or June 1943. No trial. They just wake you up at night for interrogation. They asked questions and then returned you to your cell while they checked out your answers. The charges were true; they just couldn't prove it. They used to release people against whom they had no evidence. In detention they fed us only *lugaw* (rice gruel) and salt. I lost 40 or 50 pounds before they finally released me."[55] In the view of a *viajera* from San Nicolas, "The Japanese were very strict, especially when the *makapilis* [informer-traitors] pointed you out! Suspected guerrillas the Japanese caught were usually executed."[56]

Other informants recall some positive personal interactions. The occupiers were in a position to exchange rice and other commodities for things they wanted, or even to give it as a kindness—understood by some Tagalogs as *awa*, an act of compassion toward one's fellow man. One woman reported that her house in San Juan was visited by a Japanese man carrying peanuts. She feared that he was there to arrest them, but in fact he 'wanted to barter the peanuts for the electric stoves that we made during the war.' In another incident, a 'Japanese officer was kind to us. While I was washing by the river, he dropped a [small] sack of rice purposely near me. This happened several times.'[57] Several workers from Tondo recall special meals they were given as a part of their daily work as *kargadores* on the piers of South Harbor, apparently a carryover from earlier Japanese labor practices. Some workers brought ripe mangos and bananas to the pier in order to exchange them with Japanese soldiers for rice.[58]

Doing Well in a Bad Economy

In the midst of a general social process of immiseration, a few did well, even very well. First were wholesalers and retailers whose stocks of goods were neither looted during December 1941 nor confiscated later by the Japanese military. Japanese, Chinese, American, and British merchants were all devastated by one or both of these eventualities, but Indian dry goods sellers prospered early in the war. Others who were very fortunate included "Filipinos engaged in the drug business . . . and owners of jewelry shops for whose diamonds and precious stones there has been a great demand by those who fear inflation," although declining stocks of goods to sell eventually ended their special advantages. Another group that profited consisted of Filipinos, "many of whom [were] former politicians," who made fortunes as the favored suppliers of the Japanese military and "Japanese subsidized companies." They provided all manner of capital goods, electric wires and motors, and construction tools obtained through buy and sell or worse. "The prices which the Japanese [were] willing to pay [were] fantastic."[59]

The System Collapses

Buy and Sell

Making a self-organized (and self-exploiting) living in the "informal sector" had constituted the livelihood of many Manilans even in peacetime. Obtaining goods and seeking buyers, often by setting up a vending point on the street near streetcar and jitney stops, going door to door, or otherwise traveling, was called *compra y venta* in Spanish, which became "buy and sell" in Philippine English and eventually in urban Tagalog slang as well. As the occupation wore on and the economy worsened, the number of those involved rocketed upward; when things grew desperate, "buy and sell" also became a euphemism for the one-way process of selling off one's own household goods.

During the war, the wife of a sometime Tondo tailor and bottle recycler was 'occupied in buy and sell, walking around. I met people with clothes and other things in hand. This is a cue that they want to sell. I would buy.' A stock of clothes could be taken to the nearby provinces and sold or bartered directly for rice or sold to buyers at certain markets. 'People from Ilocos came to the

Bambang market [near Palomar in Sta. Cruz] and Divisoria to buy clothes. Those Ilocano buyers exchanged clothes for rice in the provinces.' Another Tondo woman was also engaged in buying and selling secondhand clothes during the occupation. She agrees that 'there were many buyers at Bambang Street who came there from different provinces.' The clothes they preferred were known as "genuine" (pronounced "jen-wine"), meaning U.S.-made and implying quality.[60] The Bambang marketplace became the informal center of Manila's black market, a place for fencing the loot of a thousand burglaries and robberies and simultaneously a marketplace for legitimate commerce, "a flea market, used-garment bargain sale, arts and crafts display, antique show, and trading center . . . , the last-chance haven for those dying of starvation, where goods and comestibles were bought and sold without questions asked."[61]

Buy and sell was common all over the city, not just in poor inner neighborhoods. In suburban San Juan, one store employee lucky enough to retain his job was eventually reduced to working only half days for half salary. "We were not desperately hard up," he said, "but I engaged in buy and sell, taking secondhand clothes to Infanta, Quezon, on the remote east coast. With what I got for the clothes, I bought rice and brought it back to San Juan. I had no family connections in Infanta, but life was hard here, and there was so much rice there."[62]

Families that had quality possessions, such as dry goods, furniture, and appliances, could sell them in a severe pinch, and many did just that. The Filipina widow of one Sergeant Seale reported, 'My husband was a British citizen, a master sergeant with 32 years of service in the U.S. Army. . . . He gave me some money before he left [with his cadets for the fighting in Bataan], so I bought sacks of rice and corn. Then, little by little, we sold some of our furniture and clothes to buy rice. Fortunately, as an army family, we had plenty of clothes to continue selling, jewelry too. Also we got 2,000 pesos from Mr. Blouse of the Batangas-Tayabas Bus Co. My husband was with him in prison [camp], and [Mr. Blouse] sent a letter to his daughter asking her to help me. But before long, our regular diet was *lugaw*. On this diet the baby died. We lacked milk, better food, and medicine. Still, we were fortunate. Others had only the *ubod* [pith] of banana stalks and leaves from *kamote* vines.'[63] Likewise, a woman from San Nicolas recalled that her firstborn died in 1943, 'when we had no milk, only rice water.'[64] Formerly fairly affluent families were not immune.[65]

By June 1943 the food price index had tripled, and it was about to begin a much steeper climb. Pacita Pestaño-Jacinto records that by then many things were in such disastrously short supply that the grocer "caters only to the rich." Neither canned nor powdered milk was readily available now; rather, "the cow and the carabao have come into their own. The milkman (*lechero*) is again an institution." The former salary man "must sell his possessions if he wishes to live."[66]

Over time we can see pain in the inflation index. By July 1942 nominal food prices had doubled, but they remained at or below 250 (index 1941 = 100) through April 1943. Thereafter the ascending arc of prices steepened, reaching 300 in June, 650 in October, and 1,540 in December 1943 after a typhoon. The families of ordinary laborers often spent two-thirds of their income on food in more normal times; they had little possibility of keeping up in this environment. By August 1944 (just before U.S. air raids in September) the food price index had passed 11,000. It reached 33,600 in October and then tripled again by December, by which time there was little food to buy at any price.[67] Small wonder that ordinary theft and robbery were now completely out of hand.

Valentin Semilla, once a university forestry instructor, held a series of positions with the Philippine National Railroad (PNR) and lived with his family in Gagalangin. As the Japanese entered Manila, they evacuated to Los Baños, then moved again to Cabuyao and finally Biñan, Laguna, before returning to Gagalangin in June 1942. 'During the war,' he said, 'I was afraid and sick. I did not go back to work for the PNR, and they didn't call again.' Under duress, the family sold jewelry first and then their prized piano. Then 'Mrs. Semilla did some timber deals in Bikol, and from August 1943 until September 1944, she was the principal support of the family.' She did this 'through a buy-and-sell business in jewelry and other items.' With the urban food system in collapse and prices escalating rapidly, the family had to rely on selling personal and household effects in order to eat.[68]

Making Do

Many people tried to cope by growing their own food. Within Manila itself, gardening was hardly routine; in densely inhabited districts, it was only possible in supplemental amounts. But as conditions worsened many undertook to grow a basic food supply as a matter of survival. The family of a construction laborer was living in the Sta. Lucia section of San Juan when the war broke out; a year of hardships convinced them to move along the city fringe to the Ortigas Estate (near what is now called Little Baguio). One member recalled, 'During the war we planted *kamote* and *kamoteng kahoy* [cassava] for food and cut *zacate* for horse owners. We did not know where to get rice, so we ate cassava, *kamote*, and corn grits, which we ground with a *gilingan* [a circular stone grinder].'[69] Members of another family had been living on monthly remittances from their husband and father from California, which stopped with the outbreak of war. Fortunately, the family lived in the Salapan neighborhood of San Juan, where 'there was plenty of grass with goats, carabao, and cows. On the far side there were many *santol* and *kamatsile* [fruit] trees. We caught shrimp [and fish] in the river. . . . Everyone built their homes high and used the *silong* for firewood gathered from the riverbank and for raising pigs and chickens. We also planted papaya and *kamoteng kahoy*.'[70]

In the more densely inhabited areas cassava was out of the question; despite its yield, it required too much space and took too long to grow. *Kamote* was the leading subsistence crop of choice. One Tondo *cochero* said his family ate *lugaw* with *kamote* during the entire war. Another family, with the husband away in the U.S. Navy and son dead in a prisoner of war camp, set about growing *kamote* in the backyard and 'gave the *kamote* tops to the maid.' In addition, they used the pith of the banana stalk (*ubod*, usually fed to pigs) as a carbohydrate source. The family of a city carpenter who lost his job, although his wife did fairly well at buy and sell, ate only twice a day for the duration of the war because that was all he could afford. The family also grew some Chinese cabbage.[71]

For the affluent, things were often considerably better. One Tondo patriarch had a rice field in Santa Cruz, Laguna, which kept the family supplied with rice, his daughter-in-law recalls: "We were well off, and did not plant *kamote*." At the same time, she reports that cooked dog and cat meat were sold along the street in Tondo. "Before the war," she said, "people were afraid of dogs. During the war, dogs were afraid of people." In suburban San Juan, some residential properties were large enough to support raising considerable food-stuffs. One grand home "had a tennis court and a big garden. We had chickens and eggs, goats, and slaughter hogs to make bacon and ham—we salted the meat. We also grew string beans on the fence around the tennis court. [And] we got regular shipments of rice from our farm [in Nueva Ecija] thanks to the caretaker there."[72]

～

The worst was yet to come. For ordinary folks in San Juan, coconut meat now became a staple. Civil servant turned shopkeeper Luciano Salanga continues his story: "Starting in September 1944, Americans began flying over in droves. The Japanese forces increasingly began to stock food for their own use. It became difficult to get food for the store, and the prices soared. Coconuts became the main foodstuff given the lack of rice. My wife and I would travel from San Juan to the Divisoria Market with a pushcart once a week for 100 coconuts, sugar, and other things. It was no joke to push the loaded cart up-slope all the way from Divisoria to the Santa Mesa Bridge. [At the same time] others had to sell everything in order to eat, even furnishings and clothes."[73]

Like many others, a cigar worker and his family in Tondo also ate roasted coconut meat, which they called *kastanyog*, from *castaña* (Spanish for chest-nuts) and *niyog* (Tagalog for coconut)—lowly coconut meat aspiring to the luxurious flavor of roasted chestnuts, black humor about their dire situation. They also grew *kamote* in front of the house and ate cassava and bananas, one of the fruits still available.[74]

Their other major food item at this point was *sisid* rice, which ranks among the most revolting and degrading memories of wartime hardship. Today having eaten it is like a badge of honor. Air raids had sunk a number of merchant vessels in South Harbor, and as food deprivation became severe, people began diving onto these wrecks to recover sacks of rice carried as cargo. (In Tagalog, *sisid* means "to dive headfirst or swim underwater.") Nearly everyone in Tondo remembered surviving on it or, conversely, being well enough off to avoid having to eat it.

A civil servant recalls, 'We ate *sisid* rice that we got underwater from a sunken ship. My son dove and got the sack of rice. Because of *sisid* rice, Tondo folks did not starve. . . . People from other parts of Manila came to Tondo to buy *sisid* rice.' A carpenter from the Bankusay neighborhood emphasized the difficulty of recovering the sacks: 'It had been underwater a year and a half. My close friend . . . had men to dive, but not me. Not everyone can dive, only those with strong lungs. Many became deaf, and blood oozed from their ears.' The same fisherman who traded shrimp for rice said, 'You had to cook *sisid* rice and let the foul smell evaporate. Still, it smelled terrible. I couldn't eat it, so we sold it. Two days a week we ate normal cooked rice, *kanin*. Three days we ate *lugaw*, budgeting our rice supply to avoid having to eat *sisid* rice.' Most families cooked *sisid* rice with garlic, but this hardly concealed the stench.[75] With a certain earthiness, two *kargadores*, father and son, recalled that their family certainly did eat *sisid* rice during the occupation: 'You fry the *sisid* rice. Put in a little ginger, a little garlic, then it's okay. But if you break wind, people [a block away] on Sande Street would pass out.'[76]

The Japanese occupation also brought the necessity of eating dried maize to many Manilans who were unused to it. (Cebuanos do not generally think of eating corn as a hardship, but many ordinary Manilans and Tagalogs were unaccustomed to it except as a fresh vegetable.) A Tondo vegetable seller recalled eating *binatog*, maize kernels steamed until they puff, which is normally eaten with grated coconut and a bit of salt. Doreen Fernandez celebrates *binatog* and *kastanyog*, both sold by street vendors, as "imaginative solutions to privation." A circular migrant farming in Bulacan who came to Manila during the dry season every year to work construction recalled, 'We had no food problem in Bulacan. We had a rice shortage but had *kamote*, *gulay* (vegetables), and maize. We did not [normally] eat maize before the war, only when rice was short.'[77] Author-historian Nick Joaquin reconstructs his family diet during "The Great Hunger," which began, in his account, with the typhoon of November 1943: "Mongo, dried fish and coconuts became luxuries and we lived on *kangkong*; ate gruel mixed with crushed corn; and sweetened our ersatz coffee (corn roasted black and then ground) with precious black *panocha* shavings."[78]

Before the war, rats were sometimes caught by the hundreds in village rice fields in pest drives occasionally celebrated by roasting and consuming the rodents. In the city there were also rat drives, but these were conducted by the public health service, and the animals' remains were incinerated, not consumed. During the later stages of the Japanese occupation, however, all commensal animals were targeted. Not only was the consumption of dogs, cats, horses, and other domesticated animals accelerated—as had occurred during the siege of 1898—but rats and mice also became survival food items. The two *kargadores* recalled an acquaintance who trapped mice and rats, cleaned and skinned them, and then 'sold the body and the legs on the sidewalk.' The Lichaucos recall a moment late in the war when rats were sold as "STAR meat ('rats' spelled backwards) in the markets," and Mañalac describes a starving beggar eating raw a freshly skinned rat.[79] Relatively few urban dogs survived the war.

Starvation

Mortal starvation now stalked the city. Ending this crisis remains today the most strongly defensible of American motives for the land invasion of the Philippines.[80] The war deepened in 1944, and conditions in Manila deteriorated rapidly after June, as the Japanese moved to defend the islands against the American-Allied invasion. They brought in more troops, haggard and emaciated from fighting in New Guinea and elsewhere. Unfortunately for the civilian population, many arrived without adequate rations, and, although the Japanese military controlled corn and rice collection in the Cagayan Valley, by November 1944 it could no longer ship from there to Manila because of American submarines.[81] In such cases, the Japanese military often ordered troops to provision themselves in the field by taking foodstuffs from local inhabitants—even in Okinawa, where the inhabitants were considered Japanese.

As early as May 1944, President Jose P. Laurel had ordered the confiscation of "hoarded" rice stocks, but little of the rice seized subsequently showed up for distribution to civilians. In June the *kempeitai* started seizing rice hoarded in Manila.[82] At the end of September, Felipe Buencamino III noted that "the houses of many rich landowners were searched," a "desperate move." In Ermita-Malate, a household-by-household search by small army units took place in early October. In San Juan, the Japanese went from house to house and took food, including fowls and livestock. Now there was real hardship and want (*kapós*). Starvation. 'People simply died in the streets. We are not heartless, but there is a strong instinct of self-preservation.'[83] Most people now lived on *kangkong* greens or *kamote*, if they could afford it. Many subsisted on coconuts. You cut the coconut meat into squares and broil them. This diet left people bloated, with beriberi and sores on the upper body, malnourished.[84]

In Quiapo some people who had been subsisting on *kangkong* and tiny bits of rice now found their nutrition in rectangles of "bread" made of cassava flour and "smeared with *matamis na bao*"—a sweetened coconut paste.

Now "on the sidewalks hobbled the wretched wrecks of humanity—beggars of all ages—cadaverous in their tattered rags or jute sacks, gathering the last morsel of food, with the smell of their festering, putrid leg ulcers . . .permeating the air."[85] By August residents of Ermita-Malate found starving beggars at their doors, both elderly and young. To carry a sack of rice in the open was to invite attack by starving children. By mid-October, "the streets [were] full of starving people who storm the gates of houses insistently, desperately begging for rice, for a little soup . . . , for anything."[86] In November 1944 Mayor Leon G. Guinto organized pushcarts to remove the dead from the streets and sidewalks. By December the corpses of the dead were carted away in trucks every day. Now "bands of ragged clawing men and young children" were looting warehouses even in full daylight. This was not a form of popular protest, like the bread riots in early modern European cities—which would have drawn a rapid Japanese military response—but rather the most desperate form of spontaneous collective action.[87]

〜

Even caring intact families lost children to hunger and malnutrition. A. V. H. Hartendorp reports an acquaintance who "lost five small grandchildren, one after the other. They did not appear ill . . . only very thin, but then something seemed to go wrong with their breathing, and they died."[88] In Tondo a carpenter and cigar-worker couple was trying to cope: "Our first child was a boy born in 1941. He died of infantile beriberi when he was two. I was breast-feeding. I'm afraid it hurt the child. The second was a girl, born in 1942. We were surprised that she did not also get beriberi. Late in the war, many died on the sidewalks of starvation. Most were children and old people from here."[89] In many normally comfortable families there was now nothing to eat, starvation: "During those last days of World War II we ate mostly *kamote*. Earlier we had rice. In the last days there were many emaciated people. Some fell down and died. We bought some *sisid* rice but couldn't eat it. We grew *kamote* in the yard, but no *kamoteng kahoy*—it takes time. *Kamote* produce tubers quickly."[90]

> We were married on July 14, 1940, and soon rented a bungalow in Gagalangin. Our first child was a girl born in September 1941. Our second was a boy, Victor Jr., born in 1943. Victor died from malnutrition in September 1944 when the Japanese were transporting food supplies to Japan. In November, I evacuated my

family to Macabebe, Pampanga [his wife's home place], because rice was now extremely scarce in Manila. . . . Before an air raid, a *ganta* of rice cost 500 pesos. After the air raid, the price soars to 700. The rationale for this increase was that it would not be long until the Philippines would be liberated and so, if the Philippines got rid of the Japanese, then the money issued by them would be worthless.[91]

During 1944–45 in San Juan, we depended on *darak*, rice bran, that we ate fried. Our tea was boiled *banaba* leaves and *trigo*, wheat. Our third child was not able to drink milk, but he survived. We also grew *kamote* and tried anything. But we did not go to the province to look for food. When I went to the office [in the San Juan public market], I found people dying on the sidewalk. On my way home at the end of the day, they were dead and being taken away in a pushcart. When there was a siren, I ran home, sentry or no sentry. . . . It was hard then, yes. We had to endure. Even now [in the depression of 1985] it is just the same. We have to content ourselves with what we have and depend on the Lord. . . . We just had to endure.[92]

Conditions were also deteriorating in the Santo Tomas internment camp. "Benny" del Carmen, who had regularly brought supplemental food to his former boss in the camp, reported, "Late in the war, when the U.S. invaded Leyte, I was shut out of UST by the Japanese. To cope with this, I rented a house . . . across from the western corner of UST. From there, I could see the shanties of the internees [on the grounds]. Our new arrangement was that I would light a cigarette about midnight and use it to make an up-and-down signal to let them know. Then I would go to the fence at 12:30 with foodstuffs in a sack tied to a rope, and they would pull it over. It was terrifying. When they got it, they would make a circular sign with a cigarette—'OK, we got the goods.' Towards the end there was no rice . . . and we ate only coconuts."[93]

The Second Evacuation and Final Battle

The second mass evacuation of the city, unlike the first, was motivated by the subsistence crisis. The occupation rationing system for rice in Manila had proved stable, if increasingly inadequate, through mid-1943, as Ricardo Jose points out, but not thereafter.[94] By February 1944 the minister of agriculture was warning that Manila should be depopulated to avoid starvation. The price of a sack of rice in the city shot up from 400 pesos at the end of May to 2,700 in mid-September and 5,000 in early October, a reflection of the lack of food stocks and of confidence in the future value of the occupation scrip. Many people left between August and November; some of the wealthy fled to their second homes in Baguio. In September Mayor Guinto urged nonresidents to evacuate. In October President Laurel laid off most government workers and made plans to move his now skeletal administration out of Manila. Others left the city because they could no longer obtain minimal food for themselves and their families and because they saw elements of the Japanese military digging in for battle.[95] Very little of the much diminished rice harvest of November and December 1944 found its way into the city. The Japanese military was looking to its own supply with little obvious concern for the civilian population. Even the minute rations obtained through the formal neighborhood associations, now collapsing, had ended by the start of November.

A further late stream of evacuees developed in early February 1945. Tondo residents had been told to get out of the way of the shelling to come, and the Japanese announced that sections of the district would be burned. After the Allied landings in Lingayen Gulf much of the remaining Japanese army withdrew from Manila, preferring to delay the end and tie down the Americans as long as possible in the mountains, but the marines remained in the city.[96] A majority of our Tondo interviewees evacuated, quite a few at the last minute; most of the San Juan folk did not.

One of the Tondo families that chose not to evacuate was that of the fish broker Leoncia Buzon. Her eldest son, a physician and fellow fish broker, was killed in the yard by shrapnel. One of the San Juan families that did leave went to Tarlac where it had relatives.

> Around the tenth of September 1944 [I took the children, and] we managed to get passage on a charcoal-powered truck to evacuate to Gerona, Tarlac [where my great-aunt took care of us]. Filipino guerrillas surrounded the place. There we bartered old clothes for chicken, fish, and vegetables. Unlike the city, there was no *gutom* [hunger] there. My husband and father stayed on in San Juan. In one incident, my husband was saved by a special card. He was about to be bayoneted but managed to show the card, which said in Japanese that he was a good man. This incident really scared everyone. The next trip to Gerona in October, they joined us. My husband said there was no rice in Manila and now the *kamotes* were gone and there was no organized transport to bring food from the provinces.[97]

One Tondo family evacuated to Bamban, Tarlac; then Baliuag, Bulacan; and later Talavera, Nueva Ecija, where they stayed the longest: 'My husband's work in Talavera was cutting grass for the carabao. I helped in harvesting fruits and vegetables . . . and gathering *kangkong* from the river. *Walang gutom doon*. There was no hunger there.'[98] Another family 'evacuated to Pangasinan on November

20, 1944, because food was very scarce. We arrived during harvest time. At that time even if you have money, there was no food to buy in the city.'[99]

Almost inevitably law and order broke down further. In addition to the robberies, burglaries, and even killings by both desperate people and ordinary outlaws, there was political violence. In Tondo "near the end of the war there were rumors that the Japanese would conduct a 'zoning' operation here and kill all the people.[100] We were afraid of the *makapili*, and there was a *makapili* here. We believed he would signal for the attack with a whistle." So they killed the *makapili*, armed themselves, and fled to nearby Balut Island.[101] Avoiding a rumored *zona* or being fingered by an informant as a guerrilla was a good reason to evacuate.

In the end, General Yamashita's decision not to declare Manila an open city when he withdrew in late December 1944, but instead to turn it into a funeral pyre involving the deliberate brutalization and slaughter of thousands of civilians, created casualties on a scale that during February likely overshadowed the effects of starvation. We will never have an exact accounting, but an estimate by Benito Legarda Jr. of 100,000 killed and an equal number wounded and maimed—the rough equivalent of an early nuclear weapon detonation—fits the evidence.[102]

LIBERATION AND RELIEF

In the memories of survivors, the arrival of the American forces provided heaven-sent relief from privation and starvation. The authorities, both the new Filipino administration and the American soldiers, took immediate action; this did not bring immediate general prosperity, but it did create an environment in which relief food could be obtained and commerce reestablished.

Araceli Evangelista, whose family had survived with relatives in Gerona, Tarlac, recalled, "As the American forces neared, moving through Paniqui and Gerona, the Japanese evacuated. They asked farmers to bring potatoes to town to be purchased. A huge mound of *kamotes* was accumulated and paid for with worthless mimeo money. The Americans arrived in amphibious vehicles and thousands of jeeps. They had handsome tomatoes like apples versus the Japanese with dirty pushcarts with no money and no food. We didn't have milk or bread. Suddenly there were candies, chocolate, milk, and eggs. Befriend an American, and he will give you a huge can of dried eggs or a loaf of bread. Eureka!"[103]

A short story by C. V. Pedroche captures this euphoric moment. A pregnant woman is beset with a specific craving for bread, for the smell of fresh bread, but there is none and has been none for ordinary people for several years. Suddenly soldiers arrive, and the deepest craving is satisfied. New life can proceed. There is hope. "'Thank God,' she said, 'for the smell of bread.'"[104]

In the midst of the fighting, food was distributed to the civilian population in the city. Guerrillas handed out cans of sardines to survivors hunkered down in Quiapo, then Americans and guerrillas distributed "small packages of California rice, cans of beans and sardines"; later there would be corned beef and salmon. Now the Philippine Civil Affairs Unit (PCAU) for the emergency distribution of food and clothing went to work. With the internees liberated, a PCAU office was installed near UST, under sporadic Japanese artillery fire, distributing "rice, corn, canned meat, powdered egg[s] and sardines."[105]

"For days and weeks the people ate nothing but canned goods and rationed California rice," Agoncillo records.[106] Navy wife Rosalia Fortaleza said, "We got coupons or tickets. If there was any distribution, we would send my nephew to get our 'ration.'" Filoteo S. Tuason, the wartime fish dealer, set up a distribution center and store in Tondo: "For three or four years, I was distributing relief goods assigned to me by PCAU. I had small profits, but there was a great demand for the relief goods that were mainly foodstuffs."[107] From San Juan, Luciano Salanga, the wartime storekeeper, reported, "With the Japanese in Intramuros finally subdued in February 1945, I received a message to help start up the civil government. . . .We started the Emergency Control Administration [ECA] to handle Philippine civilian units. The Army gave us wheat, flour, and other goods, and we distributed them. To facilitate the distribution, certain stores were authorized to handle PCAU goods and sell them cheap. Our store was one of those. We were known here. My wife handled the store after I went back to work for the government."[108]

In addition to the PCAU, there were informal channels of food distribution through both black marketeers and the American military. The remembered experiences range from the innocent to the venal. In one case, 'There was an American army barracks on this hill [outside San Juan] called High Hill Horseshoe. They showed movies. Right after the war, my husband and his friends collected dirty clothes from the U.S. soldiers and washed them. They were paid in cash, rice, soap, laundry soap, and cigarettes. There were also barracks at Wack Wack golf course in 1945. Filipinos trooped to the barracks when the soldiers had breakfast—mashed potatoes, corned beef, and hamburger. The Americans were very kind. My eldest daughter, then eight years old, did that.'[109]

Many others report going to U.S. Army camps looking for canned goods or employment and being overwhelmed by the sight of veritable mountains of things after years of privation. Several storage depots were established on the margins of the city. To many these became major sources of immediate employment and goods. One man became a *kapatas* and took a crew of men to an army depot on Balut Island in northern Tondo. There, '[W]e loaded food on Army trucks for distribution.'[110] A woman recalled, 'I sold cooked food in front of our house. . . . And we engaged in buy and sell—of clothes, milk, noodles, and other things smuggled out by contacts from the depot at the Tondo church. In 1946 we were able to build the present house out of my husband's earnings from buy and sell.'[111] There was also outright theft of relief goods. "The looting," Agoncillo concludes, "was 'democratic,' for the partici-pants belonged to all classes of society."[112]

~

In the meantime, the metropolitan water supply remained precarious for a while. The Novaliches Reservoir had been retaken in February and was supply-ing drinking water as before, but its capacity was simply not enough to meet demand in a ruined city now absorbing growing numbers of civilian returnees, refugees, and military personnel: "The whole south side of Manila was with-out water except that provided by army tank trucks and shallow wells." The Ipo Dam in eastern Bulacan, also vital to Manila, remained in the hands of the Japanese (who wired it for demolition) until Filipino and American forces reestablished control on May 17.[113]

After a few months a revived victory gardens program got under way. By June 1945 the ECA was distributing almost 50,000 kilos of garden seeds: beans, soybeans, Chinese cabbage, eggplant, radishes, and others. These were given out in two-gram lots in the city and by the pound to farmers in nearby Rizal and Central Luzon. Marking the new normal, child peanut vendors hawked their wares near the center of the city and the new rice crop looked magnificent in some of the nearest irrigated fields (Figure 11.1).[114] And, having had more than three years for the fish population to recover, the trawler catch in Manila Bay immediately after the war was phenomenal.[115] Slowly, the com-plex and many-stranded urban provisionment system was being resurrected and extended. The story of the successful reconstruction of this system, whose long piece-by-piece rise and halting demise have been traced in this volume, lies beyond our scope.

FIGURE 11.1 Rice production renewed, July 1945, just south of Manila. (USNA II, RG18-AO-371-21, box 159)

EPILOGUE

WORLD WAR II had devastating consequences for the Philippines. The extreme destruction of Manila and other cities, shattered physical infrastructure and productive facilities, loss of life and talent in the general massacre and wastage of civilians, and the economic sundering of an effective Filipino civil service put long-term hurdles in the way of achieving real independence and economic progress. The horrendous brutality of the occupation and the mortality directly attributable to military action, especially in February 1945, overshadow some of the less dramatic consequences of the war. These were significant nevertheless.

Over the previous century the people of Manila had developed a successful system of provisionment, from production and processing or importation of foodstuffs through transport and distribution. It was expandable, from feeding a quarter of a million people to almost a million. It was resilient against all manner of catastrophes, including natural disasters (floods, droughts, plant and animal diseases), limited wars (1898–99), and price fluctuations. It was flexible enough to accommodate changes in regime (from Spanish to American to Commonwealth) and changes in taste, technology, commodity production, ethnicity of vendors, and inflation rates. In short, it worked. Just enough of the battered provisioning system was resilient and continued to function, if not to keep the very young and old alive, then at least to stave off the worst panic flight.

Then the system collapsed. Despite the increasingly desperate ingenuity of Manilans, what remained could not feed the shrunken population that survived and stayed in the city through late 1944. The systemic collapse suggests, in the end, just how intricate and fragile the structure was; it took over a century to build and less than three years to effectively destroy. The collapse of both the food supply and occupation currency from late 1943 to early 1945 put the most ordinary basic nutrition out of reach for the majority of Manilans. Despite inventive and desperate attempts at subsistence and substitution, most suffered a drastically impoverished diet. Many thousands left Manila searching for subsistence, and further thousands fled as local fighting began in February 1945.[1]

Uncounted thousands perished of starvation during these months, selected for death by their lowly positions in the economic system and their lack of sustaining social connections. In the Battle of Manila a great many of the best educated also died of military actions. But few at this end of the spectrum starved, since many middle-class residents had possessions they could trade for food. With the normal economy completely broken down, the least affluent segment in society was the most vulnerable. Without possessions to sell, many evacuated, turned to theft and robbery, or simply and quietly perished, like the entire family of "Salvador" in late 1944.

> Outside Quiapo Church . . . stood a young beggar with distended belly, patches of baldness on his scalp, rusty tin can in hand, clad in pieces of gunny sack. Supported by a crooked staff taller than he, he seldom moved, but just stood there facing the plaza, unmindful of the flies buzzing around [the] . . . ulcers on his swollen legs. He smiled when he recognized me as one of his former alley playmates before the war. We used to call him "Luga" (draining ear), although his first name was Salvador. They were a poor working family to begin with, and as he stood there he whispered to me that his family had disappeared. Three days later I went back to see him to give him some *castanyog* but found him lying on the pavement, partly covered with old newspapers, flies buzzing around him as he joined the ranks of other beggars and paupers in death.[2]

To fathom the significance of feeding Manila, we have only to gaze on the alternative, on the megacity unfed.

NOTES

A comprehensive bibliography, including consular reports, may be found at the website of the University of Wisconsin, Center for Southeast Asian Studies, http://seasia .wisc.edu/home-page/about-cseas/faculty-staff/dan-doeppers/.

ABBREVIATIONS

AAAG   *Annals of the Association of American Geographers*
ABFMS   *Agricultural Bulletin of the Federated Malay States*
ACCJ   *American Chamber of Commerce Journal* (Manila)
AEH   *Australian Economic History*
AHR   *American Historical Review*
ARBAg   Annual Report of the Bureau of Agriculture
ARBAI   Annual Report of the Bureau of Animal Industry
ARBPI   Annual Report of the Bureau of Plant Industry
ARCC   Annual Report of the Collector of Customs
ARCIR   Annual Report of the Collector of Internal Revenue
ARMB   Annual Report of the Municipal Board, City of Manila
ASEAN   Association of Southeast Asian Nations
BAHC   *Bulletin of the American Historical Collection* (Manila)
Balanza   Title varies: Islas Filipinas [Hacienda], *Balanza General del Comercio de las Islas Filipinas, Balanza Mercantil del Comercio de las Islas Filipinas, Estadistica Mercantil del Comercio Exterior de las Islas Filipinas, Estadistica General del Comercio Exterior de las Islas Filipinas*
BIA   United States, Bureau of Insular Affairs records, USNA II
Biba   Bigabang Bayan (National Rice Granary, 1944)
BIES   *Bulletin of Indonesian Economic Studies*
BNM   Biblioteca Nacional, Madrid
BPS   *Bulletin of Philippine Statistics* (continues PSR)
BR   Blair, Emma H., and James A. Robertson, eds. *The Philippine Islands, 1493–1898.* 55 vols. Cleveland: A. H. Clark, 1903–7.

Bulletin   Title varies: *Manila Daily Bulletin, Manila Bulletin, Bulletin Today*
CEA   Civilian Emergency Administration
Census 1903   United States, Bureau of the Census. *Census of the Philippine Islands, 1903.* 4 vols. Washington, D.C.: Government Printing Office, 1905.
Census 1918   Philippine Islands, Census Office of the Philippine Islands. *Census of the Philippines, 1918.* Manila: Bureau of Printing, 1920.
Census 1939   Philippines (Commonwealth), Commission of the Census. *Census of the Philippines, 1939.* Manila: Bureau of Printing, 1940.
Chronicle   *Manila Chronicle*
CIJ   *Commerce and Industry Journal* (continued by PJC)
CR   *Commerce Reports*
CRUSFC   *Commercial Relations of the United States with Foreign Countries*
CVB   The Netherlands, *Consulaire en andere Verslagen en Berichten* [consular reports]
DCTR   United States, *Daily Consular and Trade Reports*, 1912–14
ECA   Emergency Control Administration
EEH   *Explorations in Economic History*
ENSO   El Niño southern oscillation
FAO   United Nations Food and Agriculture Organization
Gaceta   *Gaceta de Manila*
GR   *Geographical Review*
Herald   *Philippines Herald* (Manila)
HRAF   Human Relations Area Files
HSBC   Hong Kong and Shanghai Bank
Hs. of Com.   United Kingdom, House of Commons Papers, a.k.a. Parliamentary Papers (Commons) [consular reports]
IJAS   *International Journal of Asian Studies*
Inquirer   *Philippine Daily Inquirer* (Manila)
IR   *Ilocos Review*
JAAA   *Journal of the American Asiatic Association*
JAE   *Journal of Asian Economics*
JAMS   *Journal of the Association of Military Surgeons of the United States*
JAS   *Journal of Asian Studies*
JAVMA   *Journal of the American Veterinary Medical Association*
JCPT   *Journal of Comparative Pathology and Therapeutics*
JEH   *Journal of Economic History*
JH   *Journal of History*
JHG   *Journal of Historical Geography*
JIH   *Journal of Interdisciplinary History*
JNL   *Journal of Northern Luzon*
JPS   *Journal of Philippine Statistics*
JSEAH   *Journal of Southeast Asian History*
JSEAS   *Journal of Southeast Asian Studies*
JSS   *Journal of the Siam Society*
KITLV   Koninklijk Institut voor Taal-, Land- en Volkenkunde
MJTG   *Malayan Journal of Tropical Geography*

MS   *Municipal Sanitation*
MT   *Manila Times*
NARIC   National Rice and Corn Corporation
NHI   *National Historical Institute*
NYT   *New York Times*
PA   *Philippine Agriculturist*
PAgR   *Philippine Agricultural Review*
PC   Philippines Constabulary
PCAU   Philippine Civil Affairs Unit
PEF   *Progreso Economico de Filipinas*
PFP   *Philippines Free Press*
PFR   *Philippine Finance Review*
PFY   *Philippine Fisheries Yearbook*
Philippines Yearbook   Title varies: *Philippines Yearbook, Philippines Herald Year Book,*
   Philippine Yearbook (1933–34 to 1940–41)
PICAU   variation of PCAU
PJAg   *Philippine Journal of Agriculture*
PJAI   *Philippine Journal of Animal Industry*
PJC   *Philippine Journal of Commerce* (continues CIJ)
PJS   *Philippine Journal of Science*
PMC   Philippine Manufacturing Corporation
PNA   Philippine National Archives, Records Management and Archives Office
PNHB   *Peking Natural History Bulletin*
PNL   Philippine National Library
PNR   Philippine National Railroad
PPG   *Progress in Physical Geography*
PQCS   *Philippine Quarterly of Culture and Society*
PR   *Philippine Review*
PRO   United Kingdom, Public Records Office, Kew, London
PRRA   Philippine Relief and Rehabilitation Administration
PS   *Philippine Studies*
PSacra   *Philippiniana Sacra*
PSR   *Philippine Statistical Review* (continues SBPI)
QBBPW   *Quarterly Bulletin, Bureau of Public Works*
RCUS   United States, *Reports of the Consuls of the United States*
RGGPI   United States, War Department, Bureau of Insular Affairs, *Report of the Gov-*
   ernor General of the Philippine Islands
RIMA   *Review of Indonesian and Malaysian Affairs*
RPC   United States, War Department, Bureau of Insular Affairs, Philippine Com-
   mission, *Report of the Philippine Commission*
SBPI   Philippine Islands, Bureau of Commerce, *Statistical Bulletin of the Philippine*
   Islands
SCR   United States, Bureau of Foreign and Domestic Commerce, *Supplement to*
   Commerce Reports (continues DCTR)
SEAS   *Southeast Asian Studies*

TIBG   *Transactions of the Institute of British Geographers*
UMJEAS   *University of Manila Journal of East Asiatic Studies*
USNA II   United States National Archives II, College Park, Maryland
UST   University of Santo Tomas

PREFACE

1. William Cronon, personal communication, 22Aug2000.

2. D. F. Doeppers, "Migration to Manila, 1893–1903: Changing Gender Representation, Migration Field, and Urban Structure" and "Migrants in Urban Labor Markets," both in D. F. Doeppers and Peter Xenos, eds., *Population and History: The Demographic Origins of the Modern Philippines* (Madison and Quezon City: 1998), 139–79, 253–63.

3. The works of geographers Albert Kolb, Martin Lewis, and Marshall McLennan stand as important partial exceptions.

INTRODUCTION: WHY PROVISIONMENT?

1. Wendell Cox, "The Evolving Urban Form: Manila," newgeography.com, accessed June 28, 2012.

2. Mike Davis, *Planet of Slums* (London: 2006).

3. Mark Jefferson, "The Law of the Primate City," GR 29 (Apr39): 226–32; Daniel F. Vining Jr., "The Growth of Core Regions in the Third World," *Scientific American* 252.4 (Apr85): 42–49; T. G. McGee and Ira M. Robinson, eds., *The Mega-urban Regions of Southeast Asia* (Vancouver: 1995). The largest such countries tend to have more than one megacity.

4. My *Manila, 1900–1941: Social Change in a Late Colonial Metropolis* (New Haven, Conn., and Quezon City: 1984) demonstrates how the timing of Philippine export business cycles and Manila construction cycles were related to business cycles in the United States, while my "Metropolitan Manila in the Great Depression: Crisis for Whom?" JAS 50.3 (Aug91): 511–35, reprinted in David Ludden, ed., *Capitalism in Asia: Perspectives on Asia, Sixty Years of the "Journal of Asian Studies"* (Ann Arbor: 2004), demonstrates the folly of thinking that this was a simple predictable relationship.

5. For want of space, two portions of this project have been or are being published separately: D. F. Doeppers, "Lighting a Fire: Home Fuel in Manila, 1850–1945," PS 55.4 (2007): 419–47; and an essay on the domestic production of alcohol.

6. See the introductory essay by William T. Rowe in *Hankow: Commerce and Society in a Chinese City, 1796–1889* (Stanford, Calif.: 1984) on dismantling older formalist Weberian ideas about Chinese and other Asian cities.

7. A massive wall was a critical innovation that allowed a few port cities in Southeast Asia to be held with relatively small numbers of European artillerymen and troops. With the deployment of large rifled cannon and pointed cylindrical projectiles after the 1860s, the military advantage of such elevated defenses largely vanished. Most of Manila's relict wall still stands, but the walls around Hanoi and Surabaya were removed as useless.

8. The simple notion of a "dual city" does not work well at most levels of society and geography in the Manila context. On attempts to separate the town of Binondo

from Manila in the 1820s, see Ruth de Llobet, "Orphans of Empire: Bourbon Reforms, Constitutional Impasse, and the Rise of Filipino Creole Consciousness in an Age of Revolution" (PhD diss., University of Wisconsin–Madison, 2011), 256–60.

9. Xavier Heutz de Lemps, "Shifts in the Meaning of 'Manila' in the Nineteenth Century," in Charles J-H. Macdonald and Guillermo M. Pesigan, eds., *Old Ties and New Solidarities: Studies on Philippine Communities* (Quezon City: 2000); Robert R. Reed, *Colonial Manila: The Context of Hispanic Urbanism and Process of Morphogenesis* (Berkeley, Calif.: 1978), 45–48; *Map of the City of Manila*, 1:80,000, 1943, Republic of the Philippines, Office of the Mayor, *City Gazette* (16Feb43), back cover.

10. Linda A. Newson, *Conquest and Pestilence in the Early Spanish Philippines* (Honolulu: 2009), 130–32; Doeppers, *Manila, 1900–1941*, 45.

11. Jared Diamond, *Collapse: How Societies Choose to Fail or Succeed* (New York: 2005).

12. The historical geography of this process and the enormous quantities supplied are addressed in Doeppers, "Lighting a Fire," 419–47.

13. William Cronon, *Nature's Metropolis: Chicago and the Great West* (New York: 1991), quote 384–85.

14. Peter Boomgaard and David Henley and their group at the Koninklijk Institut voor Taal-, Land- en Volkenkunde (KITLV) have been in the forefront in addressing this situation. See Peter Boomgaard and David Henley, eds., *Smallholders and Stock-breeders: Histories of Foodcrop and Livestock Farming in Southeast Asia* (Leiden: 2004). A notable exception is R. D. Hill, *Rice in Malaya: A Study in Historical Geography* (Kuala Lumpur: 1977).

15. See Erik Swyngedouw, *Social Power and the Urbanization of Water* (Oxford: 2004) (on Guayaquil, Ecuador); Johann Heinrich von Thünen, *Von Thünen's Isolated State*, Carla M. Wartenberg, trans., Peter Hall, ed. (Oxford: 1966, German editions 1826 and 1842); and, among others, Rhoads Murphey, "How the City Is Fed," in *Shanghai: Key to Modern China* (Cambridge: 1953), 133–64; Richard Peet, "Influences of the British Market on Agriculture and Related Economic Development in Europe before 1860," TIBG 56 (Jul72): 1–20; Kenneth Kelly, "Agricultural Change in Hoogly, 1850–1910," AAAG 71.2 (Jun81): 237–52 (on Calcutta); and Gareth Shaw, "Changes in Consumer Demand and Food Supply in Nineteenth-Century British Cities," 11.3 JHG (Jul85): 280–96.

16. Charles Tilly, "Food Supply and Public Order in Modern Europe," in Charles Tilly, ed., *The Formation of National States in Western Europe* (Princeton: 1975), quote 455; Roger Horowitz, Jeffrey M. Pilcher, and Sydney Watts, "Meat for the Multitudes: Market Culture in Paris, New York City, and Mexico City over the Long Nineteenth Century," AHR 109.4 (Oct2004): quote 1055. See also Lillian M. Li and Alison Dray-Novey, "Guarding Beijing's Food Security in the Qing Dynasty: State, Market, and Police," JAS 58.4 (Nov99): 992–1032; James L. McClain, John M. Merriman, and Ugawa Kaoru, eds., *Edo and Paris: Urban Life and the State in the Early Modern Era* (Ithaca, N.Y.: 1994); and Steven L. Kaplan, *Provisioning Paris: Merchants and Millers in the Grain and Flour Trade during the Eighteenth Century* (Ithaca, N.Y.: 1984).

17. De Llobet, "Orphans of Empire," quote 171; Jaime Vicens Vives, *An Economic History of Spain*, Frances M. López-Morillas, trans. (Princeton, N.J.: 1969), quote 644.

18. Norman G. Owen, *Prosperity without Progress: Manila Hemp and Material Life in the Colonial Philippines* (Berkeley and Quezon City: 1984), quote 64; Josep M. Fradera, "The Historical Origins of the Philippine Economy: A Survey of Recent Research of the Spanish Colonial Era," AEH 44.3 (Nov2004): 310.

19. My sincere thanks go to Chantal Oudkerk Pool, who translated the Dutch-language materials, and to Prof. Otto van den Muijzenberg, who brought their existence to my attention.

20. Public education is an infrastructural requirement not treated here. See Glenn A. May, *Social Engineering in the Philippines* (Westport, Conn.: 1980); Doeppers, *Manila, 1900–1941*; and Erin P. Hardacker, "The Impact of Spain's Educational Decree on the Spread of Philippine Public Schools and Language Acquisition," *European Education* 44.4 (Winter 2012): 8–30.

21. I owe a debt to Charles Minard, who invented a cluster of thematic mapping techniques in the 1850s while investigating the meat supply of Paris. See Arthur Robinson, *Early Thematic Mapping in the History of Cartography* (Chicago: 1982), 145, 152.

22. A note on the testimony in these oral accounts is in order. During 1985 over 100 senior citizens consented to be interviewed on their work careers and family lives in Manila during the 1930s and World War II. Most of the interviews were conducted by Loreto Seguido and myself working as a team. About 20, evenly divided, were conducted by one or the other of us working alone. Those interviewed by Seguido are so noted. Most of the interviewees lived or had lived in either Tondo or San Juan and represented survivors from the full spectrum of urban Filipino society. Depression conditions prevailed in 1985, and tension was high in the last full year of Ferdinand Marcos's rule. There was good reason to believe that introducing a recording device would complicate the interview process. Further, I was operating on a tight research budget with no room for transcriber-typists. As a result, the original transcript of each interview is composed of my on-the-spot longhand record of phrases and sentences in English or Tagalog. The notes were written in the presence of the interviewee, and in all cases the interview was understood to be for the record. From these notes I have reconstructed a prose narrative in English. In this volume such reconstructions are rendered in single rather than double quotation marks. Double quotation marks indicate a direct quote of verbatim material.

23. Ken De Bevoise, *Agents of the Apocalypse: Epidemic Disease in the Colonial Philippines* (Princeton, N.J.: 1995) quotes 164; Daniel F. Doeppers and Peter Xenos, eds., *Population and History: The Demographic Origins of the Modern Philippines* (Madison and Quezon City: 1998), 4.

24. There is also a fascinating set of questions about the aesthetics and ethnography of food and cuisine, but for the most part I leave this to others better qualified than myself. See, for example, Doreen G. Fernandez, *Tikim: Essays on Philippine Food and Culture* (Pasig: 1994); Gilda Cordero-Fernando, *Philippine Food and Life* (Pasig: 1992); Gene R. Gonzalez, *Cocina Sulipeña: Culinary Gems from Old Pampanga* (Makati: 1996); and Felice Prudente Sta. Maria, *The Governor-General's Kitchen: Philippine Culinary Vignettes and Period Recipes, 1521–1935* (Pasig: 2006).

25. Francisco de Solano, "An Introduction to the Study of Provisioning in the Colonial City," in Richard P. Schaedel, Jorge E. Hardoy, and Nora Scott Kinzer, eds.,

Urbanization in the Americas from Its Beginnings to the Present (The Hague: 1978), quote 99.

CHAPTER 1. THE MANILA RICE TRADE IN THE AGE OF SAIL

1. Newson, *Conquest and Pestilence*, 115.

2. On the international rice trade and the Philippines, see Benito Legarda Jr., *After the Galleons: Foreign Trade, Economic Change, and Entrepreneurship in the Nineteenth-Century Philippines* (Madison and Quezon City: 1999); Leon A. Mears et al., *Rice Economy of the Philippines* (Quezon City: 1974); Hill, *Rice in Malaya*. On changing production areas, see Marshall S. McLennan, *The Central Luzon Plain: Land and Society on the Inland Frontier* (Quezon City: 1980); Albert Kolb, "Die Reislandschaft auf den Philippinen," *Petermanns Geographische Mitteilungen* 86 (1940): 113–24; and Albert Kolb, *Die Philippinen* (Leipzig: 1942).

3. I rely on Vito C. Santos, *Vicassan's Pilipino-English Dictionary* (Manila: 1978/1986), as well as the works of Serrano Laktaw, Retana, and others noted in the glossary.

4. See David Henley, "Natural Resource Management and Mismanagement: Observations from Southeast Asian Agricultural History," in Greg Bankoff and Peter Boomgaard, eds., *A History of Natural Resources in Asia* (Houndsmills: 2007), 19–37; Pierre Gourou, *The Peasants of the Tonkin Delta: A Study of Human Geography*, 2 vols. (New Haven: 1955, French ed., 1936); Clifford Geertz, *Agricultural Involution* (Berkeley: 1963); John Wise, "Account of the Philippine Islands," in *Centenary of Wise and Company in the Philippines, 1826–1926* (n.p.: n.d.), 78.

5. Newson, *Conquest and Pestilence*, 141–43, 254.

6. Rafael Díaz Arenas, *Report on the Commerce and Shipping of the Philippine Islands*, Encarnacion Alzona, trans. (Manila: 1979/1838), 24.

7. Newson, *Conquest and Pestilence*, 121, quote 116.

8. Luis Alonso Alvarez, "The Spanish Taxation System and the Manila Food Market: Indications of an Early Commercialized Economy," *Kasarinlan* 14.2 (1998): 5–20.

9. On land-grabbing by some friar estates and a primer on how to bribe royal officials in these matters, see Dennis Morrow Roth, *The Friar Estates of the Philippines* (Albuquerque: 1977), 105–6, 98.

10. Ibid., quote 149. See also Nicholas P. Cushner, *Landed Estates in the Colonial Philippines* (New Haven: 1976), chaps. 4 and 7. Cushner argues that the landed estates must be studied in "relation to local and urban markets" (123n30, quote 68).

11. Roth, *Friar Estates*, quote 149.

12. Cushner, *Landed Estates*, quote 1. Cf. Owen, *Prosperity without Progress*, 130; John A. Larkin, *The Pampangans: Colonial Society in a Philippine Province* (Berkeley: 1972), 25–26.

13. *Cedulario de Manila*, Nicholas Cushner, Helen Tubangui, and Domingo Abella, eds. (Manila: 1836/1971), decrees of December 4, 1784, and February 7, 1798, 132–35, 193–94. Cf. "Ordinances Enacted by the Audiencia of Manila," October 15 and December 7, 1598, in Emma H., Blair and James A. Robertson, eds., *The Philippine Islands, 1493–1898*, 55 vols. (Cleveland: 1903–7), 10:304–10. Provisionment and the regulation of food prices was also a leading issue in early local governance in Latin

America. See John C. Super, *Food, Conquest, and Colonization in Sixteenth-Century Spanish America* (Albuquerque: 1988), 39.

14. Defined both geomorphologically and by human population density, the Luzon Central Plain includes the smooth lowland portions of Bulacan, Pampanga, Tarlac, Nueva Ecija, and Pangasinan. See maps 2 and 3 in Larkin, *The Pampangans.*

15. Among the growing literature, see Legarda, *After the Galleons*; Alfred C. McCoy and Ed. C. de Jesus, eds., *Philippine Social History: Global Trade and Local Transformations* (Quezon City and Honolulu: 1982); and Owen, *Prosperity without Progress.*

16. Legarda, *After the Galleons*, chap. 5.

17. *Eotechnic* is used here in the sense developed by Lewis Mumford in *Technics and Civilization* (New York: 1934/1963).

18. In the 1920s, Joseph Arthur LeClerc claimed that "three-fourths of the rice crop of the Philippines is produced by transplanting." J. A. LeClerc, *Rice Trade in the Far East*, Trade Promotion Series, no. 46 (Washington, D.C.: 1927), quote 56. Wise describes routine transplanting around Manila in 1837 in "Account of the Philippine Islands," 77.

19. Lucien M. Hanks, *Rice and Man* (Honolulu: 1992); Toshiyuki Miyata, "Tan Kim Ching and Siam 'Garden Rice': The Rice Trade between Siam and Singapore in the Late Nineteenth Century," in A. J. H. Latham and Heita Kawakatsu, eds., *Intra-Asian Trade and the World Market* (London: 2006), 115.

20. For areas of broadcast cultivation, see Jose S. Camus, "Rice in the Philippines," PAgR 14.1 (1921): 23. Cf. R. D. Hill, "The Cultivation of Perennial Rice, an Early Phase in Southeast Asian Agriculture?" JHG 36 (2010): 215–23.

21. Percy Hill and Kilmer O. Moe, "The Cultivation of Rice," UMJEAS 9.4 (Oct60): 125–27 (originally published in 1917). Others say that carabao begin regular work at age five or six. *Census 1903*, 4:130.

22. See Robert E. Huke, *Agroclimatic and Dry-Season Maps of South, Southeast, and East Asia* (Los Baños: 1982).

23. McLennan, *Central Luzon Plain*, 71; Francisco Calalang, *History of Bulacan* (Quezon City: 1971), 90; Glenn A. May, *Battle for Batangas: A Philippine Province at War* (New Haven: 1991), 23–24; D. H. Grist, *Rice*, 4th ed. (London: 1965), 136–38; Santos, *Vicassan's Pilipino-English Dictionary*, 1169; Frederick Wernstedt and Joseph E. Spencer, *The Philippine Island World: A Physical, Cultural, and Regional Geography* (Berkeley: 1967), unpaginated photos. On the selection and cutting of rice for seed, see Ma. Concepcion Herrera Vda. de Umali, *The Tayabas Chronicles: The Early Years (1886–1907)*, Nita Umali Berthelsen, trans., Karen Berthelsen Cardenas, ed. (Pasig: 2002), 22.

24. Fr. Manuel Buzeta and Fr. Felipe Bravo, *Diccionario Geográfico, Estadístico, Historico de las Islas Filipinas*, 2 vols. (Madrid: 1850–51), 2:443; Isabelo de los Reyes, "Malabon Monograph," in *El Folk-Lore Filipino*, Salud C. Dizon and Maria Elinora P. Imson, trans. (Quezon City: 1994, Spanish ed., 1889), quote 528; PNA, Fincas Urbanas, Tambobong, 1891; PNA, Contribución Industrial, Manila, 1892; Consul Pauli, "Manila," Hs. of Com., 1880, v. 74, C2632, quote 1368.

25. De los Reyes, *El Folk-Lore Filipino*, 528–31, 535. For a ranking of urban centers, see D. F. Doeppers, "The Development of Philippine Cities before 1900," JAS 31.4 (Aug72): 787, revised to combine Malabon and Navotas.

26. Buzeta and Bravo, *Diccionario Geográfico.*

27. On the 1823 medal, see Blair and Robertson, *The Philippine Islands*, 52:309. On the canal plans, see the "Plano Topografico" by Tomas Cortes (1846), in Edgardo J. Angara, José Maria A Cariño, and Sonia P. Ner, *Mapping the Philippines: The Spanish Period* (Cubao, Quezon City: 2009), 150–51.

28. On the Canal de la Reina, see Jose Montero y Vidal, *El Archipiélago Filipino y las Islas Marianas, Carolinas, y Palaos: Su Historia, Geográfia, y Estadística* (Madrid: 1886), 305; and "Who Must Pay for Dredging," MT, 30Aug12, 5.

29. Agustin de la Cavada, Mendez de Vigo, *Historia, Geográfica, Geologica, y Estadística de Filipinas*, 2 vols. (Manila: 1876), 2:466–67. In the general registration of vessels in 1904–5, almost all *bancas* were 5 to 15 tons while 84 percent of *cascos* were above this range, mostly between 16 and 55 tons. RPC 1904, pt. 3, 581–621; 1905, pt. 4, 139–53. The *cavan* is a dry measure of volume; weight varies by moisture content.

30. There were also 50 *cascos* based in the Bulacan waterway towns of Hagonoy and Meycauayan, almost 50 in Sesmoan and nearby Pampangan river towns, and 22 in Bacoor, Cavite. A few *cascos* were based at San Fernando in Pampanga, coastal Bataan, San Isidro in Nueva Ecija, and Cavite-San Roque. Isabelo de los Reyes argued that the number of *cascos* based in Malabon had fallen from 145 in 1872 due to a crisis in international sugar markets and the fact that many of Malabon's warehouses were empty for that reason. De los Reyes, *El Folk-Lore Filipino*, 532–33. See also PNA, Contribucíon Industrial, Manila, 1892; and Manila, 1896.

31. PNA, Contribución Industrial; see table 1.1.

32. A number of the small steamers were known as "Ynchausti boats" before Yangco purchased them and were used to carry passengers between Cavite and Manila. Santiago V. Alvarez, *Recalling the Revolution: Memoirs of a Filipino General*, Paula Carolina S. Malay, trans. (Madison and Quezon City: 1992), 20. Teodoro Yangco succeeded his father in this business. Arsenio Manuel, *Dictionary of Philippine Biography*, 2 vols. (Quezon City: 1955–70), 1:481–85.

33. The brothers fled to Hong Kong during the Revolution and subsequent war, where they and Luis Yangco were active in conducting the foreign affairs of the Aguinaldo government. Capt. John R. M. Taylor, *The Philippine Insurrection against the United States, a Compilation of Documents with Notes and Introduction*, 5 vols. (Pasay City: 1971–73), 2:487, 496; 3:197. See also Luisa Fernandez Lichauco, *Family Recollections* (Manila: 1991), 8–13; and Cornelia Lichauco Fung, *Beneath the Banyan Tree: My Family Chronicles* (Hong Kong: 2009), 29–40.

34. *Gaceta*, 23Oct1861; PNL, Historical Data Papers (1950), Bulacan Province, Hagonoy, Barangay Pugad, 71. On the bandits/pirates, see Greg Bankoff, *Crime, Society, and the State in the Nineteenth Century Philippines* (Quezon City: 1996), 61–71; and John Foreman, *The Philippine Islands*, 2nd ed. (New York: 1899), 267.

35. J. de Man, *Recollections of a Voyage to the Philippines*, E. Aguilar Cruz, trans. (Manila: 1984/1875), quote 74; Ruperto C. Santos, "The Town and Church of Pasig," PSacra 83 (1993): 343–47; Henry T. Ellis, *Hong Kong to Manilla and the Lakes of Luzon in the Philippine Isles in the Year 1856* (London: 1859), 84; Serafin D. Quiason, "The Tiangui: A Preliminary View of an Indigenous Rural Marketing System in the Spanish Philippines," PS 33.1 (1985): 22–34; Hugo H. Miller, *Economic Conditions in the Philippines* (Boston: 1913), 331.

36. Buzeta and Bravo, *Diccionario Geográfico*, quote 2:397; Islas Filipinas, Comisión Central de Estadística de Filipinas [second notebook], 1855, 67. See also de la Cavada, *Historia Geográfica*, 1:55–57; Montero y Vidal, *El Archipiélago Filipino*, 312; and Santiago Ugaldezubiaur, *Memoria Descriptiva de la Provincia de Manila* (Madrid: 1880), 38.

37. Buzeta and Bravo, *Diccionario Geográfico*, 1:210–11, quote 194. The entrepreneur who owned this mill circa 1850 was Bartolome Barretto, born in Macao and settled in Manila circa 1846. Earlier, in 1836, the Real Sociedad Economia de Amigos del Pais noted the operation of "machines for hulling rice by steam power . . . introduced by Don Eulogio de Otaduy." BR 52:310, quote 317. See also Wise, "Account of the Philippine Islands," 107; and Díaz Arenas, *Report on the Commerce and Shipping of the Philippine Islands*, 30–32. Otaduy/Otadui also acted as agent for the British-China firm Jardine Matheson. Carol Matheson Connell, *A Business in Risk: Jardine Matheson and the Hong Kong Trading Industry* (Westport and London: 2004), 6, 30; Legarda, *After the Galleons*, 158, 229, 258–59; RPC 1904, pt. 1, 596.

38. Roth, *Friar Estates*, 100–16, 97–98.

39. Buzeta and Bravo, *Diccionario Geográfico*, 1:389; Rhina Alvero-Boncocan and Dwight David Diestro, *Nineteenth Century Conditions and the Revolution in the Province of Laguna* (Quezon City: 2002), 10–33; "Friar Lands Irrigation Systems in Nearby Provinces," ACCJ 12.7 (Jul32): 5, 16.

40. Frederick H. Sawyer, *Inhabitants of the Philippines* (London: 1900), 72–74. A British engineer involved in the construction of irrigation works, Sawyer lived in the Philippines during 1877–92.

41. Timoteo J. Pascual and Liwayway P. Guillermo, in collaboration with D. H. Soriano, *Morong's 400 Years* (Manila: 1978), 32–62, quote 61; *Tanay Tercentenary Souvenir, 1640–1940, and the Towns of Rizal Province* (Manila: 1940), 117.

42. In Manila there were price quotations for Pampanga *palay* in October 1861, and Cavada reports 40,000 *cavans* of clean rice equivalent shipped from Pampanga to Manila circa 1870. Jose Felipe del Pan identifies a highly appreciated rice variety from Cavite. Jose Felipe del Pan, *Diccionario de la Administracion del Comercio y de la Vida Practica en Filipinas* (Manila: 1879), vol. 1, "arroz."

43. "Estado de la agricultura filipina en 1887," BNM, ms. 19.218. On relative productivity in the inner zone, Regino Garcia identifies Imus, Cavite, as having the highest yields and Pampanga the lowest due to its sandy soils. Regino Garcia, "The Cultivation of Rice," *Census 1903*, 4:93. Montero y Vidal cites Cavite as providing *arroz fino* to the city. Montero y Vidal, *El Archipiélago Filipino*, 308, 315. A generation later Hugo Miller finds no surplus rice production in Cavite. Hugo Miller, *Commercial Geography: The Materials of Commerce for the Philippines* (Manila: 1911), 15.

44. Gourou was concerned about the economic fate of the artisanal millers if power milling were to become widespread in the Red River delta. Pierre Gourou, *Les paysans de delta tonkinois: Etude de geographie humaine* (Paris: 1936), 518–23.

45. PNA, "Estadísticas de Manila, 1896, Pandacan, Ganaderia."

46. Ugaldezubiaur, *Memoria Descriptiva*, 38; Charles M. Conner, *Rice Culture in the Philippines* (Manila: 1912), 26–27.

47. Alvin J. Cox and A. S. Arguelles, "Soils of the Island of Luzon," PJS 9A.1 (Feb14): 25–27.

48. Silverio Apostol, "Report on Rice Cultivation in Zambales and Pangasinan," PAgR 2.5 (May09), quote 270–71. This system is described for Pangasinan in Díaz Arenas, *Report on the Commerce and Shipping of the Philippine Islands*, 49, and for the 1930s in an interview with historians and Pangasinan natives Rose Mendoza Cortes and Napoleon Casambre, 20Jul97. On the *rakem*, see Harold Conklin, *Hanunóo Agriculture* (Rome: 1957), 113–17; Teresita R. Maquiso, *Nueva Ecija's Material Culture*, Pt. 1 (Muñoz, Nueva Ecija: 1985), 17; and Francesca Bray, *The Rice Economies: Technology and Development in Asian Societies* (Berkeley: 1986), 20–21.

49. Henry T. Lewis, *Ilocano Rice Farmers: A Comparative Study of Two Philippine Rice Barrios* (Honolulu: 1971), quote 58–61. Lewis describes Ilocos awned varieties as being intermediate between the *Oryza sativa indica*, i.e., the most common rices of Southeast Asia, and the *O. s. japonica* physiological clusters. On these clusters, see "Tracing Rice Genomes Reveals Cross-Breeding," NYT, 14Jun2011, D3. Cf. Silverio Apostol, "Rice Growing in the Philippines," PAgR 3.11 (Nov10): 625–38; McLennan, *Central Luzon Plain*, 256; Conner, *Rice Culture*, 25–26; H. O. Jacobson, *Rice in the Philippines Islands* (Manila: 1914), 5–6; and Basilio R. Bautista, "The General Practice of Lowland Rice Farming," PJAg 8.1 (1937): 105–18.

50. Rosario Mendoza Cortes, *Pangasinan, 1801–1900* (Quezon City: 1990), quote 53; PNA, Erección de Pueblos Pangasinan, 1874–1897, folios 19–27B, February 3, 1873.

51. The agent of this introduction was apparently a Spanish Dominican. Rafael Magno, "A Historical Retrospect on the Town of Mangaldan (1600–1898)," PSacra 8.23 (1973): 352–55; "A Unique Irrigation Water Hoist," QBBPW 3.1 (Apr14): 41; Napoleon Casambre, interview with the author, 20Jul97.

52. On Vietnam, Mexico, and antiquity, see P. Guilleminet, "Une industrie annamite: Les norias du Quang-Ngai," *Amis de Vieux Hue, Bulletin* 13 (1926): 97–232, partially available in English through the HRAF; William E. Doolittle, "Noria Technology in Mexico," *International Journal of Molinology* 59 (1999): 8–13; Joseph Needham and Wang Ling, *Mechanical Engineering*, in Joseph Needham, ed., *Science and Civilization in China*, vol. 4, pt. 2 (Cambridge: 1965), 356–62; Rudolph P. Hommel, *China at Work* (New York: 1937), fig. 175; and Grist, *Rice*, 33–40.

53. Lewis, *Ilocano Rice Farmers*; Emerson B. Christie, "Notes on Irrigation and Cooperative Irrigation Societies in Ilocos Norte," PJS 9.2 (Apr14): 99–113; and discussions with Willem Wolters, Netherlands, May 1999.

54. This considerable flow arrived in 567 shipments from more than 40 ports. "Clean rice equivalent" simply reduces the volume of *palay* by 50 percent to its approximate postmilling volume.

55. Coastal trade deliveries to Manila in *arroz* equivalent terms were 412,111 *cavans* in 1853—a year of strong exports—versus only 234,912 *cavans* in 1854. Islas Filipinas, Comisión Central de Estadística de Filipinas. The 2-*cavan* average is suggested in Owen, *Prosperity without Progress*, 133, supported by calculations based on H. O. Jacobson, "Consumption of Rice in the Philippine Islands," PAgR 8.4 (1915): 283–85. The rate would be higher for a working adult male: 2.5 *cavans*. Camus, "Rice in the Philippines," 12.

56. Rice varieties differ in texture, aroma, flavor, and color. Some may be pale red or dark, especially after cooking; some are yellowish; and some highly desired varieties are intensely white. Colored varieties were less valued and less common in commerce than for local use. In one classification, 13 of 312 lowland varieties were colored versus 82 of 441 upland varieties. Few upland varieties were awned. H. O. Jacobson, "Methods Used to Improve Rice Culture in the Philippines," PAgR 8.3 (1915): 190–92.

57. Del Pan, *Diccionario de la Administracion*, "arroz"; Fr. Andres Carro, OSA, *Vocabulario Iloco-Español*, 2nd ed. (Manila: 1888), 160; Jose de Bosch, "Philippine Islands-Sual," Oct. 1, 1857, Hs. of Com. Papers, 1859, v. 30, session 2, 375. *Malagkit* also came from Cavite and some *mimis* was grown there. In Pampanga, *mimis* and *milagrosa* were broadcast and left unirrigated. All three were considered first class. Regino Garcia, "The Cultivation of Rice," in *Census 1903*, 4:94; M. Cunanan, "Cultivation of Rice in Pampanga," in *Census 1903*, 4:96; Hill and Moe, "Cultivation of Rice," 131; Silverio Apostol, "Report on Rice Cultivation"; Camus, "Rice in the Philippines," 11.

58. On the standard *cavan*, see Gregorio Sancianco y Goson, *The Progress of the Philippines: Economic, Administrative, and Political Studies*, Encarnacion Alzona, trans. (Manila: 1975; Spanish ed., 1881), 178–79. According to Owen, before about 1860 the "cavanes de provincia" used in Bikol were "about 39 percent larger than the standard 'cavan del Rey' or 'cavan de Manila.'" Owen, *Prosperity without Progress*, 137n37. Laguna, Masbate, and Zambales all reported some later use of nonstandard *cavans*, but in general the standardization was successful. *Census 1903*, 4:448–49. Adoption of the decimal metric system only began to unify weights and measures in Spain itself after 1858. Vicens Vives, *Economic History of Spain*, 692.

59. *Bayones*, *sacos*, *cestos*, and *bultos* were all dry containers or bundles of indefinite volume. The first two are bags. See "Glossary."

60. Mean rainfall at Manila during January–May 1866–1912 was 200 mm. The 40 mm for these months in 1871 was the second lowest of the epoch. Fr. Jose Coronas, SJ, *Extraordinary Drought in the Philippines, October 1911 to May 1912* (Manila: 1912), 4. In the end, 45 ports in the outer zone sent 547 rice shipments to Manila in this year.

61. William Gifford Palgrave, Hs. of Com., 1878, v. 73, C1950, 259.

62. Francis A. Geologo reports a tendency for babies to have been conceived during the cooler dry season months of December–March based on parish records for San Jose, Batangas, 1767–1903. Conversely, 51 percent of local mortality occurred during the four peak rainy months, June–September. Francis A. Geologo, "Marriage Patterns in the Philippines during the 19th Century: Data from the Parish Records," paper presented at the conference Asian Population History, Taipei, January 1996. Cf. Owen, *Prosperity without Progress*, 126–27; and Peter Xenos and Shui-Meng Ng, "Nagcarlan, Laguna: A Nineteenth Century Parish Demography," in D. F. Doeppers and Peter Xenos, eds., *Population and History: Demographic Origins of the Modern Philippines* (Madison and Quezon City: 1998), 201.

63. "Crops Planted and Harvested," PAgR (Oct08–May09).

64. Pangasinan reports for November, *Gaceta*, 5 and 7Dec1862.

65. During 1861–62 the only *Gaceta* price quotations for rice arriving from the inner zone came precisely during September and October.

66. Geographer Rhoads Murphey maintained that the stock on hand in prerevolutionary Shanghai amounted to so little relative to consumer demand that the price was affected within hours of a rain that would even briefly halt the flow of arrivals. Rhoads Murphey, "How the City Is Fed," in *Shanghai: Key to Modern China* (Cambridge, Mass: 1953), 143–45.

67. Cushner, *Landed Estates*, quote 44, referring to San Pedro Tunasan, Laguna.

68. Luciano P. R. Santiago, "The Last Hacendera: Doña Teresa de la Paz, 1841–1890," PS 45.3 (1998): quote 354; Luciano P. R. Santiago, personal communication, 31Jul2008.

69. See notices in *El Comercio* such as "Se aquilan dos buenas bodegas," 25Nov1873; "Bodegas," 4Mar1885, quote 4; and "Nómina de los propietarios," 5Jul1887. See also *Witton's Manila and Philippines Directory, 1903* (Manila: 1903).

70. All this was insured against fire. The *sacos* held 101,764 *picos arroz*. "Almacenes de Deposito," *El Comercio*, 11Jan1892; Philippine Supreme Court case 411, *Donaldson, Sim and Co. v. Smith Bell and Co.*, 23Apr02 (lawphil.net/judjuris/juri1902). On Yangco's *almacen*, see Lorelei D. C. de Viana, *Three Centuries of Binondo Architecture, 1594–1898: A Socio-historical Perspective* (Manila: 2001), 128–29.

71. I rely here on Norman Owen's notes from price-current circulars of the major houses.

72. This occurred in 1851, 1855, and 1857. Legarda, *After the Galleons*, table 13; *Gaceta*, 1861–1862; "Pangasinan," *Gaceta*, 5 and 7Dec1862; Norman Owen, personal research files, "Peele Hubbell" (1850–1857) and "Merchant Circulars: Jardine Matheson," containing "price-current" circulars from many Manila merchant houses (1860–72), summarized in his *Prosperity without Progress*, app. E. I am extremely grateful for the opportunity to use these files.

73. Legarda, *After the Galleons*, 158–60. On the emergence of a broad synchronization of rice prices in the riverine and coastal areas of China centered on the Yangtze Delta, see Thomas G. Rawski and Lillian M. Li, eds., *Chinese History in Economic Perspective* (Berkeley: 1992), especially the introduction and Yeh-chien Wang's "Secular Trends of Rice Prices in the Yangzi Delta, 1638–1935." By Wang's reckoning, 1862 was a year of high rice prices in the delta, almost double the long-term average (table 1.1). Cf. Loren Brandt, "Chinese Agriculture and the International Economy, 1870s–1930s: A Reassessment," EEH 22 (1985): 168–93.

74. Dutch policy finally changed when a rice export ban was imposed in 1911–12, 1914, and 1917–20. Peter Boomgaard, "The Welfare Services in Indonesia, 1900–1942," *Itinerario* 10.1 (1986): 67–68.

75. Howard Dick and Peter J. Rimmer aptly call them an "archipelago" in "Islands: Java and Luzon," in *Cities, Transport, and Communication: The Integration of Southeast Asia since 1850* (New York: 2003), 136.

76. Owen, *Prosperity without Progress*, 95, 131, on Pasacao; McLennan, *The Central Luzon Plain*, 240, 248, 259, on Nueva Ecija. Pongol was a coastal sandbank just west of the Vigan center.

77. In 1872 Pangasinan sent almost 260,000 *cavans*, up from 201,000 in 1862. In a report of May 28, 1856, de Bosch says that Pangasinan had exported "in the present year" 185,000 to 190,000 *cavans* to Manila and 60,000 to 65,000 to China. He believed

that more could be shipped if the price went up. "Philippine Islands (Sual)," Hs. of Com., 1857, v. 16, C2201, 530. Some 216,000 *cavans* were shipped in 1868. Consul George Thorne Ricketts, "Philippine Islands," Hs. of Com., 1870, v. 64, C115, 218. In that year, Pangasinan also sent 117,306 *cavans arroz* equivalent to Ilocos and Pampanga.

78. Owen, *Prosperity without Progress*, 133–34. Pasacao and the Camarines continued to send occasional rice shipments to regional markets such as Iloilo. *Gaceta*, 1Jul1869 and 3May1870.

79. Camiling, Paniqui, and Gerona were reassigned to the new province of Tarlac in 1873. Cortes, *Pangasinan, 1801–1900*, 2. Camiling had been included in the parish of Bayambang, Pangasinan.

80. Jose P. Apostol, "The Ilocanos in Zambales," JH 4 (1956): 3–15; Ed. C. de Jesus, *The Tobacco Monopoly in the Philippines: Bureaucratic Enterprise and Social Change, 1766–1880* (Quezon City: 1980), 182.

81. *Gaceta*, 18May1861, 12Oct1861, 20Oct1861, 24May1862, 1Jun1862, 14Jun1862, and 6Jul1862; Legarda, *After the Galleons*, 164.

82. *Census 1903*, 4:87; Legarda, *After the Galleons*, table 12. Mears et al. give a conversion of 56 kg for the contemporary *cavan* of clean rice and 44 kg per *cavan* of *palay* at 14 percent moisture content. Mears et al., *Rice Economy*, 23. The *Statistical Bulletin of the Philippine Islands* uses a conversion of 57.5 kilos for the older *cavan* (Philippine Islands, Bureau of Commerce, SBPI 1925, 94). Sanciancos's higher figures seem aberrant. Sanciano y Goson, *Progress of the Philippines*; Norman Owen, personal communication, 25Jul2009. On export shipments from Manila, see *Gaceta*, 13Jul1862, 3.

CHAPTER 2. PALEOTECHNIC MARVELS AND RICE PRODUCTION DISASTERS, 1876–1905

1. *Paleotechnic* refers to the industrial complex of iron, coal, and steam and the myriad applications and implications of their use. Mumford, *Technics and Civilization*.

2. James Francis Warren, *At the Edge of Southeast Asian History* (Quezon City: 1987), 7. Nicholas Loney observed steam gunboats replacing the heavy sail *faluas* throughout the archipelago in 1861. Even before this, Filipinos were organizing an effective defense of some coastal areas. "Philippines," Hs. of Com., 1862, v. 58, C3054, 345.

3. On similar changes in Iloilo, see Alfred W. McCoy, "A Queen Dies Slowly: The Rise and Decline of Iloilo City," in Alfred W. McCoy and Ed. C. de Jesus, eds., *Philippine Social History: Global Trade and Local Transformations* (Quezon City and Honolulu: 1982), 308–11. The telegraph system was a major accomplishment of the late Spanish regime.

4. Foreman, *The Philippine Islands*, quote 319.

5. Norman G. Owen, "Americans in the Abaca Trade: Peele, Hubbell and Co., 1856–1875," in Peter W. Stanley, ed., *Reappraising an Empire: New Perspectives on Philippine-American History* (Cambridge, Mass.: 1984), 215. The last full year cargoes were reported in detail was 1881.

6. Factoría was a river market center in southernmost Nueva Ecija.

7. On *palay Malabon*, see Consul Pauli, "Manila," Hs. of Com., 1880, v. 74, C2632, 1373–74. On Pangasinan shipments, see Fr. Jose Ma. Gonzalez, OP, *Labor*

Evangelica y Civilizadora de los Religiosos Dominicos en Pangasinan (Manila: 1946), 68–72.

8. Peter Xenos, "The Ilocos Coast since 1800: Population Pressure, the Ilocano Diaspora, and Multiphasic Response," in D. F. Doeppers and Peter Xenos, eds., *Population and History: The Demographic Origins of the Modern Philippines* (Madison and Quezon City: 1998), 47–50.

9. University of Santo Tomas Archives and Manila Archdiocesan Archives, Luzon parish *Libros de Entierros*; Renato Lopez, "History of Sta. Barbara in Pangasinan during Spanish Times," *Ilocos Review* 16 (1984): 10–11; RCUS, 1882, 9:28, 382. The same period witnessed four years of drought in neighboring Indonesia. On the form of diffusion, see Matthew Smallman-Raynor and Andrew D. Cliff, "The Philippine Insurrection and the 1902–4 Cholera Epidemic, pt. 1: Epidemiological Diffusion Processes in War; pt. 2: Diffusion Patterns in War and Peace," JHG 24.1–2 (Jan and Apr98): 69–89, 188–210.

10. Nueva Ecija sent out 42,500 *cavans* of rice equivalent in 1870, three-fourths in *palay* form, but *palay de Factoría* went unmentioned in *El Comercio*'s price reports as late as 1877. De la Cavada, *Historia, Geográfica*, 2:466.

11. "Estado de la agricultura filipina en 1887," BNM, ms. 19.218. Sailing vessels from Vigan landing with handicrafts and commodities to trade for rice feature prominently in oral accounts collected in Dagupan in 1969. As a result of this trade, a whole neighborhood of Ilocano residents sprang up in Dagupan's wharf district. D. F. Doeppers, "'Ethnic Urbanism' and Philippine Cities," AAAG 64.4 (Dec74): 553.

12. The database for 1891 is drawn from daily reports in *El Comercio*. These give one-word cargo descriptions, exclude secondary cargo items, often list only the province of origin, and contain small gaps that cannot readily be checked.

13. McLennan, *Central Luzon Plain*, 119. Prices for rice from Ilocos were given in 7 of 22 reports in *El Comercio* in 1891. Rice shipments from Ilocos Norte continued in 1906 and were also recorded from Zambales to both Ilocos Sur and La Union. RPC, 1906, pt. 1, 270, 479.

14. On Subic, see RPC 1904, pt. 1, 669–70.

15. Hs. of Com., 1892, v. 84, 1032, 9–10; 1893/94, v. 96, 1289, 16.

16. Owen points to the Hong Kong and Shanghai Bank (HSBC) as mobilizing "mostly British and German" capital to finance the railroad. Norman G. Owen "Fomento and the Free Market: The 19th Century Philippine Economy," in Ma. Dolores Elizalde, Josep M. Fradera, and Luis Alonso, eds., *Imperios y Naciones en El Pacífico* (Madrid: 2001), quote 1:143. A construction contract was awarded in 1887 to a British group headed by Edmund Sikes Hett. In 1888 the board of directors included the chairman of the Madras Railway Company, a member of the London committee of the HSBC, and other persons connected to British railroad and banking interests, as well as Smith Bell. Spain guaranteed an 8 percent annual return for 99 years. This company failed and was replaced by another British capital group in 1890.

17. William Stigand, "Manila," 1893, Hs. of Com., 1893/94, v. 96, 1289, quote 17.

18. Alexander R. Webb, "Projected Railways in the Philippine Islands," RCUS, 1888, v. 26, 94, quote 412; "El Primer Ferro-Carril en Filipinas," *El Resumen*, 29Mar1891; RPC 1900, 73 and 1904, pt. 3, 223. A third investor group, Speyer Brothers of London,

began acquiring a financial interest in the railway in 1902. In 1905 the U.S. Congress authorized interest guarantees on Philippine railway construction bonds at 4 percent for 30 years. The Speyer group won the bid for these guarantees and then reincorporated in the United States. Arturo G. Corpuz, *Colonial Iron Horse: Railroads and Regional Development in the Philippines, 1875–1935* (Quezon City: 1999), 53.

19. Foreman, *Philippine Islands*, quotes 319.

20. Advertisements in *El Comercio*, 25Oct–6Dec1891. Succeeding Peele Hubbell, Warner Blodgett also operated the British flag steamships *Diamante*, *Esmeralda*, and *Zafiro* on the Hong Kong–Xiamen–Manila run.

21. On capacity, see PNA, Contribución Industrial; and Garcia, "Cultivation of Rice."

22. Otto van den Muijzenberg, *The Philippines through European Lenses* (Quezon City: 2008), quotes 87. Warner Barnes used 470 shares in Luzon Rice Mills as loan collateral for the Manila branch of the HSBC, a loan still outstanding in 1916. Hong Kong and Shanghai Banking Corporation, "Statement of Condition, February 27, 1909," USNA II, BIA records, RG350, file 553; "Inspector's Report on the Manila Branch, October 14, 1916," HSBC Group Archives, London, LOH II, 124, folder Ig2. I thank Wigan Salazar for several references here.

23. *Under Four Flags: the Story of Smith Bell & Company in the Philippines* (Bristol: [1970s]), 22, 27–28, 42, quote 28. Bautista was part of Bayambang municipality until 1900. The number of Chinese in Bautista increased from 8 in 1895 to 137 in 1903. PNA, Padron de Chinos, Pangasinan, 1895; *Census 1903*, 2:257–58, 840–44.

24. The railway company resumed control on April 20, 1900. RPC 1904, pt. 3, 223 and facing 229. Mortgage records and a list of Smith Bell postings point to a mill at Calumpit operated by Smith Bell during 1905–16 and another at Dagupan that operated during roughly the same years but had closed by 1916. Legarda, *After the Galleons*, table 19, 329; PNA, Contribución Industrial, Tarlac, 1893–94, 1892–98, 1896–98, B1, and Pangasinan, 1896–97; "List of Smith Bell Employees," Ifor B. Powell papers, Special Collections, School of Oriental and African Studies, London; HSBC Group Archives, London, LOH II, 124, folder Ig2; Mears, *Rice Economy*, 127–28; *Census 1903*, 4:522–25; "Report of the Governor of Pangasinan," in ms. RPC 1907–8, 2 and 5.

25. Numerous provinces still had none, apparently including Nueva Ecija, Zambales, Bataan, Cavite, Tayabas, and Ilocos Sur. This is confirmed for Zambales in RPC 1904, pt. 1, 671.

26. PNA, Contribución Industrial, Pangasinan, 1896–98, and Batangas, 1894–97 (1); Reynaldo C. Ileto, "Hunger in Southern Tagalog, 1897–1898," in *Filipinos and Their Revolution: Event, Discourse, and Historiography* (Quezon City: 1998), 101.

27. PNA, Contribución Industrial; Conner, *Rice Culture*, 26. Cf. Foreman, *Philippine Islands*; Garcia, "Cultivation of Rice," 4:91; and Edgar Wickberg, *The Chinese in Philippine Life, 1850–1898* (New Haven: 1965), 103.

28. One water mill owner was Joaquin Arnedo, a rich landowner in Sulipan, Apalit, Pampanga. Gonzalez, *Cocina Sulipeña*.

29. Foreman, *Philippine Islands*, quote 319; PNA, Contribución Industrial; Cortes, *Pangasinan, 1901–1986*, 58.

30. Siok-hwa Cheng, *The Rice Industry of Burma, 1852–1940* (Kuala Lumpur: 1968), 95. Cf. LeClerc, *Rice Trade in the Far East*, 62–63.

31. The grain mill reproduced here in figure 2.2 was featured in ads of 24Jan and 17Jun1882. The graphic was copied from Ransomes' catalogs of April and July 1879 (University of Reading, Rural History Centre, Ransomes file G 97). See D. R. Grace and D. C. Phillips, *Ransomes of Ipswich: A History of the Firm and Guide to Its Records* (Reading: 1975), 4–7; and Klinck, "American vs. European Trade in the Philippine Islands," RCUS, 1882, vol. 8, 25½, 299–304.

32. *El Comercio*, 28Dec1896, 14Jan1898, 31Jan1898. Jose de Leon (1867–1939) later became one of the "planter millionaires . . . who profited enormously from their investment in the booming sugar industry" of the 1910s–30s. He became president of Pasudeco sugar milling and was eventually gunned down in a dispute over a financial split between millers and planters. Larkin, *Pampangans*, quote 309; "Slayers' Victims Noted," *Tribune*, 13Jul39, 11.

33. A few were from Germany, and milling equipment from there later became important. Wigan Salazar, "British and German Passivity in the Face of the Spanish Neo-mercantilist Resurgence in the Philippines, c. 1883–1898," *Itinerario* 21.2 (1997): 124–53.

34. Hs. of Com., 1889, v. 80, 494, 6, quote from Shelmerdine in Iloilo. *El Comercio* ran ads for the products of G. Buchanan, 4Jan1871, 22Apr1882, 12Jul1891; F. H. Sawyer, "Máquinaria para limpiar palay," 30Jan1891, and "Ingenio economico para arroz," 15Nov1891; and Barlow and Wilson 9Nov1873–23Sep1882. See also ads for Punzalan and Wilson in *El Resumen*, 5Jun1891; and for Fred Wilson & Co. in *La Vanguardia*, 21Jan22.

35. Well past 1900 rice for local consumption in the Mekong Delta was still husked by hand. An increase in small upcountry power mills there and in Thailand came at the end of World War I. They had appeared a decade earlier in Burma. Gourou, *Les paysans*, 518–23; Pierre Brocheux, *The Mekong Delta: Ecology, Economy, and Revolution, 1860–1960* (Madison: 1995), 68–69, 232n43; Norman G. Owen, *The Rice Industry of Mainland Southeast Asia, 1850–1914* (monograph), JSS 59 (Jul71): 119n60.

36. Ian Brown, *Economic Change in South-East Asia ca. 1830–1980* (Kuala Lumpur: 1997), 205.

37. The oil, or fat, in rice bran also contains vitamins A and E. Augustus P. West and Aurelio O. Cruz, "Philippine Rice-Mill Products with Particular Reference to the Nutritive Value and Preservation of Rice Bran," PJS 52.1 (Sep33): 32.

38. An English translation of Dr. M. Koeniger's "Ueber das Auftreten von Beriberi in Manila, 1882 und 1883," *Deutsches Archiv für Klinische Medezin* 34 (1884), was kindly provided by Ken De Bevoise (quote ms. 3). See also De Bevoise, *Agents of the Apocalypse*, chap. 5; Melinda S. Meade, "Beriberi," in Kenneth F. Kiple, ed., *The Cambridge World History of Human Disease* (Cambridge: 1993), 606–12; Vernon L. Andrews, "Infantile Beriberi," PJS 7B.2 (Apr12): 67–88; and A. J. Hermano and G. Sepulveda Jr., "The Vitamin Content of Philippine Foods, III," PJS 54.1 (May34): 61–74.

39. Cheng, *Rice Industry of Burma*, 208n23.

40. De Bevoise, *Agents of the Apocalypse*, 118–41. See also Owen, "Rice Industry," 97–100, 109–11.

41. Per capita rice production also fell dramatically in Java during the late 1880s and early 1890s. This affected Surabaya, the "busiest port and largest city" in the Dutch East Indies. H. W. Dick, *Surabaya, City of Work: A Socioeconomic History, 1900–2000* (Athens: 2002), quote 42. See also Anne Booth, "Rice Production in Nineteenth-Century Java," RIMA 19.1 (Winter 1985): 86, 93, 98–99.

42. Legarda, *After the* Galleons, 157–78.

43. Sawyer, *Inhabitants of the Philippines*, quote 130.

44. John R. Arnold, "Philippine Islands," SCR, 1914, v. 2, 80a, quote 3.

45. Xenos, "Ilocos Coast since 1800."

46. On Philippine population growth, see Doeppers and Xenos, *Population and History*, chap. 1; and Owen, *Prosperity without Progress*, 132.

47. Jean Philippe Hens, CVB, 1888, quote 1049 concerning 1887.

48. For instances of loss of land because of the loss of work animals, see Marshall McLennan, "Changing Human Ecology on the Central Luzon Plain: Nueva Ecija, 1705–1939," in Alfred W. McCoy and Ed. C. de Jesus, eds., *Philippine Social History: Global Trade and Local Transformations*, 57–90 (Quezon City and Honolulu: 1982), 77. For land uncultivated for lack of work animals long after the second wave, see Pablo Tecson, "Agricultural Conditions in Tarlac Province," PAgR 1.7 (Jul08): 301–4.

49. Legarda, *After the Galleons*, 157–78; de Bosch, "Sual," Hs. of Com. 1870, 64, C115, 218.

50. Wang, "Secular Trends of Rice Prices," especially table 1.1; Consul B. Robertson, "Canton" Commercial Report, 1876, Hs. of Com., v. 84, C1857, 3–4.

51. Specialists tend to restrict the term *El Niño* to the Ecuador-Peru region and use ENSO to refer to the broader manifestation involving the western Pacific and beyond.

52. See C. S. Ramage, "El Niño," *Scientific American* 254.6 (June 1986): 76–83; William H. Quinn et al., "Historical Trends and Statistics of the Southern Oscillation, El Niño, and Indonesian Droughts," *Fishery Bulletin* 76.3 (1978): 663–78; Quinn and Neal, "Historical Record of El Niño Events"; and Neville Nicholls, "Historical El Niño/Southern Oscillation Variability in the Australasian Region," in Henry F. Diaz and Vera Markgraf, eds., *El Niño: Historical and Paleoclimatic Aspects of the Southern Oscillation* (Cambridge: 1992).

53. The standard year designation in Quinn and Neal, "Historical Record of El Niño Events," is based on events in the southern hemisphere. South of the equatorial zone this relates to the northeast monsoon, whereas in the Central Luzon rice heartland it is the southwest monsoon six months later that produces most of the rain. One must add this lag in order to get the calendar year most appropriate to the possibility of Philippine drought.

54. Full records for Manila are available for the years 1865–98, 1900–1940, and 1951–75. Rainfall data were also consulted for Dagupan (1902–39, 1951–75), San Isidro, Nueva Ecija (1888–91, 1893–95, 1897, 1903, 1905–6, 1908–13), and Bolinao (1886–97). "Climatology," RPC, 1900, v. 4, 192, 211–14; ASEAN [Association of Southeast Asian Nations], *The ASEAN Compendium of Climatic Statistics* (Jakarta: 1982); Pablo Tecson, "Agricultural Conditions in the Province of Nueva Ecija," PAgR 1.2 (Feb08): 88–90; Tecson, "Agricultural Conditions in Tarlac Province," 301–4; Percy Hill and Kilmer O. Moe, "The Cultivation of Rice," UMJEAS 9.4 (Oct60): 112; and SBPI, 1920, 26–31.

55. I am combining the "strong" and "moderate+" categories in Quinn and Neal, "Historical Record of El Niño Events."

56. We are fortunate to have Quinn and Neal's careful recalibration of El Niño intensity (ibid., table 32.1).

57. See Huke, *Agroclimatic and Dry-Season Maps.*

58. This section is informed by consultation with geographer Robert Huke, April 24, 2000.

59. Larkin cites this as the cause of "an estimated loss of 40 percent over the previous year" in the combined rice and sugar harvest of Pampanga in 1909 (*Pampangans*, quote 208).

60. Quinn et al., "Historical Trends." The relationship between El Niños and drought is stronger than Quinn and his colleagues suspected since some moderate to strong El Niños that produced no drought in Java nevertheless had severe consequences in some parts of the Philippines (e.g., 1871 in Central Luzon and 1911–12 in Cebu).

61. Quinn and Neal, "Historical Record of El Niño Events," 638.

62. *Maize*, a term from Arawak, is a close approximation of the word used by Filipinos and many others: *mais* or *maiz*. It avoids confusion with *corn* as that term used in British English to mean generic grain. Jonathan D. Sauer, *Historical Geography of Crop Plants: A Select Roster* (Boca Raton: 1993), 228–36.

63. Alexander R. Webb, "The Philippine Islands," August 1, 1888, RCUS, 1889, v. 29, 101, quote 185; Alexander R. Webb, "Agricultural Products of the Philippines," August 16, 1888, RCUS, 1889, v. 31, 109, 277; "Vegetables and Fruits Prohibited," MT, 14Jan08; Fernandez, *Tikim*, 5–7.

64. Robert M. Zingg, "American Plants in Philippine Ethnobotany," PJS 54.2 (Jun34): quote 270. On gender-selective transmission, see Rafael Bernal, "The Mexican Heritage in the Philippines," *Africa, Asia, Latin America* 1 (19Apr65): 9–18. Joseph E. Spencer, "The Rise of Maize as a Major Food Crop Plant in the Philippines," JHG 1.1 (Jan75): 1–16; and Peter Boomgaard, "Maize and Tobacco in Upland Indonesia, 1600–1940," in Tania Murray Li, ed., *Transforming the Indonesian Uplands: Marginality, Power, and Production* (Amsterdam: 1999), 45–78.

65. Newson, *Conquest and Pestilence*, quote 47; G. H. Grist and Syed Abdul Rahman, "The Cultivation of Tenggala Padi," ABFMS 9.1 (Jan–Mar21): 19. In Spain maize used for animal feed and bread had become an important commercial crop by the mid-eighteenth century. Vicens Vives, *Economic History of Spain*, 512.

66. Islas Filipinas, Comisíon Central de Estadística de Filipinas [second notebook], 67; "Estado de la agricultura filipina en 1887," BNM, ms. 19.218. Records for Ilocos Sur, Albay, and Camarines Sur are missing from this source.

67. RPC 1912, quote 22; "Corn to Take Place of Rice," MT, 6Jan12, 7; Glenn A. May, "The Business of Education in the Colonial Philippines, 1909–1930," in Alfred W. McCoy and Francisco Scarano eds., *Colonial Crucible: Empire in the Making of a Modern American State* (Madison: 2009).

68. Miller, *Economic Conditions*, quote 52. Most maize varieties took 110 to 120 days to mature. J. F. Boomer, "Philippine Islands," SCR, October 27, 1917, 80a, 7, 21.

69. Lino L. Dizon, *Mr. White: A "Thomasite" History of Tarlac Province, 1901–1913* (Tarlac City: 2003), 136–40; "Cheap and Clean, 'Tiempo' Advocates Corn Bread," MT, 10Jul07, 6.

70. E.g., Catalino Valdezco offered a hand mill in the *El Comercio* Saturday supplement, 8Jan1881; and Labhart y Comp. ran ads in the issues of 17Jan1881 and 5Jan1884.

71. Alexander R. Webb, "Agricultural Products of the Philippines," RCUS, 1889, v. 31, 109, quote 277. The La Castellana ad ran in *El Comercio* on 1Feb1895. Early shipment and price data is from the *Gaceta* and later data from the column "Arribos" in *El Comercio*, 1Jul1894–28Apr1895. Cf., "Rice Famine," PAgR 4.10 (Oct11): 530–31. De la Cavada, in *Historia, Geográfica*, gives approximations of maize production by province in 1870, as well as arrivals in Manila by coastal shipping (52,000 kilos).

72. Pierre van der Eng, "Cassava in Indonesia: A Historical Re-appraisal of an Enigmatic Food Crop," SEAS 36.1 (Jun98): 3–31, esp. 28; Pierre van der Eng, "Trends in Indonesia's Food Supply, 1880–1995," JIH, 30.4 (2000): 598–600; "Cassava Possibilities," CIJ 9.1 (Jan33): 12–13.

73. Hens, CVB, 1888 (on 1887), quote 1049. Mallari describes village raids by landless "vagabonds and contrabandists" during times of rice shortages and high prices. Francisco Mallari, SJ, "The Eighteenth Century Tirones," PS 46.3 (1998): 293–312, quote 294.

74. Hens, "Rapport Annuel. Manilla," CVB, 1881, quote 46.

75. Peter C. Smith [Xenos], "Crisis Mortality in the Nineteenth Century Philippines: Data from Parish Records," JAS 38.1 (Nov78): quotes 75.

76. Felix M. Montemayor, *Anak Apo na Alaminos* (n.p.: 1983), 50; Emilio S. Tolentino and Tomas P. Maddela, "A Brief History of the Province of Nueva Vizcaya," Felipe S. Cortez, trans., JNL 2.2 (Jan72): 10–11.

77. C. R. B. Pickford, "Cebu," January 16, 1879, Hs. of Com., 1878–79, v. 72, C2421, quote 1607; George MacKenzie, "Manila," January 10, 1879, Hs. of Com., 1878/79, v. 70, C2285, quote 574–75; *Under Four Flags*, 22.

78. Maize had long since become the ordinary staple in Cebu, western Bohol, and parts of eastern Negros. The *Balanza* for 1879 reports the import of 44 small steam-powered mills for maize.

79. Hs. of Com., 1893–94, v. 96, 1289, quote 19–20; G. T. Ricketts, "Philippine Islands," Hs. of Com., 1870, v. 64, C115, 216; Buzeta and Bravo, *Diccionario Geográfico*, 1:548. Shipments data from the *Gaceta*, 8Aug and 12Nov1861 and 8Jan1862. Cf. Michael Cullinane and Peter Xenos, "Growth of the Population of Cebu during the Spanish Era," in D. F. Doeppers and Peter Xenos, eds., *Population and History: The Demographic Origins of the Modern Philippines* (Madison and Quezon City: 1998).

80. Vice Consul Gray, "Yloilo," Hs. of Com., 1878–79, v. 72, C2421, quote 1666–67. Over half of the direct rice imports to Iloilo in 1878 were from Japan. Cf. Gray, "Iloilo," Hs. of Com., 1880, v. 74, C2632, 1354.

81. One longs for a deeper historiography of this event such as that available for the weather-triggered famine in the Semarang Residency of Java in 1849–50. Significantly, "there were no food riots or demonstrations, no attacks on granaries, and no general overt unrest" in the Semarang area and the rural food crisis did not become a disaster for the urban food supply. Charles Tilly says something similar of Western Europe. Robert Elson, "The Famine in Demak and Grobogan in 1849–50: Its Causes and Circumstances," RIMA 19.1 (Winter 1985): 39–85, quotes 67 and 64, respectively; Tilly, "Food Supply and Public Order, 390–91.

82. Java-Madura experienced three years of drought from 1883 to 1885. In Thailand both 1884 and 1885 were bad years. H. Warington Smyth, *Five Years in Siam from 1891 to 1896*, 2 vols. (London: 1898), 2:267, 277; Hs. of Com., 1884, v. 80, C3964, 564; RPC 1900, pt. 4, 192; Coronas, *Extraordinary Drought*, 4.

83. Peter Boomgaard speculates that this may have happened in Java during the 1880s depression. On the "engrossing of lands by mestizos," see Wickberg, *Chinese in Philippine Life*, quote 143.

84. Quotes from Hs. of Com., 1892, v. 84, 1032, 11; and E. H. Rawson-Walker (quoting Acting Vice Consul William Sloan Fyfe), Hs. of Com., 1897, v. 93, 1932, 9. On the tariff of 1891, see Salazar, "British and German Passivity," 124–53.

85. Hs. of Com., 1892, v. 84, 1032, 9–10, and 1893/94, v. 96, 1289, 16. On the change of tariff effective July 20, 1895, see RCUS, 1896, v. 50, 561. RCUS, 1889, v. 31, 109, 367; and Mears et al., *Rice Economy*, chap. 2.

86. Nick Joaquin, *Manila, My Manila* (Makati: 1999), 199.

87. Lichauco, *Family Recollections*, 12–13; Ileto, "Hunger in Southern Tagalog, 102–6.

88. [Lt. Aime Ernest Motsch,] *The Diary of a French Officer on the War in the Philippines, 1898*, Marietta Enriquez de la Haye Jousselin, trans. (Manila: 1994), quotes 50, 58–59, 83, 87. Cf. *French Consular Dispatches*, Maria Luisa T. Camagay, trans. (Quezon City: 1997), 61.

89. Jose Roca de Togores y Saravia, *Blockade and Siege of Manila in 1898* (Manila: 2003/1908), 104, quotes 44, 94.

90. Richard Brinsley Sheridan, *Filipino Martyrs: Story of the Crime of February 4, 1899* (London and New York: 1900), republished in *Eyewitness Accounts in 1900*, vol. 3 of *Centennial Collection of Filipino Heroes* (Manila: 1998), quotes 20, 31.

91. "Arroz," *El Comercio*, 6Feb00. Imports were minimal in the three previous quarters. *El Comercio*, 7Apr1899.

92. Motsch, *Diary of a French Officer*, 97; Hs. of Com., "Philippine Islands," 1900, v. 97, 2436, Harford on 1899, 3; Phelps Whitmarsh, "Conditions in Manila," *Outlook* 63.18 (16Dec1899): 922–23, quote 917; Sheridan, *Filipino Martyrs*, 20.

93. This event also saw a considerable precipitation shortfall in Indonesia, much of Australia, India, and southern Africa.

94. May, *Battle for Batangas*, chap. 9; Peter Lewis, *Foot Soldier in an Occupation Force: The Letters of Peter Lewis, 1898–1902* (Manila: 1999), 115, 121–24.

95. Ho Ping-yin, "A Survey of Sino-Philippine Trade," PJC 10.6 (Jun34): 14.

96. "Arrivals," *El Comercio*, 18Nov01–8Mar02. After March 1902, the information was no longer routinely published. Cf. Firth, "Philippine Islands 1902," Hs. of Com. 1903, v. 79, 3044," 7–8.

97. On locust swarms in or following dry years and the efforts of agricultural bureaucracies to counter them, see C. R. Jones and D. B. Mackie, "The Locust Pest," PAgR 6.1 (Jan13): 5–22; and H. C. Pratt, "Locust Work in the Philippines, A Review," ABFMS 2.1 (Aug13): 69–70. On typhoon damage, see Norman Owen, "A Subsistence Crisis in the Provincial Philippines, 1845–1846," *Kinaadman* 8 (1986): 35–46.

98. K. R. Briffa et al., "Influence of Volcanic Eruptions on Northern Hemisphere Summer Temperature over the Past 600 Years," *Nature*, 4Jun98, 450–55; Clive Oppenheimer, "Climatic, Environmental, and Human Consequences of the Largest Known

Historic Eruption: Tambora Volcano (Indonesia), 1815," PPG 27.2 (2003): 230–59; RPC 1911, 27, 66.

99. "Report of the Governor of Pangasinan," ms. RPC, 1907–8, 5. On the effect of the railroad in the Central Plain, see Dick and Rimmer, "Islands," 145–46.

100. Benito Legarda Jr., personal communication, 6Feb2012. Legarda's *After the Galleons* remains one of the critical texts in Philippine economic history.

CHAPTER 3. THE MANILA RICE TRADE TO 1941

1. The transformation is abundantly documented in Marshall McLennan's *Central Luzon Plain* and "Changing Human Ecology."

2. A. H. Wells, F. Agcaoili, and R. T. Feliciano, "Philippine Rice," PJS 20.3 (Mar22): 356–57.

3. Cortes, *Pangasinan, 1901–1986*, 70–79.

4. Among others, see "Who Is Supreme in Nueva Ecija, the Gang or the Government?" PFP, 14Jan22. The Bureau of Lands bureaucracy was chronically understaffed.

5. Philippine National Archives. *Finding Aid, Erecciones*, vol. 5: *Pampanga-Zamboanga* (Manila, Philippine National Archives, 1989l), 23, 26, 67.

6. Such properties included estates in Guimba and Quezon belonging to the family of Casimiro Tinio (7,300 hectares in 1930 in large parcels). Martin Tinio Jr. recalls a total closer to 20,000 hectares plus a large cluster of others belonging to Casimiro's brother, Gen. Manuel Tinio in Licab, Sto. Domingo, Talavera, and Laur. "Statement Showing the Names of Persons or Companies Owning 500 Hectares or More of Agricultural Land," PFR 3.1–5 (Feb–Jun30): 26–27 each installment; Martin Tinio Jr., personal communication, 5Nov07.

7. By 1914 there were "not less than 130 [threshing] outfits in the Islands." H. O. Jacobson, *Rice in the Philippine Islands*, quote 6; "'Case' Maquinas Trilladoras de Arroz," *La Vanguardia*, 7Jan22, 2; Brian Fegan, "Entrepreneurs inVotes and Violence: Three Generations of a Peasant Political Family," in Alfred W. McCoy, ed., *An Anarchy of Families: State and Family in the Philippines* (Madison and Manila: 1993), 56, 58, 72.

8. "Modern Machinery Is Hernandez Cry," MT, 27Jul19.

9. John R. Arnold, "Philippine Islands," SCR, 1914, v. 2, 80a, 3–4, quote 3; McLennan, "Changing Human Ecology," 74.

10. PNA, Contribución Industrial, Nueva Ecija, 1894–97, 28Feb1895, quote; McLennan, *Central Luzon Plain*, 259. These records reveal that there were numerous small mills run by steam or animal power, but none is identified as a rice mill. On market centers, see Miller, *Economic Conditions*, 331.

11. Servicio Agronómico de Filipinas report for 1894. Nueva Ecija produced 8.0 percent of the national crop versus 17.7 for Pangasinan, 8.9 for Bulacan, 6.2 for Pampanga, 5.4 for Ilocos Sur, 5.3 for Iloilo, 4.9 for Laguna, and 4.2 for Capiz. "Estado demostrativo de la extensión cultivada de paláy y de la cosecha del mismo . . . 1894," PNA, Construcción de Casas, 1836–1898, SDS 11128 (kindly brought to my attention by Jean Paterno); *Census 1903*, 4:219; PAgR (Feb09): 113; PAgR (May11): 248; Hill and Moe, "Cultivation of Rice," 93.

12. Dick and Rimmer, "Islands," 141, drawing on McLennan, *Central Luzon Plain*, 187–89, 198, 326n52.

13. "The Need for Diversification in the Philippines," PA, 16.9 (Feb28): 561.

14. John A. Larkin, *Sugar and the Origins of Modern Philippine Society* (Berkeley: 1993), 96–97. The railroad showed a profit for 1920 and 1921, but by 1923 a decrease in freight tonnage had begun. Vanessa Jane Glynn, "Railroad Policy and Administration in the Philippines in the American Period, 1898–1924" (MA thesis, University of the Philippines, 1987), 185–90.

15. Wells, Agcaoili, and Feliciano, "Philippine Rice," 360–61.

16. Sawyer, *Inhabitants of the Philippines*, quote 17; Owen, "Fomento and the Free Market," quote 137; Dick and Rimmer, "Islands," 138.

17. James J. Halsema, *E. J. Halsema, Colonial Engineer: A Biography* (Quezon City: 1991), 61, 123; "Rice Arrivals in and Shipments from Manila," for July and August 1941, JPS 1.3–4 (Sep–Oct41): 152–53, 149–50; Dick and Rimmer, "Islands," 145.

18. Doeppers, *Manila, 1900–1941*, 39–41, fig. 5; "Rice Is Hurt by Trancazo," MT, 27Nov18; van der Eng, "Food for Growth," 602–3.

19. Bray, *Rice Economies*, 49, 148–55.

20. Akira Takahashi, *Land and Peasants in Central Luzon* (Honolulu: 1970), 49–52, 54–57; McLennan, *Central Luzon Plain*, 257; SBPI, 1929, 12, 59; Willem Wolters, personal communication, 14May99. Although it was not a common practice nationally, double cropping was widespread in relatively sandy lower Pampanga in 1907 and 1938. It was rare in Pangasinan and Nueva Ecija in 1938. Pablo Tecson, "Agricultural Conditions in the Province of Pampanga," PAgR 1.10 (Oct08): 432–37; and *Census 1939*, 2:1394–96.

21. "Delinquent Landowners in the Payment of Water Charges for 1933 of the Angat River Irrigation System," *La Opinión*, 17Nov34, 3–26.

22. Willem G. Wolters, "Population Growth, Irrigation, and Increasing Cropping Frequency in Java and Luzon," manuscript, 1999, 16–20; "A Record to Be Ashamed Of," PFP, 8Aug31, 2–3, 38; "Irrigation Activities of the Bureau of Public Works," QBBPW 6.2 (Jul17); "The Filipino Point of View: Irrigation," MT weekly ed., 20Apr19, 5; "Irrigation Is Doubled," *Herald*, 12Jan27; SBPI, 1929, 12, 59; Donald W. Fryer, *Emerging Southeast Asia: A Study in Growth and Stagnation* (New York: 1970), 178–80. The capacity for substantial second cropping finally came in the 1970s with the construction of the Pantabangan Dam in Nueva Ecija.

23. McLennan, *Central Luzon Plain*; Benedict J. Kerkvliet, *The Huk Rebellion: A Study of Peasant Revolt in the Philippines* (Berkeley: 1977).

24. The Philippines Constabulary was a "paramilitary police apparatus," a national instrument in the establishment and maintenance of colonial control and later of the indigenous elite order. Alfred W. McCoy, *Policing America's Empire: The United States, the Philippines, and the Rise of the Surveillance State* (Madison: 2009), quote 126.

25. David R. Sturtevant, "Epilog for an Old 'Colorum,'" *Solidarity* 3.8 (Aug68): 10–18; David R. Sturtevant, *Popular Uprisings in the Philippines, 1840–1940* (Ithaca: 1976), 183–92, esp. 188n26; Milagros C. Guerrero, "The Colorum Uprisings, 1924–1931," *Asian Studies* 5.1 (Jan67): 65–78, esp. 75.

26. Two generations of scholars have studied these sociopolitical phenomena. See, e.g., Kerkvliet, *Huk Rebellion*, esp. chap. 1; Brian Fegan, "The Social History of a Central Luzon Barrio," in Alfred W. McCoy and Ed. C. de Jesus, eds., *Philippine Social*

History: Global Trade and Local Transformations (Quezon City and Honolulu: 1982; Guerrero, "Colorum Uprisings"; and Vina A. Lanzona, *Amazons of the Huk Rebellion: Gender, Sex, and Revolution in the Philippines* (Madison: 2009).

27. "Internal Combustion Engines Important in [U.S.] Implement Exports," CR, 15Jul29, 160–62. Small waterwheel mills persisted in some hilly out-of-the-way places, especially in Laguna, Tayabas, and Albay. SBPI, 1918, no. 1, 45. Cf. Peter Timmer, "Choice of Technique in Rice Milling on Java: a Reply," BIES 10.1 (Mar74): 121–26.

28. PFP, Jan–Feb22, Spanish section; Philippines (Commonwealth), *Report of the Rice Commission to the President of the Philippines*, 1936, appendix 6, 40–50.

29. Robert E. Huke, *Shadows on the Land: An Economic Geography of the Philippines* (Manila: 1963), 237–39; Manila Railroad Co., *Report of the General Manager*, 1918, 1920–22; Mears et al., *Rice Economy*, 92, 127–28.

30. Cheng, *Rice Industry of Burma*, 93–95, quote 93.

31. Philippines (Commonwealth), *Report of the Rice Commission*, app. 6, 40–50.

32. Wong Kwok-Chu, *The Chinese in the Philippine Economy, 1898–1941* (Quezon City: 1999), 177; Allister MacMillan, ed., *Seaports of the Far East: Historical and Descriptive Commercial and Industrial Facts, Figures, and Resources*, 3rd ed. (London: 1926), 387, 368, 382; *Trabajo*, Feb30, 9; Jose E. Velmonte, "Palay and Rice Prices," PA 25.5 (Oct36): 394.

33. Henry Fraser and A. T. Stanton, "The Etiology of Beriberi," PJS 5B.1 (Feb10): 55–61, quote 59.

34. H. Campbell Highet, "Beriberi in Siam," PJS 5B.1 (Feb10): 73–79, quotes 75.

35. Ibid., quotes 77.

36. Victor Heiser, *An American Doctor's Odyssey* (New York: 1936), 205–10; RPC 1913, 109, quote 127; Jose Albert, "Treatment of Infantile Beriberi with the Extract of Tiqui-Tiqui," PJS 10B.1 (Jan15): 81–85.

37. M. Gutierrez and F. O. Santos, "The Food Consumption of 104 Families in Paco District, Manila," PJS 66.4 (Aug38): 397–416.

38. Aaron Rom O. Moralina, "State, Society, and Sickness: Tuberculosis Control in the American Philippines, 1910–1918," PS 57.2 (Jun09): 207; Bonnie McElhinny, "Producing the A-1 Baby: Puericulture Centers and the Birth of the Clinic in the U.S.-Occupied Philippines, 1906–1946," PS 57.2 (Jun09): 219–60. Manila mayor Juan Posadas Jr. (1934–39) sponsored a rapid expansion of puericulture centers. *Tribune*, 19Nov39, 6. The national recorded total of beriberi deaths during 1924–33 was 140,200 infants and 47,500 adults, and beriberi continued to be a leading cause of rural child mortality.

39. Schistosomiasis, another debilitating and potentially fatal disease, was discovered in the 1930s in some rice-growing locales ranging from eastern Leyte and northwestern Samar to Surigao. Caused by a parasitic blood fluke with a snail as intermediate host, it was not a factor in the areas routinely supplying rice to Manila in our period. Marcos A. Tubangui and Antonio M. Pasco, "Geographical Distribution, Incidence, and Control of *Schistosomiasis japonica* in the Philippines," PJS 74.4 (Apr41): 301–27.

40. Rainfall in June 1911 was also insufficient. The decline is calculated from Bureau of Agriculture production estimates for July 1911 through June 1912 as measured against

the five-year mean for the two preceding and two following years plus 1915–16. SBPI, 1918, no. 1, 56–57.

41. A second cluster of high rice import years emerged in Java during 1910–11.

42. Surely this was much better for the international rice trade than if they had consistently coincided. In the Philippines, the five worst years in terms of high levels of imports were 1903, 1912, 1902, 1904, and 1905, while the years of greatest net imports to Java and Madura were, in declining order, 1921, 1910, 1917, 1918, and 1922.

43. Java had its worst years in 1921–22, as well as experiencing high levels of imports in 1929. The five heaviest years of rice imports in each area during our period do not overlap—with or without a lag for the Philippines.

44. D. F. Doeppers, "The Philippines in the Great Depression: A Geography of Pain," in Peter Boomgaard and Ian Brown, eds., *Weathering the Storm: The Economies of Southeast Asia in the 1930s Depression* (Leiden and Singapore: 2000), 53–82.

45. "Rice Harvest for Batangas," MT, 30Jul19, quote 7; "Big Corn Crop to Reduce Rice Price," MT, 28Jul19.

46. "Enemies," MT, 27Jul19, weekly section, 5; "Sale by Sack [rather than *cavan*] Is Continuing," MT, 30Jul19, 1; "May Deport Chinese Exporters of Rice," MT, 1Aug19; and "Profiteers Stopped by Firm Action of Yeater," MT, 1Aug19.

47. See for example, Tilly, "Food Supply"; and Li and Dray-Novey, "Guarding Beijing's Food Security."

48. On the change of tariff, see RCUS, 1896, v. 50, 561; and 1889, v. 31, 109, 367.

49. "Review of Philippine Commerce for Calendar Year 1903," JAAA 4.6 (Jul04): quote 178.

50. RPC 1912, 22. Cf. Owen, *Prosperity without Progress*, 133; and ms. ARCC, 1919, 23.

51. Victor Buencamino, *The Memoirs of Victor Buencamino* (Mandaluyong: 1977), quote 238. Cf. Fegan, "Entrepreneurs in Votes and Violence," 66–67, 75. Victor's first cousin, Narcissa, married Jose de Leon, owner of a large rice estate centered in Talavera, Nueva Ecija. In the 1930s his brother, Felipe Jr., was the "boss of Nueva Ecija politics," a member of the National Assembly, and some thought on his way to the speakership until 1939. *Foto News*, 15May39: quote 46.

52. Alfredo Abes, "Rice Marketing Methods," CIJ 1.5 (May25): esp. 7. Cf. his more moderate "Are Local Rice Traders Profiteers?" CIJ 5.9 (Sep29): 6–7, 16. On rice policy made under the cloak of nationalism, see Percy A. Hill, "Conditions in the Rice Industry," PJC 10.12 (Dec34): 8–9, 28. On the lack of literature, see Velmonte, "Palay and Rice Prices," 382.

53. In addition to a term on the Philippine Commission, by 1922 Singson Encarnacion of the Compañia Mercantile de Filipinas had been an assemblyman and senator. In the 1930s he was a delegate to the Constitutional Convention and president of the Insular Life Association, the Philippine Guaranty Company, and the Balintawak Estate Co.

54. See "May Seize Rice in Tarlac Town" and "Government Will Import Rice from Saigon," both in *Tribune*, 21Sep35; Mears et al., *Rice Economy*, 9–11; and Fegan, "Entrepreneurs in Votes and Violence," 66–67.

55. V. Buencamino, *Memoirs*, quotes 243; "Big Rice Dealers Are in Quandary," *Tribune*, 17Oct36, quote. Cf. "Saigon Rice: Reduced Price of Imported Cereal Effective Today," *Tribune*, 15Sep36; "Import More Rice," *Tribune*, 23Sep36.

56. Mears et al., *Rice Economy*, quote 10; "Rice Mill, Bodegas Will Be Constructed," *Tribune*, 2Sep37; "Plan Modern Rice Central," *Tribune*, 11 and 17Aug38, 14. On the Dutch administration's attempt to intervene in the rice trade of Indonesia during 1917–20, see Boomgaard, "Welfare Services in Indonesia," 69.

57. Ian Coxhead, personal communication, 9Oct2002; Mears et al., *Rice Economy*, quotes 9–11. We need to know more about any licensing required for rice imports by private traders.

CHAPTER 4. CHANGING COMMERCIAL NETWORKS
IN THE RICE TRADE

1. Wickberg, *Chinese in Philippine Life*, 135, quote 63; Edgar Wickberg, "The Chinese Mestizo in Philippine History," JSEAH 5 (Mar64): 62–100.

2. One author surmises, "[I]t was very likely that the identities of the Chinese mestizos were multiple, fluid, and ambiguous." Richard T. Chu, *Chinese and Chinese Mestizos of Manila: Family, Identity, and Culture, 1860s–1930s* (Leiden and Boston: 2010), 276.

3. Russell & Sturgis received partial shipments from Pangasinan totaling 1,300 *cavans*—hardly a major presence. During 1862 Peele Hubbell and Russell & Sturgis each imported one major cargo of *arroz*.

4. Owen, "Americans in the Abaca Trade."

5. I am indebted to Norman Owen for generously sharing his research file concerning the "Peirce Family Papers," in particular George Peirce's ledger from his years in Albay, conserved in the Special Collections of the Stanford University Library. Peirce later lost everything in the trade depression of 1873–78. Ibid., 205–6.

6. On the *mayorazgo*, see *Teresa de la Paz and Her Two Husbands: A Gathering of Four Families* (Manila: 1996), 32–35; and Legarda, *After the Galleons*, 311, 315–18, and photo on 219–20.

7. Maria Teresa Colayco, *A Tradition of Leadership: Bank of the Philippine Islands* (Manila: 1984), 24–27, 210.

8. In Manila by 1800, "creole" was more an ethnic than a racial concept. Llobet, "Orphans of Empire," 25.

9. Aguirre y Cia. received a 20,000-peso loan from the Carriedo fund in 1862 using the sugar refinery as collateral. Alexander E. W. Salt, "Francisco de Carriedo y Peredo," PJS 8D.3 (Jun13): 181. See also Clarita T. Nolasco, *The Creoles in Spanish Philippines* (Manila: 1970), 54, 67; and *Gaceta*, 13Apr1862.

10. Legarda, *After the Galleons*, 219–20 (with photo), 314. Padilla exported 4,000 *cavans* of clean rice from Sual to Macao in 1838 and made further foreign trading ventures, including the sale of rice in Amoy. PNA, Erección de Pueblos Pangasinan, 1814–1883, folios 156–57. Cortes, *Pangasinan, 1801–1900*, 53–54; Rose Cortes, personal communication, 10Sep98; PNA, Mercados Publicos, Manila 1856–1901; "Espediente para subastar la obra . . ."; Martin Tinio Jr., personal communications, 4Oct07, 22Oct07, and 5Nov07.

11. Farren, "Philippine Islands," May 1, 1856, Hs. of Com., 1856, v. 57, quote 25.

12. Otto van den Muijzenberg, *Four Centuries of Dutch-Philippine Relations, 1600–2000* (Manila: 2000), 38–39; "Philippine-Dutch Social Relations, 1600–2000," *Bijdragen tot de Taal-, Land- en Volkenkunde* 157.3 (2001): 485–90. The 1861 *Guia de Forasteros* lists Petel as a French commercial house (132), but non-Spanish Europeans in Manila were often called "French."

13. From the outer zone as a whole, 22 Chinese are named as rice consignees.

14. Jose de Bosch, "Philippine Islands (Sual)," Hs. of Com., 1857 session 1, v. 16, C 2201, quote 530.

15. "Croquis Militar de la Parte de Pangasinan, 1830," in Angara, Cariño, and Ner, *Mapping the Philippines*, 122–23.

16. Cortes, *Pangasinan, 1801–1900* 54; Lopez, "History of Sta. Barbara," 112–14; de Bosch, "Philippine Islands (Sual)," Hs. of Com., 1857 session 1, v. 16, C 2201, 530.

17. De Bosch, "Philippine Islands (Sual)," Hs. of Com., 1857 session 1, v. 16, C 2201, 529–35; Magno, "Historical Retrospect," 340.

18. Owen, "Americans in the Abaca Trade," 220–25. My understanding of this system also owes much to conversations with Willem Wolters during 1999.

19. "Philippine Islands (Sual)," Hs. of Com., 1857 session 1, v. 16, C 2201, 529–35, quote 530.

20. Spanish letter in the collection of Don Peterson. I am indebted to Mr. Peterson for his assistance and to Courtney Johnson for translating the colloquial phrasing. Don Peterson, *Mail and Markings of Private Business Firms of the Spanish Philippines* (Eden: 1998). On the impact of telegraphy, see Stephen C. Lockwood, *Augustine Heard and Company, 1858–1862: American Merchants in China* (Cambridge: 1971), 104–5.

21. Arrechea organized rice shipments from Sual to Shanghai and Ningpo in the early 1860s, apparently in combination with Augustine Heard. Legarda, *After the Galleons*, 164; PNA, Calamidades Publicas—Naufragios 1843–1893, S172–S174B; Marciano R. de Borja, *Basques in the Philippines* (Reno: 2005), 92–93.

22. First-cousin marriage, or "family arranged marriage," was fairly common among people of property, a social device intended to avoid the division of assets among heirs. There were at least three such marriages among the mestizo Tuasons. *Teresa de la Paz*; Fegan, "Entrepreneurs in Votes and Violence," 55.

23. See Colayco, *Tradition*, 27, 209–10; Ninotchka Rosca, "The House That Ayala Built," in *Fookien Times Philippines Yearbook, 1975*, 158–63, 169; De Borja, *Basques*, 124–27; and *Gaceta*, 5 and 30Apr1871.

24. De Bosch, "Philippine Islands (Sual)," Hs. of Com., 1857 session 1, v. 16, C 2201, quote 532–34.

25. Vice consul Jose de Bosch, "Trade and Commerce of Sual for the Year 1866," Hs. of Com., 1867–68, v. 68, 3953, quotes 223.

26. The orthography of names was not standardized in nineteenth-century Manila. I employ here the form usually given in the *fincas urbanas* records and contemporary press. To her descendants, Cornelia's name is better spelled Laochangco. Cornelia Lichauco Fung, *Beneath the Banyan Tree: My Family Chronicles* (Hong Kong: 2009), 6–8, 18–19, 228; Cornelia Lichauco Fung, personal communications, 17Sep2007; 13Apr2009, and 16Apr2009.

27. Cornelia Lichauco Fung, personal communications, 12May and 31Oct2009, conveying reports by Rose Mendoza and transcripts: PNA, Informaciones Matrimoniales; Protocolos Manila 429 and 429 1873 Tomo 2 F. Dujua, SDS 20258, No. 613, S743B–744B (November 10, 1873).

28. *Lichauco Family Reunion, 1991* (n.p.: 1991), quote 3. Although Cornelia was a Filipino Chinese mestiza and all three of her husbands were Hokkien immigrants, there were no unions with Chinese in the next three generations of Lichaucos. Cf. De Viana, *Three Centuries of Binondo Architecture*, 234; Larkin, *Sugar*, 26; Wickberg, *Chinese in Philippine Life*, 94–96, 112; Alexander R. Webb, "The Sugar Industry of the Philippines," RCUS, 1889, v. 31, 109, 371–80.

29. *El Comercio*, 11May1881 and 29Aug1882, 3; Cornelia Lichauco Fung, *Beneath the Banyan Tree*; Luisa Fernandez Lichauco, *Family Recollections* (Manila: 1991); *Lichauco Family Reunion, 1991* (n.p.: 1991), 3, 16; PNA, Contribución Industrial, Manila 1892–97, S149; Fincas Urbanas, Quiapo 1890.

30. Colayco, *Tradition*, 80–82, 211; *El Comercio*, 27Nov1877, 1Mar1884, 20Aug1886, 28Jun1894; PNA, Servidumbre domestica, B2; Fincas Urbanas, Tondo 1891 and Santa Cruz 1891; *Guia oficial de las Islas Filipinas para 1895* (Manila: 1895), 192.

31. Severo Tuason (d. 1874) was the eldest son of José Maria Tuason. Fincas Urbanas, Caloocan-Navotas; "Hacienda de Maysilo," *Democracia*, 21May01, 2; *La Fábrica de Cerveza de San Miguel, Golden Jubilee, 1890–1940* (Manila: 1940); *Teresa de la Paz*, 55–56; Legarda, *After the Galleons*, 220, 277–78, 318. Doña Barbara Padilla owned two major Binondo *accesorias*, which were destroyed in the storm of 22Oct1882. De Viana, *Three Centuries*, 227.

32. "Relacion de los individuos. . . ," *Gaceta*, 4Jan1887; *El Comercio*, 24Feb1881, 9Dec1881, 28Dec1876, 30Dec1876; van den Muijzenberg, *Four Centuries*, 39; van den Muijzenberg, "Philippine-Dutch Social Relations," 485–86; Otto van den Muijzenberg *Philippines through European Lenses* (Manila: 2008), 76.

33. See Wickberg, *Chinese in Philippine Life*, e.g., 100–103; Owen, *Prosperity without Progress*, 173–74; Legarda, *After the Galleons*; and McLennan, *Central Luzon Plain*.

34. Wolters, undated conversation; W. G. Huff, "Bookkeeping Barter, Money, Credit, and Singapore's International Rice Trade," EEH 26.2 (Apr89): 161–89; Wickberg, *Chinese in Philippine Life*, 96–97, 100. Cf. McLennan, *Central Luzon Plain*.

35. De Bosch, "Philippine Islands (Sual)," Hs. of Com. 1857, quote 534; G. T. Ricketts, "Philippine Islands [1868]," Hs. of Com. 1870, v. 64, C115, 213.

36. Camps was a counselor of the Banco Español Filipino (1868–70, 1875–77). Colayco, *Tradition*, 211.

37. José Joaquin served as *alcalde mayor tercero* (one level below a vice mayor) of the province of Manila in 1862 and was managing director of the Banco Español Filipino in 1868–73 and 1876–84. *Gaceta*, 15Nov1862, 4; *El Comercio*, 14Sep1877, 8Nov1889, 3; Colayco, *Tradition*, 59n7, 210; De Borja, *Basques*, 131–37; *Manuel del Viajero en Filipinas* (Manila: 1875), 82.

38. In 1890 the Inchaustis were paying very substantial property taxes on their distillery in Tanduay, on five houses in Quiapo, and on a *camarin* and other substantial buildings along the Muelle San Fernando and on the Ysla de Romero. Fincas Urbanas.

39. Legarda, *After the Galleons*, 316, 330; Kunio Yoshihara, *Philippine Industrialization: Foreign and Domestic Capital* (Quezon City and Singapore: 1985), 113–14; William Gervase Clarence-Smith, "The Impact of 1898 on Spanish Trade and Investment in the Philippines," in Charles J-H. Macdonald and Guillermo M. Pesigan, eds., *Old Ties and New Solidarities: Studies on Philippine Communities* (Quezon City: 2000), 241.

40. G. T. Ricketts, "Philippine Islands [1868]," Hs. of Com. 1870, v. 64, C115, 217; N. Loney, "Philippines," Hs. of Com. 1862, v. 58, 3054, 342; McCoy, "A Queen Dies Slowly."

41. David Sturtevant estimates "El Porvenir" at 4,000 hectares. In 1930 it sprawled over parts of Tayug, Natividad, San Quintin, and Santa Maria in Pangasinan. There were numerous disputes over its boundaries. Land-grabbing by the powerful was a major issue triggering agrarian revolts in the 1920s and 1931, including that at Tayug. Sturtevant, *Popular Uprisings*, 182–92, esp. 188n26; lawphil.net/judjuris/juri1957/apr1957; "Statement Showing Names of Persons," PFR 3.5 (Jun30): 26–27. See also Philippine National Archives, *Finding Aid, Erecciones*, 5:23.

42. Wickberg, *Chinese in Philippine Life*, 135.

43. The same occurred in abaca exports. Owen, "Americans in the Abaca Trade," 226.

44. On the Roxas estates in Batangas, see Connolly, *Church Lands and Peasant Unrest*, 39, 71. On Pedro P. Roxas and agrarian relations on these estates, see May, *Battle for Batangas*, 21, 33–34, 73.

45. Outside the Pangasinan stream only one Chinese merchant was a notable player in the outer zone rice supply system in 1881: Jose Baura, who received cargoes from Zambales totaling 8,070 *cavans*.

46. De los Reyes, *El Folk-Lore Filipino*, quote 534–35.

47. Mariano later became the guardian of the persons and property of Alejandro's minor daughters. *Amparo Nable Jose, et al., Standard Oil Co. of New York, and Carmen Castro v. Mariano Nable Jose et al.*, Philippine Supreme Court case 7397, December 11, 1916 (lawphil.net/judjuris/juri1916). The Nable Jose family is the major basis for Wickberg's statement that the mestizos of Dagupan and Calasiao remained "the preeminent traders of their region." Wickberg, *Chinese in Philippine Life*, quote 135n19.

48. *Gaceta*, 8Apr1881; Wickberg, *Chinese in Philippine Life*, 135; PNA, Elecciones de Gobernadorcillos, Binondo 1837–98, SDS 14501, S 243, and Vecindario de Manila, Binondo Mestizos 1881 and 1884. Dagupan informants in 1969 recalled Mariano Nable Jose at the top of the local elite in the early twentieth century. PNA, Protocolos Pangasinan 1896 and Contribución Industrial, Pangasinan 1881–97, 1892–97, 1896–98; Dagupan congressman Angel Fernandez, Teofilo P. Guadez, and Amado Ll. Ayson, interviews with the author, April–May 1969; Restituto C. Basa, *The Story of Dagupan* (Dagupan: 1972), 12–13; John Bancroft Devins, *An Observer in the Philippines* (Boston: 1905), 82.

49. Among commercial houses, Smith Bell was first in sugar exports in 1881 and 1886 and first in abaca in 1886. Peele Hubbell was first and third in abaca exports in 1881 and 1886, respectively, and second in sugar exports in both years. Legarda, *After the Galleons*, table 19; *Under Four Flags*, 22.

50. PRO, FO 72/1890, British consuls and vice consuls. Later there was a notice for his substantial *camarin* and "beautiful property" along the river in Dagupan, as well as

24,000 *cavans* of rice on hand. "Dagupan," *El Comercio*, 3Nov1894. I thank Norman Owen for information on Heald, personal communication, 26Aug99. Cf. Ifor Powell, "The British in the Philippines in the American Era, 1898–1946," BAHC 9.2 (April–June 1981): 11, 18.

51. Owen, "Americans in the Abaca Trade," 227–29.

52. *El Comercio*, 22Feb1890.

53. *Lichauco Family Reunion*, 15. Later Amparo Nable Jose, a daughter of the late Alejandro and stepdaughter of Luisa, managed a tenanted estate of around 1,000 hectares located near Tayug, Pangasinan. Known as the Hacienda Hermanas Nable Jose (expropriated in 1959), it began as a section of El Porvenir. These estates were acquired by Donato Nable Jose and Macario Lichauco, respectively, apparently prior to 1885, in settlement of a gambling debt owed them by a Spaniard (lawphil.net/judjuris/juri1965/jul1965).

54. Lichauco, *Family Recollections*. Luisa is listed in *El Comercio* as a generous contributor to the funds for the family of the deceased commander of the Manila Police (4Jul1893) and for the erection of the Rizal Monument in Manila (6Feb02).

55. Nolasco, *Creoles*, 60–66. On Yangco's great house on the river in Binondo, see Fernando N. Zialcita and Martin I. Tinio Jr., *Philippine Ancestral Houses (1810–1930)* (Quezon City: 1980), 56.

56. PNA, Contribución Industrial, Ilocos Sur, 1891–95, 1893–94, 1896–97, 1894–98, 1896–98; *Rosenstock's Press Reference Library, Philippine Edition* (Manila: 1913), 116, 113; PNA, Padrones de Chinos, 1894. Encarnacion, who died a decade later, was also handling cargoes from the Ilocos ports of Vigan, Caoayan, Currimao, and several others in La Union in 1881 and 1891. I am indebted to Rose Marie Mendoza for supplying leads to much of this information. A failed suit against Sy Quia's estate by the grandchildren of his probable first wife in China is found in *Sy Joc Lieng, et al. v. Petronila Encarnacion et al., Philippine Reports* 16, 4718 (March 19, 1910), 145–287, affirmed by the U.S. Supreme Court in 1913 in 228 U.S. 335. Chu, *Chinese and Chinese Mestizos of Manila*, 228, 305–12. See also Alfonso Felix, ed., *The Chinese in the Philippines (1770–1898)* (Manila: 1969), 2:117–204; and lawphil.net/judjuris/juri1920/mar1920.

57. Separately the two owned and contracted out numerous indigo-processing facilities in four municipalities and between them owned three coastal sailing vessels.

58. Thirty years later, however, there were no Sys among the nine Chinese founders of the Tutuban Rice Exchange. Philippines (Commonwealth), *Report of the Rice Commission*, 1936, 29.

59. See Chu, *Chinese and Chinese Mestizos of Manila*. Sy Tay's wife was Chinese. When he died in 1901, the executor of his estate and guardian of his children was Sy Giang, a major entrepreneur in alcoholic beverages. Sy Tay's business included insurance, and he had become the owner of substantial houses in Intramuros, Ermita, and San Miguel districts. Philippine Supreme Court cases *John B. Early v. Sy Giang*, September 5 and 22, 1905, 1889, and 1890; *John B. Early and Edward N. White v. Sy Giang*, September 5, 1905, 2027 (lawphil.net/judjuris/juri1905); *Sy Joc Lieng, et al. v. Petronila Encarnacion, et al.* (lawphil.net/judjuris/juri1920/mar1920).

60. See Lockwood, *Augustine Heard*, 104–5.

61. The *Dagupan* has not turned up in surviving records, but Mariano Nable Jose registered another small steamer in 1897–98.

62. Manuel, *Dictionary of Philippine Biography*, 1:482–83.

63. D. F. Doeppers, "Destination Selection and Turnover among Chinese Migrants to Manila in the Nineteenth Century," JHG 12.4 (Oct86): fig. 4, 395.

64. Mariano Nable Jose ended up in debt to the Standard Oil Company. Eventually his riverfront property and storage bodegas were bought by the Spanish tobacco giant Tabacalera. "List of Vessels," in RPC 1904, pt. 3, 581–615; RPC 1905, pt. 4, app. C, 139–53. *Amparo Nable Jose, et al., Standard Oil of New York, and Carmen Castro v. Mariano Nable Jose, et al.*, Philippine Supreme Court case 7397, December 11, 1916 (lawphil.net/judjuris/juri1916).

65. In addition to Yap Tico, these were Serafin Tejuco, Yap Sioco, and Uy Lianfun. Of the 54 arrivals, 39 were described as rice and the other 15 as *con general*—surely also rice.

66. Wickberg, *Chinese in Philippine Life*, 103.

67. "Movimiento del Puerto," *El Comercio*, 4Nov01–8Mar02; *Under Four Flags*, 34.

68. Todd Lucero Sales, "Chino Bravo: The Story of Don Pedro Singson Gotiaoco," *Southwall* (Cebu) 1.3 (May–Jun2005): quotes 92–97; Mike Cullinane, personal communication, 7Sep05, including notes on an interview with prewar businessman Victoriano Go in Cebu City, 20May76. Fe Susan Go (Victoriano's daughter) confirms the substance of this interview (personal communication, 21Aug08). See also Rodrigo C. Lim, *Who's Who in the Philippines, Chinese Edition* (Manila: 1930), 153–54, 208.

69. In 1891 Yap Siocco organized the import of at least three cargoes from Saigon and advertised rice for sale that had arrived on a Norwegian freighter employed by Smith Bell.

70. On Yap Tico, see Wong, *Chinese in the Philippine Economy*, 47; McCoy, "A Queen Dies Slowly," quote 310; Wickberg, *Chinese in Philippine Life*, 85, 87–88; Alfred W. McCoy, "Ylo-ilo: Factional Conflict in a Colonial Economy, Iloilo Province, Philippines, 1937–1955" (PhD diss., Yale University, 1977), 1:87; and *El Comercio*, 26Jan and 6Dec01. Yap Seng was born in 1874 and came to the Philippines in 1888. Lim, *Who's Who in the Philippines, Chinese Edition*, 174, 178.

71. Consul Meerkamp, CVB, 1896 (on 1895), 177.

72. "Rice Hoards in City Big," MT, 8Aug19, 1.

73. Walter Robb, "Cabanatuan: Rocky Ford," in *Filipinos: Pre-war Philippines Essays* (Victoneta Park: 1963/1939), quote 95. On the Luzon Rice Mills mortgage in 1916, see Hong Kong and Shanghai Banking Corporation, "Inspector's Report on the Manila Branch, October 14, 1916," HSBC Group Archives, London, LOH II, 124, folder Ig2, reference courtesy of Wigan Salazar.

74. In addition to the mill established by Faustino and his brothers, there was an operating pool of 40,000 pesos put up mostly by three aunts (daughters of Cornelia Laochanco, sisters of Faustino's late father). An additional 5,000 pesos was put up by Mariano Nable Jose. When the manager neglected to provide a timely accounting and terminate the partnership, one of the investor aunts sued. *Eugenia Lichauco, et al. v. Faustino Lichauco*, Philippine Supreme Court case 10040, January 31, 1916 (lawphil

.net/judjuris/juri1916), quotes. *Amparo Nable Jose, et al. v. Mariano Nable Jose, et al.*, Philippine Supreme Court case 7397, December 11, 1916 (lawphil.net/judjuris/juri1916).

75. Daniel F. Asuncion, "A Study of Marketing Rice in Nueva Ecija," PA 21.3 (Aug32): 177–93, based on a 1929–30 survey of 620 rice land owners in twelve municipalities.

76. Velmonte, "Palay and Rice Prices," quotes 384, 390. Velmonte reports, "The large mills usually have their own separate clientele among growers, built up in the course of many years."

77. Martin Tinio Jr., personal communication, 11Nov2007. Basilia was the third wife and widow of General Manuel Tinio, former governor of Nueva Ecija and director of lands in 1913–16. Manuel and brother Casimiro were founders of two of the largest private collections of landholdings in Nueva Ecija.

78. Chinese owned 11 of 15 large bonded rice warehouses in Nueva Ecija in 1935. Filipinos owned 17 of 26 smaller facilities (processing less than 30,000 *cavans*). Velmonte, "Palay and Rice Prices," 407.

79. In 1904 the three Manila mills were owned by Chua Toco, Cue Jua Dy, and Co Piaco. Chua Toco succeeded Jose Garcia Chua Pinco in the 1890s, and Co Piaco succeeded Marcelo Nubla, a Philippine Chinese lawyer. Wong, *Chinese in the Philippine Economy*, 75, 134; RPC, ARMB, 1902–3, 225; *Rosenstock's Manila City Directory*, Oct–Dec 1904, 225; *Rosenstock's Manila City Directory*, Jul–Dec 1906, 306. Chinese comprised at least 59 of 66 "rice dealers" listed in *Rosenstock's Manila City Directory* for 1904, 223–25. At the sarisari store level in 1896–97, Chinese operated a dense network of 94 stores in Tondo, 42 in Santa Cruz, 40 in Intramuros, etc. PNA, Contribución Industrial, Manila, 1896, 2, 5, and 1896–97.

80. The exchange soon added maize and mongos/mung beans. Wong, *Chinese in the Philippine Economy*, 44–45, 160–62, 175–81; Alfredo Abes, "Rice Marketing Methods," CIJ 1.5 (May25): 4, 7; "Rice Distribution in the Philippines and the Tutuban Rice Exchange," PJC 12 (Apr36): 13, 36; Nemesio Lontoc, "The Present Rice Industry in the Philippines," UMJEAS 3.3 (Apr54): 299.

81. Wong names Co Leco in Sta. Rosa, Nueva Ecija, and Tan Sio in Cabanatuan as examples of urban rice dealers who invested in such upstream linkages in the 1910s and were still in business in 1935. Wong, *Chinese in the Philippine Economy*, 64. See also Velmonte, "Palay and Rice Prices," 407.

82. Wong, *Chinese in the Philippine Economy*, 39–40, 44–45, 133, 140, 152–53, 177, 201, quotes 44, 153; McLennan, *Central Luzon Plain*, 258. See also ads in *Rosenstock's Directory of China and Manila*, 1915; *Rosenstock's Directory of China and Manila*, 1916; and MT, 29Jul19, 9. See Yu's ads for Saigon rice in *El Comercio*, 15Nov and 4Dec01; *Ang Bayang Pilipino*, 28May14, 10; and *Bulletin*, 21Feb23, Chinese supplement, 4.

83. "Rice Hoards in City Big," 1. On the effects of inflation and deflation in the World War I cycle, see Doeppers, *Manila, 1900–1941*, 39–44. Yu was treasurer of the Philippine Chinese General Chamber of Commerce in Manila during 1907–10 and 1919–20 and a founding director of the China Banking Corporation.

84. Both companies adapted to the evolving legal environment by incorporating circa 1919–20, and both had multiple interests. Both men served as heads of the Chinese Chamber of Commerce. MacMillan, *Seaports of the Far East*, 400–401; Wong,

Chinese in the Philippine Economy; Wickberg, *Chinese in Philippine Life*, 87–88; Chu, *Chinese and Chinese Mestizos of Manila*, 257; Lim, *Who's Who in the Philippines, Chinese Edition*, 166; "Rice Hoards in City Big," 1.

85. Burma was an exception. See Cheng, *Rice Industry of Burma*, 82–85, 92, 228–29.

86. See S. Sugiyama and Linda Grove, eds., *Commercial Networks in Modern Asia* (Richmond: 2001), especially the commentaries by Ian Brown and Anthony Reid, 251–56, 261–64, respectively.

Chapter 5. Vegetables, Fruit, and Other Garden Produce

1. *Ulam* is pronounced *oo* as in the English in *loose*.

2. Doeppers, *Manila, 1900–1941*, 40.

3. Charles Wilkes, *Narrative of the United States Exploring Expedition during the Years 1838, 1839, 1840, 1841, 1842* (Philadelphia: 1845), 5:301–2; J. de Man, *Recollections of a Voyage*, 4; Hens, CVB, 1888, quote 1050. *Mabulo* is a Philippine evergreen with a dark red, fuzzy fruit used for its medicinal, including antidiarrheal, properties.

4. De Bevoise, *Agents of Apocalypse*, 58.

5. Doreen Fernandez, "The 10 Most Popular Filipino Dishes," *People*, 16Apr78, 8; Edilberto N. Alegre, "Poor Man's Fare," in Doreen G. Fernandez and Edilberto N. Alegre, *Sarap: Essays on Philippine Food* (Manila: 1988), 213; F. Lamson-Scribner, *List of Philippine Agricultural Products and Fiber Plants* (Manila: 1904); G. C. Lugod, "Wild Plants Used as Vegetables," in James E. Knott and Jose R. Deanon Jr., eds., *Vegetable Production in Southeast Asia* (Los Baños: 1967), 342–47; Elmer D. Merrill, *A Dictionary of Plant Names of the Philippine Islands* (Manila: 1903), 159. *Kangkong* is *Ipomoea aquatica* Forskal (or Hans); *malunggay* is *Moringa oleifera* Lam.

6. De Bevoise, *Agents of Apocalypse*, 58.

7. Gutierrez and Santos, "Food Consumption of 104 Families," 409, quote 398.

8. De Bevoise, *Agents of Apocalypse*, 56–57.

9. Super, *Food, Conquest, and Colonization*, 23. Cf. William Stigand, "Manila," Hs. of Com. 1893/94, v. 96, no. 1289; Foreman, *Philippine Islands*, 374.

10. Von Thünen, *Von Thünen's Isolated State*, 9.

11. Kenneth Kelly, "Agricultural Change in Hoogly, 1850–1910," AAAG 71.2 (Jun81): 237–52.

12. Buzeta and Bravo, *Diccionario Geográfico*, alphabetical entries; Cushner, *Landed Estates*, 41.

13. Buzeta and Bravo, *Diccionario Geográfico*, 2:393–94, quote 421. Bruce Cruikshank, *Spanish Franciscans in the Colonial Philippines, 1578–1898*, 5 vols. (Hastings: 2003), 2:349.

14. James H. Shipley, "Report on the Agricultural Conditions along the Line of the Manila and Dagupan Railroad," 9Jun02, RPC, pt. 1, 637–38.

15. Wilkes, *Narrative*, 5:301; Buzeta and Bravo, *Diccionario Geográfico*; Cruikshank, *Spanish Franciscans*.

16. Montero y Vidal, *El Archipiélago Filipino*, 308; "Cosas de Provincias," *Gaceta*, 1Jun1891.

17. Francisco Rodriguez, "Memoria general de la Provincia de Cavite en el año 1892," in Michael C. Francisco, trans., *Cavite en Siglo 19* (Cavite: 2002), 126, 130; Soledad

Borromeo, "El Cadiz Filipino: Colonial Cavite, 1571–1896" (PhD diss., University of California, Berkeley, 1973), 73; "Estado de la agricultura filipina en 1887," BNM, ms. 19.218; *Census 1903*, 4:125.

18. Santiago, "The Last Hacendera," 340–60; *Teresa de la Paz*; *Barretto v. Tuason, Philippine Reports*, v. 50, 23Mar26, 888–971; "Hacienda de Maysilo," *Democracia*, 21May01; Vicens Vives, *Economic History of Spain*, 626–27, 644–45. I also benefited from a memorable conversation with Benito Legarda Jr., Ang Gubat, Sampaloc, 30Jul96.

19. PNA, Estadistica Manila, Terrenos Agricolas, 1896. These included ten of twelve localities in the then huge expanse of Caloocan. See also "Statement Showing the Names of Persons," 26–27.

20. Benito Legarda Jr., interview with the author, 7Aug2008.

21. PNA, Estadistica de Manila, Terrenos Agricolas, 1896.

22. Victor S. Clark, "Labor Conditions in the Philippines," [U.S.] *Bulletin of the Bureau of Labor* 58 (May05): 806; RPC 1904, pt. 1, 590; Lamson-Scribner, *List of Philippine Agricultural Products*; J. R. Deanon Jr. and J. M. Soriano, "The Legumes," in James E. Knott and Jose R. Deanon Jr., eds., *Vegetable Production in Southeast Asia* (Los Baños: College of Agriculture, University of the Philippines, 1967), 66, 86–88.

23. Horatio de la Costa, SJ, *Jesuits in the Philippines, 1581–1768* (Cambridge: 1967), 133, 208, 514; Newson, *Conquest and Pestilence*, 124.

24. Aniano Eladya, "Report on the Chinese Market Gardening System," PAgR 20 (1927): 255–59. They may also have used night soil—human excrement—but it is not mentioned in this report.

25. See James Blaut, "The Economic Geography of a One-Acre Farm on Singapore Island," MJTG 1 (1953): 37–48.

26. Eladya, "Report on Chinese Market Gardening," 255–59; Francisco Agcaoili, "Some Vegetables Grown in the Philippine Islands," PJS 11A.3 (May16): 91–100.

27. De Bevoise, *Agents of Apocalypse*, 209–10n32; James A. LeRoy, *Philippine Life in Town and Country* (New York: 1905), 96.

28. Murphey, "How the City Is Fed," quotes 136–37. Cf. Yoshinobu Shiba, "Ningpo and Its Hinterland," in G. William Skinner, ed., *The City in Late Imperial China* (Stanford: 1977), 391–439, esp. 425.

29. Gourou, *Peasants of the Tonkin Delta*, 417.

30. Walter Buchler, "Sewage Treatment in the Philippine Islands," MS 6.5 (May35): 144; RPC 1904, pt. 1, 124; "Problems of a City Health Officer," PFP, 3Aug40, 24; Eladya, "Report on Chinese Market Gardening," 255–59.

31. Edwin Bingham Copeland, "Spanish Agricultural Work," PAgR 1.8 (Aug08): 308–11.

32. Luciano P. R. Santiago, personal communication, 31Jul2008.

33. Zingg, "American Plants in Philippine Ethnobotany," 224.

34. Foreman, *Philippine Islands*, 356.

35. Ugaldezubiaur, *Memoria Descriptiva*, 37–39; Francisco X. Baranera, SJ, *Handbook of the Philippine Islands*, Alexander Laist, trans. (Manila: 1899), 59; RPC 1904, pt. 1, 590; RPC 1906, pt. 1, 422; "Provincia de Bataan," *Gaceta*, 11Apr1861. The 1886–87 survey reports 3.7 million bunches of leaves grown in Pasay. "Estado de la agricultura filipina en 1887," BNM, ms. 19.218; PNA, Estadistica de Manila, Terrenos Agricolas, 1896.

36. Sawyer, *Inhabitants*, quote 227–28.

37. Maria Luisa T. Camagay, *Working Women of Manila in the 19th Century* (Quezon City: 1995), 28–33; Islas Filipinas, Comisión Central de Estadística de Filipinas, *20 cuaderno*, 1855; "La Buyera," *Ilustracion Filipina*, 15Jun1859, 61–64.

38. Francis A. Geologo, "The Philippines in the World Influenza Pandemic of 1918–1919," PS 57.2 (Jun2009): 261–92; Doeppers, *Manila, 1900–1941*, 41, 157n16.

39. Heiser, *American Doctor's Odyssey*, 112–13; *Commercial Handbook of the Philippine Islands* (Manila: 1924), 97; PJC 9.11 (Nov33): 14. Cf. D. F. Doeppers, "Metropolitan Manila in the Great Depression," 521–25; and Anthony Reid, "From Betel-Chewing to Tobacco-Smoking in Indonesia," JAS 44.3 (1985): 529–47.

40. Pangasinan native Napoleon Casambre was a major source on this topic, personal communication, 26Jan2005.

41. "Sabong," *Foto News*, 15Jul37, 22; M. L. Jovellanos, *Traditional Veterinary Medicine in the Philippines* (Bangkok: 1992), 25, 28.

42. Another was splitting massive quantities of fuelwood. Doeppers, "Lighting a Fire"; Bruce Cruikshank, "Commercial Patterns in the Province of Tayabas, c. 1820," manuscript, 22May2014, quote 6; A. F. Byars, "Coconuts in Laguna and Tayabas," PAgR 1.12 (Dec08): 516. During World War I there was a great boom in industrial coconut oil extraction for export—for munitions. Doeppers, *Manila, 1900–1941*, 24–25.

43. Doeppers, ""Lighting a Fire."

44. Minami Teisuke, "Report on the Philippine Islands, 1886," trans. and partially reprinted in Josefa M. Saniel, *Japan and the Philippines, 1868–1898* (Quezon City: 1969), 304–15, esp. 312. On this, see also Motoe Terami-Wada, "A Japanese Take Over of the Philippines," BAHC 13.1 (January 1985): 7–28, esp. 14.

45. Greg Bankoff, "Horsing Around: The Life and Times of the Horse in the Philippines at the Turn of the Twentieth Century," in Peter Boomgaard and David Henley, eds., *Smallholders and Stockbreeders: Histories of Foodcrop and Livestock Farming in Southeast Asia* (Leiden: 2004), 236; RPC 1906, pt. 3, 92.

46. RPC 1905, pt. 1, 485; *Census 1903*, 4:240–41; PSR 3.4 (1936): 318.

47. Robb, "Sunrise in Manila," 3.

48. Von Thünen, *Von Thünen's Isolated State*, 9.

49. Guano was advertised in Tagalog media in the 1910s for use as fertilizer on *zacate* grounds and was imported from at least the 1890s through the 1930s. "Guano para Zacate," *El Renacimiento*, 6Mar09; *El Comercio*, 1Apr01; *Taliba*, 15Feb10, 3; "Native Guano Sold in Manila," CIJ 1.4 (Apr25); "Guano Industry," CIJ 6.8 (Sep30): 9; "Guano," PJC 11.12 (Dec35): 18. In the 1890s the Cebu research station conducted fertilizer experiments using local bat guano. Copeland, "Spanish Agricultural Work," 308–11.

50. Ugaldezubiar, *Memoria Descriptiva*, 37–39; Fr. Felix de Huerta, *Estado Geográfico, Topográfico, Estadístico, Histórico-Religioso de la Santa y Apostólica Provincia de S. Gregorio Magno de Religiosos Menores Descalzos en las Islas Filipinas* (Binondo: 1865); "Forage Investigations in the Philippine Islands," PAgR 1.2 (Feb08): 71–82; PNA, Estadistica de Manila, Terrenos Agricolas, 1896; PNA, Prestacion Personal de Manila, mid-1880s.

51. Islas Filipinas, Comisión Central de Estadística de Filipinas, *20 cuaderno*, 67.

52. Of 412 hectares of *zacatales* within the city limits in 1918, just 9 remained in 1939. *Census 1918*, 3:397–98; SBPI, *1932*, 122–23; *Census 1939*, 2:1570; PJAI 2.4 (Jul–Aug35): 309.

53. "Copra Meal," PFP, 8Jul16, 33.

54. Bankoff, "Horsing Around," 240–41; Copeland, "Spanish Agricultural Work," 317.

55. William G. Clarence-Smith, "Diseases of Equids in Southeast Asia, c. 1800–c. 1945: Apocalypse or Progress?" in Karen Brown and Daniel Gilfoyle, eds., *Healing the Herds: Disease, Livestock Economies, and the Globalization of Veterinary Medicine* (Athens: 2010), 139. Cf. "Forage Investigations," 76, 79; "Investigations in the Philippines," DCTR, 14Feb12, 38, 676–77; Greg Bankoff, "A Question of Breeding: Zootechny and Colonial Attitudes toward the Tropical Environment in the Late Nineteenth-Century Philippines," JAS 60.2 (May2001): 413–37.

56. Liberate Tuaño, interview, 26Feb85. Others fed forage grass (*damo*) delivered from Cavite.

57. Cox and Arguelles, "Soils of the Island of Luzon," 16–24.

58. Mrs. Natividad Samio de Gamboa, interview, 15Apr85.

59. Filoteo S. Tuason, interview with Loreto Seguido, 24May85.

60. Arturo Bautista, interview, 4May85. See also Maria Cristina Blanc Szanton, *A Right to Survive: Subsistence Marketing in a Lowland Philippine Town* (University Park: 1972).

61. V. Buencamino, *Memoirs*, quote 17–18.

62. A. P. Laudico, "In the Grip of Usury," pt. 1: "Market Vendors," *Graphic*, 18Jan34, 10–11, 53–55; *Anti-usury Bulletin*, various issues, 1935–36; John T. Macloed, "Among the Usurers," in *The Sliding Scale and Other Philippine Sketches* (Manila: 1910), 71–80.

63. Garlic is sometimes used to treat elevated blood pressure and fever in humans. See Jovellanos, *Traditional Veterinary Medicine*, 20–21, 27.

64. Pedro Serrano Laktaw, *Diccionario Tagálog-Hispano* (Manila: 1914), 555; Juan de Noceda and Pedro Sanlucar, *Vocabulario de la Lengua Tagala* (Manila: 1860), 121; Sta. Maria, *Governor-General's Kitchen*, 114; Merrill, *Dictionary of Plant Names*, 122; Santos, *Vicassan's Pilipino-English Dictionary*, 820.

65. Sta. Maria, *Governor-General's Kitchen*, quotes 251.

66. See store licenses in PNA, Contribución Industrial, Nueva Ecija, 1893–99 (R2); Bataan, 1892–97; and Batangas, 1894–96 (2).

67. De la Cavada records only 136,000 kilos of *cebollas* received in Manila during 1870. De la Cavada, *Historia*, 1:385. See also Manuel Sastrón, *Batangas y su Provincia* (Malabon: 1895).

68. Stuart McCook, "'The World Was My Garden': Tropical Botany and Cosmopolitanism in American Science, 1898–1935," in McCoy and Scarano, eds., *Colonial Crucible*, 499–507.

69. Aniano Eladya, "Bermuda Onion," PJAg 6.2 (1935): 175; Pedro A. Rodrigo, "Preliminary Studies on the Use of Bermuda Onion as a Rotation Crop with Lowland Rice." PJAg 7.4 (1936): 317–25; Aniano Eladya, Adriano M. Orgas, and Francisco de Jesus, "The Economic Aspect of Growing Bermuda Onion in the Philippines," PJAg

7.4 (1936): 353–66; Harvey V. Rohrer, "The Philippine Market for Fresh Vegetables," CR, 9Dec29, 614.

70. Willem Wolters, personal communication, 4Sep2000; Willem Wolters, "Channel Organization and Collective Action in Philippine Onion Marketing," in Huub de Jonge and Willem Wolters, eds., *Commercialization and Market Formation in Developing Societies* (Saarbrucken: 1993), 94–115.

71. On Taal and Bauan, see Buzeta and Bravo, *Diccionario Geográfico*; Sastrón, *Batangas*, 101, 110, 116, 187; and RPC 1906, pt. 1, 193.

72. Mongo, or *balatong*, is *Phaseolus mungo* L. according to Merrill, *Dictionary of Plant Names*, 174.

73. RPC 1904, pt. 1, 590; "Precios Mercados," in PNA, Servidumbre Domestica 6, January 13, 1887. In the survey of 1886–87, Pangasinan reported raising 20,000 *cavans* of mongos; Batangas and Ilocos Norte also supplied some to Manila. "Estado de la agricultura filipina en 1887," BNM, ms. 19.218.

74. *Gaceta*, "Provincia de Batangas," 7Sep1862; *Gaceta*, Dec1880–Nov1881; Cavada, *Historia Geográfica*, 1:384–95. Cf. F. C. Kingman and E. D. Doryland, "The Principal Forage Crops of the Philippines." PAgR 10.3 (1917): 268–69.

75. "Report on Rice Culture in Zambales and Pangasinan," PAgR 2.5 (May09): 269; BPS 1–2 (1939): 66–67; Camus, "Rice," 37; *Census 1939*, 2:1545, 1169.

76. H. D. Gibbs and F. Agcaoili, "Soja-Bean Curd, an Important Oriental Food Product," PJS 7A.1 (Feb12): 47–53; H. D. Gibbs and F. Agcaoili, "Some Filipino Foods," PJS 7A.4 (Dec12): 398; Chu, *Chinese and Chinese Mestizos of Manila*, 50.

77. Sauer, *Historical Geography of Crop Plants*, 36–41.

78. Ugaldezubiaur, *Memoria Descriptiva*; Cavada, *Historia Geográfica*, 1:386; Felix Keesing, *Ethnohistory of Northern Luzon* (Stanford: 1962), 52, 65–66.

79. Henry T. Ellis, *Hong Kong to Manilla and the Lakes of Luzon in the Philippine Isles in the Year 1856* (London: 1859), 65; Margherita Arlina Hamm, *Manila and the Philippines* (New York and London: 1898), 62.

80. "Estado de la agricultura filipina en 1887," BNM, ms. 19.218. Records for Ilocos Sur, Zambales, Albay, and Camarines Sur are missing.

81. "Agriculture, Trade and Industry," PR 6.4 (Apr–May21): 222.

82. Agcaoili, "Some Vegetables," 94; "Ubes de Cebu," *El Comercio*, 26Jan1897.

83. Vicens Vives, *Economic History of Spain*, 513, 652–53.

84. This provides some (local) support for Josep M. Fradera's argument that the Spanish colonial government played a major role in assigning lands to specific commercial crops—an argument that for the colony as a whole may be overstated and misses the ability of local people to evade. Fradera, "Historical Origins of the Philippine Economy," 311–12. Cf. Cruikshank, "Commercial Patterns," 5.

85. Elsewhere these are known as *cajel* today. "Estado de la agricultura filipina en 1887," BNM, ms. 19.218. Both fruits were also reported as being grown in small quantities in Cavite and Basilan and in larger quantities in Calamianes for shipment to Iloilo. In one classification scheme, the botanist Elmer D. Merrill (*Dictionary of Plant Names*, 136) grouped a wide variety of Philippine citrus forms together as *Citrus aurantium* Linn, but he classified the *dayap*, *limon*, Lukban/*suha*, and a few others as separate species. Later the *kalamansi* was also recognized as separate.

86. May, *Battle for Batangas*, 42–47, 276–77; Sastrón, *Batangas*, 180–81, 191; RPC 1906, pt. 1, 193; RPC 1911, 6; Manuel, *Dictionary of Philippine Biography*, 1:268–74.

87. William S. Lyon, "Commercial Orange Production," PAgR 1.2 (Feb08): 101; Mariano Cruz, "Orange Cultivation in Batangas Province," PAgR 2.6 (Jun09): 311–16; A. H. Wells, F. Agcaoili, and Maria Y. Orosa, "Philippine Citrus Fruits," PJS 28.4 (Dec25): 453–527.

88. P. J. Wester, "The Situation in the Citrus District of Batangas," PAgR 6.3 (Mar13): 127.

89. Cruz, "Orange Cultivation," 313. See also E. Arsenio Manuel and Magdalena Avenir Manuel, *Dictionary of Philippine Biography* (Quezon City: 1995), 4:559–70; H. D. Gibbs and F. Agcaoili, "Philippine Citrus-Fruits," PJS 7A.6 (Dec12): 403–15; Lyon, "Commercial Orange Production," 101; RPC 1902, pt. 1, 642; and ms. ARBAg 1923, in ms. RGGPI, 1923, 3:103–8.

90. These are *Citrus nobilis Loureiro*, *Citrus sinensis Osbeck*, and *Citrus grandis Osbeck*, respectively.

91. Francisco Ignacio Alcina, SJ, *Historia de las Islas e Indios de Bisayas*, Ms.: 1668, Fr. Cantius J. Kobak, OFM and Fr. Lucio Gutierrez, OP, trans., "On the Many Different Varieties of Oranges . . .," PSacra 88 (Jan95): 155–71. The seventeenth-century Jesuit missionary Alcina was a remarkable scientific generalist who reported on natural and agricultural phenomena observed during a long residence in the Visayas. See Huerta, *Estado Geográfico*, 73, 77–79, 111–12.

92. *Census 1903*, quote 4:140. See also Wells, Agcaoili, and Orosa, "Philippine Citrus Fruits." Green in commerce, it can be orange when fully ripened.

93. Alfred W. Crosby, *The Columbian Voyages, the Columbian Exchange, and Their Historians* (Washington, D.C.: 1987).

94. Only Nueva Vizcaya and Benguet reported *patatas* in the survey of 1886–87. Benguet reported 2,000 *cavans*, with some sold in Manila, La Union, and Pangasinan. "Estado de la agricultura filipina en 1887," BNM, ms. 19.218. *El Comercio*, 10Dec1873 and 2Jan1889.

95. Foreman, *Philippine Islands*, quote 355; Sawyer, *Inhabitants*, quote 260.

96. Martin W. Lewis, "Agricultural Regions in the Philippine Cordillera," GR, 82.1 (Jan92): 32. Zingg reports white potatoes grown at 600 to 1,200 meters in "American Plants," 239.

97. RPC 1904, pt. 2, quotes 67–68; Martin W. Lewis, *Wagering the Land: Ritual, Capital, and Environmental Degradation in the Cordillera of Northern Luzon, 1900–1986* (Berkeley: 1992), 95–98.

98. Robert R. Reed, *City of Pines: The Origins of Baguio as a Colonial Hill Station and Regional Capital* (Berkeley: 1976; Baguio: 1999); Warwick Anderson, *Colonial Pathologies: American Tropical Medicine, Race, and Hygiene in the Philippines* (Durham: 2006), chap. 5, "The White Man's Psychic Burdens."

99. Reed, *City of Pines*; Halsema, *E. J. Halsema*, 190.

100. Richard Hooley, however, maintains that the American regime failed to establish an effective program of agricultural extension work and research results were slow to find farm application. "American Economic Policy in the Philippines, 1902–1940: Exploring a Dark Age in Colonial Statistics," JAE 16 (2005): 471.

101. Charles L. Cheng and Katherine V. Bersamira, *The Ethnic Chinese in Baguio City and in the Cordillera* (Baguio: 1997), 143–44, 170, 185–87.

102. Pedro A. Rodrigo, "A Survey of the Vegetable Industry in Baguio and the Trinidad Valley, Mountain Province, Luzon," PJAg 4 (1933): 179–99, quote 188. (This was not the only anthropogenic soil.)

103. Halsema, *E. J. Halsema*, 175, 220, 274; Manila Railroad Company, *Report of the General Manager, 1930*, 50; "Ah Gong's," *Bulletin*, 11May35; Lewis, *Wagering the Land*, 95–98.

104. Lewis, "Agricultural Regions," 41. Renato Rosaldo, "The Years of War, 1941–45," in *Ilongot Headhunting, 1883–1974: A Study in Society and History* (Stanford: 1980), 120–34.

105. See advertisements in *El Comercio*, 20Jan and 3Sep1881, quotes; notice "Cebollas de Bombay," 20Dec1880, *Balanza*; "Splendid American Onions," MT 18Feb1899; and RPC 1904, pt. 2, 66. Philippine production was reportedly less than a million kilos in 1918 and less than three million in 1938. *Census 1918*, 3:404–5; *Census 1939*, 2:1561.

106. Hs. of Com. 1899, v. 98, 2281, "Trade of Amoy," 13.

107. Pedro A. Rodrigo, "Garlic Culture in the Philippines," PJC 11.12 (Dec35): 13–14; Henry T. Lewis, *Ilocano Rice Farmers: A Comparative Study of Two Philippine Rice Barrios* (Honolulu: 1971), 57–58; "Onions, Potatoes, Garlic, and Groceries," PJC 9.12 (Dec33): 14; Rohrer, "Philippine Market," 615; "Facts and Figures about Philippine Trade with China," PJC 11.3–5 (May–Mar35): 18.

108. Walter Robb, "I Weep for the Chinese," *Harper's*, Sep50, 59. Cantonese speakers made up 10 percent of Manila's Chinese community in the 1890s.

109. Rohrer, "Philippine Market," CR, 9Dec29, quote 614; Shinzo Hayase, "Japanese Goods and Their Dealers under the U.S.-Philippine Free Trade System," read in manuscript, 2002; Shinzo Hayase, "A Study of Early Popular Consuming Society: The Philippines and Japanese Goods under American Colonial Rule," paper presented at the Seventh International Philippine Studies Conference, Leiden, June 2004; *Philippine Yearbook*, 1936–37, 167.

110. Rohrer, "Philippine Market," quote 614.

111. Wong Chen-ta, "Vegetable Farming in Hong Kong, a Study in Agricultural Geography" (PhD diss., University of Hong Kong, 1971), 247–49; Kenneth Ruddle and Gongfu Zhong, *Integrated Agriculture-Aquaculture in South China* (Cambridge: 1988); Wong Chor-yee, "Proto-industrialization and the Silk Industry of the Canton Delta, 1662–1934" (PhD diss., University of Wisconsin–Madison, 1995).

112. Alfred Lin, personal communication, 26Feb2002. See also Alfred H. Y. Lin, *Rural Economy of Guangdong, 1870–1937: A Study of the Agrarian Crisis and Its Origins in Southernmost China* (London and New York: 1997), 77. Lin suggests that some of the produce may have come to Hong Kong from Swatow since immense quantities of vegetables and eggs were sent out from there. Hs. of Com. 1905, v. 88, 3455, "Trade of Swatow 1904," 8.

113. W. G. Huff, *The Economic Growth of Singapore: Trade and Development in the Twentieth Century* (Cambridge: 1994), 155.

114. He later went on to a useful, if controversial, career in world public health with the Rockefeller Foundation. For an incisive critical view of Heiser, see Anderson, *Colonial Pathologies*.

115. Heiser, *American Doctor's Odyssey*, quote 104, also 169; Victor G. Heiser, "The Outbreak of Plague in Manila during 1912," PJS 8B.2 (Apr13): 115.

116. Simone Bijlard, "Battered Beauties: A Study of French Colonial Markets in Cambodia," *Newsletter of the International Institute of Asian Studies* (Leiden), no. 59 (Spring 2012): 4–5.

117. RPC 1903 pt. 1, 651; 1904 pt. 1, 242; 1913, 111–12. The three markets still accounted for 81 percent of total collections two decades later. ARCIR 1919, 32.

118. "Vegetables and Fruits Prohibited," MT, 14Jan08. On vendors, see Camagay, *Working Women*.

119. Thomas W. Jackson, "Sanitary Conditions and Needs in Provincial Towns," PJS 3B.5 (Nov08): 432–33; "Hordes of Big Green Flies," *Tribune*, 29Jun30, 5.

120. Tomas Confesor, "To Mayor Posadas," *The Critic*, 6Nov34, quote 7; Carlos Quirino, *Amang: The Life and Times of Eulogio Rodriguez Sr.* (Quezon City: 1983), 85; "Problems of a City Health Officer," 24.

121. "Colorum Markets Imperil Health of City Residents," *The News (Behind the News)*, 25Jun39, 3; "That Divisoria Market Row," 10–11; "Avert Fight at Board Session," *Bulletin*, 22Jan31.

122. Heiser, *American Doctor's Odyssey*, quote 112–13.

123. Julius G. Voight, "Fruit Culture in the Philippines," RCUS, v. 12, 1884, no. 41½, quote 776.

124. J. Bartlett Richards, "Philippine Economic Conditions, Annual Report for 1936," in Miguel R. Cornejo, ed., *Cornejo's Commonwealth Directory of the Philippines, 1939* (Manila: 1939), 636.

125. Harvey V. Rohrer, "Market for United States Fresh Fruits in the Philippines," CR, 29Apr29, 278–79. Oranges were the leading Chinese fruit export to Manila in the late 1930s at 2 to 3 million kilos per year. Philippines (Commonwealth), Department of Agriculture and Natural Resources, Statistics Division, BPS, 1939, 1–2 qtrs., 183.

126. Gibbs and Agcaoili, "Philippine Citrus Fruits," PJS (1912): 413–14; Wells, Agcaoili, and Orosa, "Philippine Citrus Fruits," PJS 28.4 (Dec25): 516–20. See "Ah Gong's" and "International Cold Stores," *Bulletin*, 11May35 and 17Jan31, respectively; Philippines Cold Stores ad, PFP, 6Jan08.

127. "Oranges," CIJ 6.10 (Oct30): quote 15. See also "Sunkist," *Taliba*, 10Dec41, 31.

128. The ad appeared in *El Comercio*, 20Dec1880. Many similar ads were placed by the La Malagueña, La Marina, and El Globo stores, e.g., on 12Dec1890, 10Nov1894, 6Dec1894, and 26Dec1896. See also Isaac Elliot, "Oriental Market for Dairy Products and Fruits," RCUS, 1895, v. 48, 177, 205.

129. Nick Joaquin, ed., *Intramuros* (Manila: 1988), 45; Hs. of Com. 1904, v. 101 pt. 2, "Trade and Commerce of the Consular District of Portland," 7, 22; American Grocery and Ah Gong's ads for fresh grapes and apples, *Bulletin*, 11May35; "American Apples for the Philippines," CR, 1Jun15, 987; CR, Daniel Moriarity, "International Trade in Apples," CR, 26Jul26, 217; Rohrer, "Market for United States Fresh Fruits," 278–79.

130. Vicens Vives, *Economic History of Spain*, 649–51, quote 649.

131. Rohrer, "Market for United States Fresh Fruits," quote 280. See also "Uvas Frescas," Almacen de Binondo, *El Renacimiento*, 3–5Jan05.

132. Harvey V. Rohrer and Carl H. Boehringer, "Importance of the Philippine Market for Canned Foodstuffs," CR, 2Sep29, 623–29, quote 625; Rohrer, "Merchandising American Products in the Philippines," 19May30, 404; Elliott, "Oriental Market," RCUS, 1895, v. 48, 177, 205; E. A. Burlingame Johnson, "American Fruits in the Orient," Amoy, RCUS, 1899, v. 61, 229, 313–14; "Philippine Tariff and the Food Supply," JAAA 3.1 (Feb03): 16; Hamm, *Manila and the Philippines*, 16.

CHAPTER 6. FISHING AND AQUACULTURE

1. Gonzalez, *Cocina Sulipeña*, 40–43.
2. Gutierrez and Santos, "Food Consumption," 405, 414.
3. Gibbs and Agcaoili, "Soja-Bean Curd," 47–53; Gibbs and Agcaoili, "Some Filipino Foods," 398. *Taho* is derived from the Hokkien *tahu*.
4. Gutierrez and Santos, "Food Consumption," quote 400. The season of the survey was October to March.
5. Super, *Food, Conquest, and Colonization*, 29–31.
6. Carl O. Sauer, *Agricultural Origins and Dispersals: The Domestication of Animals and Foodstuffs* (Cambridge: 1969), 24–32.
7. On the central role of fish in the dietary of Southeast Asians, see Antonio de Morga, "Report on Conditions in the Philippines," Manila, June 8, 1598, in Emma H. Blair and James A. Robertson, eds., *The Philippine Islands, 1493–1898*, 55 vols. (Cleveland: 1903–7), 10:85; Anthony Reid, *Southeast Asia in the Age of Commerce, 1450–1680*, vol. 1 (New Haven: 1988)28–29; *Philippines, Nutrition Survey of the Armed Forces, a Report by the Interdepartmental Committee on Nutrition for National Defense* (Washington, D.C.: 1957).
8. At Obando, just north of the city, there is a statue that is the focus of a famous waterborne procession known in 1850 and later, to Jose Rizal, as Our Lady of Salambao. Buzeta and Bravo, *Diccionario Geográfico*; Jose Rizal, *Noli me Tangere*, Soledad Lacson-Locsin, trans. (Manila and Honolulu: 1997), 39.
9. Fish caught by hook and line with the use of lanterns contributed significantly to the commercial flow sent to Manila from southern Luzon in the 1930s. Agustin F. Umali, "The Fishery Industries of San Miguel Bay," PJS 63.2 (Jun37): 229, 244; Agustin F. Umali, "The Fishery Industries of Ragay Gulf," PJS 65.3 (Mar38): 179, 191–92.
10. Paula Carolina Malay, "Some Tagalog Folkways," UMJEAS 6.1 (Jan57): 69–88, quote 75. On crabs, see Alvin Seale, "Preservation of Commercial Fish and Fishery Products in the Tropics," PJS 9D.1 (Feb14): 4. On mussels, see Deogracias V. Villadolid, "The Fisheries of Lake Taal, Pansipit River, and Balayan Bay, Luzon," PJS 63.2 (Jun37): 193.
11. Deogracias V. Villadolid, "Kanduli Fisheries of Laguna de Bay, Philippine Islands," PJS 54.4 (Aug34): quotes 546–47; F. Landa Jocano and Carmelita E. Veloro, *San Antonio, A Case Study of Adaptation and Folk Life in a Fishing Community* (Quezon City: 1976), 89; "No Epidemic Here," *Bulletin*, 9May29. On pollution, see Xavier Heutz de Lemps, "Waters in Nineteenth Century Manila," PS 49.4 (2001): 488–517.
12. Antonio de Morga, *Sucesos de las Islas Filipinas*, in Emma H. Blair and James A. Robertson, eds., *The Philippine Islands, 1493–1898*, 55 vols. (Cleveland: 1903–7), 16:96; Benjamin F. Matthes, *Boegineesch-Hollandsch Woordenboek and Ethnographischen Atlas* (The Hague: 1874), atlas, pl. 13a, fig. 1, on fish traps in Indonesia.

13. In 1862 *diliman* arrived primarily from Mindoro and Nasugbu, Batangas. In the 1930s, it was gathered in Laguna. Deogracias V. Villadolid and Mamerto D. Sulit, "A List of Plants Used in Connection with Fishing," PA 21 (Jun32): 26.

14. See PNA, Estadisticas Manila 1896–1898, Navotas, Estado Urbano-Agricola-Comercial, 1896; PNA, Vecindario Provincia de Manila, Navotas, 1877–80.

15. The jumbled *baklad* tax registration numbers for Manila Bay end at 511, suggesting that the surviving record is incomplete. Cristina Bernabe copied the individual tax records. PNA, Corrales de Pesca Manila, Junta de Obras del Puerto de Manila, producto del arbitrio del corrales de pesca, January–June 1882.

16. The *baklad* tax was sharply graded: 2 pesos for the common half-meter category, 12.50 for 2.5 meters, and 40 for 3 meters, proceeding by 15-peso increments up to 115 for 8 meters.

17. Villadolid, "Fisheries of Lake Taal," 215, 217, 222–25, quotes 222. PNA, Distrito de Morong, Relacion del numero de Corrales de pesca plantados en . . . Laguna de Bay," July 15, 1887.

18. PNA, Corrales de Pesca Cavite, "Relacion nominal de las licencias de corrales de pesca in la Bahia de esta Capital . . . ," 1892; Rodriguez, "Memoria general de la Provincia de Cavite," 124.

19. Teodoro G. Megia, "An Oceanographic Study of the Fisheries of Manila Bay," PFY (1953): 104–7.

20. Ragay Gulf abuts Camarines Sur and southern Quezon provinces. Umali, "Fishery Industries of Ragay Gulf"; Umali, "Fishery Industries of San Miguel Bay," 236; Agustin F. Umali, "The Fishery Industries of Southwestern Samar," PJS 54.3 (Jul34): 377, 382; Vicente C. Aldaba, "Manila Bay," PJS 47.3 (Mar32): 405.

21. Fortunata Buzon de Guzon and her sister Josepha, interviews, 6 and 17May85; PNA, Vecindario Provincia de Manila, Tondo Mestizos, 1889 and 1891; PNA, Corrales de Pesca Manila, Junta de Obras, March 1882, no. 343, and January 1883, no. 86. In the 1880s only four *cabezas* listed their occupation as *pescador* out 119 *cabecerias* of *naturales* and mestizos in Tondo. PNA, Vecindario Provincia de Manila, Tondo; PNA, Corrales de Pesca, Cavite, 1892.

22. Greg Bankoff, "Spanish Attitudes toward the Tropical Forests of the Philippines, 1863–1898," in Greg Bankoff and Peter Boomgaard, eds., *A History of Natural Resources in Asia: The Wealth of Nature* (Houndsmills: 2007).

23. Another notable *baklad* owner was shipping entrepreneur Luis R. Yangco, who owned three deep-water corrals off Tondo's Bankusay neighborhood in the 1890s.

24. *Census 1939*, 4:429, 485; USNA II, photo RG18-AA, box 179, file 12.

25. See "Los corrales de pesca en bahia su reglamentacion," *El Renacimiento*, 5Jan03. On the local competition for licenses, see Donn V. Hart, *Securing Aquatic Products in Siaton Municipality, Negros Oriental Province, Philippines* (Manila: 1956), 34–37.

26. Alexander Spoehr, "Change in Philippine Capture Fisheries: An Historical Overview," PQCS 12.1 (Mar84): 35. On the 1906 dynamite ban, see Shinzo Hayase, "Meiji-ki Manira-wan no Nippon-jin Gyomin," in Akimichi Tomoya, ed., *Kaijin no Sekai* (Tokyo: 1998), 343–68. An English translation, "Japanese Fishermen in Manila Bay during the Meiji Period (1868–1912)," was kindly provided by the author. See also "Dynamiters of Fish Caught," "Dynamite Is Stolen," and "Dynamite Causes Alarm in

Bataan," MT, 16Aug12, 30Jul19, and 26Mar24; "Dynamites Again Used for Fishing," *Herald*, 14Jan27; "Dynamite Users Arrested," and "Three Fishermen Killed in Dynamite Blast," *Tribune*, 22Aug34 and 13Jul39; P. L. Jusay, "Dynamite! Fish Enemy No. 1," *Sunday Tribune Magazine*, 31Mar40, 16, 45; and Umali, "Fishery Industries of Ragay Gulf," 178.

27. "Fish Industries Help Increase Town Funds," *Tribune*, 29Sep36, 8; Aldaba, "Manila Bay," 412; Spoehr, "Change in Philippine Capture Fisheries," PQCS 12.1 (Mar84): 32.

28. Damaso Tria, interview with the author, 2Mar85.

29. Porfirio R. Manacop and Cipriano Menguito, "Philippine Commercial Fishing Craft and Gear," PFY (1953): 199–202; David L. Szanton, *Estancia in Transition*, rev. ed. (Quezon City: 1981), 26–33.

30. Porfiro R. Manacop, "The Sexual Maturity of Some Commercial Fishes Caught in Manila Bay," PJS 59.3 (Mar36): 384.

31. Agustin F. Umali, "The Japanese Beam Trawl Used in Philippine Waters," PJS 48.3 (Jul32): 389–410; Umali, "Fishery Industries of Ragay Gulf," 183–84, Umali, "Fishery Industries of San Miguel Bay," 254; Guillermo J. Blanco and Felix Arriola, "Five Species of Philippine Shrimps," PJS 62.2 (Feb37): 219–20; Zoilo Alviar, "Japanese Fishing in the Philippines," CIJ 7.6 (Jun31): 9, 13; "Slump in Fish Market Expected," *Tribune*, 31Jan39, 3.

32. Faustino Lichauco was a grandson of Cornelia Laochanco and nephew of Luisa Lichauco viuda de Nable Jose, leading suppliers of domestic rice to the city in the 1860s and 1890s, respectively. He had served in Japan for the Foreign Affairs Committee of the aspiring Aguinaldo government during the Philippine Revolution. Sr. Marissa (Luisa Lichauco, daughter of Tomas), interview with the author, 27Jul2008; Marcial P. Lichauco, *Dear Mother Putnam: A Diary of the Second World War in the Philippines* (n.p.: 1996/1949), 3–4, 61, 184, 213; Lichauco, *Family Recollections; Lichauco Family Reunion*, 26.

33. Their numbers grew from 17 vessels and 39 crewmen in 1903 to 31 vessels and 150 crewmen in 1912.

34. Serafin D. Quiason elaborates on the scandal of Filipinos serving as "dummies" for Japanese boat owners when the rules were changed (about 1934–35) to require Philippine capital participation. Serafin D. Quiason, "The Japanese Community in Manila, 1898–1941," PHR (1970): 206.

35. These paragraphs are based on the English translation of Hayase, "Meiji-ki Manira-wan no Nippon-jin Gyomin." I am very grateful for the opportunity to have read the English manuscript version.

36. Heraclio R. Montalban and Claro Martin, "Two Japanese Fishing Methods Used by Japanese Fishermen in Philippine Waters," PJS 42.4 (Aug30): 465–71.

37. Villadolid, "Kanduli Fisheries," 545–52; Vicente C. Aldaba, "The Kanduli Fishery of Laguna de Bay," PJS 45.1 (May31): 29–39; Andres M. Mane and Domiciano K. Villaluz, "The Pukot Fisheries of Laguna de Bay," PJS 69.4 (Aug39): 397–413; *Tanay Tercentenary*, photo facing 65.

38. Villadolid, "Fisheries of Lake Taal"; Deogracias V. Villadolid, "Methods and Gear Used in Fishing in Lake Taal and the Pansipit River," PA 20 (Feb32): 571–79;

Hilario A. Roxas and Santos B. Rasalan, "The Mullet Fishery of Naujan, Mindoro,"
PJS 59.2 (Feb36): 259–73; Hilario A. Roxas, "Our Capacity of Producing Sufficient
Supply of Fish Problematical—but Not Impossible," in *Commercial and Industrial
Manual of the Philippines, 1940–1941* (Manila: 1941), 870–72.

39. Umali, "Fishery Industries of Ragay Gulf," 176–79; Umali, "Fishery Industries
of San Miguel Bay." 236; Umali, "Fishery Industries of Southwestern Samar," 377, 382;
Heraclio R. Montalban, "Investigations on Fish Preservation at Estancia, Panay" PJS
42.2 (Jun30): 328–31. The manufacture of ice in Manila began by at least 1870.

40. A. W. Herre and J. Mendoza, "Bangos Culture in the Philippine Islands," PJS
38:4 (Apr29): 451–510, quote 451; Gutierrez and Santos, "Food Consumption," 398.

41. Sastrón, *Batangas*, 95, 113, 201, 244; "Fish Pond Industry," PJC 16.1 (Jan40): 23.

42. *200 Years of Las Piñas* (Manila: 1962), quote 13.

43. See Cavada, *Historia Geográfica*, 2:442; Larkin, *Pampangans*, 302; and Artemas L.
Day, "Difficulties Encountered in the Culture of the Bangos, or Milkfish, in Zambales
Province," PJS 10D.5 (Sep15): 307–8.

44. Philippine Supreme Court case 3898, *City of Manila v. Tomas Cabangis*, February 18, 1908 (lawphil.net/judjuris/juri1908); "Fishpond Ordered Removed by Court,"
Tribune, 3Feb39, 3; Calalang, *History of Bulacan*, 92.

45. Manuel L. Roxas, "The Manufacture of Sugar from Nipa Sap," PJS 40.2
(Oct29): 185–229. See also McLennan, *Central Luzon Plain*, 17; Guillermo J. Blanco and
Deogracias V. Villadolid, "Fish-Fry Industries of the Philippines," PJS 69.1 (May39):
69–100; Wallace Adams, Heraclio R. Montalban, and Claro Martin, "Cultivation of
Bangos in the Philippines," PJS 47.1 (Jan32): 1–38, esp. 26; and *Census 1939*, 4:366, 481,
503. Ayala holdings in Pampanga equaled 3,392 hectares. "Statement Showing the
Names of Persons or Companies Owning 500 Hectares or More of Agricultural Land,"
PFR 3.4 (May30): 26–27. On renting fishponds, see PNL, Historical Data Papers (ca.
1950), Bulacan Province, Hagonoy, Barrio San Roque, 48.

46. *Census 1939*, 4:360, 483, 491. In 1953, 41 holdings in Orani, Bataan, were distributed thus: less than 1–5 hectares ten, 6–10 fourteen, 11–25 eleven, and more than 25 six.
By contrast, the largest in Las Piñas was 8 hectares. Fishponds produced approximately
11 percent of total fish caught or raised in the country during 1948–52. The proportion
was much higher in Manila. PFY (1953): 165–69, 119. Cf. *Nandaragupan: The Story of a
Coastal City and Dagupan Bangus* (Dagupan: 2005), 176–215.

47. PNL, Historical Data Papers, Bulacan: Hagonoy and Paombong on swampland conversion.

48. On the algal mat and diatom food sources, see Ling Shao-Wen, *Aquaculture in
Southeast Asia: A Historical Overview* (Seattle: 1977), 46–49. On the Chinese polycul-
ture system, see Ruddle and Zhong, *Integrated Agriculture-Aquaculture*; and Wong,
"Proto-Industrialization."

49. On the new fishponds at Sariaya, see "Tayabas," in RPC 1906, 466. On the
economic decline of nipa wine-producing areas under the new internal revenue
requirements, see J. S. Hord, ARCIR, in RPC 1904, pt. 3, esp. 671–72; and "Pangasinan," RPC 1906, 416.

50. See the photo "Three O'Clock in the Morning at a Bangus Landing," *Employee*,
1939, 211.

51. Herre and Mendoza, "Bangos Culture," 461; "Salt Water Intrusion," *Inquirer*, 1Feb98, 14; Coronas, *Extraordinary Drought*, 4.

52. Vicente C. Aldaba, "The Dalag Fishery of Laguna de Bay," PJS 45.1 (May31): 41–60; Aldaba, "Fishing Methods in Laguna de Bay," 1–28; Gutierrez and Santos, "Food Consumption," 398.

53. Ling, *Aquaculture*, quote 22; Hilario A. Roxas and Agustin F. Umali, "Fresh-Water Fish Farming in the Philippines," PJS 63.4 (Aug37): 435–65.

54. "Tilapia Takes the Philippines by Storm," in PFY (1953): 26–27; Herminio R. Rabanal, "Aquaculture in the Philippines (1898–1998)," in Rafael D. Guerrero III, ed., *100 Years of Philippine Fisheries and Marine Science* (Los Baños: 1998), 70–115, quote 87; Loreto Pastor, "Dagupan Bangus Gets Bad Image," *Bulletin*, 15Jul81, 23.

55. Florencio Talavera and Leopoldo A. Faustino, "Edible Mollusks of Manila." PJS 50.1 (Jan33): 9.

56. Domiciano K. Villaluz, "Oyster Farming," PJS 65.2 (Apr38): 303–11; Rabanal, "Aquaculture," 76–77; PNA, Estadisticas Manila, 1896–1898, Navotas, Estado Urbano-Agricola-Comercial 1896. Oysters were later displaced by the green mussel, or *tahong* (*Mytilus smaragdinus*). Local shellfish were made worthless by the proliferation of "red tide" phytoplankton in Manila Bay. "Red Tide Worsening," *Inquirer*, 24Jun92; "30 'Red Tide' Areas Placed under State of Calamity," *Inquirer*, 5Jul92. Something like a red tide occurred off Bataan in 1908. "Peridinium" (editorial), PJS 3A.3 (Jun08): 187–88. During these outbreaks, small fish, gulls, and terns disappeared from the affected area, and many stalls of Manila fishmongers were vacant. See Greg Bankoff, "Societies in Conflict: Algae and Humanity in the Philippines," *Environment and History* 5 (1999): 1–27.

57. Rabanal, "Aquaculture," 110.

58. Ms. RPC 1909, ARMB, quote 352.

59. Buzeta and Bravo, *Diccionario Geográfico*, 1:362; Sastrón, *Batangas*, 14. Data were compiled from the *Gaceta* and Islas Filipinas, Comisión Central de Estadística de Filipinas, *20 cuaderno*, 1855, 67. On supply from Taguig, see Cavada, *Historia Geográfica*, 1:55, 172. On the 1930s, see Philippines (Commonwealth), Department of Labor, "Fact Finding Survey Report, 1936," typescript, quote 61.

60. On shipping dried, smoked, and fresh fish in ice to Manila by train, see Umali, "Fishery Industries of Ragay Gulf," esp. 176–79, 195. Cf. "Fishing in Estancia," PFP, 16Aug30, 47; and Szanton, *Estancia*, 34–36.

61. Interviews by Loreto Seguido with Filoteo S. Tuason, and Brigida Alcaire-Pahit, 24May85. Tuason began as a teacher and then worked as an administrative employee before selling groceries on consignment in the Paco public market in 1934–41.

62. The *bulong*, or *bulungan*, system of whispered private bidding was and is widespread. Rod Cabrera, "Navotas: The Smell Remains the Same," *Philippine PANORAMA*, 21May78, 12–13, 34; *Nandaragupan*, 8–9. On the *bulong* system in rice and hog sales, see the index to this volume.

63. Geraldo N. Santiago and Eduarta Gagahasin, interviews, 22Apr85.

64. PNA, Contribución Industrial, Manila 1892–97 and 1897; "Sucesos," *Gaceta*, 4Aug1891; de los Reyes, "Malabon Monograph," quote 534–35.

65. The *o*'s in *bagoong* are sounded separately. Zoilo Alviar, "The Fishing Industry in the Ilocos Provinces," CIJ 7.9 (Sep31): 24.

66. Gibbs and Agcaoili rated the following for protein content: *bagoong*, 8–13 percent; *bagoong alamang*, 15.6 percent; and *patis*, 4.5 percent. Gibbs and Agcaoili, "Some Filipino Foods," 387–88. See also *Philippines, Nutrition Survey*, 87–93.

67. E. W. Taylor, "Ipon Fisheries of Abra River," PJS 14.1 (Jan19): 127–30; Alviar, "Fishing Industry in the Ilocos," 7; "Bagoong," CIJ 6.7 (Jul30): 14; Eliseo Nazareno, "'Bagoong' and Other Preserved Fish," CIJ 9.1 (Jan33): 6; Guillermo J. Blanco, "Fisheries of Northeastern Luzon and the Babuyan and Batanes Islands," PJS 66.4 (Aug38): 513–14; Umali, "Fishery Industries of Southwestern Samar," 383–85; Angeles S. Santos, *Ang Malabon* (Malabon: 1975); "Bagoong and Patis Manufacturers," in PFY (1953): 174–75; Edilberto N. Alegre, "Poor Man's Fare," in Doreen G. Fernandez and Edilberto N. Alegre, *Sarap: Essays on Philippine Food*, 210–21 (Manila: 1988), 218–19.

68. A rare ad for "Bagoong Makabayan" appeared in *El Renacimiento*, 2Sep09.

69. Fernandez, *Tikim*, quote 63.

70. Kevin McIntyre, ""Eating the Nation: Fish Sauce in the Crafting of Vietnamese Community" (PhD diss., University of Wisconsin–Madison, 2001). See also J. Guillerm, *L'Industrie du Nuoc-Mam en Indochine* (Saigon: 1931).

71. The spare handful of studies on fish sauce are highly relevant to the nutrition, health, and work capacity of ordinary people in the region. On the resource base for *patis*, see S. V. Bersamin, "Fluctuations in the Planktonic Population of the Estuaries of Navotas and Malabon," PJS 86.4 (1957): 339–57.

72. Brian Fegan, personal communication, 22Jun95; Umali, "Fishery Industries of Ragay Gulf"; Cavada, *Historia Geográfica*, 1:180 on Laguna; *Libertas*, 5Feb00; Pablo Fernandez, OP, and Jose Arcilla, SJ, "Memorial on the Province of Bataan," *PSacra* 7.21 (1972): 467, 479–481 (originally written in 1887). Orani and Balanga each had a dealer in this trade. See PNA, Contribución Industrial, Bataan, 1892–97 and 1893–96.

73. Zosimo Montemayor, "Mañgabol Fisheries of Bayambang, Pangasinan," PA 16.2 (Jul27): 73–79, quotes 74.

74. Umali, "Fishery Industries of Ragay Gulf," quotes 195; Umali, "Fishery Industries of Southwestern Samar," 383–84; Blanc Szanton, *A Right to Survive*, 30–32.

75. In the 1850s, "in the sitio named Omboy are the meat and fish markets of Binondo situated in a large warehouse with dividers. The market . . . operates at night" (Omboy is derived from *umbuyan*). Buzeta and Bravo, *Diccionario Geográfico*, quote 2:239; Islas Filipinas, Comisión Central de Estadística de Filipinas, *20 cuaderno*, 1855, 25; Miguel Ma. Varela, SJ, personal communication, 27Jun74. On the barrio name, Incendio, see *El Comercio*, 8Jan1887.

76. *El Comercio*, 16May1891; Agustin F. Umali, "The Cast Net as a Deep-Water Fishing Appliance in Manila Bay," PJS 51.4 (Aug33): 555–65.

77. On the smoking and drying of fish for the Manila market, see Claro Martin, "Methods of Smoking Fish around Manila Bay," PJS 55.1 (Sep34): 79–90; and Umali, "Fishery Industries of Ragay Gulf," 177. As a prominent food of poor workers, see "The Bluffers" (cartoon), *The Independent*, 16Aug19.

78. Eric Tagliacozzo, "A Necklace of Fins: Marine Goods Trading in Maritime Southeast Asia, 1780–1860," IJAS 1.1 (2004): 23–48, drawing on James Francis Warren, *The Sulu Zone, 1768–1898: The Dynamics of External Trade, Slavery, and Ethnicity in the Transformation of a Southeast Asian Maritime State* (Quezon City: 1985/1981); Díaz

Arenas, *Report on the Commerce and Shipping of the Philippine Islands*, 12–13; David E. Sopher, *The Sea Nomads* (Singapore: 1977).

79. Cruikshank, "Commercial Patterns," 5–6.

80. Edmund Roberts, *Embassy to the Eastern Courts of Cochin-China, Siam, and Muscat* (New York: 1837), quote 51–52; Cavada, *Historia Geográfica*, cuadros 19, 20; *Balanza*, 1857–94; Jose S. Domantay, "Philippine Commercial Holothurians," PJC 10.9 (Sep34): 5–7. Cavada cites 23,000 kilos of domestic shark's fin arriving in the city circa 1870 (1:387).

81. Leonard Blussé, "In Praise of Commodities: An Essay on the Cross-Cultural Trade in Edible Bird's-Nests," in Roderich Ptak and Dietmar Rothermund, eds., *Emporia, Commodities, and Entrepreneurs in Asian Maritime Trade, c. 1400–1750* (Stuttgart: 1991), 326–27, drawing on Warren, *Sulu Zone*; Chu, *Chinese and Chinese Mestizos of Manila*, 162; Canuto G. Manuel, "Beneficial Swiftlet and Edible Birds' Nest Industry in Bacuit, Palawan," PJS 62.3 (Mar37): 379–91.

82. Richards, "Philippine Economic Conditions," 635.

83. Dennis Morrow Roth, *The Friar Estates of the Philippines* (Albuquerque: 1977), 74; "Commissioned to conduct a census of Tondo Province's tributary populations [was the] *oidor* of the Real Audiencia of Manila . . .," July 10, 1741, in Luis C. Dery, *Pestilence in the Philippines: A Social History of the Filipino People, 1571–1800* (Quezon City: 2006), 200–201. Cf. Alcina (1668), *Historia*, in PSacra 111 (Sep–Dec2002): 559–69; and Reid, *Southeast Asia*, 1:28–29.

84. "Mineral Production of the Philippine Islands," DCTR, 8Mar13, 55, 1197; BPS 1939, 1–2 qtrs., 186; "Salt," PJC, Dec34, 17; Calalang, *History of Bulacan*, 87–88. Cf. Blanc Szanton, *Right to Survive*, 45–49.

85. Alvin J. Cox and T. Dar Juan, "Salt Industry and Resources of the Philippine Islands," PJS 10A.6 (Nov15): 373–401; Umali, "Fishery Industries of Southwestern Samar," 385–86; Guillermo L. Ablan, Jose R. Montilla, and Basilio M. Martin, "The Salt-Making Industry of Northwestern Luzon, PJS 72.3 (Jul40): 319–29. Cf. PNA, Servidumbre Domestica 6, "Precios, Mercados," January 13, 1887; Herre and Mendoza, "Bangos Culture," 473.

86. There were also other methods of local production. Andrea Yankowski, "*Asinan*: Documenting Bohol's Traditional Method of Salt Production," PQCS 35.1–2 (Mar–Jun2007): 24–47.

87. Doeppers, "Lighting a Fire," 419–47.

88. Gerrit Knaap and Luc Nagtegaal, "A Forgotten Trade: Salt in Southeast Asia, 1670–1813," in Roderich Ptak and Dietmar Rothermund, eds., *Emporia, Commodities, and Entrepreneurs in Asian Maritime Trade, c. 1400–1750*, 127–57 (Stuttgart: 1991), 127–57; McIntyre, "Eating the Nation"; Gerard Henry Sasges, "Contraband, Capital, and the Colonial State: The Alcohol Monopoly in Northern Viet Nam, 1897–1933" (PhD diss., University of California, Berkeley, 2006).

89. De Jesus, *Tobacco Monopoly*, 22.

90. *El Comercio* reported the arrival of tinned sardines (8May1872), and a store advertised sardines in tomato sauce and oil (31Dec1873). "Philippine Tariff and the Food Supply," 16; *Monthly Summary of Commerce and Finance of the United States for 1914*,

Bureau of Foreign and Domestic Commerce, U.S., 21:1153; Hayase, "Meiji-ki Manira-wan no Nippon-jin Gyomin." English translation.

91. Harvey V. Rohrer, "Fishing and Fish Consumption in the Philippines," CR, 5Jan31, 39–40; Felix Franco, "Towards Self-Sufficiency in Fishery Products," PJC 10.9 (Sep34): 8–9; Valeriano K. Luz, "The Philippines as a Market for Canned Goods," PJC 10.6 (Jun34): 11–12; "Leading Imports of the Philippines," in *Philippine Yearbook*, 1936–37, 172. See ads for imports, "Sardinas P.C.C.," PFP, 18Feb22; and "Mayon" sar-dines by Daido Boeki Kaisha, in *Philippine Yearbook*, 1933–34, 86.

92. John G. Butcher, "The Marine Animals of Southeast Asia; Towards a Demo-graphic History, 1850–2000," in Peter Boomgaard, David Henley, and Manon Osse-weijer, eds., *Muddied Waters: Historical and Contemporary Perspectives on Management of Forests and Fisheries in Island Southeast Asia* (Leiden: 2005), 63–96.

93. David Henley and Manon Osseweijer, "Introduction: Forests and Fisheries in Island Southeast Asia; Histories of Natural Resource Management and Mismanage-ment," in Peter Boomgaard, David Henley, and Manon Osseweijer, eds., *Muddied Waters: Historical and Contemporary Perspectives on Management of Forests and Fisheries in Island Southeast Asia* (Leiden: 2005), 20–21.

Chapter 7. "Generations of Hustlers"

1. Doeppers, *Manila, 1900–1941*, 73.

2. Sauer, *Agricultural Origins*.

3. "Ordinances of the Audiencia, 1598," in Emma H. Blair and James A. Robert-son, eds., *The Philippine Islands, 1493–1898*, 55 vols. (Cleveland: 1903–7), 10:305–9, quote 306. Pigs, chickens, and carabao featured importantly in the competitive feast-ing of late pre-Hispanic power centers. Laura Lee Junker and Lisa Niziolek, "Food Preparation and Feasting in the Household and Political Economy of Pre-Hispanic Philippine Chiefdoms," in Elizabeth A. Klarich, ed., *Inside Ancient Kitchens: New Directions in the Study of Daily Meals and Feasts* (Boulder: 2010), quote 18.

4. Alcina (1668), *Historia*, in PSacra III (Sep–Dec 2002): quote 583; Wickberg, *Chinese in Philippine Life*, quote 113, giving a reference from the 1820s. Ruth de Llobet cites a trial in 1813 in which creole "perpetual aldermen" were accused of "arbitrary and tyrannical abuses" of the *abasto y carnes* system of urban supply. Llobet, "Orphans of Empire," 171.

5. Looking to possible Mexican antecedents, the meat supply of Mexico City and Cuernavaca was long organized by an *abasto* contractor who supplied beef and mutton at a fixed price for the duration of the contract. This system gave way to a freer com-merce in meat in the 1780s. Ward Barrett, "The Meat Supply of Colonial Cuernavaca," AAAG 64.4 (Dec74): 525–40; Horowitz, Pilcher, and Watts, "Meat for the Multitudes," 1054–83.

6. C. O. Levine, "Egg Production in China a Giant Industry," PEF 1.2 (May23): 35–36; Morga, *Sucesos*, 16:90.

7. Copeland, "Spanish Agricultural Work," 318.

8. De la Cavada, *Historia Geografica*, 1:180; Sawyer, *Inhabitants*, 197.

9. Buzeta and Bravo, *Diccionario Geográfico*, 1:446; De la Cavada, *Historia Geográ-fica* 1:179–80; Montero y Vidal, *El Archipiélago Filipino*, 308; *Census 1903*, 4:236. On

market prices, see Hs. of Com., 1872, v. 60, C501, 208 and 1878, v. 73, C1950, 252–53. On shipments, see Sastrón, *Batangas*, 251; and Governor J. Losada "Batangas," in RPC 1906, pt. 1, 194.

10. This is interpolated from Doeppers, *Manila, 1900–1941*, fig. 7.

11. RCUS, C. H. Cowan, 1893, 43:159, 574.

12. On hybrid breeding stock, see "Livestock and Poultry in the Philippine Islands," PAgR 4.9 (Sep11): 492–94; ms. AR Department of Agriculture, 1921, 87; and ms. ARBAg, 1924, 285.

13. Estefano C. Farinas, "Avian Pest, a Disease of Birds Hitherto Unknown in the Philippine Islands," PJAg 1.4 (1930): 311.

14. Ms. ARBAI, 1930, quote 11; Sta. Maria, *Governor-General's Kitchen*, quote 256. *Tinola* begins with lightly sautéing pieces of chicken with ginger, garlic, and onions. These are then boiled together. Later cubes of green papaya or squash are added and, finally, *sili* pepper leaves.

15. Mrs. Jessie Lichauco (widow of son Marcial), interview with the author, 27Jul2008; and Lichauco, *Family Recollections*, 40.

16. See an ad for the Nagoya Poultry and Supply Co. in the *Manila Carnival, Commercial and Industrial Fair, Commercial Handbook*, 1932, 53. The Japanese poultry farms turned up in a 1940 survey of 1,408 Japanese enterprises listed in *Manira Nihon Shogyo Kaigisho Tsuho* 54 (15Jan41): 66, reported by Hayase in "A Study of Early Popular Consuming Society."

17. "Chickens," CIJ 8.2 (Feb32): 15; Carlos Quirino, *Philippine Tycoon: The Biography of an Industrialist, Vicente Madrigal* (Manila: 1987), 73; Juan Rangasajo, "Progress of the Local Poultry Industry," PJC 10.4 (Apr34): 13–14. For 1933 statistics, see M. R. Montemayor, "Half-Century of Livestock Raising," in *A Half-Century of Philippine Agriculture* (Manila: 1952), 292; and Carlos X. Burgos, "Poultry Shows and Poultry Judging Contests," PJAI 2.3 (May–Jun35): 225–37.

18. PSR 4.1–2 (1937): 199–200; BPS 6.1–2 (1939): 144; *Census 1939*, 2:1111; Raol V. Baron and Conrado A. Valdez, "The Contributions of Veterinary Medicine to the Poultry Industry," in Mauro F. Manuel et al., eds., *A Century of Veterinary Medicine in the Philippines, 1898–1998* (Quezon City: 2002), 89–93.

19. Herrera vda. de Umali, *Tayabas Chronicles*, quote 112.

20. A survey of Manila workingmen's families concluded that about 2 percent of their animal protein came from eggs. Gutierrez and Santos, "Food Consumption," 400.

21. Heiser, *An American Doctor's Odyssey*, quote 169. Cf. "Trade of Swatow 1904," Hs. of Com., 1905, v. 88, 3455, 8.

22. "Prices, Imports, and Exports of Agricultural Produce," Hs. of Com., 1903, v. 82, Cd. 1616, 167.

23. Ms. ARBAI, 1930, quotes 11–12.

24. On feeds, see "Las Gallinas Pueden Poner Huevos sin la Ayuda de los Gallos," PFP, 10May30, 54–55; F. M. Fronda, "The Philippine Poultry Industry," *United States Egg and Poultry Magazine* 35 (Oct29): 26; Villadolid, "Fisheries of Lake Taal," 192; and Levine, "Egg Production in China," 36.

25. "Urge Tariff on Rice, Eggs," *Tribune*, 24Feb31, 1; "Tariff Duty on Meat and Eggs Imported," CIJ 8.8 (Aug32): 14. On imports and expanding local egg production, see

Gregorio San Agustin and Santiago Y. Rotea, "Philippine Egg Production," PJAI 3.6 (Nov–Dec36): 479–95; "Las Gallinas," 55; PSR 4.1–2 (1937): 198–201; BPS 6.1–2 (1939): 144, 183; BPS 6.3 (1939): 67; and Rangasajo, "Progress," 13. During 1931–34, a breakdown is provided. Fresh imported duck eggs evidently exceeded imported hen's eggs 5.0 to 4.5 million dozen in 1932 plus an additional 334,000 dozen preserved duck eggs. San Agustin and Rotea, "Philippine Egg Production," 490–91. By 1934 it was 1.3 million dozen hen's eggs to 0.5 million dozen duck eggs, including only 19,000 dozen preserved duck eggs. "Facts and Figures about Philippine Trade with China," PJC 11.3 (Mar35): 27.

26. On efforts to improve the counting, see ms. ARBAI, 1933, 24. On provincial and local patterns of egg production, see Census 1939, 2:1129–40 and "Rizal," 3:1499–1500; Carlos X. Burgos, "Commercial and Industrial Development of Livestock in the Philippines," PJAI 1.4 (Jul–Aug34): 211; and Montemayor, "Half-Century," 292.

27. Edwin S. Cunningham, "The Egg and Poultry Industry in China," PJC 10.4 (Apr34): 5–6; "The Egg Industry in the Philippines," CIJ 1.11 (Nov25): 5, 10, 12.

28. On the need for more developed marketing organizations, see "Would Build Up Egg Business," Bulletin, 4May26, 9. On the potential viability of a provincial egg-drying business that would export to the United States, see "Egg Industry in the Philippines," 12.

29. "Tandang Maria," Maria Balboa vda. Reyes, interview, 14May85. Cf., Fronda, "Philippine Poultry Industry," 27, 70; Calalang, History of Bulacan, 93; J. F. Boomer, "Poultry Industry in the Philippines," DCTR, 16Oct14, 285; and Teddy Santo Domingo, "Bulacan's Egg Basket," Sunday Times Magazine, 19Jan47, 3.

30. "₱13,000 Lost in Binondo Fire," Tribune, 30Dec39. Cf. Graphic, 12Jan39, 2Mar39, and 25Jul40. At last report, eggs from Bulacan and Batangas were selling for about 9 percent more than those from China. For the official figures on egg supply by province, see PSR 4.1–2 (1937): 198; BPS 1.1–2 (1939): 144; and BPS 1.4 (1939): 67.

31. Wilkes, Narrative of the United States Exploring Expedition, quote 5:301. In Spanish, pateros means "caring for ducks."

32. Buzeta and Bravo, Diccionario Geográfico, 2:397; Islas Filipinas, Comisión Central de Estadística de Filipinas, 20 cuaderno, 1855, 67; Montero y Vidal, El Archipiélago Filipino, 99–100; "The Balut Industry of Pateros," ACCJ 9.1 (Jan29): 8–9; "Balut and Salted Eggs," CIJ 7.9 (Sep31): 14; Ugaldezubiaur, Memoria Descriptiva, 38 (on Taguig and Pasig).

33. Now Bulacan farmers encourage "itinerant herders of brown Peking ducks to run their armies on the harrowed fields before planting" to reduce the number of golden kuhol snails, a major pest. Brian Fegan, "Philippines [Bulacan] Fieldwork," unpublished report, 21Jan95.

34. RPC 1906, pt. 1, quote 422.

35. De la Cavada, Historia Geográfica, cuadro 19; Aldaba, "Fishing Methods in Laguna de Bay," PJS 45.1 (May 1931): 24–26. See also the essay on street food in Fernandez, Tikim, 10–11.

36. "Duck Raising Advocated Here," Bulletin, 20Nov24, 8; Walter Robb, "Balut," in Filipinos: Pre-war Philippines Essays (Victoneta Park: 1963/1939), 48–53; on Pateros, C. W. Edwards, "The Livestock Industry of the Philippines," PAgR 9.2 (1916): 148; on

Hagonoy, "The Balut Industry," CIJ 7.4 (Apr31): 7. See also *Census 1939*, 2:1141; and Carlos X. Burgos, "The Duck Industry of the Philippines," PAgR 17.2 (1924): 87–100, photo V.

37. In 1934 the Purity Bakery next to the Quinta Market advertised *balut*, salted eggs, and fresh duck (and chicken) eggs from Pateros. *La Opinión*, 19, 21, 28Nov34.

38. Wilkes, *Narrative of the United States Exploring Expedition*, quote 5:301; Ellis, *Hong Kong to Manila*, 64; Harford, "Trade and Commerce of the Philippine Islands, 1899," Hs. of Com., 1900, v. 97, 2436, 3; Sawyer, *Inhabitants*, 183, 189–90, 220; "Snipe Season," advertisement in PFP (7Aug26): 6; Sta. Maria, *Governor-General's Kitchen*, 254; Hamm, *Manila and the Philippines*, 16. On ducks, see "Zambales," in RPC 1904, pt. 1, 665. On quail, see John T. Zimmer, "A Few Rare Birds from Luzon and Mindoro," PJS 13D.5 (Sep18): 223.

39. Herrera, *Tayabas Chronicles*, 72, 151, 182.

40. In the rich cuisine of Sulipan, Pampanga, fat grain-fed doves or pigeons and also quail were "braised in their juices with butter, wine and brandy" and cooked in a pie. Gonzalez, *Cocina Sulipeña*, quote 6, 72–73; Census 1939, 2:1131–41.

41. Gonzalez, *Cocina Sulipeña*, quote 50, 46–47. Cf. De la Cavada, *Historia Geográfica*, 57; Carlos X. Burgos, "Turkey Raising in the Philippines," PJAI 1.6 (Nov–Dec34): 379–91; and Sta. Maria, *Governor-General's Kitchen*, 254.

42. The pork:beef ratio is for 1915–19. Ms. ARMB in ms. RPC 1909, 1:341; ARCIR 1915–20.

43. This compares to 6,183 and 6,012 in 1853 and 1854, respectively. Islas Filipinas, Comisión Central de Estadística de Filipinas, *20 cuaderno*, 1855, "Resumen," 1853, 1854.

44. Iloilo sent more than 1,000 hogs into interisland commerce in 1860, but these would have gone mainly to Negros and/or Bikol. N. Loney, Hs. of Com., 1862, v. 58, 3054, 344. In 1870 Manila received a net number of 7,808 hogs by coastal shipping. De la Cavada, *Historia Geografica*, 1:384–95.

45. "Matadero Publico," in Islas Filipinas, *Boletin de Estadística de la Ciudad de Manila* 1.1–12 (1895–96). Cavite also fit this rainy season pattern.

46. Adults age 21 to 59 were obliged to fast. Fr. John Schumacher, SJ, personal communication, April 9, 1998.

47. "Ordinances of the Audiencia," 10:308–9. In 1884 and 1885, an equally pronounced second nadir of pork consumption developed in September–November. The data on supply are from official monthly and annual reports in the *Gaceta* (1869–72) and *El Comercio* (including 1884–86 and 1891).

48. Dra. Helen Mendoza, personal communication, 15Sep99.

49. Oscar Evangelista, personal communication, 19Oct99. "El Lucero," *El Comercio*, 27Feb1895, quote. Cf. *El Comercio*: "Almacen," *El Comercio*, 29Oct1875; and "Bakalao superior de Noruega," *El Comercio*, 7Jan1884. *Bakalaw* was also eaten at other times by those who could afford it. It was an occasional Sunday menu entrée of the Café del Recreo and Café Universal during 1887–91. On the import of *bakalaw*, see the *Gaceta*, 2Jun, 17Aug, and 21Sep1861; and the annual *Balanza* for 1854–66 and 1891–94. The total reported volume increased materially in each of these last four years, nearly doubling to 300,000 kilograms.

50. "Estado de la ganaderia filipina en 1887," BNM, ms. 19.218.

51. In 1822 more than a dozen Chinese registered in Manila gave their occupation as *porquero*, "one who takes care of hogs," or in this case, "deals in" them. These men resided in the old Parian and Binondo in the city and at the provisioning centers of Malabon and Pasig. PNA, Padron de Chinos Ynfieles, 1822. The 1890s data are from the PNA, Contribución Industrial, by province.

52. Otto Scheerer, "Batanes Report," in RPC 1908, 273.

53. "Estado de la ganaderia filipina," BNM, ms. 19.218. Bulacan and Pampanga failed to answer, and the records of some other likely provinces, such as Ilocos Sur, Isabela, Zambales, and the major Bikol provinces, have been lost.

54. For Luzon provinces the reported incidence was about 38 percent of the rate for 1903: 7.3 versus 19.1 per 100. For swine numbers, I used Montero y Vidal, *El Archipiélago Filipino*, 329, 342, 361, augmented by De la Cavada. Likely both figures were underreported. De la Cavada presents a comparison of the 1870 population returns of the ecclesiastical and civil statistical reports (neither a proper census). I used the larger figure in each case. See Michael Cullinane, "Accounting for Souls," in D. F. Doeppers and Peter Xenos, eds., *Population and History: The Demographic Origins of the Modern Philippines* (Madison and Quezon City, 1998), 321; and De la Cavada, *Historia Geografica*, 1:372; 2:337.

55. At 57,400, Cebu Province reported the largest swine total in the survey of 1886–87. "Estado de la ganaderia filipina," BNM, ms. 19.218.

56. *Census 1903*, 4:236; De la Cavada, *Historia Geografica*, 1:180; "Matadero Publico"; Sawyer, *Inhabitants*, 136; Montero y Vidal, *El Archipiélago Filipino*, 308; Sastrón, *Batangas*, 37. The incidence of swine would have been even less in the Muslim areas had an enumeration been taken there. Losada, "Batangas," 194. Laguna reported more than 15,000 hogs in the 1886–87 survey but failed to indicate the number sent out. "Estado de la ganaderia filipina en 1887," BNM, ms. 19.218.

57. PNA, "Estadistica de Manila," bundles 1896 and 1896–98, Ganaderia. The four urban districts are Sampaloc, Pandacan, Sta. Ana, and Ermita.

58. H. J. Andrews ads *El Comercio*, 1Jun1897 and 22Dec02.

59. Quoted from a sausage ad in *El Comercio*, 6Dec1894. See also ads in El Comercio, 20Apr1891, 27Apr1891, and 7Jan01. On the German Sausage Factory, see Wigan Maria Walther Tristan Salazar, "German Economic Involvement in the Philippines, 1871–1918" (PhD diss., School of Oriental and African Studies, University of London, 2000), 193–95; and "Max Is Back," MT, 28Sep08, 10.

60. Some 395 and 150 hogs were landed in Manila from China in 1891 and 1901–2, respectively, and 184 from Japan in 1894. *Balanza*, 1854–94; RPC 1901–2, pt. 1, 381. Saigon exported almost 500 hogs in 1898 and 10,000, 8,000, and 5,351 during 1900–2. Hs of Com.: 1899, v. 99, "Trade of Saigon," 6; 1901, v. 82, "Trade of French Indo-China," 15; 1902, v. 107 and 6; 1903, v. 77, "Trade and Commerce of Cochin-China," 6. Cf. Santiago Y. Rotea, "Preliminary Report on the Manufacture of Hams under Philippine Conditions," PJAI 3.4 (Jul–Aug36): 307; and "Philippine Tariff and the Food Supply," 16.

61. David Moberly, "Hog Cholera," PAgR 1.2 (Feb08): 91–94; V. Buencamino, *Memoirs*, 69.

62. Nicolas S. Sevilla, "The Slaughter of Livestock and Meat Inspection in the Philippines," PJAI 2.2 (Mar–Apr35): 140; Wickberg, *Chinese in Philippine Life*, 109.

63. Sawyer, *Inhabitants*, 184; "La trichina," *El Comercio*, 10Jun1881; Gregorio San Agustin, "Meat Inspection in the Philippines," PJAI 3.6 (Nov–Dec36): 519–20.

64. "Discussion of the Paper by Dr. Thomas W. Jackson," PJS 3B.5 (Nov08): 439–42.

65. PAgR 1 (Jan08): quote 52; W. R. L. Best, "Notes on Swine Breeding in the Philippines," PAgR 7.3 (Mar14): 119–20; Estefano C. Farinas, "Some of the Most Important Diseases Affecting Our Hog Industry," PJAI 3.4 (Jul–Aug36): 332–34; Benjamin Schwartz, "Helminth Parasites of Hogs," PJS 27.2 (Jun25): 227–33. Jagor noted that pigs in Bikol fed on human wastes, a practice also common in Manila in the nineteenth century and in lowland Luzon provinces in the early 1900s. Fedor Jagor, *Travels in the Philippines* (Manila: 1965), 112 (original German edition 1873); Jackson, "Sanitary Conditions," 433; "Animals Dying," *Varsity News*, 21Sep20; Lewis, *Wagering the Land*, 30.

66. On avoidance, see geographer Frederick J. Simoons, *Eat Not This Flesh: Food Avoidance from Prehistory to the Present*, rev. ed. (Madison: 1994). See also Jacinto C. Antonio, "A Survey of Pork Marketing in the City of Manila," PJAI 6.5 (Nov–Dec37): 450; and Farinas, "Some of the Most Important Diseases," 328–33.

67. During 1901–7, hog cholera was found in most provinces and may have been long standing. Moberly, "Hog Cholera," PAgR 1.2 (Feb08): 91–94. See also (Jan08): 50.

68. RPC 1901–2, pt., 1, quote 381.

69. Edwards, "Livestock Industry," quote 145; ms. ARBAg, 1916, quote 49. See also William F. Nydegger and Corinne Nydegger, *Tarong: An Ilocos Barrio in the Philippines* (New York: 1966), 29.

70. On slaughter weights, see ms. ARBAI, 1933, 165; PSR 4.4 (1937): 616–17; and Farinas, "Some of the Most Important Diseases," 333.

71. "Comparemos," *El Comercio*, 3May1872. Highly discrepant livestock numbers were also reported in Spain itself during the nineteenth century. Vicens Vives, *Economic History*, 654.

72. Ms. ARBAg, 1921, 15; ms. ARBAI, 1930, quote 41; ms. ARBAI, 1933, quote 24, also in PJAI 1.4 (Jul–Aug34): 231–32. For the inflated numbers, see SBPI 10 (1927): 50–51; and 12 (1929): 64–65. Thereafter, see PSR 4.4 (1937): 624–25; and JPS 1.3 (Sep41): 161–62. Tutuban, Velasquez, and Pritil are all in Tondo.

73. On the growth of the indigenous middle class, see Doeppers, *Manila, 1900–1941*, chap. 3.

74. On hog dealers, see "Poultry and Meat," PJC 9.5 (May33): quote 14; PJC 8.1 (Jan32): 15; and Antonio, "Survey of Pork Marketing."

75. Interviews in the Velasquez-Nepomuceno neighborhood were conducted with Ricardo Almario, 18Apr85; Dr. Federico Rubio, 18 and 21Apr85; and Mrs. Gliceria Clemente, 21Apr85. Not all *viajeros* operated on borrowed capital.

76. Walter Robb, "Lipa's Classic Poverty," ACCJ 10.10 (Oct30): quote 6; PAgR 14.2 (1921): pl. 6; PAgR 15.2 (1922): pl. 4; Robert Reed, "Corn," in Robert E. Huke, ed., *Shadows on the Land: An Economic Geography of the Philippines* (Manila: 1963), 261.

77. At the Agricultural School in Muñoz, Nueva Ecija, two-thirds of the feed of the school's hogs by weight consisted of bran, 12 percent of *binlid* (broken rice), and another 12 percent of table scraps. Corn grains, slaughter scraps, copra meal, fishmeal,

coarse salt, and slaked lime were minor components. Antonio Hernandez, "Hog Raising under Muñoz Conditions," PJAI 3.3 (May–Jun36): 206–7.

78. On the origins of hogs slaughtered in the city in 1935, see Farinas, "Some of the Most Important Diseases," 334. For provincial reports of hogs exported and imported, see PSR 3.4 (1936): 362, 331. Similar patterns characterized 1936–37. BPS 1–2 (1939): 128, 130.

79. *Census 1939*, 2:1111, 1124–36.

80. Earlier members of the elite of colonial Mexico City routinely ate mutton "stewed or pit roasted, often with chilies in marinade (*adobo*)," and this may have carried over among some in Manila. Horowitz, Pilcher, and Watts, "Meat for the Multitudes," quote 1066. Cf. De la Cavada, *Historia Geografica*, 1:384–95; J. F. Boomer, "Philippine Islands," SCR, 1916, 2:80a, 8; and Resil B. Mojares, "Deciphering a Meal," in Jonathon Chua, ed., *Feasts and Feats: Festschrift for Doreen G. Fernandez* (Quezon City: 2000), 190.

81. Data are from Cesar Majul in William G. Clarence-Smith, "Lebanese and Other Middle Eastern Migrants in the Philippines," in Akira Usuki, Omar Farouk Bajunid, and Tomoko Yamagishi, eds., *Population Movement beyond the Middle East: Migration, Diaspora, and Network* (Osaka: 2005), quotes 131; and *Census 1903*, 4:335.

82. V. Buencamino, *Memoirs*, 8; Nick Joaquin, *Manila, My Manila*, 199; BPS 1.4 (1939): 117; "Matanza clandestina," *El Comercio*, 31Jan1895; Lewis, *Wagering the Land*, 85.

83. Likewise *carnero* ordinarily refers to sheep in Spanish, but Noceda and Sanlucar (*Vocabulario*) give *kambing*, or "goat," as its Tagalog equivalent. Sawyer, *Inhabitants*, 220; Max L. Tornow, "The Economic Condition of the Philippines," *National Geographic* (Feb1899): 40–41; "House Warming at Mariveles," MT, 16Jan08; ARMB, 1911, 85, 82. On *kaldereta*, see Gonzalez, *Cocina Sulipeña*, 70–71; and Clarence-Smith, "Lebanese," 115–43.

84. "Matanza clandestina," *El Comercio*, 31Jan1895 and "Matanza," 9Aug1881. Lewis, *Wagering the Land*, 30–31, 85; Simoons, *Eat Not This Flesh*, 204–22; Gourou, *Peasants of the Tonkin Delta*, 2:458–59.

85. Fr. John Schumacher, SJ, personal communication, 1Mar2000; Enrique T. Carlos and Loinda R. Baldrias, "Veterinary Medicine and Animal Welfare," in Mauro F. Manuel et al., eds., *A Century of Veterinary Medicine in the Philippines, 1898–1998* (Quezon City: 2002), quote 152.

86. Little Cabra, or Goat Island, lies at the northern end of Lubang.

87. Morga, *Sucesos*, quotes 90 and 90n77. Cf. William Henry Scott, *Barangay: Sixteenth-Century Philippine Culture and Society* (Quezon City: 1994), 45–46; Montero y Vidal, *El Archipiélago Filipino*, 329, 342, 361; Antonio Peña, "Agricultural Conditions in the Philippines," PAgR 14.2 (1921): 157–58, and 15.2 (1922): 145–47; Sastrón, *Batangas*, 37; *Census 1903*, 4:236; and PSR 3.4 (1936): 363.

88. Nydegger and Nydegger, *Tarong*, quotes 29; *Census 1939*, 2:1126, 1136. In addition to the depredations of goats, the Ilocos hills were also thoroughly cleared of woody species for use as fuel in curing tobacco.

89. Herrera, *Tayabas Chronicles*, 22, 75 on *mayas*; Sta. Maria, *Governor-General's Kitchen*, quote 257; Alegre, "Poor Man's Fare," 212–13. From Ilocos consumption of "beetles, locusts, crickets, and certain ants" was reported, but there was nothing in

Manila like the ninth-month market for edible beetle grubs in Hanoi. Nydegger and Nydegger, *Tarong*, 29; Georges Boudarel and Nguyen Van Ky, *Hanoi: City of the Rising Dragon*, Claire Duiker, trans. (Lanham and Oxford: 2002), 5.

90. Otherwise this was a rural phenomenon. Patricio N. Abinales, "Let Them Eat Rats: The Politics of Rodent Infestation in the Postwar Philippines," PS 60.1 (Mar2012): 93n4.

CHAPTER 8. BEEF, CATTLE HUSBANDRY, AND RINDERPEST

1. Despite the use of the term *cattle* here, it is well to keep in mind that the bio-logic form of domesticated bovines went through substantial change during our period. An early version of this section was published as "Beef Consumption and Regional Cattle Husbandry Systems in the Philippines, 1850–1940," in Peter Boomgaard and David Henley, eds., *Smallholders and Stockbreeders*, 307–24. I thank the KITLV Press for permission to use this material. Parts of this chapter have benefited from critical readings by Matthew Turner and William G. Clarence-Smith and by discussions at the conference Science, Disease, and Livestock Economies at St. Antony's College, Oxford University, June 23, 2005, and an extensive discussion with R. D. Hill in Hong Kong, July 1, 2000.

2. Jollibee's began operating in Manila in 1978, McDonald's in 1981. See "Fast-Food, the Fast Money-Makers," *IBON Facts & Figures*, 31Jan84.

3. R. Gonzalez y Martin, *Filipinas y Sus Habitantes* (Bejar: 1896), 81. Cf. Anderson, *Colonial Pathologies*, 43.

4. De Bosch, "Sual," Hs. of Com., 1857, v. 16, 531; Jagor, *Travels*, 208; W. G. Palgrave, Hs. of Com., 1878, v. 73, C1950, 262.

5. Consul G. T. Ricketts, "Philippine Islands," Hs. of Com., 1872, v. 60, C501, 208. The nominal value of carabao doubled during 1853–62. Nicholas Loney, "Philippines," Hs. of Com., 1862, v. 58, 3054, 343.

6. Acting Consul George Mackenzie, "Manila," Hs. of Com., 1878–79, v. 70, C2285, 577, quote 589. Sales of imported textiles were down for the same reason.

7. Martine Barwegen, "Browsing Livestock History: Large Ruminants and the Environment in Java, 1850–2000," in Peter Boomgaard and David Henley, eds., *Smallholders and Stockbreeders: Histories of Foodcrop and Livestock Farming in Southeast Asia* (Leiden: 2004), 283–305.

8. Colin P. Groves, "Domesticated and Commensal Mammals of Austronesia and Their Histories," in Peter Bellwood, James J. Fox, and Darrell Tryon, eds., *The Austronesians: Historical and Comparative Perspectives* (Canberra: 1995), 152–63; Scott, *Barangay*, 46; Alcina (1668), *Historia*, in PSacra 96 (Sep–Dec97): 502–5, 512–13, 524–25.

9. Scott, *Barangay*, 201, 287.

10. Pedro Chirino, *Relacion de las Islas Filipinas*, in Emma H. Blair and James A. Robertson, eds., *The Philippine Islands, 1493–1898*, 55 vols. (Cleveland: 1903–7), 12, 188; Bray, *Rice Economies*, 48. Cf. Karen M. Mudar, "Patterns of Animal Utilization in the Holocene of the Philippines," *Asian Perspectives* 36.1 (Spring 1997): 74–75, 100.

11. Junker and Niziolek, "Food Preparation," 17–53. On various prohibitions of carabao slaughter, see *Gaceta de Manila*, 1Jul1870 and 2Nov1875: and RPC 1904, pt. 2, 234.

12. J. J. O'Brien, SJ, *The Historical and Cultural Heritage of the Bicol People*, 2nd ed. (Naga City:1968), 183; Owen, *Prosperity without Progress*, 144, 167–69. An "*estancia*" with 2,000 head on the small island of Ambil was the only one of "some importance" in Mindoro Province in 1886–87. "Estado de la ganaderia filipina en 1887," BNM, ms. 19.218.

13. De Bosch, "Sual," Hs. of Com., 1859 session 2, v. 30, C2579, quote 752.

14. De la Cavada, *Historia Geográfica*, 1:180, 204; Montero y Vidal, *El Archipiélago Filipino*, quote 308; "Estado de la ganaderia filipina en 1887," BNM, ms. 19.218. Cf. Stanton Youngberg, "The North to South Movement of Animals on the Island of Luzon," PAgR 5.12 (Dec12): 653–59.

15. See table 8.2. The survey mentions 13 cattle *estancias* in Masbate-Ticao. "Estado de la ganaderia filipina en 1887," BNM, ms. 19.218.

16. Islas Filipinas, "Estado de importación y exportación del Puerto de Manila, 1818" (Archivo Franciscano Ibero-Oriental 111/4).

17. "Estados sobre el Movimiento comercial de Filipinas con Emuy," *Gaceta*, 8Mar1861; de Bosch, "Sual," Hs. of Com., 1857, v. 16, C2201, 531; De la Cavada, *Historia Geográfica*, 1:387. Oscar Evangelista, personal communication, 2Jul2000.

18. Noceda and Sanlucar, *Vocabulario*.

19. On deer keeping, consumption, and the trade in venison and deer hides, see Peter Boomgaard, "The Age of the Buffalo and the Dawn of the Cattle Era in Indonesia, 1500–1850," in Peter Boomgaard and David Henley, eds., *Smallholders and Stockbreeders: Histories of Foodcrop and Livestock Farming in Southeast Asia* (Leiden: 2004), 276–77.

20. Lewis E. Gleeck Jr., *Americans on the Philippine Frontiers* (Manila: 1974), 187.

21. E. N. Anderson, *The Food of China* (New Haven: 1988), 145.

22. Yldefonso de Aragon, "Descripción geográfica y topográfica de la ysla de Luzón," in Horatio de la Costa, SJ, ed., *Readings in Philippine History* (Manila: 1965), 158–59.

23. Isagani R. Medina, *Cavite before the Revolution (1571–1896)* (Quezon City: 1994); Bankoff, *Crime, Society*, 79–81; Pauli, "Manila," Hs. of Com., 1880, v. 74, C2632, 1370; Blair and Robertson, *Philippine Islands*, 40:215n; RPC 1906, pt. 2, 203; "Stolen Carabao," MT, 6Apr03.

24. My thanks go to Rose Marie Mendoza, Tina Bernabe, and Dennis Santiago, who systematically combed the *contribución industrial* in search of the sparse records for the *especulador de vacuno* (cattle buyer) and other categories.

25. Six were named Cabrera, five Ylagan, three Atienza, and three Punzalan.

26. Matthew Turner, personal communication, 5Feb2001.

27. John E. Rouse, *The Criollo: Spanish Cattle in the Americas* (Norman: 1977), 11, 50–53.

28. This was an advantage of being located in Wallacea, the broad biological transition zone between most of Southeast Asia and Australia–New Guinea. Peter Boomgaard, *Frontiers of Fear: Tigers and People in the Malay World* (New Haven: 2001), chap. 2; James C. Scott, "Agricultural Conditions in Masbate," PAgR 3.6 (Jun10): 355–58.

29. Super, *Food, Conquest, and Colonization*, 28–32.

30. Cushner, *Landed Estates*, 37, 42, 121n56, quotes 39 and 121n7; De la Costa, *Jesuits in the Philippines*, 275–76; Roth, *Friar Estates*, 177n7; Dennis Morrow Roth "Philippine

Forests and Forestry, 1565–1920," in Richard P. Tucker and J. F. Richards, eds., *Global Deforestation and the Nineteenth Century World Economy* (Durham: 1983), 31.

31. Michael Cullinane, personal communications 3Sep99 and 2Aug2005.

32. Doeppers, "Lighting a Fire," 426–28. On Burias the Spanish firm of Gutierrez Hermanos established a pastured estate of more than 11,000 hectares, which was said to have had 20,000 cattle on the eve of World War II. "Statement Showing the Names," 26–27.

33. Jagor, *Travels*, quote 64.

34. See Dery, *From Ibalon to Sorsogon*, 97–99; Owen, *Prosperity without Progress*, 168; and "Estado de la ganaderia filipina en 1887," BNM, ms. 19.218.

35. McLennan, *Central Luzon Plain*, 206–9, quote 156. *Parang* is a mixed open prairie; a *cogonal* is mainly composed of *cogon/Imperata*. Wernstedt and Spencer, *Philippine Island World*, 102–4.

36. Scott, "Agricultural Conditions in Masbate," 355–58; Edwards, "Livestock Industry," 142–44; Stanton Youngberg, "The Fresh Beef Supply of the Philippine Islands," PJAg 1.4 (1930): 306–7; De la Cavada, *Historia Geográfica*, 1:378, 2:351, 413.

37. Ellis, *Hong Kong to Manila*, 96. In the 1886–87 survey, Jalajala municipality reported one *estancia* with 920 cattle, horses, carabao, and hogs. "Estado de la ganaderia filipina en 1887," BNM, ms. 19.218, Morong.

38. Bankoff, "Horsing Around," 240.

39. Josue Soncuya, "Castración de Ganado Mayor," PEF 1.3 (June 1923): 31. Castration of goats is known from the seventeenth century. Alcina (1668), *Historia*, in PSacra 96 (Sep–Dec97): 516–17.

40. "Livestock and Poultry in the Philippine Islands," PAgR 4.9 (Sep11): quote 484.

41. Soncuya, "Castración," 31; De la Cavada, *Historia Geográfica*, 2:465; David C. Kretzer, "Castration of Cattle on the Range," PJAI 1.2 (Mar34): 77–83. Male carabao used as plow animals were routinely neutered in the 1930s. *Census 1939*, 2:1102.

42. Peter Boomgaard, personal communication, 29Jan2003.

43. Buencamino, "Driving Skeletons to the Meat Market," PJC 9.12 (Dec33): 8; "Livestock and Poultry," 471; G. W. Peters, *Picturesque Manila* (Manila: 1899), unpaginated, photo of mounted cowboys.

44. Buzeta and Bravo, *Diccionario Geográfico*, 1:362, 370, 374 and 2:436; Montero y Vidal, *El Archipiélago Filipino*, 308, 316.

45. Matthew Turner, personal communication, 5Feb2001.

46. See RPC 1900, pt. 4, 14; "Need for Diversification," 560–63.

47. See Pauli, "Manila," Hs. of Com., 1880, v. 74, C2632, 1370; Pauli, Hs. of Com. 1880, v. 75, C2713, 1851; and Sawyer, *Inhabitants*, 198. On the Batangas fair, see Blair and Robertson, *Philippine Islands*, 52:321 (livestock prizes); "Fiestas Publicas," *El Comercio*, 2Dec1872, 6Feb1875, 16Jan1891; and C. W. Edwards, "The Batangas Livestock Show," PAgR 6 (Mar13): 131–36.

48. On cattle types, see Youngberg, "Fresh Beef Supply," 303, 307. See also Crisologo Atienza, "Batangas: The Premier Province in Animal Production," PJC 13.12 (Dec37): 13, 16; ms. ARBAI, 1930, 16, and 1933, 162; and Juan J. Angeles, "Marketing Cattle and Beef," PJC 14.5 (May38): 13. Sawyer says Ilocos had fine cattle in the 1890s, but they were sold to upland peoples rather than in Manila. Sawyer, *Inhabitants*, 251.

49. Rodney J. Sullivan, *Exemplar of Americanism: The Philippine Career of Dean C. Worcester* (Ann Arbor: 1991), 215–23; Ronald Edgerton, "Americans, Cowboys, and Cattlemen on the Mindanao Frontier," in Peter W. Stanley, ed., *Reappraising an Empire: New Perspectives on Philippine-American History* (Cambridge: 1984), 171–97; SCR, J. F. Boomer, "Philippine Islands," 1916, 2:80a, 8; "Like the 'Great West' of America," PFP, 11Mar22, 2. See also Kretzer, "Castration," 77–83; and Hilario G. Fusilero, "Spaying Heifers on the Ranch," PJAI 2.4 (Jul–Aug35): 297–301.

50. Ms. ARBAI, 1933, in PJAI 1.4 (Jul–Aug34): quote 231.

51. Both were built in 1851, fell in the great earthquake of 1863, and were rebuilt or refurbished at the turn of the century.

52. Ms. RPC, ARMB, 1907–8, 341, 352; ARCIR, 1919, 32.

53. Horowitz, Pilcher, and Watts, "Meat for the Multitudes," 1054–83.

54. Antonio, "Survey of Pork Marketing," pl. 2; Wickberg, *Chinese in Philippine Life*, 109.

55. Michael Cullinane, personal communication, 3Feb2004.

56. Mackenzie, "Manila," Hs. of Com., 1878–79, v. 70, C2285, 589. Late 1894 and early 1895 saw arrests of clandestine operators at sites in Paco, San Nicolas, and Santa Cruz. *El Comercio*, "Matanza clandestina," 2Nov1894, 9, 26 31Jan1895; and 4Feb1895.

57. By 1902 the slaughter fee had been converted into a charge per kilogram of dressed meat.

58. "La trichina," *El Comercio*, 10Jun1881; San Agustin, "Meat Inspection," 519–20.

59. Sevilla, "Slaughter of Livestock," 139–46; San Agustin, "Meat Inspection," 517–23; "Matadero Publico"; "De la Board of Health," *El Comercio*, 3Dec00, 4; RPC 1902, pt. 1, 137; RPC 1903, pt. 1, 650; RPC 1904, pt. 1, 221; RPC 1905, pt. 1, 608; ARBAg, 1911–12; PAgR 5.12 (December 1912): xx; V. Buencamino, *Memoirs*, 69, 133; E. A. Rodier, "The Relation of Domestic Animals to Human Health," PJAg 1.4 (1930): 445–50; ms. ARBAI, 1930, 67–69; ms. ARBAI, 1933, 119–20; "Meat Fights Meat," *Graphic*, 6Apr32, 12–13, 45; "City Plans New Slaughterhouse," *Tribune*, 9Oct36, 2; "Problems of a City Health Officer," PFP, 3Aug40, 24.

60. See Cronon, *Nature's Metropolis*, 225–35, on Chicago as "porkopolis."

61. Filomeno G. Domingo, "Marketing Cattle and Beef," PJC 9.5 (May33): 9; V. Buencamino, *Memoirs*, 133.

62. This pattern of commercial behavior dates back to the 1840s at least. Owen, "Americans in the Abaca Trade," 217.

63. Ms. ARBAI, 1933, 20. See also the ads "The Family Will Enjoy a Change," International Cold Stores," MT, 25Sep12; "The 'Cut's' the Thing," Philippine Cold Stores, MT, 26Sep12; and "Prime Australian Roast Beef," Philippines Cold Stores, *Bulletin*, 31Jan25.

64. The map includes only the number of head slaughtered in Manila by province of origin. The reader will have to make an appropriate mental adjustment for the much higher average dressed weight of animals from Batangas and Bukidnon.

65. Vicente Ferriols, "Foot and Mouth Disease at the City Slaughterhouse," PJAI 2.5 (1935): 349–55. Angeles, "Marketing Cattle and Beef," 19.

66. Donald G. McNeil Jr., "Rinderpest, Scourge of Cattle, Is Vanquished" NYT, 28Jun2011, D1, 4.

67. De Bevoise, *Agents of Apocalypse*, 158–59.

68. *Hungerford's Diseases of Livestock*, 9th ed. (Sydney: 1990), 386–87; *Black's Veterinary Dictionary*, Geoffrey P. West, ed., 16th ed. (London and Totowa: 1988), 124–25. See also Vicente Ferriols, J. D. Generoso, and A. B. Coronel, "Rinderpest in the Philippines," PJAI 10.3 (1950): 289–306; H. E. Keylock, "The Control of Rinderpest in a Large Dairy Herd in Shanghai," JCPT 46.3 (Sep33): 149–58; H. E. Keylock, "Cattle-Plague in China," JCPT 22.3 (Sep09): 193–213; S. Anderson, "An Outbreak of Cattle Plague in China," *Indian Medical Gazette* 36 (Sep01): 327–28; and Harold Brown, "Rinderpest," PNHB 4.2 (1929): 87–94.

69. Dominik Hünniger, "Policing Epizootics: Legislation and Administration during Outbreaks of Cattle Plague in Eighteenth-Century Northern Germany as Continuous Crisis Management," and Peter A. Koolmees, "Epizootic Diseases in the Netherlands, 1713–2002: Veterinary Science, Agricultural Policy, and Public Response," both in Karen Brown and Daniel Gilfoyle, eds., *Healing the Herds: Disease, Livestock Economies, and the Globalization of Veterinary Medicine* (Athens: 2010), 76–91 and 19–41, respectively.

70. Karen Brown, "Conclusion," in Karen Brown and Daniel Gilfoyle, eds., *Healing the Herds: Disease, Livestock Economies, and the Globalization of Veterinary Medicine* (Athens: 2010), 270. Other important bovine diseases, including foot-and-mouth disease and anthrax, also reached the Philippines. See Jose Ll. Reyna, "First Anthrax Outbreak in Leyte," PJAI 3.5 (1936): 397–410; and "Threat of an Anthrax Outbreak," PJAI 3.4 (Jul–Aug36): 249–60. The Leyte outbreak left seventeen humans dead.

71. H. W. Smyth, *Five Years in Siam*, app. 6, 2:285–87, quote 285.

72. Martine Barwegen, personal communication, 10Mar2003; Martine Barwegen "For Better or Worse? The Impact of the Veterinarian Service on the Development of the Agricultural Society in Java in the Nineteenth Century," in Karen Brown and Daniel Gilfoyle, eds., *Healing the Herds: Disease, Livestock Economies, and the Globalization of Veterinary Medicine* (Athens: 2010), 92–107.

73. C. W. Daniels, "The Outbreaks of Rinderpest in Selangor, 1903 and 1904," *Journal of Tropical Veterinary Medicine* 2 (1907): 159–62.

74. *The Merck Veterinary Manual*, 9th ed. (Whitehouse Station: 2005), 619–20. Almost 1,300 "sheep" were imported in 1890. The category was then relabeled "sheep and goats," and the numbers jumped to 6,400 and 9,800 in 1891 and 1894, respectively. See *Balanza* for the years indicated; Sawyer, *Inhabitants*, 220; and Jagor, *Travels*, 112, 208.

75. J. Ph. Hens, CVB, 1888 (concerning 1887), quotes 1049. Hens's mention of a "southeast" monsoon is either a reference to the general wind direction in May or a misstated reference to the main southwest monsoon of June–September.

76. Gines Geis Gotzens, *Una epizootic en Filipinas* (Manila: 1888), 16–17.

77. Larkin, *Pampangans*, 207; De Bevoise, *Agents of Apocalypse*, 158–59.

78. De Bevoise, *Agents of Apocalypse*, 159.

79. Foreman, *Philippine Islands*, 391; De Bevoise, *Agents of Apocalypse*, 159–60; Paul Rodell, personal communications, 21Oct99 and 27Apr2000, including copies of PNA documents Estadistica, Zambales, ganados, 1886, and "Memorias," Zambales, 1892; Jose Vicente Braganza, "Alaminos," *Ilocos Review* 10 (1979): 96. Rodell emphasizes cattle as

wealth in Zambales. Paul Rodell, "La Iglesia Filipina Independiente, 1902–1910" (PhD diss., State University of New York, Buffalo, 1992), 69–72, 142.

80. Annual data on slaughter cattle origins are from *El Comercio*, "Matanza de reses," 1Feb1886, 15Jan1887, and 9Jan1888.

81. Cattle from Batangas slaughtered in Manila increased from 3,900 in 1885 to 4,350 in 1887.

82. Sastrón, *Batangas*, 227; Rodriguez, "Memoria general de la Provincia de Cavite," 152.

83. Vicente Ferriols, "A Brief Resume of Rinderpest Control Work in the Philippines," PJA 1.4 (1930): 393.

84. *El Comercio*, "Matanza," 11Jul1888; P. K. A. Meerkamp, CVB, 1890 (on 1889), quote 7–8.

85. Jovellanos, *Traditional Veterinary Medicine*.

86. "La Epizootica," *El Comercio*, 11Jan1891.

87. J. Ph. Hens, CVB, 1888 (on 1887), quotes 1049; Angel K. Gomez, "Eradication and Control of Rinderpest in the Philippine Islands," JAVMA 113.857 (Aug48): quote 113; Barwegen, "For Better or Worse?"

88. *Merck Veterinary Manual*, 620.

89. De Bevoise, *Agents of Apocalypse*, 161–63; Owen, *Prosperity without Progress*, quote 181.

90. RPC 1900–1901, pt. 2, quote 93; *Census 1903*, 4:236, 373–76.

91. Scheerer, "Batanes Report," 273.

92. Ileto, "Hunger in Southern Tagalog," 113–15; De Bevoise, *Agents of Apocalypse*, 158–61; May, *Battle for Batangas*, 72–73, 202.

93. Youngberg, "North to South Movement," quote 654; Tecson, "Agricultural Conditions in Tarlac Province," 301; "Provincial Reports," PAgR 3.10 (Oct10): 589; Tornow, "Economic Conditions," 40.

94. Philippine Commission hearings, RPC 1900–1901, pt. 2, 194, 210, 278, quotes 24–25, 237.

95. Clarence-Smith, "Diseases of Equids," 132–34.

96. De Bevoise, *Agents of Apocalypse* esp. 161–63, 235–37. See also May, *Battle for Batangas*, 266–67.

97. This policy had unforeseen consequences in China itself, as in Guangxi Province (west of Canton) a British customs official reported, "A new cause of local disaffection is the disappearance of the farmers' cattle: it is said tens of thousands have been taken away to supply the want now felt in Manila—the number is probably exaggerated, but the fact is fact. The agriculturists are alarmed as they lose their ploughing oxen etc.—mostly carried off by brigands and soldiers and sold to American agents for the Philippines." Hart to Campbell, April 12, 1903, in Chan Xiafei and Han Rongfang, eds., *Archives of China's Imperial Maritime Customs, Confidential Correspondence between Robert Hart and James Duncan Campbell, 1874–1907*, 4 vols. (Beijing: 1992), 3:724. This reference was kindly provided by Dan Meissner.

98. "Trade and Commerce of the Philippine Islands, 1898," Hs. of Com. 1899, v. 101, 2319, 3; Director of Agriculture, "The Animal-Disease Problem," PAgR 1.5

(May08): 186–93; CRUSFC, 1907, vol. 1, "Amoy," 349; "Zafiro's Big List," MT, 14Jan08; "Zafiro Ties Up," MT, 24Feb08.

99. In 1903, 1,805 water buffalo were purchased in the Shanghai area, of which 1,370 were alive and accepted by the insular purchasing agent following temporary immunization. A further 429 soon died, leaving just 941 to be distributed. "Review of Philippine Commerce," 179–80; Meerkamp, CVB, 1903, 958–59; Meerkamp, CVB, 1904, 1167; Meerkamp, CVB, 1905, 834; Montemayor, "Half-Century of Livestock Raising," 287. Sale believes many fewer carabao were finally imported. Pedro S. Sale, "History of the Importations of Carabaos," PJAI 2.5 (Sep–Oct35): 345–47.

100. By 1902 some French colonists were planning to establish cattle ranches in southern Annam "with a view to supplying the American demand in the Philippines, a certain quantity being already exported thither." Little, "Report on the Affairs of Indo-China 1902," Hs. of Com., 1904, v. 98, 3117, 6; Henri Brenier, *Essai d'Atlas Statistique de l'Indochine Française* (Hanoi: 1914), 190–92.

101. Brocheux, *Mekong Delta*, 98–99.

102. Charles G. Thomson, "Report on the Animal Industry of Indo-China," PAgR 1.5 (May08): 197–98.

103. "Cattle Had Rinderpest," MT, 22Jul07, 15; Director of Agriculture, "Animal-Disease Problem," 186–93.

104. Fung, *Beneath the Banyan Tree*; Lichauco, *Family Recollections*, 4, 17–23, 34; *Rosenstock's Manila City Directory*, 1921, 495. See articles in MT: "Bids ₱60,000 for Talim Quarry" and "City Can Do Cheaper Work," both 17Jun07; and the column "Municipal Board Notes," 22Jun07 and 16Dec15. See also Michael Cullinane, *Ilustrado Politics: Filipino Elite Responses to American Rule, 1898–1908* (Quezon City: 2003), 132.

105. On Soriano, see "Cattle Are Infected," MT, 30Sep08. On Smith Bell, see "Marine and Shipping," MT, 2Feb10.

106. Youngberg, "North to South Movement," 653–59, quote 654.

107. V. Buencamino, *Memoirs*, 107–8. See also Larkin, *Pampangans*, 258–61.

108. RPC 1911, 170–71.

109. The first law banning the importation of diseased cattle took effect on January 1, 1907, but it had little immediate impact.

110. Soriano was well connected in his own right, but starting in 1920 his partner in the importing business was veterinarian Victor Buencamino, whose brother, Assemblyman Felipe Buencamino Jr., became the "boss" of Nueva Ecija politics and a nationally powerful figure. See Fegan, "Entrepreneurs in Votes and Violence," 54–55, 66.

111. "Cattle Had Rinderpest," (on Lichauco); "Cattle Are Infected" (on Soriano); "Long-Drawn Battle against Rinderpest," PFP, 12Sep08, 3. Between 1904 and 1921, some 20 cattle and livestock dealers were listed at least once in the annual *Rosenstock's Manila City Directory*. Besides Soriano and Lichauco, these included Barretto and Co. in 1904–6, Jose Flameño in 1904–11, and Gregorio Olegario in 1911.

112. Arthur Stanley, "Notes on an Outbreak of Cattle-Plague in Shanghai," *Journal of Hygiene* 2.1 (1902): 44; *Lichauco Family Reunion*, 23. On Sisiman, see PAgR 4.10 (October 1911): 544. From 1911 to 1916, Lichauco transported cattle in a vessel that he renamed *Sisiman*. Fung, *Beneath the Banyan Tree*, 71.

113. ARBAg, 1910–11, in PAgR 5.1 (Jan12): 18–19; ARBAg, 1911–12, in PAgR 5.12 (Dec12): xx; ARBAg, 1911–12, in PAgR 4.10 (Oct11): 544. A weaker form of quarantine in the port of Manila was in place in 1908, with imported animals restricted to the owners' corrals until released following a health inspection. PAgR 1.1 (Jan08): 51; "Veterinary Work of the Bureau of Agriculture," PAgR 1:11 (Nov08): 447–49.

114. ARBAg, 1911–12, in PAgR 5.12 (Dec12): xviii–xix; "Inspection of Cattle of City," MT, 25Sep12.

115. George S. Baker, "Cattle Importation," PAgR 5.12 (Dec12): 660–61.

116. See *Black's Veterinary Dictionary*, 125; Ferriols, Generoso, and Coronel, "Rinderpest," 29; "Hits Million Mark Again," PFP, 14Jun13, 2–3; and ms. ARBAg 1919, 38.

117. V. Buencamino, *Memoirs*, 134–35, quote 135.

118. Ms. ARBAg, 1918, 118; SBPI, 1928, no. 11, 58. On the pressure to restart live cattle imports from Hong Kong, see DCTR, vol. 17, no. 23, 28Jan14, 357; vol. 17, no. 38, 14Feb14, 605; vol. 17, no. 125, 28May14, 1166; vol. 18, no. 231, 2Oct15, 30; and vol. 18, no. 241, 14Oct15, 201.

119. Two early graduates of the University of the Philippines' College of Veterinary Science, Todulo Topacio and Ildefonso Patdu, assisted in this work. V. Buencamino, *Memoirs*, 112.

120. Brown, "Rinderpest."

121. Ms. ARBAg, 1916, 40–47; 1917, 35; 1918, 108–10; 1921, 26; 1933, 135. See photos in USNA II, RG350-P-Am-9-4 thru 7, box 5, file 4 and PJAI 1.6 (Nov–Dec34): 468. See also Antonio Peña, "Cattle Raising in the Philippines," CIJ 6.7 (Jul30): 7.

122. Ferriols, Generoso, and Coronel, "Rinderpest," 303; ms. ARBAI, 1933, in PJAI 1.4 (Jul–Aug34): quote 223.

123. In August 1918, 105 cattle, "badly infected with foot and mouth disease of a virulent type," arrived from Saigon for Lichauco. Ms. ARBAg, 1918, 133. From Hong Kong 100 cattle were consigned to Go Tamco Hermanos. "Loongsang to Bring Cargo of Livestock," MT, 30Aug19, 2.

124. Import sources and figures are from reports in PAgR, 1911–13; ms. ARMB, 1910; ms. ARCIR, 1913; and ms. ARBAg, 1916–21. China provided almost 3,000 head in 1916.

125. *Lichauco & Co. v. Apostol and Corpus*, 19628, *Philippine Reports* 44 (4Dec22): 138–165 (lawphil.net/judjuris/juri1922).

126. V. Buencamino, *Memoirs*, 131–38, quotes 134, 137; ms. ARBAg, 1919, 39.

127. J. Collas, "Manila's Meat Supply," PFP, 13Sep30, 20–21; J. Collas, "The Battle of Beef," PFP (20Sep30): 51–52.

128. Half the beef consumed in Manila in 1929 arrived frozen from abroad, especially Australia. Collas, "Manila's Meat Supply," 20.

129. "Carnes refrigerada de Australia," *El Comercio*, 1Aug1899.

130. Ms. ARBAI, 1933, quote 19. See also "Carabao Meat," *Tribune*, 16Sep37. During 1937–39, Manila's slaughter carabao came from Masbate (26 percent), Bikol (19 percent), Batangas (11 percent), and Mindoro (11 percent). PSR 4.4 (1937): 615; BPS 1–2 (1939): 141; BPS 3 (1939): 63. On the end of imports, see "Local Beef Market," CIJ (Jul27): 7; ms. ARBAI, 1930, 7–9; and "The Economics of Hard Times," PFP, 14Feb31, 30.

131. Mauro F. Manuel et al., eds., *A Century of Veterinary Medicine in the Philippines, 1898–1998* (Quezon City: 2002).

132. De la Cavada, *Historia, Geográfica*, 1:370, 378; 2:337, 351, in each case taking the larger of the reports of population by province. *Census 1903*, 2:123, 4:235, 239; *Census 1939*, 2:45, 1117, 1120, 1135, combining the numbers of animals on and off farms.

CHAPTER 9. FLUIDS OF LIFE

1. On the grotesquely divergent mortality rates in the otherwise quite similar port cities of Hamburg and Bremen when the political elites of one decided to invest in a sanitary water supply and the other did not, see John C. Brown, "Coping with Crisis? The Diffusion of Waterworks in Late Nineteenth-Century German Towns," JEH 48.2 (Jun88): 307–18.

2. Swyngedouw, *Social Power and the Urbanization of Water*; Xavier Heutz de Lemps, "Une urgence de cent cinquante annes: La construction de lamene deau de Manille (1733–1882)," in Denis Bocquet and Samuel Fettah, eds., *Resaux techniques, modernisation urbaine, et conflits de pouvoir (XVIIIe. XXe. siecle)* (Rome: 2004), read in Spanish manuscript, 2001; William Gervase Clarence-Smith, *Cocoa and Chocolate, 1765–1914* (London: 2000). See also Santiago Artiaga and M. Mañosa, "A Brief of Account of the Spanish Projects to Supply Manila with Potable Water," *Unitas* 11.5 (Nov32): 263–87.

3. Artiaga and Mañosa, "Brief of Account," 264–66.

4. Manuel, *Dictionary of Philippine Biography*, 1:482. Eleven *bancas* carrying jars of water to the city were recorded in a one-week sample of Pasig River traffic in 1853. Islas Filipinas, Comisión Central de Estadística de Filipinas, *20 cuaderno*, 1855, 67.

5. Buzeta and Bravo, *Diccionario Geográfico*.

6. Massimo Livi-Bacci, *Population and Nutrition: An Essay on European Demographic History*, Tania Croft-Murray, trans. (Cambridge: 1991), 38.

7. Alvin J. Cox, George W. Heise, and V. Q. Gana, "Water Supplies in the Philippine Islands," PJS 9.4 (Jul14): 273–74. Although it attributed the situation to a different cause, *El Comercio* ran statistics for 1876 showing that Manila had the highest overall death rate among lowland provinces. "La trichina," *El Comercio*, 10Jun1881. Cf. Gerry Kearns, "Death in the Time of Cholera," JHG 15.4 (Oct89): 425–32.

8. These themes are developed by Xavier Heutz de Lemps in "Une urgence." I appreciate the opportunity to read this essay in the original Spanish manuscript. Cf. Salt, "Francisco de Carriedo y Peredo," 194, quote 173n22; Consul Wilkinson, "Manila," 1882, Hs. of Com., 1883, v. 72, 513; Sawyer, *Inhabitants*, 15–16; Harry F. Cameron, "Municipal Water Supply," QBBPW 1.3 (Oct12): 16–18; and Dolores Romero Muñoz, "The Supply of Fresh Water to Manila," in *Manila, 1571–1898: Occidente en Oriente* (1998, English edition), 241–45. Artiaga and Mañosa review the designs that led to the Carriedo system in "Brief of Account," 263–87.

9. The homes of Gonzalo Tuason and Luis R. Yangco were connected in 1891 and those of Cornelia Laochanco and Joaquin Inchausti in 1893. *El Comercio*, "Servicio de Aguas," 15Jan1891 and 1Aug1893, plus 11Jan1890 on the volume of flow. Cf. Dick, *Surabaya*, 166–68, 175–76.

10. Heiser, *American Doctor's Odyssey*, quote 121. Buzeta and Bravo, *Diccionario Geográfico*, "Tambobong," 2:443; "Nuestro Grabados," *El Comercio*, 29Jul1882, supplement, 1; Frank Lewis Minton, "How the 'Tigbalang' Fought the Waterworks," ACCJ (Apr29): 9, 14–15; RPC 1902 pt. 1, 99; Brown, "Coping with Crisis?" 308.

11. Report by Superintendent of Government Laboratories Paul C. Freer, March 14, 1904, in ARMB, 1903–4, 162–64, quote 163. Cf. RPC 1902, pt. 1, 99, 268, 329–30.

12. O. F. Williams, "Health of Manila," RCUS, 1899, v. 60, no. 225, quote 295; Sawyer, *Inhabitants*, 184; ARMB, in RPC 1905, pt. 1, 494; "Meant It for Bipeds," MT, 5Jan07, 1; Heiser, *American Doctor's Odyssey*, 111, 118–20, 125, 130. Cf. Report of the Commission of Public Health, in RPC 1903, pt. 2, 94; and a discussion of charcoal home filters in Sta. Maria, *Governor-General's Kitchen*, 164.

13. Among many admonitions concerning water, see Joseph A. Guthrie, "Some Observations While in the Philippines," JAMS 13 (1903): 148. According to Worcester, at least 4,386 died in the city during this two-year epidemic, over 1,000 in July 1902 alone, and more than 105,000 in the provinces. Dean C. Worcester, *History of Asiatic Cholera in the Philippine Islands* (Manila: 1909), 20, 25.

14. Advertising quotes are from MT 14Feb08; and *El Renacimiento*, 11May09. See also MT, 21May07, 8Jul07, 13Jul07, 29Jul07, 14Jan08, and 28Sep08; *Razón*, 10Mar10; Cox, Heise, and Gana, "Water Supplies," 343; MacMillan, *Seaports of the Far East*, 393; and Rufino Abriol, "Amoebic Abscess of the Liver among Filipinos," PJS 12.3 (May17): 121–46.

15. Paul S. Sutter, "Tropical Conquest and the Rise of the Environmental Management State: The Case of the U.S. Sanitary Efforts in Panama." In Alfred W. McCoy and Francisco Scarano, eds., *Colonial Crucible: Empire in the Making of a Modern American State* (Madison: 2009), 317–26.

16. Anderson, *Colonial Pathologies*, chap. 1 ("American Military Medicine Faces West"); Erik Larson, *Devil in the White City* (New York: 2003), 138.

17. Leopoldo A. Faustino et al., *Manila Water Supplies* (Manila: 1931), 14–15; Leopoldo A. Faustino, "The Water Supply of Manila from Underground Sources," PJS 45.2 (Jun31): 119–49.

18. James Case supervised construction of the Montalban Dam. He was appointed city engineer in 1903 and soon became the chief engineer of city sewer and water construction. Following completion of the dam, Governor-General Forbes made him director of an expanded Bureau of Public Works. After 1910 he became general manager of the Cuban Construction and Engineering Co., which was responsible for building Havana's drainage works. Later he worked rebuilding war-torn Europe and eventually received the Laurie prize of the American Society of Civil Engineers. Halsema, *E. J. Halsema*, 30–31, 49, 331n23–24.

19. On the right of the constabulary to guard streams on private lands, see "City Water Supply Must Be Protected," MT, 24Aug19.

20. Nancy Peluso, "Territorializing Local Struggles for Resource Control," public lecture, University of Wisconsin–Madison, 12Sep97. Owning or buying standing coconut trees without owning the land is sometimes recognized in the Philippines.

21. It is possible that a filtration system was not built because the Municipal Board overspent in 1913 and then had to practice stringent economy in 1914. RPC 1914, 32.

22. "Let's Raise a Monument to Major Case," PFP (24Dec08): 3.

23. RPC 1913 quotes III, 122; Charles Burke Elliott, *The Philippines to the End of the Commission Government* (Indianapolis: 1917), 202–3; Cox, Heise, and Gana, "Water Supplies," 274–85, 342–44, esp. 283–85; George W. Heise, "Notes on the Water Supply of the City of Manila," PJS 11A.1 (January 1916): 1–13; Heiser, *American Doctor's Odyssey*, 121–32; "Report of the City of Manila," in RPC 1901–2, pt. 1, 88; RPC 1912, 34–35. Cf. the report by Superintendent of Government Laboratories Paul C. Freer, March 14, 1904, in ARMB 1903–4, 162–64. Another possibility for typhoid and dysentery transmission is food handled by a chronic carrier. A large study found this to be highly unusual. O. Garcia and A. Vazquez-Colet, "Bacteriological Examination of Stools of Food Handlers in Manila," PJS 24.6 (Jun24): 735–41.

24. The urban population grew by at least 12 percent in five years. Doeppers, *Manila, 1900–1941*, 45.

25. A very different water system was built in Shanghai's International Settlement. This took water directly from the highly polluted Huangpu River, treated it in settling tanks with a coagulant to remove "fine particles of solid matter," and passed it through multiple sand and gravel filtration beds. Bartlett Yung, "The Water Supply of Greater Shanghai," *Journal of the Association of Chinese and American Engineers* 11.3 (Mar30): quote 34.

26. Previously it was known as the City Department of Sewer and Waterworks Construction, 1912–21, and before that the Division of Sewer and Waterworks Construction in the city Department of Engineering and Public Works. The new board initially included the mayor as president, plus the president of the elected Manila Municipal Board, the elected governor of Rizal Province, the director of the Insular Bureau of Public Works, and three others. By 1926 it included the same officers plus the Manila city engineer and the city treasurer, as well as a public member—often foreign businessman and Manila "booster" Horace B. Pond.

27. Gideon joined the Philippine civil service in 1903 and had been water supply and sewers superintendent in the city's Department of Engineering and Public Works since at least 1911.

28. Metropolitan Water District, "Extension of Water Supply, Preliminary Report to the Board of Directors," by A. Gideon, Manager, August 1922, with geological reports, quotes 40 and 61 (copy in possession of the author). In 1921 the board included Mayor Ramon J. Fernandez, Municipal Board president Ramon R. Papa, Andres Gabriel, engineer Jose Paez of Rizal (the best-educated engineer of his day and first Filipino director of the Bureau of Public Works), and appointees Pond, Vicente P. Genato, and politician Felipe Buencamino Jr. The assistant manager on July 1, 1921, was Federico J. Muñoz. *Official Roster of Officers and Employees in the Civil Service of the Philippine Islands*, 1921.

29. "Manila's Water Supply," 11. Cf. ACCJ (May32): 5, map.

30. George W. Heise, "Water Supplies in the Philippine Islands II," PJS 10A.2 (Mar15): 149; RGGPI, 1925, 14–15; "Water Lack Serious," *Bulletin*, 17May26; "City Water Getting Low," *Bulletin*, 15May29; Novaliches Dam, World's Largest Earth Barrier," *Bulletin*, 14Jun29; "The Typhoid Outbreak," *Herald*, 12Jan27; "Annual Report,

Metropolitan Water District"; ms. RGGPI, 1928, 8–11; ms. RGGPI, 1932, table 1; Faustino et al., *Manila Water Supplies*; P. I. de Jesus and J. M. Ramos, "Effect of Filtration on the Sanitary Quality of the Water of the Metropolitan Water District," PJS 59.4 (Apr36): 455–71.

31. An attempt by Sakdals to blow up the water main was rendered unsuccessful when repairs took less than 24 hours. Doeppers, "Metropolitan Manila," 532. On the leak, see "Water Supply Normal Today," *Tribune*, 13May36.

32. Work on a seven-kilometer tunnel extending southward from the dam site was under way in 1929. A. D. Alvir, "A Geological Study of the Angat-Novaliches Region," PJS 40.3 (Nov29): 392.

33. "Move to Prevent Water Shortage," *Tribune*, 2Dec38; Philippines (Commonwealth), Metropolitan Water District, *Report of the Metropolitan Water District*, 1936, and *Report of the Metropolitan Water District*, January 1-June 30, 1939, 15–16.

34. Faustino et al., *Manila Water Supplies* quote 5; Faustino, "Water Supply," 130, pls. 2–3.

35. A *banga* is a large pottery water jar, long an essential item of kitchen furnishing. Mrs. Estevan Munarriz and daughters Dra. Natividad and Rosario Munarriz, interview with the author, 12Mar85.

36. Mr. and Mrs. Antonio Sumbillo, interview, 14May85.

37. Lei Zhang, dissertation project description, "Water Lords: Water Carriers in Beijing, 1900–1937," Syracuse University, 2014.

38. There were "labor market conspiracies" in which male natives of the city dominated the skilled production trades vis-à-vis migrants, and migrants dominated skilled and unskilled construction work in the 1890s. Doeppers, "Migrants in Urban Labor Markets," 253–64.

39. Swyngedouw, *Social Power and the Urbanization of Water*.

40. Faustino et al., *Manila Water Supplies*; Commonwealth of the Philippines, *Report of the Metropolitan Water District*, 1936, 1938, and 1939; Worcester, *History of Asiatic Cholera*, 167.

41. Alfred W. McCoy, *The Politics of Heroin*, rev. ed. (Chicago: 2003), 6; Grace Elizabeth Hale, "When Jim Crow Drank Coke," NYT, 28Jan2013.

42. Ads for *Aguas gaseosas* plants and home delivery published in *El Comercio* include "Fábrica de Aguas Gaseosas de D. Juan Caro y Mora," 12Feb1891, quote; "Fábrica de Aguas Minerales, A. G. Siegert," 15Dec1891; "A. S. Watson Co., Ltd.," 11Apr1891; "de B. Roca," 1Oct1894; and "Luzon Aerated Water Manufactory," 3Oct01. See also the ads "El Rosario," *La Opinión*, Suplemental Ilustrado, 5Mar1888; "La Concepcion," *Rosenstock's Manila City Directory*, 1904–16; and *Razón*, 18Feb–15Oct10. Caro y Mora, a Spaniard, also operated a pharmacy. *El Comercio*, 9Sep1890.

43. Herrera, *Tayabas Chronicles*, 92.

44. In the early decades of the twentieth century, ice was delivered at a cut rate to the homes of insular and city employees by the city cold storage and ice plant. "Special Ice Rate to Gov't Employees," MT, 19Aug19, 6.

45. See the ad "Aguas gaseosas: soda, limonada, zarzaparrilla, tonica," A. G. Siegert, Quiapo, *El Comercio*, 15Dec1891; and ads and listings in *Rosenstock's Manila City Directory*, 1904, 1906, 1908; See also Salazar, "German Economic Involvement," 190–92.

Other pioneers in this industry were the Botica Zobel and Carlos Plitt (d. 1884); the latter advertised "Soda Americana" in *El Comercio*, 18Sep1880.

46. See "La Zarzaparrilla de Bristol," *El Comercio*, 9Aug1881 and 5Jan1884. Initially, this product was carried by the "Botica de Manuel G. Mendieta," itself frequently advertised in *El Comercio*, e.g., on 12Jan1872.

47. In Hong Kong in 1871 Watson was advertising "soda water, lemonade, and other aerated waters." Later it was active in Xiamen as the Aerated Water and Ice Co. In Manila, see ads in El Comercio for "Gran Fábrica, A. S. Watson Co." and "Botica Inglesa," 21April1881, 21Nov1881, 23Sep1882, and 7Dec1890; "Lime Juice," 22Jun1885; and "Cuidado con las Aguas," 11Mar1896. Cf. the ads "Water Analysis" in MT, 18Feb1899 and 12Feb08 (whole page); *El Grito del Pueblo*, 13Aug03; and *Rosenstock's Manila City Directory*, Oct–Dec 1904, 1915, 1931–32.

48. Bautista also became a substantial Tondo landowner, buying properties offered to him. Daughter-in-law Victorina de la Cruz Bautista, interview, 1Jun85. For ads, see *Witton's Directory, 1903*; *Rosenstock's Manila City Directory*, various editions, 1906–37; *Ang Kapatid ng Bayan*, 12Sep06, 2; *Razón*, 15Oct10; and Manila Carnival, *Commercial and Industrial Fair, Commercial Handbook*, 1931, 51.

49. *El Heraldo de la Revolución*, 29Sep–3Nov1898.

50. See ads in *Rosenstock's Manila City Directory*, 1915, 1916; and "Be Cautious" and "Magingat . . ." in PFP, 4Jan30. For Royal ads, see *Trabajo*, 28Feb23; MT 28Nov28; PFP, 7–28Jul28; *Sampaguita*, 30Mar30; and *Tribune*, 13May36. For a "Tru-Orange" ad, see *Herald*, 22Feb39, 16. "Ice Cream and Coke" were listed as the final course of a restaurant's Christmas dinner offering circa 1915. Luning B. Ira, Isagani Medina, and Nik Ricio, *Streets of Manila* (Manila: 1977), 112.

51. Gutierrez and Santos, "Food Consumption," 408–9.

52. Frederick J. Simoons, "The Traditional Limits of Milking and Milk Use in Southern Asia," *Anthropos* 65 (1970): 547–93; Norman Kretchmer, "Lactose and Lactase," *Scientific American* 227.4 (Oct72): 70–78.

53. Simoons, "Traditional Limits of Milking"; Paul Wheatley, "A Note on the Extension of Milking Practices into Southeast Asia during the First Millennium A.D.," *Anthropos* 60 (1965): 577–90; Anderson, *Food of China*, 145; Morga, *Sucesos*, 90; Frederick J. Simoons, "New Light on Ethnic Differences in Adult Lactose Intolerance," *American Journal of Digestive Diseases* 18.7 (July 1973): 602–8.

54. Fr. Alcina describes Chinese as making "quantities of salted cheeses" from carabao milk and selling them locally in seventeenth-century Cebu. Alcina (1668), *Historia*, in PSacra 96 (Sep–Dec97): 528–29. *Kesong* is derived from the Spanish, *queso*.

55. Sampaloc and Santa Cruz recorded the largest number of male *lecheros* among six districts of the city in the mid-1880s, Tondo was not included. PNA, Prestacion Personal, Manila.

56. "La Lechera," *Ilustracion Filipina*, 1Aug1860, quotes, 178–79, reprinted in Sta. Maria, *Governor-General's Kitchen*, 167.

57. Adel P. den Hartog, *Diffusion of Milk as a New Food to Tropical Regions: The Example of Indonesia, 1880–1942* (Wageningen: 1986), 79, 84.

58. "Lecheros," *El Comercio*, 5Jun1889, quote. In this, Manila was like nineteenth-century New York City. Norman Schaftel, "A History of the Purification of Milk in

New York," in Judith Walzer Leavitt and Ronald L. Numbers, eds., *Sickness and Health in America* (Madison: 1978), 276–82.

59. Musgrave and Richmond calculated the infant death rate at 50 percent or more. W. E. Musgrave and George F. Richmond, "Infant Feeding and Its Influence upon Infant Mortality in the Philippine Islands," PJS 2.4 (Aug07): 361, quote 370; W. E. Musgrave, "Infant Mortality in the Philippine Islands," PJS 8B.6 (Dec13): 459–67, quote 462. The calculated death rate was dependent on birth and death reports that were routinely incomplete. See "Cocoanut Oil Tabooed Now," MT, 22May07, 1; and "Getting after Impure Food Dealers," MT 23May07, 4 (editorial). The latter makes explicit some lessons learned from the packinghouse scandals in Chicago.

60. Alex R. Webb, "The Philippine Islands," RCUS, 29, 101, January 1889, quote 182.

61. See ads for Eagle Brand in MT, 26Sep08; and Bear/*Oso* Brand in *El Renacimiento*, 9Dec03, *La Vanguardia*, 10–31Jan22, and signs on the Escolta and Plaza Moraga. *231 Views of Manila and the Philippine Islands* (Manila: ca. 1910), 41, 43.See also ads for Milkmaid in MT, 2Jan13; Dragon Brand, in MT, May–Jun07; and Royal Brand, Swiss Eagle, *Lipang Kalabaw*, 19Oct07.

62. By 1916 the "Nestlé and Anglo-Swiss Milk Co." was advertising "Señorita," "leche natural esterilizada," in PFP, 1Jul16, 28. Nestlé later took over the Milkmaid brand and promoted it with full-page ads and large signs. See the Nestlé ad "Progress Towards Health," *Tribune*, 5Oct30, 6.

63. Musgrave and Richmond, "Infant Feeding," 363; Francisco Agcaoili, "The Composition of Various Milks and Their Adaptability for Infant Feeding," PJS 8A.3 (Jun13): 143–44. In the *Bulletin*, 1935, there were weekly ads for Carnation, Bear Brand, and Nestlé. Advertising for Mayon milk by Daido Boeki Kaisha was concentrated in the Tagalog media including *Taliba*, 25May40. In 1934 and 1935, about 16 million kilos of evaporated and sweetened condensed milk was imported. The United States supplied 86 percent of the total in 1934 and 69 percent a year later. The Netherlands supplied 24 percent in 1935 and led in the evaporated and condensed categories in 1936. Richards, "Philippine Economic Conditions, 636; "Leading Imports," 166; Luz, "Philippines as a Market," 12.

64. Hartog, *Diffusion of Milk*, 92, 102, and chap. 6.

65. Ibid., quote 76–77.

66. Musgrave and Richmond, "Infant Feeding," 366–68, 381–82; Jose Albert, "Treatment of Infantile Beriberi," 81–85. On cigarette advertising, see Doeppers, "Metropolitan Manila," 522–25.

67. In 1984 Nestlé agreed to a marketing code demanded by the European and American boycott movement, one that limits public advertising to poor Third World mothers and stopped the company from undercutting breast-feeding. Hartog, *Diffusion of Milk*, 213.

68. "Appalling Ignorance Kills Manila's Babes," MT, 10Jul12, 9. The committee included Drs. Proceso Gabriel, Luis Guerrero, and W. E. Musgrave.

69. "Wage Campaign for Good Milk," MT, 23Sep12, quotes 1; "Manila Milk Is Condemned," MT, 18Jul12, 1.

70. Schaftel, "History of the Purification of Milk," 286–88.

71. "La Mision de la Gota de Leche," *Renacimiento Filipino* (21Nov10): 3–11; "La Protección de la Infancia, Gota de Leche," *El Renacimiento*, 16Mar09; *Sunday Tribune Magazine*, 26Sep37, 4.

72. "Fresh Milk Supply," 282. Cf. Agcaoili, "Composition of Various Milks," photo following 149; RPC, 1913, 108; and Doeppers, *Manila, 1900–1941*, 170n34.

73. Anonymous interview with the author, Jun85; Oscar Evangelista, personal communication, 2Apr2003. The observation concerning intestinal tuberculosis came in a conversation with Warwick Anderson, 17Oct2003, and is confirmed in B. C. Crowell, "The Chief Intestinal Lesions Encountered in One Thousand Consecutive Autopsies in Manila," PJS 9B.5 (Sep14): 454. Intestinal tuberculosis was discovered in 5.6 percent of autopsy cases in the city in 1912–14.

74. "Fresh Milk Supply," 280–82. Philippine Supreme Court case 14761, *Arce Sons and Co. v. Selecta Biscuit Co., Inc., et al.*, January 28, 1961 (lawphil.net/judjuris/juri1961). The last two facilities were producing about 35 liters a day.

75. This is calculated from *Census 1939*, 2:1139. By this accounting, goats provided 1.4 percent of the national milk supply.

76. Isaac M. Elliott, "The Oriental Market for Dairy Products and Fruits, Philippine Islands," RCUS, 1895, v. 48, 177, quote 205; PNA, Contribución Industrial Manila 1896 (5): 1892–97; "Leche Pura," *El Comercio*, 7Jan1898.

77. Ads for the Vaqueria Australiana appeared in *El Comercio*, 31Mar1898 and 19May11; MT, 18Feb1899; and *Bangon*, 31May09, 19.

78. David Taylor, "London's Milk Supply, 1850–1900: A Reinterpretation," *Agricultural History* 45.1 (Jan71): 33–38.

79. Elsewhere in Southeast Asian and Chinese port cities, Indian sojourners operated small dairies. In Singapore their animals were mostly Murrah water buffaloes with some Multani milk cattle. By 1939 there were likely more than 600 Indians in metropolitan Manila. It is possible they were served by such a dairying operation. Rudolf Wikkramatileke and Karpal Singh, "Tradition and Change in an Indian Dairying Community in Singapore," AAAG 60.4 (Dec70): 717–42.

80. "Manila Dairy Farm," MT, 25Mar1899; the Vaqueria Española, *El Comercio*, 19May11; *Rosenstock's Manila City Directory*, 1916, 1921. On the Lecheria de San Juan del Monte, see *El Comercio*, 3 and 7Oct01. See also PAgR 1 (Jan08): 21; and J. F. Boomer, "Increased Demand for Milk in Philippines," DCTR, 29Dec14, 1350–52.

81. "Model Dairy Opens in City," MT, 10Jul12, 10.

82. "Dairy Produce Dealers," in *Rosenstock's Manila City Directory* and *Manila City Directory*, 1916, 1921, 1926–27, 1933–34, 1936–37, 1941. See also "Fresh Milk Supply," 279. Perez was a "Baghdadi Jew" who also raised sheep and was the ritual slaughterer of sheep for the consumption of the city's Jewish families. Cesar Majul in Clarence-Smith, "Lebanese," quotes 131; V. Buencamino, *Memoirs*, 144.

83. By 1937 its remaining cows had been merged with those of the La Loma Dairy.

84. By contrast, there were at least 32 dairies with a total of 2,500 cows serving Bandung in 1935 in upland Java. The majority of these were run by Dutchmen. Hartog, *Diffusion of Milk*, 210–11.

85. Tomas V. Rigor, "Dairying in the Philippines," PJAI 3.5 (Sep–Oct36): 353–60, quote 355; "Fresh Milk Supply," 279–84; Rodier, "Relation," 449. In order to provide

milk for his wife circa 1920, Victor Buencamino imported an Ayrshire cow from Australia.

86. David G. Fairchild, "Breeds of Milch Cattle and Carabaos for the Philippine Islands," PAgR 4.9 (Sep11): 501–3, quote 503. Fairchild ran the U.S. Department of Agriculture's Office of Seed and Plant Introduction from 1903 to 1928. McCook, "The World Was My Garden," 501; David G. Fairchild, *The World Was My Garden: Travels of a Plant Explorer* (New York: 1938), 152–53, 225.

87. ARBAI, 1930, in ms. ARGGPI, 1930, 12; Stanton Youngberg. "Red Scindi Milch Cows Very Promising." ACCJ 12:9 (Sep32): 5; "Fresh Milk Supply," 282. See Elliott, "Philippine Islands," RCUS, 1895, v. 48, 177, 205, on the rapid deterioration of midlatitude dairy cows in the tropics. See also Wikkramatileke and Singh, "Indian Dairying Community," 717–42.

88. Herminio A. Bernas, "The Grade Sussex-Nellore as Dairy Cattle," PJAI 2.6 (Nov–Dec35): 375–77; M. R. Montemayor, "Half-Century of Livestock Raising," 285–86; Lewis Gleeck, "Hardie Dairy," in *American Business and Philippine Economic Development* (Manila: 1975), 125.

89. Jack J. Rutledge, "Applications of Reproductive Biology Technology and Indigenous Knowledge to Dairying in Southeast Asia," lecture, Friday Forum, Center for Southeast Asian Studies, University of Wisconsin–Madison, November 19, 2004. Rutledge is Professor of Animal Science and Associate Director of the University of Wisconsin Babcock Institute for International Dairy Research and Development.

90. The technical manager was Niels N. Nyborg. "Fresh Milk Supply," 283; San Agustin and Rotea, "Dairy Products"; *La Fábrica de Cerveza*; "Magnolia Announces," *Bulletin*, 22Jan31. See the ads "Leche Magnolia, ang mabuting gatas," and "Masarap na Magnolia Milk Sherbet," in *Mabuhay*, 29Apr34.

91. See the ads "Ang Yelo ay mura," for iceboxes, *Mabuhay*, 29Apr34, 9; "La Nueva Refrigeradora Electrolux," which operated on gas, *El Debate*, 12May35, 6; and "Crosley Shelvador," an electric model, *El Debate*, 12May35, 6.

CHAPTER 10. FOREIGN FASHIONS

1. In a parallel case, Luisa Fernandez Lichauco, wife of cattle magnate Faustino Lichauco, writes of trying her first American grapefruit in 1912 on a Hamburg-America steamer and immediately deciding to send a box to her husband in Manila. By the 1930s imported grapefruits were readily available in the city's upscale groceries. Lichauco, *Family Recollections*, 29.

2. Norman Owen, personal communication, 23Mar2010.

3. Crosby, *Columbian Voyages, the Columbian Exchange, and Their Historians*.

4. Fr. John Schumacher, SJ, personal communication, 21Mar2001; Ruby Paredes, "Memoir," 8Sep2000; Domingo de Salazar, "Relation of the Philippine Islands," in Blair and Robertson, eds., v. 7, 34; Reed, *Colonial Manila*, 56; Joseph E. Spencer, *Land and People in the Philippines* (Berkeley: 1952), 3.

5. Thanks to Resil Mojares for pointing out the derivation of *tinapay*. On this word as a name for rice cakes in seventeenth-century Visayan, see Alcina (1668), *Historia*, in PSacra 85 (1994): 160–63, and in Ilocano, Morice Vanoverbergh, "The Ilocano Kitchen," PJS 60.1 (May36): 4.

6. I. M. Roberts and P. Herington, "Growth in the Japanese Import Demand for Rural Products," *Quarterly Review of Agricultural Economics* 35.3 (Jul72): 208–9. On consumption, see Mears et al., *Rice Economy*, 55, 189.

7. De la Costa, *Jesuits in the Philippines*, 12; de la Costa, *Readings in Philippine History*, 67–68; Maria Lourdes Diaz-Trechuelo, "The Economic Development of the Philippines in the Second Half of the Eighteenth Century," PS 11.2 (Apr63): 201.

8. Paul P. de la Gironière, *Twenty Years in the Philippines* (Manila: 1962), 221; Legarda, *After the Galleons*, 175; Islas Filipinas, "Estado de importación y exportación del Puerto de Manila, 1818," Archivo Franciscano Ibero-Oriental, 111/4; Díaz Arenas, *Report on the Commerce and Shipping of the Philippine Islands*, 54. Bruce Cruikshank reports that in the pueblo of Tayabas in 1820, local people were buying and selling wheat, "which is then transported and sold in Santa Cruz, Laguna," possibly for transport on to Manila. Cruikshank, "Commercial Patterns."

9. *Gaceta de Manila*, Mar1861–Dec1862; Reed, *Colonial Manila*, 53; Cushner, *Landed Estates*, 43; Edmund Roberts, *Embassy to the Eastern Courts*, 53–54.

10. Spencer, "Rise of Maize," quote 13.

11. Buzeta and Bravo, *Diccionario Geográfico*, 1:29, 210; Huerta, *Estado Geográfico*, 170, 178; Cruikshank, *Spanish Franciscans*, 4:234–35, 258; Islas Filipinas, Comisión Central de Estadística de Filipinas, *20 cuaderno*, 1855, "Resumen 1853," and 67.

12. Super, *Food, Conquest, and Colonization*, 33, 35–37; Kaplan, *Provisioning Paris*.

13. Vicens Vives, *Economic History of Spain*, 648–49.

14. Foreman, *Philippine Islands*, quotes 436; "Estado de la agricultura filipina en 1887," BNM, ms. 19.218; P. J. Wester, "Situation in the Citrus District," 126; De la Cavada, *Historia Geográfica*, 2:420.

15. William Armstrong Fairburn, "The Manila Trade," in *Merchant Sail*, 6 vols. (Center Lovell: 1945–55), 4:2443–53; *Gaceta de Manila*, 1862; USNA, "Dispatches from U.S. Consuls in Manila, 1817–1899," microcopy T-43, rolls 2–3; "Consular Dispatches, Hong Kong," FM 108.

16. C. H. Cowan "Philippine Islands, Manila," American Flour, RCUS, 1894, v. 45, 165, quote 309. On Philippine exports to California, see G. T. Ricketts, "Manila Trade and Commerce, 1875," Hs. of Com. 1876, v. 75, C1486, 892.

17. On import duties, see Legarda, *After the Galleons*, 192, 199.

18. No clear pattern emerges from these data on importing Castilian flour under the protective tariffs of the 1880s and 1890s, probably because Spain itself was now a deficit producer.

19. Daniel James Meissner, "Shanghai Success: The Development of the Chinese Mechanized Flour Milling Industry, 1900–1910" (PhD diss., University of Wisconsin–Madison, 1996), quote 61; Charles Klinck, "American vs. European Trade in the Philippine Islands," RCUS, 1882, v. 8, 25½, quote 299; Oscar F. Williams, "The Philippine Islands," RCUS, 1898, v. 57, 299; Frank H. Hitchcock, *Trade of the Philippine Islands* (Washington, D.C.: 1898), 146–47; De la Cavada, *Historia Geográfica*, 1:384–95.

20. The *molino* was managed by Englishmen: R. P. Duncan in 1895 and Donaldson, Sim & Co. in 1899. See ads in *El Comercio*, 1895 and 1898; and MT, 1899. Earlier there was an ox-powered wheat mill. Comisión Central de Estadística de Filipinas, *20 cuaderno*, 1855, 17.

21. G. E. A. Cadell, "Philippine Islands, Cebu," flour markets, RCUS, 1894, v. 45, 165, quote 308. On Señorita XXX flour, see *El Comercio*, 5Apr1888, 9Jan1891, 3Jan1893, and 23Jun1895; and *La Oceania Española*, 4Mar1898. Señorita, Corona, and Mandarin, all imported brands, were routinely listed in the price reports of *El Comercio*, 24Jun–15Sep1895.

22. On the rise of the California flour industry, see Meissner, "Shanghai Success," chap. 2; and Daniel Meissner, "Bridging the Pacific: California and the China Flour Trade," *California History* 76.4 (Winter 1997–98): 82–93. On flour exports, see Horace Davis, "California Breadstuffs," *Journal of Political Economy* 2 (1893–94): 604, 608.

23. On the formation of the Northwest wheat region, see D. W. Meinig, *The Great Columbia Plain: A Historical Geography, 1805–1910* (Seattle: 1968). See ads in *El Comercio*, 19Jan1889 and 14Jan1898. See also "History of Sperry Flour Company," *The Modern Millwheel*, 2 (Sep–Dec38): 9–12; and *The Sperry Family* 1.12 (Dec17): 11–14. I am indebted to General Mills archivist Katie Dishman for these materials. On shipments, see "Trade and Commerce of Portland and District," Hs. of Com. 1899, v. 103, 2295, 37–39, and 1904, v. 101, pt. 2, 35.

24. See ads "Los Operarios Filipinos" in *El Renacimiento*, 9Mar and 11Sep09; *Razón*, 12Mar10. On the flour trade with Australia, see *El Comercio*, 1Dec00, 1; "Philippine Trade," MT, 24Feb08, 11; "Philippine Islands," SCR, 1915, v. 2, 80a, 1, 4; D. W. Meinig, *On the Margins of the Good Earth: The South Australian Wheat Frontier, 1869–1884* (London: 1962), ch. 9; and Nicholls, "Historical El Niño," 154–60.

25. Meissner, "Shanghai Success," quote 218 and chaps. 7–8. The value of flour imports to Canton, Amoy, and Swatow increased by 226 percent from 1897 to 1899. Henry B. Miller, "Flour in China," U.S. Consular Reports, 65, 247, April 1901, 524.

26. "Trust Control," *Manila American*, 1May06, quotes 1; "Philippine Foreign Commerce in 1907," PAgR 1.7 (Jul08): quote 282; "Harina de Australia," *Ang Bayang Pilipino*, 28May14, 38. On the U.S. share of Philippine imports, see "Trade of the Philippine Islands 1902," Hs. of Com., 1903, v. 79, 3044, 8; and ACCJ 4.3 (Mar24): 9.

27. On the question of business ethics, Owen reports that the nineteenth-century New England businessmen of Peele Hubbell were "honest" as regards contracts and business dealings, but they routinely attempted "mercantile collusion" in the interest of an "orderly market" for the purchase of abaca. Owen, "Americans in the Abaca Trade," 217. On Wilcox and his rapacious tactics, see Meissner, "Shanghai Success," 80–89.

28. Tornow, "Economic Condition," 45.

29. "Philippines Market for Flour," MT, 4Mar10, 10. See "Flour Merchants," in *Rosenstock's Manila City Directory*, 1905, 1911, 1916; Warner Barnes ads in MT, 18Feb1899 and 12Dec15; and "Connell Brothers Company, Inc.," MT, Investors and Settlers ed., Feb10, 57. The Connells handled flour from 15 American mills. On U.S.-Philippine trade preference policy, see Doeppers, *Manila 1900–1941*, chap. 1.

30. J. Bartlett Richards, "Philippine Economic Conditions, Annual Report for 1936," in Miguel R. Cornejo, ed., *Cornejo's Commonwealth Directory of the Philippines, 1939* (Manila: 1939), quotes 634–35. See also PFP, 6Sep19, 16; ms. ARCC in ms. ARG-GPI, 1921, 1221; ms. ARCC in ms. ARGGPI, 1924, 22; ms. ARCC in ms. ARGGPI, 1929, 14–15; SBPI, 1923, 89; SBPI, 1929, 166; CR, 3Apr17, 22, 27; *Trabajo* 3.5 (Feb23),

ad 37; "Leading Imports," 166; and *Commercial and Industrial Manual of the Philippines, 1940–1941* (Manila: 1941), 762, 804.

31. N. V. M. Gonzalez, *The Bread of Salt and Other Stories* (Seattle: 1993), 96–106. Gonzalez wrote in 1958, but he evokes an earlier time. See also C. V. Pedroche, "Thank God, She Said, For the Smell of Bread," *Philippine American* 1.3 (Nov45): 43–45.

32. Comisión Central de Estadística de Filipinas, *20 cuaderno*, 1855, 19–22; Wickberg, *Chinese in Philippine Life*, 109.

33. Doeppers, "Lighting a Fire," 435.

34. Sta. Maria, *Governor-General's Kitchen*, 133–38, quote 76.

35. William Stigand, "Manila," July 24, 1893, Hs. of Com. 1893/94, v. 96, no. 1289, quote 10; Sta. Maria, *Governor-General's Kitchen*, quote 75.

36. PNA, Contribución Industrial, Manila 1892–95 and 1895, S-1 to S-14; *Rosenstock's Manila City Directory*, Oct–Dec 1904, 154–55. See ads in *El Renacimiento*, 11Sep09, 6; *Bangon*, 30Apr09, 16; *Renacimiento Filipino*, Jul10, 16; and *Razón*, 15Oct10.

37. Gutierrez and Santos, "Food Consumption," 398–99; and Tamar Adler, "Against the Grain," review of Aaron Bobrow-Strain's *White Bread: A Social History of the Store-Bought Loaf*, *New York Times Book Review*, 1Jul2012, 19.

38. Perfecto Rivera, "Shifts from Job Printing to Bakery and Develops Profitable Business," CIJ 7.6 (Jun31): 8–9, 11, 18; *Manila Carnival, Commercial and Industrial Fair*, 1932, ad, quote 13; Manuel and Manuel, *Dictionary of Philippine Biography*, 3:413–14. Cf. MacMillan, *Seaports of the Far East*, 414.

39. In 1893 and 1894 more than a million pounds of "bread, biscuit, macaroni, vermicelli, etc.," were imported. Hitchcock, *Trade*, 25. See also Rev. Philip Wilson Pitcher, *In and About Amoy*, 2nd ed. (Shanghai and Foochow: 1912), 158, 219.

40. "El Zaragozano," *El Comercio*, 27Mar1872.

41. V. Buencamino, *Memoirs*, quote 9; Ambeth R. Ocampo, "The Malolos Banquet: Food as Historical Document," read in manuscript, 20Oct2000. See also Sta. Maria, *Governor-General's Kitchen*, 125, on the range of wheat flour pastries made at home at Christmastime and given as gifts.

42. Nick Joaquin, *Intramuros*, 29; ad, *Libertas*, 8Oct01; Zialcita and Tinio, *Philippine Ancestral Houses*, 119; Laling H. Lim, "At the Edge of the City," in Erlinda Enriquez Panlilio, ed., *The Manila We Knew* (Pasig: 2006), 21.

43. Ruby Paredes, "Memoir." I am entirely indebted to Dr. Paredes for unpacking these memories and insights.

44. Ibid.

45. Unlike the situation in Japan, the increasing popularity of bread and other flour-based products in Manila did not happen because of chronic rice shortages or a "militarization of nutrition." Katarzyna J. Cwiertka, "Popularizing a Military Diet in Wartime and Postwar Japan," *Asian Anthropology* 1 (2002): 1–30.

46. On *salabat*, see Herrera, *Tayabas Chronicles*, 18, 26, 52, 55, 166. In Manila some took *salabat* with their *bibingka* as a hot treat after midnight Christmas mass. Nick Joaquin, *Intramuros*, 29.

47. To be more thorough one should bring cold beverages into consideration—the colas and other so-called soft and energy drinks that routinely contain caffeine.

48. Robert Gardella, *Harvesting Mountains: Fujian and the China Tea Trade, 1757–1937* (Berkeley: 1994).

49. Buzeta and Bravo, *Diccionario Geográfico*, "Estado" at the end of vol. 2; "Facts and Figures about Philippine Trade with the U.S.," PJC 11.4 (Apr35): 14; Hitchcock, *Trade*, 24–25.; Ellis, *Hong Kong to Manilla*, 34.

50. Clarence-Smith, *Cocoa and Chocolate*, 10.

51. "Chocolate and Your Health," *Harvard Men's Health Watch*, 13.7 (Feb2009): quotes 2.

52. The coffee plant comes from Ethiopia and was spread in Southeast Asia via Islamic traders. "Java," a colloquial American name, relates to production on volcanic upland soils in insular Southeast Asia.

53. Marcy Norton, "Tasting Empire: Chocolate and the European Internalization of Mesoamerican Aesthetics," AHR, 111.3 (Jun2006): 660–91.

54. Manuel credits an Augustinian with helping to propagate newly arrived cacao plants in Bauan, Batangas, in 1695, while Alcina refers to cacao as early as 1668. Manuel, *Dictionary of Philippine Biography*, 1:286; Alcina (1668), *Historia*, in PSacra 85 (1994): 142–45.

55. Tomas de Comyn, *State of the Philippines in 1810*, William Walton, trans. (Manila: 1969/1821), quote 12.

56. Jose Rizal, *Noli me Tangere*, Soledad Lacson-Locsin, trans. (Manila and Honolulu: 1997), quote 65; Clarence-Smith, *Cocoa*, 18, 21, 23, 26, 166.

57. In Mexico *champorado* is a blend of cocoa and thick corn gruel. Zingg, "American Plants," 231, 260–62; Dr. Guillermo de Venecia, interview, Madison, 9Sep2003; Jagor, *Travels*, 75; Foreman, *Philippine Islands*, 354; "El Chocolate," *El Comercio*, 11Jan1871.

58. On cultivation practices, see *Census 1903*, 4:116.

59. Cushner, *Landed Estates*, 43, Cruikshank, *Spanish Franciscans*, 1:364–65; "Estado de importación y exportación del Puerto de Manila, 1818," Manila, 30Jul1819, Archivo Franciscano Ibero-Oriental 111/4.

60. Buzeta and Bravo, *Diccionario Geográfico*, 1:29–30, 203, 447. Buzeta and Bravo also list cacao as a major crop in Canlaon, Negros Oriental, and Dingras, Ilocos Norte.

61. "Precios Corriente," *El Comercio*, published every two weeks, April 28, 1888 through 1891. See also ads in *El Comercio*: "Cacao de Sibuyan y de Bondoc," 23Dec1873; "Cacao de Davao," 29Jan–1Apr1881; "Cacao superior" 27Dec1882; and "Cacao del Monte," 2Nov01.

62. "Estado de la agricultura filipina en 1887," BNM, ms. 19.218.

63. These buyers are variously described as based in San Pablo, having a warehouse in Bay, and/or authorized to buy in all the Laguna pueblos. PNA, Contribución Industrial, Laguna 1898 and 1893–95. Cf. "Cacao Legitimo de San Pablo (Laguna)," *El Comercio*, 28Apr1888.

64. Foreman, *Philippine Islands*, quote 354. George Anderson, "Cacao in the Philippines," DCTR, August 16, 1913, 191, 960; Montero y Vidal, *El Archipiélago Filipino*, 326–27, 340, 360; *Census 1903*, 4:206–7; William S. Lyon, "Cultivation of Cacao," in *Census 1903*, 4:105–16; Clarence-Smith, *Cocoa*, 246; Wernstedt and Spencer, *Philippine Island World*, 206–7, 644.

65. Clarence-Smith, *Cocoa*, quotes 94–97. See also Legarda, *After the Galleons*, 77–86; "Cacao Caracas y Guayaquil superior vende de Benjamin Schwob," *El Comercio*, 15Dec1882 (the seller was Swiss); "Cacao de Caracas, Guayaquil y Carupano, venden Muñoz Hermanos," *El Comercio*, 5Mar1891; and Jagor, *Travels*, 11.

66. This situation changed starting in the 1870s when the development of iron ore exports led to the rise of major industry and banking in the Basque region. Vicens Vives, *Economic History of Spain*, 662–63.

67. Juan Bautista Yrissary was in Manila by at least 1842. See *La Esperanza* (Manila), 3, 17, 26, and 30Dec1846; *Commercial Directory of Manila, 1901*, xxxvii; and De Borja, *Basques in the Philippines*, 128–32. The orthography of these names is inconsistent.

68. *Commercial Directory of Manila, 1901*, xxxvii; Eric Tagliacozzo, *Secret Trade, Porous Borders: Smuggling and States along a Southeast Asian Frontier, 1865–1915* (New Haven: 2005).

69. Wong Lin Ken, *The Trade of Singapore, 1819–69, Journal of the Malayan Branch of the Royal Asiatic Society* 33, pt. 4, no. 192 (1960), trade route maps. On January 11, 1872, a Spanish bark arrived in Manila from Menado following stops in Zamboanga and Iloilo with 42 Chinese passengers and an unstated cargo for "Chino M. Conling." It was surely carrying cocoa. Manuel Conling was a Christian Chinese from Xiamen. From 1856 through 1873 he was the business partner of a "naturalized Spaniard" in Zamboanga. In 1873 he was residing in Manila, where he became the (third) husband of Cornelia Laochangco, a major Manila rice dealer. Also in 1873 he took charge of the opium monopoly in Zamboanga. Many thanks to Cornelia Lichauco Fung for these materials, transcribed from PNA, Protocolos Manila 570 1873 Tomo 1/index [notary] Francisco Hernandez y Fajarnes SDS 20432 Nos. 33 S70B-74 and 36 S79–80 77, October 11 and 13, 1873, respectively. The 1881 commercial record includes vessels arriving in Manila from Callao, Peru, via Honolulu (2) and from Ternate (1), the Moluccas (2), and Makassar (2). Their cargoes likely included cocoa.

70. Clarence-Smith, *Cocoa*, 38, 44–45, 51, 53, 56; Jagor, *Travels*, 74; *Balanza* for the years indicated. Imports were 1.1 million kilos for Nov1911–Oct1912. See PAgR (Jan1912–Feb1913).

71. Clarence-Smith, *Cocoa*, quotes 11, 27; ad "Cacao sin aceite, Gche & Cia. de Dresden," *El Comercio*, 11 and 23Oct1891; A. S. Watson ad, "Christmas Grood," *El Comercio*, 30Dec1890.

72. Foreman, *Philippine Islands*, 354–55 quotes 354; Clarence-Smith, *Cocoa*, quote 90. For a description of this practice, see T. H. Pardo de Tavera, "Los Chinos," PFP (8Jul16): 30; and photos in *Nuevo Mundo* (Madrid), 22Jun1898, 5. See also de Viana, *Three Centuries of Binondo Architecture*, 66.

73. Islas Filipinas, Comisión Central de Estadística de Filipinas, *20 cuaderno*, 1855, 17, 20.

74. Café Oriental ran frequent ads in *El Comercio*. See, e.g., issues of 25Nov1873, 23Dec1873, and 14Jul1877. The La Bilbaina *fábrica* was steam powered by 1871. See, e.g., *El Comercio*, 3Jan1871, 25Oct1881, 31Mar1888 (full-page ad), 3Nov1894, and 2Nov01. La Bilbaina offered food and provisions at locations on the Escolta and in Intramuros in 1887. La Palma de Mallorca was first operated by Juan Pons on Calle Solana in Intramuros with a second venue in Quiapo. It continued on Solana under

Juan Galmes (by 1905) and featured prominently in the nostalgia for lost landmarks among the residents who survived the Japanese last stand in Intramuros in 1945. *El Comercio*: 5Mar1887, 4; 15Feb1891; *The American*, 20May1899; *El Renacimiento*, 3Jan05; *Rosenstock's Manila City Directory*, various editions; Nick Joaquin, *Intramuros*, 29, 40, 45, 68. On both, see PNA, Contribución Industrial: Manila 1892–97. Rueda Hermanos operated another steam-powered shop, La Marina. *Commercial Directory of Manila, 1901*, 54; RPC 1903, pt. 1, 604–6.

75. On the Paglinawans, see Rivera, "Shifts from Job Printing to Bakery," quote 8; Manuel and Manuel, *Dictionary of Philippine Biography*, 3:413–14; *Witton's Manila and Philippines Directory*; *Rosenstock's Manila City Directory*, various editions; and Philippine Supreme Court case 10738, *Rueda Hermanos & Co. v. Felix Paglinawan & Co.*, January 14, 1916 (lawphil.net/judjuris/ juri1916).

76. Clarence-Smith, *Cocoa*, 57, 74, 77, 81; *Rosenstock's Manila City Directory*, 1931–32, 400; ad for "Van Houten's Cocoa," MT, 25May07; ad for "Lowney's Make," MT, 1Aug07.

77. Ruby Paredes, "Memoir," quotes; Bienvenido Santos, interview, 19Apr85, quotes.

78. Clarence-Smith, *Cocoa*, 11–19, quote 14; Norton, "Tasting Empire," 660–91. Coffee and tea were heavily promoted in the United States in the 1880s by the A&P Company, soon to become one of the most successful American grocery chains.

79. Benito Legarda Jr., *The Hills of Sampaloc* (Makati: 2001), 79–80; Lewis, *Foot Soldier*, 3, 17; Michael Meyers Shoemaker, *Quaint Corners of Ancient Empires: Southern India, Burma, and Manila* (New York: 1899), 167.

80. Jean Gelman Taylor, *Indonesia: Peoples and Histories* (New Haven: 2003), 203. Coffee production came to be highly concentrated on Dutch estates in the volcanic highlands of East Java. Charles A. Fisher, *South-east Asia: A Social, Economic, and Political Geography* (London: 1964), 282.

81. Legarda, *After the Galleons*, 302.

82. Simeon Luz, "Cultivation of Coffee in the Municipality of Lipa, Province of Batangas," in *Census 1903*, 4:78–84; shipments listed in the *Gaceta*, Dec1880–Nov1881. See also "Precios Corriente," El Comercio, 9Aug1881 and 28Apr1888; and ads for "La Balear," *El Comercio*, 21August1880 and 10Feb1881.

83. William Gervase Clarence-Smith and Steven Topik, eds., *The Global Coffee Economy in Africa, Asia, and Latin America, 1500–1989* (Cambridge: 2003), 7.

84. Wickberg, *Chinese in Philippine Life*, 105; Hitchcock, *Trade*, 139. Later exports went to Spain. *El Comercio*, 1Feb1887 and 30Apr1888.

85. [Nick Joaquin] Quijano de Manila, "Towards a Coffee Renaissance," PFP, February 21, 1953, quote 10.

86. Luz, "Cultivation," in *Census 1903*, 4:78–84; Sastrón, *Batangas*, 18–19; De Comyn, *State of the Philippines*, 12. Buzeta and Bravo mention coffee as the second-ranking product in Calaca and third in Balayan, Rosario, San Juan de Bocboc, and Batangas municipality plus Maragondon in upland Cavite. It is difficult to verify that these represent a careful ordering.

87. Montero y Vidal, *El Archipiélago Filipino*, 326–27; Jagor, *Travels*, 77. The agricultural survey of 1886–87 lists coffee production as 46,500 (unstated units of volume)

in Batangas, 15,000 in Cavite, and 3,750 in Laguna. Basilan's authorities estimated their production at 200,000, but the data are far from precise. "Estado de la agricultura filipina en 1887," BNM, ms. 19.218. See also Pickford, "Trade and Commerce of Cebu, 1878," Hs. of Com. v.72, C2421, 1607.

88. Rodriguez, "Memoria general de la Provincia de Cavite," 2–79, 117–83.

89. Many went to nearby Mindoro, others to the frontier zone between Tagalogs and Bikolanos. On Batangas society and conditions, see May, *Battle for Batangas*; on outmigration, see Miller, *Economic Conditions*, 293. In Manila in 1895 wholesale prices were routinely reported for coffee coming from Batangas, Cavite, Zamboanga, and Cebu, usually with the notation of very small supplies. "Precios Corriente," *El Comercio*, 24Jun–15Sep1895.

90. William Gervase Clarence-Smith, "The Coffee Crisis in Asia, Africa, and the Pacific, 1870–1914," in William Gervase Clarence-Smith and Steven Topik, eds., *The Global Coffee Economy in Africa, Asia, and Latin America, 1500–1989* (Cambridge: 2003), 101–2, 104.

91. Clarence-Smith, "Coffee Crisis," quote 115.

92. Ibid.; Abelardo Valenzuela, "Composition of Philippine Coffee," PJS 40.3 (Nov29): 349–51; Copeland, "Spanish Agricultural Work," 315; RPC 1906, pt. 2, 43.

93. Clarence-Smith, *Cocoa*, 30–32; "Problems in Reviving the Coffee Industry of Lipa" PA 16.9 (Feb28): 562–63; "Coffee Industry in Philippines," PEF 1.2 (May23): 39; Valenzuela, "Composition," 349–51; *Census 1903*, 4:325; *Census 1939*, 2:1658, 1669; SBPI, 1926, 32–33; SBPI, 1929, 51–52. Cf. Wernstedt and Spencer, *Philippine Island World*, 206–7.

94. Scheerer lived in Manila during 1882–96, where he built and later sold the La Minerva cigar factory before moving to Benguet. He then taught German in elite Japanese schools, served as vice-governor of Batanes, and finished his career teaching Philippine linguistics at the University of the Philippines. See van den Muijzenberg, *Philippines through European Lenses*, 219–35; Salazar, "German Economic Involvement," 277–78; Howard T. Fry, *A History of the Mountain Province* (Quezon City: 1983), 3, 6, 8–10; Reed, *City of Pines*, 91–92; and Manuel, *Dictionary of Philippine Biography*, 2:350–54.

95. Lewis, *Wagering the Land*, 134, 179, quotes, 27–28.

96. Conklin, "Ifugao Ethnobotany 1905–1965," *Economic Botany* 21.3 (Jul67): quote 257; W. F. Pack, "Coffee Culture in the Province of Benguet," in *Census 1903*, quote 4:85.

97. Bodegas belonging to Pedro Singson Gotiaoco in Cebu City were destroyed in the Spanish bombardment of April 7, 1898. He approximated his losses in beverage-related stores as 10,400 pesos divided among coffee and cacao at 77:23—a rare, if imperfect, relative measure. Michael Cullinane, personal communication, 20Mar2009.

98. Rivera, "Shifts from Job Printing to Bakery," 8; Clarence-Smith, "Coffee Crisis," quote 115.

99. The same public health campaign attempted to clean up the city's milk supply. "Coffee Case Hard Fought," MT 29Oct12, quote 1.

100. Gutierrez and Santos, "Food Consumption," quote 398; SBPI, 1923, 1926, and 1929.

101. [Nick Joaquin], "Towards a Coffee Renaissance," 10–11, 31, quote 10. The same was true in the mid-1930s. On caffeine content, see Valenzuela, "Composition," 350–51. See also the Ah Gong Sons' ad in *Manila City Directory*, 1936–37, 400.

102. Kristin Hoganson, "Buying into Empire: American Consumption at the Turn of the Twentieth Century," in Alfred W. McCoy and Francisco Scarano, eds., *Colonial Crucible: Empire in the Making of a Modern American State* (Madison: 2009), 248–59, quote 254.

CHAPTER 11. SUBSISTENCE AND STARVATION
IN WORLD WAR II, 1941–45

1. Among the most useful treatments of this period are Ikehata Setsuho and Ricardo Trota Jose, eds., *The Philippines under Japan: Occupation Policy and Reaction* (Quezon City: 1999); Teodoro A. Agoncillo, *Fateful Years: Japan's Adventure in the Philippines, 1941–45*, 2 vols. (Quezon City: 1965); A. V. H. Hartendorp, *The Japanese Occupation of the Philippines*, 2 vols. (Manila: 1967); Benito Legarda Jr., *Occupation '42* (Manila: 2003) Benito Legarda Jr., *Occupation: The Later Years* (Quezon City: 2007); David Joel Steinberg, *Philippine Collaboration in World War II* (Manila: 1967); and Antonio S. Tan, *Chinese in the Philippines during the Japanese Occupation, 1942–1945* (Quezon City: 1981). On the food supply crisis in particular, see Ricardo Trota Jose, "The Rice Shortage and Countermeasures during the Occupation," in Ikehata Setsuho and Ricardo Trota Jose, eds., *The Philippines under Japan: Occupation Policy and Reaction* (Quezon City: 1999), 197–214, 344–53; and Benedict Tria Kerkvliet, "Withdrawal and Resistance: The Political Significance of Food, Agriculture, and How People Lived during the Japanese Occupation in the Philippines," in Laurie Sears, ed., *Autonomous Histories, Particular Truths: Essays in Honor of John R. W. Smail* (Madison: 1993), 175–93.

2. Translations and reconstructions of conversations are shown in 'single' rather than "double" quotation marks, which are restricted to direct quotations of verbatim material. See note 22 in the introduction.

3. Steinberg, *Philippine Collaboration*, 27.

4. Gutierrez and Santos, "Food Consumption," 407.

5. Jose, "Rice Shortage," 198–99, 202; Victor Buencamino, "Food for Everybody," PFP, 20Dec41, 29.

6. Luciano Salanga, civil servant, interview, San Juan, 18Feb85.

7. See the ads "La Verdad," *Taliba*, 17Dec41; and "Pananalakay sa Himpapawid [Aerial Assault]," San Miguel Brewery, *Taliba*, 18Dec41.

8. PFP, 20Dec41, 19; Fernando J. Mañalac, *Manila: Memories of World War II* (Quezon City: 1995), 13–14; Lichauco, *Dear Mother Putnam*, 11–12; V. Buencamino, *Memoirs*, 265; Carlos Quirino, *Chick Parsons: America's Master Spy in the Philippines* (Quezon City: 1984), 9, 12.

9. Dominador del Rosario, printer, interview, 25Feb85; Rosalia Fortaleza, navy wife and teacher, interview, 22Feb85.

10. Anonymous, interview, San Nicolas, May85.

11. Margaret Sams, *Forbidden Family: A Wartime Memoir of the Philippines, 1941–1945* (Madison: 1989), 56.

12. Penang in Malaya was also declared an open city in December 1941, as was Paris the year before.

13. Felipe Buencamino III, *Memoirs and Diaries of Felipe Buencamino III (1941–1944)* (Makati City: 2003), 62.

14. Pacita Pestaño-Jacinto, *Living with the Enemy: A Diary of the Japanese Occupation* (Pasig: 1999), 14, 22, 68; Lichauco, *Dear Mother Putnam*, 3; V. Buencamino, "Food for Everybody," 28–29.

15. "Forced Migration," PFP, 20Dec41, 24–25; Mañalac, *Manila*, 9–10.

16. Upracio Cruz, printer, interview, Sampaloc, 8Jun86; Mrs. Carmen Quirolgico Bengson, interview, San Juan, 8Mar85.

17. Jesus Concepcion, carpenter and *maestro de obras*, interview, San Juan, 18Feb85.

18. Lina Kisikisi vda. de Dungo, interview, Langit, Sampaloc, 4Jun85.

19. See Steinberg, *Collaboration*, 181n15.

20. Mears et al., *Rice Economy*, 12.

21. Jose, "Rice Shortage," 199–204, 208; Mears et al., *Rice Economy*, 13. Jose describes 120 grams of rice as an average meal for one person.

22. Martin Tinio Jr., personal communication, 11Jan2008. Kerkvliet argues that farmers in a number of local autonomous zones were able to withhold foodstuffs from the external forces of landlords and the Japanese regime. Kerkvliet, "Withdrawal and Resistance," 185–89.

23. Sams, *Forbidden Family*, 77, 79.

24. See Lichauco, *Dear Mother Putnam*, 57, 111; ads in the *Tribune*, 4Sep42; and Agoncillo, *Fateful Years*, 2:534.

25. A. V. H. Hartendorp, *History of Industry and Trade of the Philippines* (Manila: 1958), 88.

26. Lichauco, *Dear Mother Putnam*, 79 (February 1943).

27. Charles Parsons, "Report on Conditions in the Philippine Islands as of June 1943," 333–67, in Charles A. Willoughby, *The Guerrilla Resistance Movement in the Philippines, 1941–1945* (New York: 1972), quote 342–43; Yoshiko Nagano, "Cotton Production under Japanese Rule, 1942–1945," in Setsuho Ikehata and Richard Trota Jose, eds., *The Philippines under Japan: Occupation Policy and Reaction* (Quezon City: 1999), 171–95.

28. Pestaño-Jacinto, *Living with the Enemy*, quote 94.

29. Filoteo S. Tuason, interview with Loreto Seguido, Tondo, 24May85.

30. Adolfo Jose, interview, Vitas, Tondo, 23Apr85. According to Felipe Buencamino III (*Memoirs*, 82), the occupiers had banned night fishing on the bay by March 1942.

31. Geraldo N. Santiago, interview, Vitas, Tondo, 22Apr85; Choson Valeriano, plumber, interview, Gagalangin, Tondo, 13May85. "Dirty" implies contaminated.

32. Kerkvliet, *Huk Rebellion*, chap. 3. Cf. Ma. Felisa A. Syjuco, *The Kempei Tai in the Philippines, 1941–1945* (Quezon City: 1988), 36; Lichauco, *Dear Mother Putnam*, 52; Quirino, *Chick Parsons*, 70, 81–82.

33. The PMC was Proctor and Gamble's Philippine Manufacturing Corporation, a landmark of northern Tondo on a spit of land at the entry point of the Vitas River into Manila Bay.

34. Arturo Bautista, carpenter, interview, Bankusay, Tondo, 4May85; Cf. Jose, "Rice Shortage," 202, 207–8, 210; Hartendorp, *Japanese Occupation*, 2:102–3, 117.

35. Lichauco, *Dear Mother Putnam*, 144–45; Rolando E. Villacorte, *Baliwag Then and Now* (Caloocan City: 1970), 89.

36. Quotes are from, respectively, Alberto Javier, *cochero*, interview, Sunog Apog, Tondo, 13May85; and Mariano Wengco, *cochero*, interview with Loreto Seguido, Tondo, 27May85. See also Villacorte, *Baliwag*, 88–90.

37. Francisco Cruz, interview, Tondo, 10Jun86.

38. Ricardo Almario, *viajero*, interview, Tondo, 18Apr85, quote; Dr. Frederico Rubio, interview, Tondo, 18Apr85; Lichauco, *Dear Mother Putnam*, 66, 84.

39. Lichauco, *Dear Mother Putnam*, 2–3; Doeppers, "Home Fuel in Manila," 419–47.

40. Martin Tinio Jr., personal communications, 22Oct and 11Nov2007; Legarda, *Occupation '42*, 25.

41. Typically, such machines took 15 minutes to start and were hot, slow, and clanky. Ken Ragland, personal communications, 11 and 16Jan2003, clarifying information in "Charcoal for Gasoline," *Graphic*, 24Oct40, 12–13, 56; and "Coco-Gas Shown Again," *Tribune*, 16Mar41, 5.

42. Quotes are from, respectively, Felicisimo R. Soldaña, *umbuyan* worker, interview, Tondo, 28May85; and Victor Santos, *umbuyan* worker, interview with Loreto Seguido, Tondo, 23May85.

43. Villacorte, *Baliwag*, 91–94.

44. Agoncillo, *Fateful Years*, 2:537.

45. Federico Pelayo, *cochero*, interview, Sunog Apog, Gagalangin, 13May85. Young *kamote* leaves are a good source of vitamin A, thiamin, and riboflavin. Lugod, "Wild Plants," 342.

46. Tan, *Chinese in the Philippines during the Japanese Occupation*, 50, 54; Antonio S. Tan, *Chinese in the Philippines, 1898–1935: A Study of Their National Awakening* (Quezon City: 1972).

47. Yung li Yuk-wai, *Huaqiau Warriors: Chinese Resistance Movement in the Philippines, 1942–1945* (Quezon City: 1996). Hartendorp, *History of Industry*, 85–86; Pestaño-Jacinto, *Living with the Enemy*, 84, 107, 145–46, 165, 199; Mañalac, *Manila*, 81, 96–97; Tan, *Chinese in the Philippines during the Japanese Occupation*, 55.

48. Luciano Salanga, interview, San Juan, 18Feb85. See also Parsons, "Report on Conditions"; and Quirino, *Chick Parsons*, 121–58.

49. Mañalac, *Manila*, 16, 40, 46–48, 64. See also "That Divisoria Market Row," 10–11.

50. Prewar ethnic friction in retail trade is analyzed in Jose R. Ablang, "Nationalizing Our Retail Trade," in *Commercial and Industrial Manual of the Philippines, 1940–1941* (Manila: 1941), 881–83. On renewed postwar friction, see, inter alia, "Chinese Stalls: To Oust or Not to Oust," *Sunday Times Magazine*, 19Jan47, 18–19.

51. Mrs. Petra Aguirre, interview, Tondo, 18Apr85.

52. Legarda, *Occupation '42*, 92, 101–3.

53. Sams, *Forbidden Family*, 88.

54. Benjamin del Carmen, interview, San Juan, 20Feb85. On the feeding of family and friends, see Sams, *Forbidden Family*, 69–71, 76; and "Fernandez, *Tikim*, 141–42.

55. Justice J. B. L. Reyes, interview with the author, Sampaloc and San Juan, 8Aug89.

56. Mrs. Natividad Samio de Gamboa, interview, San Nicolas, 15Apr85; Jesus Concepcion, *maestro de obras*, interview, San Juan, 18Feb85. In a quiet voice one San Juan worker said, "There are still *makapilis* now [in 1985]." There were informers all along, but the Makapili organization was not formed until November 1944.

57. Feliza N. Lopez, politico, interview, San Juan, 27Feb85.

58. Torribio "Ibyong" Roxas and Anselmo Roxas, interview with Loreto Seguido, Tondo, 23May85; Sixto Basa, interview, Tondo, 7Jun85. Chroniclers of the occupation often see these kinds of events as part of a "winning hearts and minds" campaign, especially in 1943. Many others had horrifying experiences and quickly came to fear the arbitrariness of Japanese actions toward civilians. See Charles A. Parsons, "1943," in Carlos Quirino, *Chick Parsons: America's Master Spy in the Philippines* (Quezon City: 1984), 121–22; Steinberg, *Collaboration*, 94; and Kiyoshi Osawa, *A Japanese in the Philippines: An Autobiography* (Tokyo: 1981), chap. 6, esp. 152–53.

59. Lichauco, *Dear Mother Putnam*, quotes 104–5; F. Buencamino, *Memoirs*, 109.

60. Quotes are from Carmen S. Samson, bottle recycling, interview, Angustia, Tondo, 1Jun85; Lucia Macaranas vda. Soriano, interview, Antonio Rivera Street, Tondo, 14May85; and Nicomedes Simundo and Felicidad Villera Simundo, interview, Vitas, Tondo, 24Apr85.

61. Mañalac, *Manila*, 87.

62. Zuellig employee Lorenzo Beroña, interview, San Juan, 21Feb85.

63. Mrs. Seale, interview, San Juan, 18Feb85.

64. Consuelo Trinidad vda. de Alisasis, interview, San Nicolas, 3May85.

65. Kerkvliet, "Withdrawal and Resistance," 181; Jose, "Rice Shortage," 208, 210; Pestaño-Jacinto, *Living with the Enemy*, 99.

66. Pestaño-Jacinto, *Living with the Enemy*, 137, 164–65, 171.

67. Cesar M. Lorenzo, "Inflation and Cost of Living in Manila during the Japanese Occupation," ACCJ 22.2 (Feb46): 18.

68. Valentin Semilla, interview, Gagalangin, 13May85, quote. See also Legarda, *Occupation: The Later Years*, 126.

69. Angelita Salvador, faith healer and construction labor, interview, San Juan, 12Mar85. On *gilingan* use in wartime, see Rosalina M. Franco-Calairo and Emmanuel Franco Calairo, *Ang Kasaysayan ng Novaliches* (Novaliches: 1997), 97.

70. Guillermo Licudine, interview, San Juan, Salapan, 28Feb85.

71. Quotes are from Mrs. Petra Aguirre, husband in the navy, interview, Tondo, 18Apr85; Alberto Javier, *cochero*, interview, Sunog Apog, Tondo, 13May85; and Antonio Bonifacio, interview, Tondo, 1Jun85.

72. Quotes are from Victorina de la Cruz Bautista, soft drink *fábrica*, interview, Angustia, Tondo, 1Jun85; and Dr. and Mrs. Jose Barcelona, San Juan, and Prof. Tigi Barcelona (son), interview, 5Mar85.

73. Luciano Salanga, interview, San Juan, 18Feb85. The one-way trip was about 3.5 miles. The Mañalac *sarisari* store in Quiapo was closed in late 1943 or early 1944 when the NARIC-cum-Biba ran out of rice and corn supplies. Mañalac, *Manila*, 48, 69–70, 89–91; V. Buencamino, *Memoirs*, 343.

74. Maria Balboa vda. Reyes, interview, Antonio Rivera Street neighborhood, Tondo, 14May85. See also Soledad H. Leynes, "Survival Meals," in Gilda Cordero-Fernando, ed., *The Culinary Culture of the Philippines* (n.p.: 1976), 195.

75. Quotes are from Augusto Martin, civil servant, interview, Tondo, 21Apr85; Arturo Bautista, *carpintero*, interview, Bankusay, Tondo, 4May85; and Geraldo N. Santiago and Eduarta Gagahasin, interview, Vitas, Tondo, 22Apr85.

76. "*Isangag mo ang sisid rice. Lagyan mo ng kaunting luya, kaunting baswang, ayos na. Kaya lang, pag-utot mo, patay 'yung nasa Sande.*" Torribio "Ibyong" Roxas and son Anselmo Roxas, abaca and tobacco classifier, *pahinante*, *kargadores*, interview with Loreto Seguido, Tondo, 23May85. For an alternate rumor about *sisid* rice, see Pestaño-Jacinto, *Living with the Enemy*, 257; and Mañalac, *Manila*, 84.

77. Quotes are from Catalina Torres vda. Valentin, interview by Loreto Seguido, Tondo, 27May85; Doreen Fernandez, *Tikim*, 3; and Baltazar Valdez, rice farmer and construction worker, interview, Tondo and Bulacan, 20May85. See also Pestaño-Jacinto, *Living with the Enemy*, 249.

78. Nick Joaquin, *Manila, My Manila*, quotes 294. *Panocha* is boiled sugarcane juice dried in half a coconut shell to make a cake of semirefined brown sugar.

79. Torribio "Ibyong" Roxas and son Anselmo Roxas, interview with Loreto Seguido, Tondo, 23May85; Fung, *Beneath the Banyan Tree*, 167; Mañalac, *Manila*, 102–3. See also Pestaño-Jacinto, *Living with the Enemy*, 212; Lichauco, *Dear Mother Putnam*, 193, 202; and Patricio N. Abinales, "Let Them Eat Rats: The Politics of Rodent Infestation in the Postwar Philippines," PS 60.1 (Mar2012): 69–101.

80. For a brief review of strategic alternatives to invasion, see Legarda, *Occupation: The Later Years*, 95–97.

81. F. Buencamino, *Memoirs*, 170, 176.

82. Quotes are from Dolores L. and Domingo Gomez, interview, Sampaloc, February 16, 1985; and Jose, "Rice Shortage," 209–11. See also Saburo Ienaga, *The Pacific War, 1931–1945: A Critical Perspective on Japan's Role in World War II* (New York: 1978), 185, 288n14.

83. F. Buencamino, *Memoirs*, 153; Rosalia Fortaleza, interview, San Juan, 22Feb85, quote.

84. Legarda, *Occupation: The Later Years*, 103; Pestaño-Jacinto, *Living with the Enemy*, 233.

85. Mañalac, *Manila*, 84–86, 97, quotes 85–86, 102.

86. Pestaño-Jacinto, *Living with the Enemy*, 221–22, 249, quote 237; F. Buencamino, *Memoirs*, "walking corpses," 172.

87. Pestaño-Jacinto, *Living with the Enemy*, quote 255; Lichauco, *Dear Mother Putnam*, 194.

88. Hartendorp, *History of Industry*, 145.

89. Asuncion "Aling Sion," Salazar, interview, Sunog Apog, Tondo, 16May85.

90. Ambassador Carlos Faustino, interview with the author, Gagalangin, 25Apr85.

91. Victor Santos, interview with Loreto Seguido, Tondo, 23May85. Approached at random, Santos turned out to be the son of Pasqual Santos, a three-time city councilor last elected with the group of Manuel de la Fuerte in December 1940 on a platform of social justice for the poor.

92. Mrs. Pilar del Rosario, clerk of court, interview, San Juan, 22Feb85. Rice bran can be a good source of vitamins A, B_1, and E. Tea made from boiling *banaba* leaves was more traditionally prescribed by midwives to counteract diarrhea.

93. "Benny" del Carmen, interview, Sampaloc, 20Feb85.

94. Jose, "Rice Shortage," 206–12, esp. 204. See also Pestaño-Jacinto, *Living with the Enemy*, 215, 220, 225–26, 234, 257. A number of farmers had already switched from rice to *kamote* and cassava because rice, harvested all at once, was too easily confiscated. Franco-Calairo and Franco Calairo, *Ang Kasaysayan ng Novaliches*, 93–105.

95. Mañalac, *Manila*, 95; F. Buencamino, *Memoirs*, 157–58. The Marcial Lichaucos sold their suburban second house in Mandaluyong in June 1944 for cash with which to buy rice. During the Battle of Manila, with liberation at hand but starvation and injury all around, they fed 100 homeless refugees with the remains of their stock. Lichauco, *Dear Mother Putnam*, 155, 171, 185, 190, 211.

96. Holding out in the Mountain Province in 1945, starving Japanese soldiers foraged exhaustively, causing the extirpation of numerous locally adapted cultivar races and spreading starvation among the indigenous people. Lewis, *Wagering the Land*, 17, 107–11; Ogawa Tetsuro, *Terraced Hell: A Japanese Memoir of Defeat and Death in Northern Luzon, Philippines* (Rutland and Tokyo: 1972).

97. Araceli Evangelista, interview, San Juan, 15Feb85; Oscar Evangelista, son, personal communication, 24Nov2002.

98. Brigida Alcaire Pahit, interview with Loreto Seguido, Tondo, 24May85.

99. Lucia Macaranas vda. Soriano, interview, Antonio Rivera Street, Tondo, 14May85. See also Legarda, *Occupation: The Later Years*, 125.

100. "Zoning" (*zona*) was surrounding a neighborhood suspected of harboring guerrillas and interrogating—or worse—everyone thus trapped. F. Buencamino, *Memoirs*, 179, 186–87, 189.

101. Anonymous, interview, Tondo, Apr85, quote. See also Pestaño-Jacinto, *Living with the Enemy*, 180, 243; Mañalac, *Manila*, 62–63; and Hartendorp, *History of Industry*, 99.

102. F. Buencamino, *Memoirs*, 188–89; Legarda, *Occupation '42*, 138, 143; Legarda, *Occupation: The Later Years*, 249–58. Syjuco, *The Kempei Tai in the Philippines*, 43–47.

103. Araceli Evangelista, interview, San Juan, 15Feb85.

104. C. V. Pedroche, "Thank God, She Said, for the Smell of Bread," 43–45.

105. Mañalac, *Manila*, quotes 148, 161; Jose, "Rice Shortage." By contrast, there were still shortages in liberated Hong Kong. Philip Snow, *The Fall of Hong Kong: Britain, China, and the Japanese Occupation* (New Haven: 2003), 297–99.

106. Agoncillo, *Fateful Years*, quote 2:885.

107. Quotes are from, respectively, Antonio Fortaleza, U.S. Navy, and Rosalia Fortaleza, teacher, interview, San Juan, 22Feb85; and Filoteo S. Tuason, interview with Loreto Seguido, Tondo, 24May85.

108. Luciano Salanga, interview, San Juan, 18Feb85. The ECA was reestablished in March 1945, taking over from PCAU the emergency relief distribution of food and clothing. Both agencies were superseded by PRRA, the Philippine Relief and Rehabilitation Administration, in December 1945, but the term PCAU, or PICAU, was still in use among the interviewees. See Mears et al., *Rice Economy*, 12–15, 340–41.

109. Angelita Salvador, construction laborer and faith healer, interview, San Juan, 12Mar85.

110. Felicisimo R. Soldaña, interview, Tondo, 28May85. "That was my income during this period, 1945–46," he said. "I did not go back to my prewar job with the city, and by 1946 I was a senior foreman."

111. Anonymous, interview with Loreto Seguido, Tondo, May85.

112. Agoncillo, *Fateful Years*, quote 2:885.

113. Legarda, *Occupation: The Later Years*, 226–28, quote 227.

114. "Procedure Is Outlined for Getting Seeds," *Free Philippines* (Manila), 20Jun45. One culinary outcome of this period was the relatively high status accorded the American canned meat product called Spam. Andrew Martin, "Today's Special," *NYT*, 15Nov2008, B1–2.

115. Butcher, "Marine Animals," 75.

Epilogue

1. Some argue that many fled from rural violence and flocked to the city, swelling the number of mouths to feed. This certainly happened, but the evidence is clear that the bulk of net migration, early and again late in the war, was from the city to the countryside.

2. Mañalac, *Manila*, 102.

GLOSSARY

This work relies on Vito C. Santos, *Vicassan's Pilipino-English Dictionary* (Manila: 1978/1986), supplemented by Pedro Serrano Laktaw, *Diccionario Tagalog-Hispano* (Manila: 1914); Leo James English, CSsR, *English-Tagalog Dictionary* (Makati, Metro Manila: 1977); Juan de Noceda and Pedro de Sanlucar, *Vocabulario de la Lengua Tagala* (Manila: 1860); Jose Felipe del Pan, *Diccionario de la Administracion, del Comercio y de la Vida Practica en Filipinas,* vol. 1, A-Con (Manila: 1879,); and W. E. Retana, *Diccionario de Filipinismos* (New York and Paris: 1921).

In modern Filipino/Tagalog orthography *k* has replaced the sometimes inapt and confusing Spanish *c.* Since they are interchangeable, *c* and *k* are listed together below. Likewise, *ts* and *ks* are now preferred to *ch* and *cc,* respectively. Note also the sometimes inconsistent orthography of Spanish *y* and *i* in names. For ease of reading, I have left the *c* or *ch* in several common words, such as *chocolate.*

Plural constructions in Tagalog are ordinarily preceded by *mga.* For ease of reading, where the Tagalog noun derives from Spanish, these are (or were during our period) pluralized by adding *es* as in *kargadores.* Others are left in the singular form, for example, *talipapa* rather than *mga talipapa.*

Commodity storage facilities are *almacen, bangán,* bodega, *kamalig, kamarin/camarin, tambobong,* and *tangkil.*

Weights, measures, and containers are *arroba, batulan, bayon, bulto, buslo, canasto, kabán/cavan, cesto, ganta, picul/pico, saco, tapayan,* and *tinaja.*

accesoria—a building containing a number of small apartments or shops in a row with apartments above or behind, often with direct outside access. Spanish. In particularly dense locales, as along the Escolta, there may be three stories. It is called a *posesion* in the *fincas urbanas* registers of 1890–91.

adobo—marinade and cooking sauce for chicken and other meats made of *suka*/nipa vinegar, garlic, black pepper, and soy sauce. Spanish and Tagalog from Mexican Spanish. Besides imparting a hallmark palette of flavors, it also preserves.

arrabal—a suburban municipal-parish territory near but outside the city walls. Spanish derived from Arabic.

arroba—Spanish unit of measure equal to about 4 gallons. Norman Owen defines it as a unit of weight equal to 25 pounds, 11.5 kilograms, or two-elevenths of a *picul* (Owen, *Prosperity without Progress*, 275).

artesian, artesian well—in Philippine English, any well with a steel pipe driven some depth into the ground. In American English, a well from which water flows of its own accord without pumping.

Azcarraga—a major and extremely wide Manila street, later renamed C. M. Recto. In much of the nineteenth century each segment of this street had its own name: Felipe II/Paseo Azcarraga, Nuevo, Gral. Izquierdo, San Bernaldo, Iriz, and Alix.

bahay na bato—literally "stone house," generally meaning an expensive, stoutly built house with a massive stone ground floor and a timber main floor often extending out beyond the ground floor. Tagalog.

baklad—general name for a fish corral or fish trap. Tagalog.

bangán—a granary, especially one made of stone and lime in the *zaguan* of a *bahay na bato*, attached to such a dwelling, or free standing. Tagalog.

Bankusay—a beach-side neighborhood and fish landing in Tondo. Elderly informants native to the neighborhood pronounce it "Bangkusáy."

batulan—a large cylindrical *cesto* or deep *canasto* for carrying cargo (del Pan, *Diccionario;* Serrano Laktaw, *Diccionario*). Tagalog.

bayawak—monitor lizard. Tagalog.

bayon—a commodity sack made of *buri* fiber and used for the transport of rice, sugar, and so on. Tagalog.

bayuhan—a rice-milling operation that utilizes a pounding motion. Tagalog.

bigas—rice that has been completely milled. Tagalog.

biyahera, biyahero. See *viajera*.

bodega—a warehouse, dock warehouse, storage structure, or storage room. Spanish and Tagalog derived from Spanish.

Bolinao Peninsula—a group of municipalities in northern Zambales that were detached and transferred to Pangasinan soon after 1900. Some authors recognize Bolinao speech as a distinct language.

bulto—a generic term used in commerce in order to simplify descriptions of *fardos* (bales, bundles), *baules* (trunks), *cajones* (boxes), and similar objects of transport. Spanish (del Pan, *Diccionario*).

buslo—a small basket used to carry eggs, fruit, tomatoes, onions, garlic, and so on. Tagalog (Spanish *canastrillo*).

carabao/*kalabaw*—water buffalo (*Bubalus bubalis*). Tagalog and English derived from Tagalog.

casco/kasco—literally a shell, such as a turtle shell. Spanish. (1) Widely used in Tagalog and Philippine Spanish for a long, flat-bottomed barge employed in rivers, canals, and other quiet waters, propelled by poling occasionally supplemented with lanteen sails and later towing; (2) more rarely, refers to the central zone of a *poblacion* or central settlement.

cavan/kabán—a dry measure of volume legally standardized at 75 liters in all provinces in 1862, a.k.a. *cavan del Rey* or *cavan de Manila* as opposed to various earlier *cavanes de provincias*. In a *cavan* of rice/*bigas* the weight varies by moisture content.

Central Plain (of Luzon)—the smooth lowland portions of Bulacan, Pampanga, Tarlac, Nueva Ecija, and Pangasinan. See maps in Larkin, *Pampangans*, 4, 7.

cesto/cesta—something woven of bamboo or rattan and used for transporting produce. There are many local types and names, including *batulan*. Spanish (del Pan, *Diccionario*).

contribución industrial—annual tax on businesses and productive facilities begun in the 1880s.

contribución urbana—tax on income from urban buildings, begun in 1878.

conurbation—a very large and complex city, one that has incorporated formerly separate towns and cities, a metropolis.

darak—rice bran normally fed to horses with molasses but also cooked and eaten as a starvation food. Tagalog.

depósito—underground reservoirs in San Juan, part of the Carriedo water system, also a storehouse. Spanish.

ENSO—El Niño/Southern Oscillation of atmospheric pressure ratios across the southern Pacific.

eotechnic—a technological complex centered on the use of wood, water, and wind as material and motive force, a term derived from Lewis Mumford's classic *Technics and Civilization*.

Escolta—the leading upscale commercial street in the central business district in Binondo.

estero—inlet, tidal creek, estuary. Spanish.

fábrica/pabrika—millworks, factory, a building for such activities. Spanish, Tagalog from Spanish.

ganta—a three-liter measure of dry grain, actually measured by a *salóp*, a cubical box for measuring out exactly three liters.

genuine—pronounced "jen-wine" or sometimes "jen-oo-wine." American or other well-made clothes and dry goods during the shortages of the Japanese occupation. Philippine English and Tagalog.

hacendero, hacendera—the owner of a hacienda, a Philippine form of *hacendado*. Philippine Spanish and Philippine English.

hijos del país—creoles, especially in the eighteenth and nineteenth centuries. Spanish.

kalesa/calesa—a two-wheeled horse-drawn taxi or gig for three passengers. Spanish and Tagalog derived from Spanish.

kamalig—a storage room or structure, especially for commercial goods or commodities, a warehouse or bodega. Tagalog.

kamarin/camarin—the nineteenth-century Philippine Spanish equivalent of *kamalig*.

kamote/camóte—the American sweet potato (as opposed to the Asian yam). Tagalog and Spanish derived from Nahuatl.

kanasta/canasta—a basket or crate. With a narrow mouth called a *canasto*. Spanish.

kangkóng—a vine of swamps and stagnant pools, the leaves and stems of which are eaten as a green vegetable. Tagalog.

kapatas—a labor gang recruiter and foreman. Tagalog derived from Spanish.
kapós—"not enough," as in an insufficient supply of food or money. (Santos, *Vicassan's Pilipino-English Dictionary*, 314), want/wanting. Tagalog.
kargadór/cargador—one who labors by carrying, a porter, stevedore. Spanish and Tagalog derived from Spanish.
karinderiya/carinderia—a down-market restaurant serving precooked food. The diminutive *karihan* is a table or stand used for selling and serving cooked food. Tagalog. The term seems to be derived from Indian *curry* and to date from the decision of hundreds of sepoys to stay on in the Tagalog area after the British East India Company military withdrew in the late eighteenth century.
karne/carne—the general word for animal meat in use in Manila today. Tagalog derived from Spanish. The former Tagalog word for meat, *lamán*, has narrowed in meaning to represent a lean piece or cut of meat without bones.
kastanyog/castanyog—roasted coconut meat, a starvation food, jokingly derived from *castaña* (Spanish "chestnuts") and *niyog* (Tagalog "coconuts"), that is, ersatz chestnuts. Tagalog and Philippine English.
katulong—a household helper. Tagalog.
kutsero/cochero—a rig driver. Spanish, Tagalog derived from Spanish.
Liberation—the American invasion and subsequent Allied victory over Japanese forces in late 1944–45 and the months of exuberant interaction between U.S. soldiers and Filipinos that followed.
lugaw—gruel or rice porridge. Tagalog and Ilocano.
Makapili—an abbreviation used for the organization and membership of the Kalipunang Makabayan ng mga Pilipino, mostly composed of ordinary Filipinos who had once belonged to the Ganap Party and before it Sakdal (1930s); radical nationalist and mostly agrarian populists who welcomed and tried to aid the Japanese forces during World War II on the premise that Japan would give them arms and other aid in the cause of real Philippine independence. Members were used by the Japanese occupation forces as guards, laborers, collectors of information, and finally armed fighters. The term is often used to mean "informer" or "traitor."
manggigiling—rice pounder, later a miller. Tagalog.
manguera—mango vendor. Spanish and Tagalog.
Manila Province, Morong, Rizal, Metro Manila—Immediately north and south of urban Manila and its expanding *arrabales* in the mid-nineteenth century were settlements and lands administered as part of a province called Tondo—including but not to be confused with the Manila district of the same name. In the 1850s–70s these municipal territories included Malabon and Navotas, Caloocan, Novaliches, and San Juan del Monte and in the south Muntinglupa, Las Piñas, Parañaque, Malibay, Pasay, and others. Still with similar boundaries north and south, this unit in the 1880s and 1890s was designated Manila Province. In the early twentieth century these same northern and southern peripheral areas were reassigned as parts of the new Rizal Province—an entity distinctly separated from the city. Despite the subsequent massive suburbanization of Manila, the new suburbs remained in Rizal, administratively separate, until some were pulled in during the Japanese occupation. They were again made part of Rizal after independence until the broader

entity of Metro Manila was instituted under martial law in the 1970s. To the east, arrangements fluctuated at the same time. In the mid-nineteenth century, San Pedro Makati, Pasig, Pateros, Taguig, the Marikina Valley, and places farther east, such as Antipolo, Taytay, and Cainta, were included in the Tondo territory. Most remained part of Tondo and were included in Manila Province during the 1880s and 1890s. The last three, however, were placed in Morong when that special provincial territory was created in 1853 out of a portion of Tondo/Manila and La Laguna. All were reunited in the new Rizal Province after 1900. After the creation of Metro Manila, Cainta, Taytay, and Antipolo, as well as San Mateo and Montalban, remained in Rizal. The others, including Marikina, Quezon City, Caloocan, Malabon, Navotas, and Valenzuela, were transferred to Metro Manila.

matadero—an abattoir. Urban Tagalog and early-twentieth-century Philippine English derived from Spanish.

merienda—light food eaten in the late afternoon. Tagalog and Philippine English from Spanish.

mimis—an especially white, nice-smelling, and expensive rice variety. The term is used in Cebuano and Ilocano, but is now little known in Manila speech.

Morong—a province created in 1853 out of an eastern portion of Tondo/Manila and La Laguna. It extended from Cainta and Taytay in the west to Jalajala in the east and was named after Morong municipality, its designated capital. Fifty years later it was absorbed into the new Rizal Province. See Manila Province.

NARIC—the National Rice and Corn Corporation, an official vehicle for operating in the markets for rice and maize.

palagad—a dry season rice crop, ordinarily much smaller than the main wet season crop. Tagalog.

palay—unhusked rice grains, also the entire plant. Tagalog, also adopted into Philippine Spanish to replace *arroz sucio*.

paleotechnic—the early industrial technology complex centered on iron, coal, and steam as material, energy source, and motive power, after Mumford, *Technics and Civilization*.

Panahon ng Hapon—the period of Japanese occupation. Tagalog.

panganay—eldest child in a sibling set, firstborn. Tagalog.

pansit—a readily cooked rice noodle dish. Tagalog derived from Hokkien.

pansiteria—an inexpensive restaurant specializing in *pansit* noodle dishes.

pantalan—a boat landing or wharf. Tagalog and Pangasinan, also sometimes Philippine Spanish derived from Tagalog.

PCAU—Philippine Civil Affairs Unit for the emergency relief distribution of food and clothing.

peacetime—pronounced "pace-tyme," the period before World War II, especially the 1930s. Philippine English and Tagalog derived from English.

petchay/petsay—Chinese cabbage. Tagalog derived from Hokkien.

picul/pico—unit of weight equivalent to approximately 60 to 63.25 kilograms or 132 to 140 pounds (Owen, *Prosperity without Progress*, 277; *Census 1903*, 4:449).

pilanderia—a place and apparatus for pounding rice, a *bayuhan*. A Spanish Filipinism. (Serrano Laktaw, *Diccionario*, 120, 122; Retana, *Diccionario*, 149).

pinawa—rice that has been cleaned but not polished, "brown rice." Tagalog.

poblacion—literally "county seat," the principal settlement and township of a municipality. Spanish.

Polo—community adjacent to Obando in far southern Bulacan, now called Valenzuela.

propetaryo—owner, proprietor. Tagalog derived from Spanish *propietario*. Today the Tagalog term *may-ari* is commonly used.

pulutan—something to eat while consuming alcohol, for example, goat, *bayawak*, or dog. Tagalog.

Quingua—municipality in southern Bulacan, now called Plaridel.

regidor—a Manila city councilman in the Spanish system.

rice terminology—*palay*/unhusked rice, *pinawa*/clean brown rice, and *bigas*/milled and polished white rice. Tagalog.

Rizal Province. See Manila Province.

salabat—tea made from ginger. Tagalog and Philippine Spanish.

sarisari store/*tienda de sarisari*—a small store with a limited range of everyday consumables. Tagalog, Philippine English, and Philippine Spanish.

sili—chili pepper, including the leaves. Tagalog derived from Nahuatl via Spanish.

silong—the ground level of a house, often open sided and used for work or storage. Tagalog.

sinigang—soup or broth flavored with sour tamarind, often containing fish and *kangkong* and/or other vegetables. Tagalog and Ilocano. In Ilocos *kamotes* are added.

sisid rice—rice that was recovered after having been under seawater for some time, eaten cooked with garlic, nauseous-smelling survival food. Tagalog.

Tabacalera—common name for the Spanish-French Compañia General de Tabacos de Filipinas.

tagulan—rainy season, in Manila the period of the southwest monsoon. Tagalog.

talipapa—makeshift and unlicensed but convenient street market for provisions. Tagalog.

tambobong—granary or barn. Tagalog. Long used as the formal name of Malabon.

tangkil—small granary or barn (Santos, *Vicassan's Pilipino-English Dictionary*), a small *bangán* or *kamalig*. Tagalog.

tapayan/*tinaja*—a large, round storage and shipping jar used primarily for liquids. Tagalog and Spanish, respectively. Not to be confused with a *tapayán*, or "bakery" (Tagalog).

Tayabas—later renamed Quezon Province and expanded to add Infanta, the east coast section.

tiangue/*tiangui*—a regular, open-air, periodic market day and location. Philippine Spanish and Tagalog derived from Nahuatl, various spellings.

time-distance—distance thought of in terms of the time required to travel rather than miles. Cost-distance is a similar concept in which the map based on miles is distorted by the cost in effort or money of transporting something from A to B.

tinapay—bread or biscuits, literally "baked dough." Tagalog and Ilocano.

tinola—a dinner soup with ginger, cut-up chicken, green papaya or squash, onions, garlic, other vegetables, and *sili* chili pepper leaves. Tagalog and Ilocano.

tsamporado/*champorado*—chocolate rice porridge. Tagalog derived from Mexican Spanish.

tsokolate/chocolate—chocolate, especially in beverage form. Tagalog and Spanish, respectively, derived from Nahuatl.

tuyo—small fish preserved whole by means of salting and drying. Tagalog.

ubod—pith, for example, of a banana stalk. Tagalog.

umbuyan—facility for smoking fish. Tagalog.

utang/mga utang —debt/debts, obligations. Tagalog.

viajera/viajero—a traveling buyer of produce or hogs. Spanish, now *biyahera* in Tagalog.

zacate/sakate—one or another of several long native grasses or Bermuda grass cut fresh for use as fodder. Philippine Spanish and Tagalog, respectively, from Nahuatl.

zaguan—the open ground floor of a great stone house, often unfloored. Spanish.

zoning/*sona*—a word used during the Japanese occupation to describe the mass roundup of guerrilla suspects by cordoning off a whole village or neighborhood, now used more broadly. Philippine English and Tagalog (*sona* or *zona*).

INDEX

An Anarchy of Families: State and Family in the Philippines
Edited by ALFRED W. MCCOY

The Hispanization of the Philippines:
Spanish Aims and Filipino Responses, 1565–1700
JOHN LEDDY PHELAN

Pretext for Mass Murder: The September 30th Movement and
Suharto's Coup d'État in Indonesia
JOHN ROOSA

The Social World of Batavia: Europeans and Eurasians in
Colonial Indonesia, second edition
JEAN GELMAN TAYLOR

Việt Nam: Borderless Histories
Edited by NHUNG TUYET TRAN and ANTHONY REID

Thailand's Political Peasants: Power in the Modern Rural Economy
ANDREW WALKER

Modern Noise, Fluid Genres: Popular Music in Indonesia, 1997–2001
JEREMY WALLACH